AN INTRODUCTION TO LINEAR ALGEBRA

L. MIRSKY
Department of Pure Mathematics,
University of Sheffield

DOVER PUBLICATIONS, INC.
New York

Published in Canada by General Publishing Company, Ltd., 30 Lesmill Road, Don Mills, Toronto, Ontario.

Published in the United Kingdom by Constable and Company, Ltd., 3 The Lanchesters, 162–164 Fulham Palace Road, London W6 9ER.

This Dover edition, first published in 1990, is an unabridged and unaltered republication of the 1972 printing of the corrected 1961 edition of the work first published at the Clarendon Press, Oxford, England, in 1955. It is reprinted by special arrangement with the Oxford University Press, Inc., 200 Madison Avenue, New York, N.Y. 10016.

Manufactured in the United States of America
Dover Publications, Inc.
31 East 2nd Street
Mineola, N.Y. 11501

Library of Congress Cataloging-in-Publication Data

Mirsky, L. (Leonid)
An introduction to linear algebra / L. Mirsky.
p. cm.
Reprint. Originally published: Oxford : Clarendon Press, 1955.
Includes bibliographical references and index.
ISBN 0-486-66434-1 (pbk.)
1. Algebras, Linear. I. Title.
QA184.M57 1990
512′.5—dc90 90-40299
CIP

PREFACE

My object in writing this book has been to provide an elementary and easily readable account of linear algebra. The book is intended mainly for students pursuing an honours course in mathematics, but I hope that the exposition is sufficiently simple to make it equally useful to readers whose principal interests lie in the fields of physics or technology. The material dealt with here is not extensive and, broadly speaking, only those topics are discussed which normally form part of the honours mathematics syllabus in British universities. Within this compass I have attempted to present a systematic and rigorous development of the subject. The account is self-contained, and the reader is not assumed to have any previous knowledge of linear algebra, although some slight acquaintance with the elementary theory of determinants will be found helpful.

It is not easy to estimate what level of abstractness best suits a textbook of linear algebra. Since I have aimed, above all, at simplicity of presentation I have decided on a thoroughly concrete treatment, at any rate in the initial stages of the discussion. Thus I operate throughout with real and complex numbers, and I define a vector as an ordered set of numbers and a matrix as a rectangular array of numbers. After the first three chapters, however, a new and more abstract point of view becomes prominent. Linear manifolds (i.e. abstract vector spaces) are considered, and the algebra of matrices is then recognized to be the appropriate tool for investigating the properties of linear operators; in fact, particular stress is laid on the representation of linear operators by matrices. In this way the reader is led gradually towards the fundamental concept of invariant characterization.

The points of contact between linear algebra and geometry are numerous, and I have taken every opportunity of bringing them to the reader's notice. I have not, of course, sought to provide a systematic discussion of the algebraic background of geometry, but have rather concentrated on a few special topics, such as changes of the coordinate system, reduction of quadrics to principal axes, rotations in the plane and in space, and the classification of quadrics under the projective and affine groups.

The theory of matrices gives rise to many striking inequalities. The proofs of these are generally very simple, but are widely scattered throughout the literature and are often not easily accessible. I have here attempted to collect together, with proofs, all the better known inequalities of matrix theory. I have also included a brief sketch of the theory of matrix power series, a topic of considerable interest and elegance not normally dealt with in elementary textbooks.

Numerous exercises are incorporated in the text. They are designed not so much to test the reader's ingenuity as to direct his attention to analogues, generalizations, alternative proofs, and so on. The reader is recommended to work through these exercises, as the results embodied in them are frequently used in the subsequent discussion. At the end of each chapter there is a series of miscellaneous problems arranged approximately in order of increasing difficulty. Some of these involve only routine calculations, others call for some manipulative skill, and yet others carry the general theory beyond the stage reached in the text. A number of these problems have been taken from recent examination papers in mathematics, and thanks for permission to use them are due to the Delegates of the Clarendon Press, the Syndics of the Cambridge University Press, and the Universities of Bristol, London, Liverpool, Manchester, and Sheffield.

The number of existing books on linear algebra is large, and it is therefore difficult to make a detailed acknowledgement of sources. I ought, however, to mention Turnbull and Aitken, *An Introduction to the Theory of Canonical Matrices*, and MacDuffee, *The Theory of Matrices*, on both of which I have drawn heavily for historical references.

I have received much help from a number of friends and colleagues. Professor A. G. Walker first suggested that I should write a book on linear algebra and his encouragement has been invaluable. Mr. H. Burkill, Mr. A. R. Curtis, Dr. C. S. Davis, Dr. H. K. Farahat, Dr. Christine M. Hamill, Professor H. A. Heilbronn, Professor D. G. Northcott, and Professor A. Oppenheim have all helped me in a variety of ways, by checking parts of the manuscript or advising me on specific points. Mr. J. C. Shepherdson read an early version of the manuscript and his acute comments have enabled me to remove many obscurities and ambiguities; he has, in addition, given me considerable help with Chapters IX and X. The greatest debt I owe is to Dr. G. T. Kneebone and Professor R. Rado with both of whom, for several years past, I have been in the habit of discussing problems of linear algebra and their presentation to students. But for these conversations I should not have been able to write the book. Dr. Kneebone has also read and criticized the manuscript at every stage of preparation and Professor Rado has supplied me with several of the proofs and problems which appear in the text. Finally, I wish to record my thanks to the officers of the Clarendon Press for their helpful co-operation.

NOTE

I HAVE to thank a number of correspondents, and especially Professor R. Rado, for drawing my attention to various minor errors and misprints; the new printing of the book has enabled me to make the necessary corrections. Furthermore, the bibliography has been brought up to date and a fairly large number of new problems (most of them collected at the end of the book) has been added. For several of these I am indebted to Dr. H. K. Farahat.

CONTENTS

PART I

DETERMINANTS, VECTORS, MATRICES, AND LINEAR EQUATIONS

PART I

DETERMINANTS, VECTORS, MATRICES, AND LINEAR EQUATIONS

I

DETERMINANTS

THE present book is intended to give a systematic account of the elementary parts of linear algebra. The technique best suited to this branch of mathematics is undoubtedly that provided by the calculus of matrices, to which much of the book is devoted, but we shall also require to make considerable use of the theory of determinants, partly for theoretical purposes and partly as an aid to computation. In this opening chapter we shall develop the principal properties of determinants to the extent to which they are needed for the treatment of linear algebra.†

The theory of determinants was, indeed, the first topic in linear algebra to be studied intensively. It was initiated by Leibnitz in 1696, developed further by Bézout, Vandermonde, Cramer, Lagrange, and Laplace, and given the form with which we are now familiar by Cauchy, Jacobi, and Sylvester in the first half of the nineteenth century. The term 'determinant' occurs for the first time in Gauss's *Disquisitiones arithmeticae* (1801).‡

1.1. Arrangements and the ϵ-symbol

In order to define determinants it is necessary to refer to arrangements among a set of numbers, and the theory of determinants can be based on a few simple results concerning such arrangements. In the present section we shall therefore derive the requisite preliminary results.

1.1.1. We shall denote by $(\lambda_1,..., \lambda_n)$ the *ordered* set consisting of the integers $\lambda_1,..., \lambda_n$.

† For a much more detailed discussion of determinants see Kowalewski, *Einführung in die Determinantentheorie*. Briefer accounts will be found in Burnside and Panton, *The Theory of Equations*, and in Ferrar, **2**, Aitken, **15**, and Perron, **18**. (Numbers in bold-face type refer to the bibliography at the end.)

‡ For historical and bibliographical information see Muir, *The Theory of Determinants in the Historical Order of Development*.

DEFINITION 1.1.1. *If* $(\lambda_1,..., \lambda_n)$ *and* $(\mu_1,..., \mu_n)$ *contain the same (distinct) integers, but these integers do not necessarily occur in the same order, then* $(\lambda_1,..., \lambda_n)$ *and* $(\mu_1,..., \mu_n)$ *are said to be* ARRANGEMENTS† *of each other. In symbols:* $(\lambda_1,..., \lambda_n) = \mathscr{A}(\mu_1,..., \mu_n)$ *or* $(\mu_1,..., \mu_n) = \mathscr{A}(\lambda_1,..., \lambda_n)$.

We shall for the most part be concerned with arrangements of the first n positive integers. If $(\nu_1,..., \nu_n) = \mathscr{A}(1,..., n)$ and $(k_1,..., k_n) = \mathscr{A}(1,..., n)$, then clearly $(\nu_{k_1},..., \nu_{k_n}) = \mathscr{A}(1,..., n)$. We have the following result.

THEOREM 1.1.1. (i) *Let* $(\nu_1,..., \nu_n)$ *vary over all arrangements of* $(1,..., n)$, *and let* $(k_1,..., k_n)$ *be a fixed arrangement of* $(1,..., n)$. *Then* $(\nu_{k_1},..., \nu_{k_n})$ *varies over all arrangements of* $(1,..., n)$.

(ii) *Let* $(\nu_1,..., \nu_n)$ *vary over all arrangements of* $(1,..., n)$, *and let* $(\mu_1,..., \mu_n)$ *be a fixed arrangement of* $(1,..., n)$. *The arrangement* $(\lambda_1,..., \lambda_n)$, *defined by the conditions*

$$\nu_{\lambda_1} = \mu_1, \quad ..., \quad \nu_{\lambda_n} = \mu_n,$$

then varies over all arrangements of $(1,..., n)$.

This theorem is almost obvious. To prove (i), suppose that for two *different* choices of $(\nu_1,..., \nu_n)$—say $(\alpha_1,..., \alpha_n)$ and $(\beta_1,..., \beta_n)$—$(\nu_{k_1},..., \nu_{k_n})$ is the same arrangement, i.e.

$$(\alpha_{k_1},..., \alpha_{k_n}) = (\beta_{k_1},..., \beta_{k_n}),$$

and so $$\alpha_{k_1} = \beta_{k_1}, \quad ..., \quad \alpha_{k_n} = \beta_{k_n}.$$

These relations are, in fact, the same as

$$\alpha_1 = \beta_1, \quad ..., \quad \alpha_n = \beta_n,$$

although they are stated in a different order. The two arrangements are thus identical, contrary to hypothesis. It therefore follows that, as $(\nu_1,..., \nu_n)$ varies over the $n!$ arrangements of $(1,..., n)$, $(\nu_{k_1},..., \nu_{k_n})$ also varies, without repetition, over arrangements of $(1,..., n)$. Hence $(\nu_{k_1},..., \nu_{k_n})$ varies, in fact, over all the $n!$ arrangements.

The second part of the theorem is established by the same type of argument. Suppose that for two different choices of $(\nu_1,..., \nu_n)$—say $(\alpha_1,..., \alpha_n)$ and $(\beta_1,..., \beta_n)$—$(\lambda_1,..., \lambda_n)$ is the same arrangement, i.e.

$$\alpha_{\lambda_1} = \mu_1 = \beta_{\lambda_1}, \quad ..., \quad \alpha_{\lambda_n} = \mu_n = \beta_{\lambda_n}.$$

† We avoid the familiar term 'permutation' since this will be used in a somewhat different sense in Chapter IX.

Then $(\alpha_1,..., \alpha_n) = (\beta_1,..., \beta_n)$, contrary to hypothesis, and the assertion follows easily.

1.1.2. DEFINITION 1.1.2. *For all real values of x the function* $\operatorname{sgn} x$ (read: signum x) *is defined as*

$$\operatorname{sgn} x = \begin{cases} 1 & (x > 0) \\ 0 & (x = 0) \\ -1 & (x < 0). \end{cases}$$

EXERCISE 1.1.1. Show that

$$\operatorname{sgn} x . \operatorname{sgn} y = \operatorname{sgn} xy,$$

and deduce that

$$\operatorname{sgn} x_1 . \operatorname{sgn} x_2 ... \operatorname{sgn} x_k = \operatorname{sgn}(x_1 x_2 ... x_k).$$

DEFINITION 1.1.3.

(i) $$\epsilon(\lambda_1,..., \lambda_n) = \operatorname{sgn} \prod_{1 \leqslant r < s \leqslant n} (\lambda_s - \lambda_r).\dagger$$

(ii) $$\epsilon\begin{pmatrix} \lambda_1,..., \lambda_n \\ \mu_1,..., \mu_n \end{pmatrix} = \epsilon(\lambda_1,..., \lambda_n).\epsilon(\mu_1,..., \mu_n).$$

EXERCISE 1.1.2. Show that if $\lambda_1 < ... < \lambda_n$, then $\epsilon(\lambda_1,..., \lambda_n) = 1$. Also show that if any two λ's are equal, then $\epsilon(\lambda_1,..., \lambda_n) = 0$.

EXERCISE 1.1.3. The interchange of two λ's in $(\lambda_1,..., \lambda_n)$ is called a *transposition*. Show that, if $(\lambda_1,..., \lambda_n) = \mathscr{A}(1,..., n)$, then it is possible to obtain $(\lambda_1,..., \lambda_n)$ from $(1,..., n)$ by a succession of transpositions. Show, furthermore, that if this process can be carried out by s transpositions, then

$$\epsilon(\lambda_1,..., \lambda_n) = (-1)^s.$$

Deduce that, if the same process can also be carried out by s' transpositions, then s and s' are either both even or both odd.

THEOREM 1.1.2. *If* $(\lambda_1,..., \lambda_n)$, $(\mu_1,..., \mu_n)$, *and* $(k_1,..., k_n)$ *are arrangements of* $(1,..., n)$, *then*

$$\epsilon\begin{pmatrix} \lambda_1,..., \lambda_n \\ \mu_1,..., \mu_n \end{pmatrix} = \epsilon\begin{pmatrix} \lambda_{k_1},..., \lambda_{k_n} \\ \mu_{k_1},..., \mu_{k_n} \end{pmatrix}.\ddagger$$

We may express this identity by saying that if $(\lambda_1,..., \lambda_n)$ and $(\mu_1,..., \mu_n)$ are subjected to the *same* derangement, then the value of

$$\epsilon\begin{pmatrix} \lambda_1,..., \lambda_n \\ \mu_1,..., \mu_n \end{pmatrix}$$

† Empty products are, as usual, defined to have the value 1. This implies, in particular, that for $n = 1$ every ϵ-symbol is equal to 1.

‡ Definition 1.1.3 implies, of course, that

$$\epsilon(\lambda_{k_1},..., \lambda_{k_n}) = \operatorname{sgn} \prod_{1 \leqslant i < j \leqslant n} (\lambda_{k_j} - \lambda_{k_i}).$$

remains unaltered. To prove this we observe that

$$(\lambda_{k_j}-\lambda_{k_i})(\mu_{k_j}-\mu_{k_i}) = (\lambda_s-\lambda_r)(\mu_s-\mu_r), \tag{1.1.1}$$

where

$$r = \min(k_i, k_j), \qquad s = \max(k_i, k_j). \tag{1.1.2}$$

Now if r, s (such that $1 \leqslant r < s \leqslant n$) are given, then there exist unique integers i, j (such that $1 \leqslant i < j \leqslant n$) satisfying (1.1.2). Thus there is a *biunique correspondence* (i.e. a one-one correspondence) between the pairs k_i, k_j and the pairs r, s. Hence, by (1.1.1),

$$\prod_{1\leqslant i<j\leqslant n} (\lambda_{k_j}-\lambda_{k_i})(\mu_{k_j}-\mu_{k_i}) = \prod_{1\leqslant r<s\leqslant n} (\lambda_s-\lambda_r)(\mu_s-\mu_r).$$

Therefore, by Exercise 1.1.1,

$$\operatorname{sgn} \prod_{1\leqslant i<j\leqslant n} (\lambda_{k_j}-\lambda_{k_i}) . \operatorname{sgn} \prod_{1\leqslant i<j\leqslant n} (\mu_{k_j}-\mu_{k_i})$$
$$= \operatorname{sgn} \prod_{1\leqslant r<s\leqslant n} (\lambda_s-\lambda_r) . \operatorname{sgn} \prod_{1\leqslant r<s\leqslant n} (\mu_s-\mu_r),$$

i.e.

$$\epsilon\begin{pmatrix}\lambda_{k_1},\dots,\lambda_{k_n}\\ \mu_{k_1},\dots,\mu_{k_n}\end{pmatrix} = \epsilon\begin{pmatrix}\lambda_1,\dots,\lambda_n\\ \mu_1,\dots,\mu_n\end{pmatrix}.$$

THEOREM 1.1.3. *Let $1 \leqslant r < s \leqslant n$. Then*

$$\epsilon(1,\dots,r-1,s,r+1,\dots,s-1,r,s+1,\dots,n) = -1.$$

The expression on the left-hand side is, of course, simply $\epsilon(1,2,\dots,n)$ with r and s interchanged. Denoting this expression by $\epsilon(\lambda_1,\dots,\lambda_n)$, we observe that in the product

$$\prod_{1\leqslant i<j\leqslant n} (\lambda_j-\lambda_i)$$

there are precisely $2(s-r-1)+1 = 2s-2r-1$ negative factors, namely,

$$\begin{matrix} (r+1)-s, & (r+2)-s, & \dots, & (s-1)-s, \\ r-(r+1), & r-(r+2), & \dots, & r-(s-1), \\ r-s. & & & \end{matrix}$$

Hence, $\epsilon(\lambda_1,\dots,\lambda_n) = (-1)^{2s-2r-1} = -1$, as asserted.

The results obtained so far are sufficient for the discussion in § 1.2 and § 1.3. The proof of Laplace's expansion theorem in § 1.4, however, presupposes a further identity.

THEOREM 1.1.4. *If $(r_1,\dots,r_n) = \mathscr{A}(1,\dots,n)$, $(s_1,\dots,s_n) = \mathscr{A}(1,\dots,n)$, and $1 \leqslant k < n$, then*

$$\epsilon\begin{pmatrix}r_1,\dots,r_n\\ s_1,\dots,s_n\end{pmatrix} = (-1)^{r_1+\dots+r_k+s_1+\dots+s_k}\, \epsilon\begin{pmatrix}r_1,\dots,r_k\\ s_1,\dots,s_k\end{pmatrix} . \epsilon\begin{pmatrix}r_{k+1},\dots,r_n\\ s_{k+1},\dots,s_n\end{pmatrix}.$$

By Exercise 1.1.1 we have

$$\epsilon(r_1,\dots,r_n)$$
$$= \operatorname{sgn}\prod_{1\leqslant i<j\leqslant k}(r_j-r_i).\operatorname{sgn}\prod_{k+1\leqslant i<j\leqslant n}(r_j-r_i).\operatorname{sgn}\prod_{\substack{1\leqslant i\leqslant k\\ k+1\leqslant j\leqslant n}}(r_j-r_i)$$
$$= \epsilon(r_1,\dots,r_k).\epsilon(r_{k+1},\dots,r_n).(-1)^{\nu_1+\dots+\nu_k}, \tag{1.1.3}$$

where, for $1 \leqslant i \leqslant k$, ν_i denotes the number of numbers among $r_{k+1},\dots,r_n$ which are smaller than r_i.

Let $r'_1,\dots,r'_k$ be defined by the relations

$$(r'_1,\dots,r'_k) = \mathscr{A}(r_1,\dots,r_k), \qquad r'_1 < \dots < r'_k,$$

and denote by ν'_i $(1 \leqslant i \leqslant k)$ the number of numbers among $r_{k+1},\dots,r_n$ which are smaller than r'_i. Then

$$\nu'_1 = r'_1-1, \quad \nu'_2 = r'_2-2, \quad \dots, \quad \nu'_k = r'_k-k,$$
$$\nu_1+\dots+\nu_k = \nu'_1+\dots+\nu'_k = r_1+\dots+r_k-\tfrac{1}{2}k(k+1),$$

and hence, by (1.1.3),

$$\epsilon(r_1,\dots,r_n) = (-1)^{r_1+\dots+r_k-\frac{1}{2}k(k+1)}\,\epsilon(r_1,\dots,r_k).\epsilon(r_{k+1},\dots,r_n).$$

Similarly

$$\epsilon(s_1,\dots,s_n) = (-1)^{s_1+\dots+s_k-\frac{1}{2}k(k+1)}\,\epsilon(s_1,\dots,s_k).\epsilon(s_{k+1},\dots,s_n),$$

and the theorem now follows at once by Definition 1.1.3 (ii).

1.2. Elementary properties of determinants

1.2.1. We shall now be concerned with the study of certain properties of *square arrays* of (real or complex) numbers. A typical array is

$$\begin{matrix} a_{11} & a_{12} & \cdot & \cdot & \cdot & a_{1n} \\ a_{21} & a_{22} & \cdot & \cdot & \cdot & a_{2n} \\ \cdot & \cdot & \cdot & \cdot & \cdot & \cdot \\ a_{n1} & a_{n2} & \cdot & \cdot & \cdot & a_{nn} \end{matrix} \tag{1.2.1}$$

DEFINITION 1.2.1. *The n^2 numbers a_{ij} $(i,j = 1,\dots,n)$ are the* ELEMENTS *of the array* (1.2.1). *The elements*

$$a_{i1}, a_{i2},\dots, a_{in}$$

constitute the i-th ROW, *and the elements*

$$a_{1j}, a_{2j},\dots, a_{nj}$$

constitute the j-th COLUMN *of the array. The elements*

$$a_{11}, a_{22},\dots, a_{nn}$$

constitute the DIAGONAL *of the array, and are called the* DIAGONAL ELEMENTS.

The double suffix notation used in (1.2.1) is particularly appropriate since the two suffixes of an element specify completely its position in the array. We shall reserve the first suffix for the row and the second for the column, so that a_{ij} denotes the element standing in the ith row and jth column of the array (1.2.1).

With each square array we associate a certain number known as its determinant.

DEFINITION 1.2.2. *The* DETERMINANT *of the array* (1.2.1) *is the number*

$$\sum_{(\lambda_1,\ldots,\lambda_n)} \epsilon(\lambda_1,\ldots,\lambda_n) a_{1\lambda_1}\ldots a_{n\lambda_n}, \tag{1.2.2}$$

where the summation extends over all the $n!$ *arrangements* $(\lambda_1,\ldots,\lambda_n)$ *of* $(1,\ldots,n)$.† *This determinant is denoted by*

$$\begin{vmatrix} a_{11} & a_{12} & \cdot & \cdot & \cdot & a_{1n} \\ a_{21} & a_{22} & \cdot & \cdot & \cdot & a_{2n} \\ \cdot & \cdot & \cdot & \cdot & \cdot & \cdot \\ a_{n1} & a_{n2} & \cdot & \cdot & \cdot & a_{nn} \end{vmatrix}, \tag{1.2.3}$$

or, more briefly, by $|a_{ij}|_n$.

Determinants were first written in the form (1.2.3), though without the use of double suffixes, by Cayley in 1841. In practice, we often use a single letter, such as D, to denote a determinant.

The determinant (1.2.3) associated with the array (1.2.1) is plainly a polynomial, of degree n, in the n^2 elements of the array.

The determinant of the array consisting of the single element a_{11} is, of course, equal to a_{11}. Further, we have

$$\begin{vmatrix} a_{11} & a_{12} \\ a_{21} & a_{22} \end{vmatrix} = \epsilon(1,2)a_{11}a_{22} + \epsilon(2,1)a_{12}a_{21} = a_{11}a_{22} - a_{12}a_{21};$$

$$\begin{vmatrix} a_{11} & a_{12} & a_{13} \\ a_{21} & a_{22} & a_{23} \\ a_{31} & a_{32} & a_{33} \end{vmatrix} = \epsilon(1,2,3)a_{11}a_{22}a_{33} + \epsilon(1,3,2)a_{11}a_{23}a_{32} + \epsilon(2,1,3)a_{12}a_{21}a_{33} + \epsilon(2,3,1)a_{12}a_{23}a_{31} + \epsilon(3,1,2)a_{13}a_{21}a_{32} + \epsilon(3,2,1)a_{13}a_{22}a_{31}$$

$$= a_{11}a_{22}a_{33} - a_{11}a_{23}a_{32} - a_{12}a_{21}a_{33} + a_{12}a_{23}a_{31} + a_{13}a_{21}a_{32} - a_{13}a_{22}a_{31}.$$

We observe that each term of the expression (1.2.2) for the determinant $|a_{ij}|_n$ contains one element from each row and one element from each column of the array (1.2.1). Hence, if any array

† The same convention will be observed whenever a symbol such as $(\lambda_1,\ldots,\lambda_n)$ appears under the summation sign.

contains a row or a column consisting entirely of zeros, its determinant is equal to 0.

A determinant is a *number* associated with a square array. However, it is customary to use the term 'determinant' for the array itself as well as for this number. This usage is ambiguous but convenient, and we shall adopt it since it will always be clear from the context whether we refer to the array or to the *value* of the determinant associated with it. In view of this convention we may speak, for instance, about the elements, rows, and columns of a determinant. The determinant (1.2.3) will be called an *n-rowed determinant*, or a determinant of *order n*.

1.2.2. Definition 1.2.2 suffers from a lack of symmetry between the row suffixes and the column suffixes. For the row suffixes appearing in every term of the sum (1.2.2) are fixed as $1,\ldots, n$, whereas the column suffixes vary from term to term. The following theorem shows, however, that this lack of symmetry is only apparent.

THEOREM 1.2.1. *Let D be the value of the determinant* (1.2.3).

(i) If $(\lambda_1,\ldots, \lambda_n)$ *is any fixed arrangement of* $(1,\ldots, n)$, *then*

$$D = \sum_{(\mu_1,\ldots,\mu_n)} \epsilon\begin{pmatrix}\lambda_1,\ldots, \lambda_n\\ \mu_1,\ldots, \mu_n\end{pmatrix} a_{\lambda_1\mu_1}\ldots a_{\lambda_n\mu_n}.$$

(ii) If $(\mu_1,\ldots, \mu_n)$ *is any fixed arrangement of* $(1,\ldots, n)$, *then*

$$D = \sum_{(\lambda_1,\ldots,\lambda_n)} \epsilon\begin{pmatrix}\lambda_1,\ldots, \lambda_n\\ \mu_1,\ldots, \mu_n\end{pmatrix} a_{\lambda_1\mu_1}\ldots a_{\lambda_n\mu_n}.$$

In view of Definition 1.2.2 we have

$$D = \sum_{(\nu_1,\ldots,\nu_n)} \epsilon\begin{pmatrix}1,\ldots, n\\ \nu_1,\ldots, \nu_n\end{pmatrix} a_{1\nu_1}\ldots a_{n\nu_n}. \tag{1.2.4}$$

Let the same derangement which changes $(1,\ldots, n)$ into the fixed arrangement $(\lambda_1,\ldots, \lambda_n)$ change $(\nu_1,\ldots, \nu_n)$ into $(\mu_1,\ldots, \mu_n)$. Then

$$a_{1\nu_1}\ldots a_{n\nu_n} = a_{\lambda_1\mu_1}\ldots a_{\lambda_n\mu_n},$$

and, by Theorem 1.1.2 (p. 3),

$$D = \sum_{(\nu_1,\ldots,\nu_n)} \epsilon\begin{pmatrix}\lambda_1,\ldots, \lambda_n\\ \mu_1,\ldots, \mu_n\end{pmatrix} a_{\lambda_1\mu_1}\ldots a_{\lambda_n\mu_n}.$$

Hence, by Theorem 1.1.1 (i) (p. 2),

$$D = \sum_{(\mu_1,\dots,\mu_n)} \epsilon\begin{pmatrix}\lambda_1,\dots,\lambda_n\\ \mu_1,\dots,\mu_n\end{pmatrix} a_{\lambda_1\mu_1}\dots a_{\lambda_n\mu_n},$$

and the first part of the theorem is therefore proved.

To prove the second part we again start from (1.2.4). Let the same derangement which changes $(\nu_1,\dots,\nu_n)$ into the fixed arrangement $(\mu_1,\dots,\mu_n)$ change $(1,\dots,n)$ into $(\lambda_1,\dots,\lambda_n)$. Then, by Theorem 1.1.2,

$$D = \sum_{(\nu_1,\dots,\nu_n)} \epsilon\begin{pmatrix}\lambda_1,\dots,\lambda_n\\ \mu_1,\dots,\mu_n\end{pmatrix} a_{\lambda_1\mu_1}\dots a_{\lambda_n\mu_n},$$

and also
$$\nu_{\lambda_1} = \mu_1, \quad \dots, \quad \nu_{\lambda_n} = \mu_n.$$

Hence, by Theorem 1.1.1 (ii),

$$D = \sum_{(\lambda_1,\dots,\lambda_n)} \epsilon\begin{pmatrix}\lambda_1,\dots,\lambda_n\\ \mu_1,\dots,\mu_n\end{pmatrix} a_{\lambda_1\mu_1}\dots a_{\lambda_n\mu_n},$$

as asserted.

Theorem 1.2.2. *The value of a determinant remains unaltered when the rows and columns are interchanged, i.e.*

$$\begin{vmatrix} a_{11} & a_{12} & \cdot & \cdot & \cdot & a_{1n} \\ a_{21} & a_{22} & \cdot & \cdot & \cdot & a_{2n} \\ \cdot & \cdot & \cdot & \cdot & \cdot & \cdot \\ a_{n1} & a_{n2} & \cdot & \cdot & \cdot & a_{nn} \end{vmatrix} = \begin{vmatrix} a_{11} & a_{21} & \cdot & \cdot & \cdot & a_{n1} \\ a_{12} & a_{22} & \cdot & \cdot & \cdot & a_{n2} \\ \cdot & \cdot & \cdot & \cdot & \cdot & \cdot \\ a_{1n} & a_{2n} & \cdot & \cdot & \cdot & a_{nn} \end{vmatrix}.$$

Write $b_{rs} = a_{sr}$ $(r, s = 1,\dots, n)$. We have to show that $|a_{ij}|_n = |b_{ij}|_n$. Now, by Theorem 1.2.1 (ii) and Definition 1.2.2,

$$|b_{ij}|_n = \sum_{(\lambda_1,\dots,\lambda_n)} \epsilon\begin{pmatrix}\lambda_1,\dots,\lambda_n\\ 1,\dots,n\end{pmatrix} b_{\lambda_1 1}\dots b_{\lambda_n n}$$
$$= \sum_{(\lambda_1,\dots,\lambda_n)} \epsilon(\lambda_1,\dots,\lambda_n) a_{1\lambda_1}\dots a_{n\lambda} = |a_{ij}|_n,$$

and the theorem is therefore proved.

Exercise 1.2.1. Give a direct verification of Theorem 1.2.2 for 2-rowed and 3-rowed determinants.

Theorem 1.2.2 shows that there is symmetry between the rows and columns of a determinant. Hence every statement proved about the rows of a determinant is equally valid for columns, and conversely.

Theorem 1.2.3. *If two rows (or columns) of a determinant D are interchanged, then the resulting determinant has the value $-D$.*

Let $1 \leqslant r < s \leqslant n$, and denote by $D' = |a'_{ij}|_n$ the determinant obtained by interchanging the rth and sth rows in $D = |a_{ij}|_n$. Then

$$a'_{ij} = \begin{cases} a_{ij} & (i \neq r;\ i \neq s) \\ a_{sj} & (i = r) \\ a_{rj} & (i = s). \end{cases}$$

Hence, by Definition 1.2.2,

$$D' = \sum_{(\lambda_1,\ldots,\lambda_n)} \epsilon(\lambda_1,\ldots,\lambda_n) a'_{1\lambda_1} \ldots a'_{n\lambda_n}$$

$$= \sum_{(\lambda_1,\ldots,\lambda_n)} \epsilon(\lambda_1,\ldots,\lambda_n) a_{1\lambda_1} \ldots a_{s\lambda_r} \ldots a_{r\lambda_s} \ldots a_{n\lambda_n}.$$

But, by Theorem 1.1.3 (p. 4), $\epsilon(1,\ldots,s,\ldots,r,\ldots,n) = -1$, and so

$$D' = -\sum_{(\lambda_1,\ldots,\lambda_n)} \epsilon\begin{pmatrix} 1,\ldots, & s,\ldots, & r,\ldots, & n \\ \lambda_1,\ldots, & \lambda_r,\ldots, & \lambda_s,\ldots, & \lambda_n \end{pmatrix} a_{1\lambda_1} \ldots a_{s\lambda_r} \ldots a_{r\lambda_s} \ldots a_{n\lambda_n}.$$

Hence, by Theorem 1.2.1 (i), $D' = -D$.

Corollary. *If two rows (or two columns) of a determinant are identical, then the determinant vanishes.*

Let D be a determinant with two identical rows, and denote by D' the determinant obtained from D by interchanging these two rows. Then obviously $D' = D$. But, by Theorem 1.2.3, $D' = -D$, and therefore $D = 0$.

Exercise 1.2.2. Let $r_1 < \ldots < r_k$. Show that, if the rows with suffixes $r_1, r_2, \ldots, r_k$ of a determinant D are moved into 1st, 2nd,..., kth place respectively while the relative order of the remaining rows stays unchanged, then the resulting determinant is equal to

$$(-1)^{r_1+r_2+\ldots+r_k-\frac{1}{2}k(k+1)} D.$$

When every element of a particular row or column of a determinant is multiplied by a constant k, we say that the row or column in question is multiplied by k.

Theorem 1.2.4. *If a row (or column) of a determinant is multiplied by a constant k, then the value of the determinant is also multiplied by k.*

Let $D = |a_{ij}|_n$ be a given determinant and let D' be obtained from it by multiplying the rth row by k. Then

$$D' = \begin{vmatrix} a_{11} & . & . & . & a_{1n} \\ . & . & . & . & . \\ ka_{r1} & . & . & . & ka_{rn} \\ . & . & . & . & . \\ a_{n1} & . & . & . & a_{nn} \end{vmatrix} = \sum_{(\lambda_1,\ldots,\lambda_n)} \epsilon(\lambda_1,\ldots,\lambda_n) a_{1\lambda_1}\ldots(ka_{r\lambda_r})\ldots a_{n\lambda_n}$$

$$= k \sum_{(\lambda_1,\ldots,\lambda_n)} \epsilon(\lambda_1,\ldots,\lambda_n) a_{1\lambda_1}\ldots a_{n\lambda_n} = kD.$$

The next theorem provides a method for expressing any determinant as a sum of two determinants.

THEOREM 1.2.5.

$$\begin{vmatrix} a_{11} & . & . & a_{1r}+a'_{1r} & . & . & a_{1n} \\ . & . & . & . & . & . & . \\ a_{n1} & . & . & a_{nr}+a'_{nr} & . & . & a_{nn} \end{vmatrix}$$

$$= \begin{vmatrix} a_{11} & . & . & a_{1r} & . & . & a_{1n} \\ . & . & . & . & . & . & . \\ a_{n1} & . & . & a_{nr} & . & . & a_{nn} \end{vmatrix} +$$

$$+ \begin{vmatrix} a_{11} & . & . & a'_{1r} & . & . & a_{1n} \\ . & . & . & . & . & . & . \\ a_{n1} & . & . & a'_{nr} & . & . & a_{nn} \end{vmatrix}.$$

Denoting the determinant on the left-hand side by $|b_{ij}|_n$, we have

$$b_{ij} = \begin{cases} a_{ij} & (j \neq r) \\ a_{ir}+a'_{ir} & (j = r). \end{cases}$$

Hence, by Theorem 1.2.1 (ii) (p. 7),

$$|b_{ij}|_n = \sum_{(\lambda_1,\ldots,\lambda_n)} \epsilon(\lambda_1,\ldots,\lambda_n) b_{\lambda_1 1}\ldots b_{\lambda_r r}\ldots b_{\lambda_n n}$$

$$= \sum_{(\lambda_1,\ldots,\lambda_n)} \epsilon(\lambda_1,\ldots,\lambda_n) a_{\lambda_1 1}\ldots(a_{\lambda_r r}+a'_{\lambda_r r})\ldots a_{\lambda_n n}$$

$$= \sum_{(\lambda_1,\ldots,\lambda_n)} \epsilon(\lambda_1,\ldots,\lambda_n) a_{\lambda_1 1}\ldots a_{\lambda_r r}\ldots a_{\lambda_n n} +$$

$$+ \sum_{(\lambda_1,\ldots,\lambda_n)} \epsilon(\lambda_1,\ldots,\lambda_n) a_{\lambda_1 1}\ldots a'_{\lambda_r r}\ldots a_{\lambda_n n}$$

$$= \begin{vmatrix} a_{11} & . & . & a_{1r} & . & . & a_{1n} \\ . & . & . & . & . & . & . \\ a_{n1} & . & . & a_{nr} & . & . & a_{nn} \end{vmatrix} + \begin{vmatrix} a_{11} & . & . & a'_{1r} & . & . & a_{1n} \\ . & . & . & . & . & . & . \\ a_{n1} & . & . & a'_{nr} & . & . & a_{nn} \end{vmatrix}.$$

EXERCISE 1.2.3. State the analogous result for rows.

A useful corollary to Theorem 1.2.5 can now be easily proved by induction. It enables us to express a determinant, each of whose elements is the sum of h terms, as the sum of h^n determinants.

COROLLARY.

$$\begin{vmatrix} a_{11}^{(1)}+\dots+a_{11}^{(h)} & . & . & . & a_{1n}^{(1)}+\dots+a_{1n}^{(h)} \\ . & . & . & . & . \\ a_{n1}^{(1)}+\dots+a_{n1}^{(h)} & . & . & . & a_{nn}^{(1)}+\dots+a_{nn}^{(h)} \end{vmatrix}$$

$$= \sum_{k_1,\dots,k_n=1}^{h} \begin{vmatrix} a_{11}^{(k_1)} & . & . & . & a_{1n}^{(k_n)} \\ . & . & . & . & . \\ a_{n1}^{(k_1)} & . & . & . & a_{nn}^{(k_n)} \end{vmatrix}.$$

THEOREM 1.2.6. *The value of a determinant remains unchanged if to any row* (*or column*) *is added any multiple of another row* (*or column*).

By saying that the sth row of a determinant is added to the rth row we mean, of course, that every element of the sth row is added to the corresponding element of the rth row. Similar terminology is used for columns.

Let $D = |a_{ij}|_n$, and suppose that D' denotes the determinant obtained when k times the sth row is added to the rth row in D. Assuming that $r < s$ we have

$$D' = \begin{vmatrix} a_{11} & . & . & . & a_{1n} \\ . & . & . & . & . \\ a_{r1}+ka_{s1} & . & . & . & a_{rn}+ka_{sn} \\ . & . & . & . & . \\ a_{s1} & . & . & . & a_{sn} \\ . & . & . & . & . \\ a_{n1} & . & . & . & a_{nn} \end{vmatrix}.$$

Hence, by Theorem 1.2.5 (as applied to rows),

$$D' = \begin{vmatrix} a_{11} & . & . & . & a_{1n} \\ . & . & . & . & . \\ a_{r1} & . & . & . & a_{rn} \\ . & . & . & . & . \\ a_{s1} & . & . & . & a_{sn} \\ . & . & . & . & . \\ a_{n1} & . & . & . & a_{nn} \end{vmatrix} + \begin{vmatrix} a_{11} & . & . & . & a_{1n} \\ . & . & . & . & . \\ ka_{s1} & . & . & . & ka_{sn} \\ . & . & . & . & . \\ a_{s1} & . & . & . & a_{sn} \\ . & . & . & . & . \\ a_{n1} & . & . & . & a_{nn} \end{vmatrix},$$

and so, by Theorem 1.2.4 and the corollary to Theorem 1.2.3,

$$D' = D+k \begin{vmatrix} a_{11} & \cdot & \cdot & \cdot & a_{1n} \\ \cdot & \cdot & \cdot & \cdot & \cdot \\ a_{s1} & \cdot & \cdot & \cdot & a_{sn} \\ \cdot & \cdot & \cdot & \cdot & \cdot \\ a_{s1} & \cdot & \cdot & \cdot & a_{sn} \\ \cdot & \cdot & \cdot & \cdot & \cdot \\ a_{n1} & \cdot & \cdot & \cdot & a_{nn} \end{vmatrix} = D.$$

1.3. Multiplication of determinants

We shall next prove that it is always possible to express the product of two determinants of the same order n as a determinant of order n.

Theorem 1.3.1. (Multiplication theorem for determinants) *Let $A = |a_{ij}|_n$ and $B = |b_{ij}|_n$ be given determinants, and write $C = |c_{ij}|_n$, where*

$$c_{rs} = \sum_{i=1}^{n} a_{ri} b_{is} \quad (r, s = 1,..., n).$$

Then

$$AB = C. \tag{1.3.1}$$

We have

$$\begin{aligned} C &= \sum_{(\lambda_1,...,\lambda_n)} \epsilon(\lambda_1,..., \lambda_n) c_{1\lambda_1}...c_{n\lambda_n} \\ &= \sum_{(\lambda_1,...,\lambda_n)} \epsilon(\lambda_1,..., \lambda_n)\Big(\sum_{\mu_1=1}^{n} a_{1\mu_1} b_{\mu_1\lambda_1}\Big)...\Big(\sum_{\mu_n=1}^{n} a_{n\mu_n} b_{\mu_n\lambda_n}\Big) \\ &= \sum_{\mu_1=1}^{n} ... \sum_{\mu_n=1}^{n} a_{1\mu_1}...a_{n\mu_n} \sum_{(\lambda_1,...,\lambda_n)} \epsilon(\lambda_1,..., \lambda_n) b_{\mu_1\lambda_1}...b_{\mu_n\lambda_n}. \end{aligned} \tag{1.3.2}$$

By Definition 1.2.2 the inner sum in (1.3.2) is equal to

$$\begin{vmatrix} b_{\mu_1 1} & \cdot & \cdot & \cdot & b_{\mu_1 n} \\ \cdot & \cdot & \cdot & \cdot & \cdot \\ b_{\mu_n 1} & \cdot & \cdot & \cdot & b_{\mu_n n} \end{vmatrix}.$$

Hence, if any two μ's are equal, then, by the corollary to Theorem 1.2.3, the inner sum in (1.3.2) vanishes. It follows that in the n-fold summation in (1.3.2) we can omit all sets of μ's which contain at least two equal numbers. The summation then reduces to a simple summation over $n!$ arrangements $(\mu_1,..., \mu_n)$, and we therefore have

$$\begin{aligned} C &= \sum_{(\mu_1,...,\mu_n)} a_{1\mu_1}...a_{n\mu_n} \sum_{(\lambda_1,...,\lambda_n)} \epsilon(\lambda_1,..., \lambda_n) b_{\mu_1\lambda_1}...b_{\mu_n\lambda_n} \\ &= \sum_{(\mu_1,...,\mu_n)} \epsilon(\mu_1,..., \mu_n) a_{1\mu_1}...a_{n\mu_n} \sum_{(\lambda_1,...,\lambda_n)} \epsilon\begin{pmatrix} \mu_1,..., \mu_n \\ \lambda_1,..., \lambda_n \end{pmatrix} b_{\mu_1\lambda_1}...b_{\mu_n\lambda_n}. \end{aligned}$$

Hence, by Theorem 1.2.1 (i) (p. 7),

$$C = \sum_{(\mu_1,\ldots,\mu_n)} \epsilon(\mu_1,\ldots,\mu_n) a_{1\mu_1} \ldots a_{n\mu_n} \begin{vmatrix} b_{11} & \cdot & \cdot & \cdot & b_{1n} \\ \cdot & \cdot & \cdot & \cdot & \cdot \\ b_{n1} & \cdot & \cdot & \cdot & b_{nn} \end{vmatrix}$$

$$= \begin{vmatrix} a_{11} & \cdot & \cdot & \cdot & a_{1n} \\ \cdot & \cdot & \cdot & \cdot & \cdot \\ a_{n1} & \cdot & \cdot & \cdot & a_{nn} \end{vmatrix} \cdot \begin{vmatrix} b_{11} & \cdot & \cdot & \cdot & b_{1n} \\ \cdot & \cdot & \cdot & \cdot & \cdot \\ b_{n1} & \cdot & \cdot & \cdot & b_{nn} \end{vmatrix} = AB.$$

The theorem just proved shows how we may form a determinant which is equal to the product of two given determinants A and B. We have, in fact, $AB = C$, where the element standing in the rth row and sth column of C is obtained by multiplying together the corresponding elements in the rth row of A and the sth column of B and adding the products thus obtained. The determinant C constructed in this way may be said to have been obtained by multiplying A and B 'rows by columns'. Now, by Theorem 1.2.2, the values of A and B are unaltered if rows and columns in either determinant or in both determinants are interchanged. Hence we can equally well form the product AB by carrying out the multiplication 'rows by rows', or 'columns by columns', or 'columus by rows'. These conclusions are expressed in the next theorem.

THEOREM 1.3.2. *The equality* (1.3.1) *continues to hold if the determinant* $C = |c_{ij}|_n$ *is defined by any one of the following sets of relations*:

$$c_{rs} = \sum_{i=1}^{n} a_{ri} b_{si} \quad (r, s = 1,\ldots, n);$$

$$c_{rs} = \sum_{i=1}^{n} a_{ir} b_{is} \quad (r, s = 1,\ldots, n);$$

$$c_{rs} = \sum_{i=1}^{n} a_{ir} b_{si} \quad (r, s = 1,\ldots, n).$$

An interesting application of Theorem 1.3.2 will be given in § 1.4.1 (p. 19).

EXERCISE 1.3.1. Use the definition of a determinant to show that

$$\begin{vmatrix} a_{11} & \cdot & \cdot & \cdot & a_{1m} & 0 & \cdot & \cdot & \cdot & 0 \\ \cdot & \cdot & \cdot & \cdot & \cdot & \cdot & \cdot & \cdot & \cdot & \cdot \\ a_{m1} & \cdot & \cdot & \cdot & a_{mm} & 0 & \cdot & \cdot & \cdot & 0 \\ 0 & \cdot & \cdot & \cdot & 0 & 1 & \cdot & \cdot & \cdot & 0 \\ \cdot & \cdot & \cdot & \cdot & \cdot & \cdot & & & \ddots & \\ 0 & \cdot & \cdot & \cdot & 0 & 0 & \cdot & \cdot & \cdot & 1 \end{vmatrix} = \begin{vmatrix} a_{11} & \cdot & \cdot & \cdot & a_{1m} \\ \cdot & \cdot & \cdot & \cdot & \cdot \\ a_{m1} & \cdot & \cdot & \cdot & a_{mm} \end{vmatrix}.$$

Deduce, by means of Theorem 1.3.1, that

$$\begin{vmatrix} a_{11} & . & . & . & a_{1m} & 0 & . & . & . & 0 \\ . & . & . & . & . & . & . & . & . & . \\ a_{m1} & . & . & . & a_{mm} & 0 & . & . & . & 0 \\ 0 & . & . & . & 0 & b_{11} & . & . & . & b_{1n} \\ . & . & . & . & . & . & . & . & . & . \\ 0 & . & . & . & 0 & b_{n1} & . & . & . & b_{nn} \end{vmatrix}$$

$$= \begin{vmatrix} a_{11} & . & . & . & a_{1m} \\ . & . & . & . & . \\ a_{m1} & . & . & . & a_{mm} \end{vmatrix} . \begin{vmatrix} b_{11} & . & . & . & b_{1n} \\ . & . & . & . & . \\ b_{n1} & . & . & . & b_{nn} \end{vmatrix} .$$

1.4. Expansion theorems

1.4.1. We have already obtained a number of results which can be used in the evaluation of determinants. A procedure that is still more effective for this purpose consists in expressing a determinant in terms of other determinants of lower order. The object of the present section is to develop such a procedure.

DEFINITION 1.4.1. *The* COFACTOR A_{rs} *of the element* a_{rs} *in the determinant*

$$D = \begin{vmatrix} a_{11} & . & . & . & a_{1n} \\ . & . & . & . & . \\ a_{n1} & . & . & . & a_{nn} \end{vmatrix}$$

is defined as

$$A_{rs} = (-1)^{r+s} D_{rs} \quad (r, s = 1, \ldots, n),$$

where D_{rs} *is the determinant of order* $n-1$ *obtained when the r-th row and s-th column are deleted from D.*

For example, if

$$D = \begin{vmatrix} a_{11} & a_{12} & a_{13} \\ a_{21} & a_{22} & a_{23} \\ a_{31} & a_{32} & a_{33} \end{vmatrix},$$

then $$A_{11} = (-1)^{1+1} \begin{vmatrix} a_{22} & a_{23} \\ a_{32} & a_{33} \end{vmatrix} = a_{22}a_{33} - a_{23}a_{32}$$

and $$A_{23} = (-1)^{2+3} \begin{vmatrix} a_{11} & a_{12} \\ a_{31} & a_{32} \end{vmatrix} = a_{12}a_{31} - a_{11}a_{32}.$$

EXERCISE 1.4.1. Suppose that $|b_{ij}|_n$ is the determinant obtained when two adjacent rows (or columns) of a determinant $|a_{ij}|_n$ are interchanged. Show that if the element a_{rs} of $|a_{ij}|_n$ becomes the element $b_{\rho\sigma}$ of $|b_{ij}|_n$, then $B_{\rho\sigma} = -A_{rs}$, where A_{rs} denotes the cofactor of a_{rs} in $|a_{ij}|_n$ and $B_{\rho\sigma}$ the cofactor of $b_{\rho\sigma}$ in $|b_{ij}|_n$.

Theorem 1.4.1. (Expansion of determinants in terms of rows and columns)

If the cofactor of a_{pq} *in* $D = |a_{ij}|_n$ *is denoted by* A_{pq}, *then*

$$\sum_{k=1}^{n} a_{rk} A_{rk} = D \quad (r = 1,..., n), \tag{1.4.1}$$

$$\sum_{k=1}^{n} a_{kr} A_{kr} = D \quad (r = 1,..., n). \tag{1.4.2}$$

This theorem states, in fact, that we may obtain the value of a determinant by multiplying the elements of any one row or column by their cofactors and adding the products thus formed. The identity (1.4.1) is known as the expansion of the determinant D in terms of the elements of the rth row, or simply as the expansion of D in terms of the rth row. Similarly, (1.4.2) is known as the expansion of D in terms of the rth column. In view of Theorem 1.2.2 (p. 8) it is, of course, sufficient to prove (1.4.1).

We begin by showing that

$$\begin{vmatrix} 1 & 0 & . & . & . & 0 \\ b_{21} & b_{22} & . & . & . & b_{2n} \\ . & . & . & . & . & . \\ b_{n1} & b_{n2} & . & . & . & b_{nn} \end{vmatrix} = \begin{vmatrix} b_{22} & . & . & . & b_{2n} \\ . & . & . & . & . \\ b_{n2} & . & . & . & b_{nn} \end{vmatrix}. \tag{1.4.3}$$

Let B, B' denote the values of the determinants on the left-hand side and the right-hand side respectively of (1.4.3). We write $B = |b_{ij}|_n$, so that $b_{11} = 1$, $b_{12} = ... = b_{1n} = 0$. Then

$$B = \sum_{(\lambda_1, \lambda_2,..., \lambda_n)} \epsilon(\lambda_1, \lambda_2,..., \lambda_n) b_{1\lambda_1} b_{2\lambda_2} ... b_{n\lambda_n}$$

$$= \sum_{(\lambda_2,..., \lambda_n) = \mathscr{A}(2,..., n)} \epsilon(1, \lambda_2,..., \lambda_n) b_{2\lambda_2} ... b_{n\lambda_n}.$$

But, for any arrangement $(\lambda_2,..., \lambda_n)$ of $(2,..., n)$, we clearly have

$$\epsilon(1, \lambda_2,..., \lambda_n) = \epsilon(\lambda_2,..., \lambda_n).$$

Hence $$B = \sum_{(\lambda_2,..., \lambda_n) = \mathscr{A}(2,..., n)} \epsilon(\lambda_2,..., \lambda_n) b_{2\lambda_2} ... b_{n\lambda_n} = B',$$

as asserted.

Next, by Theorems 1.2.4 and 1.2.5 (pp. 9–10), we have

$$D = \begin{vmatrix} a_{11} & \cdot & \cdot & \cdot & a_{1n} \\ \cdot & \cdot & \cdot & \cdot & \cdot \\ a_{r1} & \cdot & \cdot & \cdot & a_{rn} \\ \cdot & \cdot & \cdot & \cdot & \cdot \\ a_{n1} & \cdot & \cdot & \cdot & a_{nn} \end{vmatrix}$$

$$= \sum_{k=1}^{n} a_{rk} \begin{vmatrix} a_{11} & \cdot & \cdot & \cdot & & a_{1k} & & \cdot & \cdot & \cdot & a_{1n} \\ \cdot & \cdot & \cdot & \cdot & \cdot & \cdot & \cdot & \cdot & \cdot & \cdot & \cdot \\ 0 & \cdot & \cdot & \cdot & 0 & 1 & 0 & \cdot & \cdot & \cdot & 0 \\ \cdot & \cdot & \cdot & \cdot & \cdot & \cdot & \cdot & \cdot & \cdot & \cdot & \cdot \\ a_{n1} & \cdot & \cdot & \cdot & & a_{nk} & & \cdot & \cdot & \cdot & a_{nn} \end{vmatrix}$$

$$= \sum_{k=1}^{n} a_{rk} \Delta_{rk}, \tag{1.4.4}$$

where Δ_{rk} is the determinant obtained from D when the kth element in the rth row is replaced by 1 and all other elements in the rth row are replaced by 0. By repeated application of Theorem 1.2.3 (p. 8) we obtain

$$\Delta_{rk} = (-1)^{r-1} \begin{vmatrix} 0 & \cdot & \cdot & \cdot & 1 & \cdot & \cdot & \cdot & 0 \\ a_{11} & \cdot & \cdot & \cdot & a_{1k} & \cdot & \cdot & \cdot & a_{1n} \\ \cdot & \cdot & \cdot & \cdot & \cdot & \cdot & \cdot & \cdot & \cdot \\ a_{r-1,1} & \cdot & \cdot & \cdot & a_{r-1,k} & \cdot & \cdot & \cdot & a_{r-1,n} \\ a_{r+1,1} & \cdot & \cdot & \cdot & a_{r+1,k} & \cdot & \cdot & \cdot & a_{r+1,n} \\ \cdot & \cdot & \cdot & \cdot & \cdot & \cdot & \cdot & \cdot & \cdot \\ a_{n1} & \cdot & \cdot & \cdot & a_{nk} & \cdot & \cdot & \cdot & a_{nn} \end{vmatrix}$$

$$= (-1)^{(r-1)+(k-1)} \begin{vmatrix} 1 & 0 & \cdot & \cdot & 0 & 0 & \cdot & \cdot & 0 \\ a_{1k} & a_{11} & \cdot & \cdot & a_{1,k-1} & a_{1,k+1} & \cdot & \cdot & a_{1n} \\ \cdot & \cdot & \cdot & \cdot & \cdot & \cdot & \cdot & \cdot & \cdot \\ a_{r-1,k} & a_{r-1,1} & \cdot & \cdot & a_{r-1,k-1} & a_{r-1,k+1} & \cdot & \cdot & a_{r-1,n} \\ a_{r+1,k} & a_{r+1,1} & \cdot & \cdot & a_{r+1,k-1} & a_{r+1,k+1} & \cdot & \cdot & a_{r+1,n} \\ \cdot & \cdot & \cdot & \cdot & \cdot & \cdot & \cdot & \cdot & \cdot \\ a_{nk} & a_{n1} & \cdot & \cdot & a_{n,k-1} & a_{n,k+1} & \cdot & \cdot & a_{nn} \end{vmatrix}.$$

Hence, by (1.4.3), $\qquad \Delta_{rk} = (-1)^{r+k} D_{rk},$

where D_{rk} denotes the determinant obtained when the rth row and kth column are deleted from D. Hence, by (1.4.4),

$$D = \sum_{k=1}^{n} a_{rk}(-1)^{r+k} D_{rk} = \sum_{k=1}^{n} a_{rk} A_{rk},$$

and the theorem is proved.

We now possess a practical method for evaluating determinants. This consists in first using Theorem 1.2.6 (p. 11) to introduce a number of zeros into some row or column, and then expanding the determinant in terms of that row or column. Consider, for example, the determinant

$$D = \begin{vmatrix} 9 & 7 & 3 & -9 \\ 6 & 3 & 6 & -4 \\ 15 & 8 & 7 & -7 \\ -5 & -6 & 4 & 2 \end{vmatrix}.$$

Adding the last column to each of the first three we have

$$D = \begin{vmatrix} 0 & -2 & -6 & -9 \\ 2 & -1 & 2 & -4 \\ 8 & 1 & 0 & -7 \\ -3 & -4 & 6 & 2 \end{vmatrix}.$$

Next, we add once, twice, and four times the third row to the second row, first row, and fourth row respectively. This leads to the expression

$$D = \begin{vmatrix} 16 & 0 & -6 & -23 \\ 10 & 0 & 2 & -11 \\ 8 & 1 & 0 & -7 \\ 29 & 0 & 6 & -26 \end{vmatrix}.$$

Expanding D in terms of the second column we obtain

$$D = -\begin{vmatrix} 16 & -6 & -23 \\ 10 & 2 & -11 \\ 29 & 6 & -26 \end{vmatrix},$$

and we can continue the process of reduction in a similar manner until D is evaluated.

EXERCISE 1.4.2. Show that $D = -532$.

The expansion theorem (Theorem 1.4.1) can be used to show that the value of the *Vandermonde determinant*

$$D = \begin{vmatrix} a_1^{n-1} & a_1^{n-2} & . & . & . & a_1 & 1 \\ a_2^{n-1} & a_2^{n-2} & . & . & . & a_2 & 1 \\ . & . & . & . & . & . & . \\ a_n^{n-1} & a_n^{n-2} & . & . & . & a_n & 1 \end{vmatrix}$$

is given by

$$D = \prod_{1 \leqslant i < j \leqslant n} (a_i - a_j). \tag{1.4.5}$$

The assertion is obviously true for $n = 2$. We shall assume that it is true for $n-1$, where $n \geqslant 3$, and deduce that it is true for n. We may clearly assume that all the a's are distinct, for otherwise (1.4.5) is true trivially. Consider the determinant

$$\begin{vmatrix} x^{n-1} & x^{n-2} & . & . & . & x & 1 \\ a_2^{n-1} & a_2^{n-2} & . & . & . & a_2 & 1 \\ . & . & . & . & . & . & . \\ a_n^{n-1} & a_n^{n-2} & . & . & . & a_n & 1 \end{vmatrix}.$$

Expanding it in terms of the first row, we see that it is a polynomial in x, say $f(x)$, of degree not greater than $n-1$. Moreover

$$f(a_2) = \ldots = f(a_n) = 0,$$

and so $f(x)$ is divisible by each of the (distinct) factors $x-a_2,\ldots, x-a_n$. Thus

$$f(x) = K(x-a_2)\ldots(x-a_n);$$

and here K is independent of x, as may be seen by comparing the degrees of the two sides of the equation. Now, by (1.4.1), the coefficient of x^{n-1} in $f(x)$ is equal to

$$\begin{vmatrix} a_2^{n-2} & . & . & . & a_2 & 1 \\ . & . & . & . & . & . \\ a_n^{n-2} & . & . & . & a_n & 1 \end{vmatrix},$$

which, by the induction hypothesis, is equal to

$$\prod_{2 \leqslant i<j \leqslant n} (a_i-a_j).$$

This, then, is the value of K; and we have

$$f(x) = (x-a_2)\ldots(x-a_n) \prod_{2 \leqslant i<j \leqslant n} (a_i-a_j).$$

We now complete the proof of (1.4.5) by substituting $x = a_1$.

The result just obtained enables us to derive identities for discriminants of polynomials. The *discriminant* Δ of the polynomial

$$x^n + a_1 x^{n-1} + \ldots + a_{n-1} x + a_n, \qquad (1.4.6)$$

whose roots are $\theta_1,\ldots, \theta_n$, is defined as

$$\Delta = \prod_{1 \leqslant i<j \leqslant n} (\theta_i-\theta_j)^2.$$

It follows that $\Delta = 0$ if and only if (1.4.6) has at least two equal roots. To express Δ in terms of the coefficients of (1.4.6) we observe that, in view of (1.4.5),

$$\Delta = \begin{vmatrix} \theta_1^{n-1} & \theta_1^{n-2} & . & . & . & \theta_1 & 1 \\ . & . & . & . & . & . & . \\ \theta_n^{n-1} & \theta_n^{n-2} & . & . & . & \theta_n & 1 \end{vmatrix} . \begin{vmatrix} \theta_1^{n-1} & \theta_1^{n-2} & . & . & . & \theta_1 & 1 \\ . & . & . & . & . & . & . \\ \theta_n^{n-1} & \theta_n^{n-2} & . & . & . & \theta_n & 1 \end{vmatrix}.$$

Carrying out the multiplication columns by columns, we have

$$\Delta = \begin{vmatrix} s_{2n-2} & s_{2n-3} & \cdot & \cdot & \cdot & s_{n-1} \\ s_{2n-3} & s_{2n-4} & \cdot & \cdot & \cdot & s_{n-2} \\ \cdot & \cdot & \cdot & \cdot & \cdot & \cdot \\ s_{n-1} & s_{n-2} & \cdot & \cdot & \cdot & s_0 \end{vmatrix},$$

where $s_r = \theta_1^r + ... + \theta_n^r$ $(r = 0, 1, 2, ...)$. Using Newton's formulae† we can express $s_0, s_1, ..., s_{2n-2}$ in terms of the coefficients $a_1, ..., a_n$ of (1.4.6), and hence obtain Δ in the desired form.

Consider, for example, the cubic polynomial $x^3 + px + q$. Here

$$\Delta = \begin{vmatrix} s_4 & s_3 & s_2 \\ s_3 & s_2 & s_1 \\ s_2 & s_1 & s_0 \end{vmatrix},$$

and it is easily verified that

$$s_0 = 3, \quad s_1 = 0, \quad s_2 = -2p, \quad s_3 = -3q, \quad s_4 = 2p^2.$$

Hence $\Delta = -(4p^3 + 27q^2)$, and thus at least two roots of $x^3 + px + q = 0$ are equal if and only if $4p^3 + 27q^2 = 0$.

Exercise 1.4.3. Show, by the method indicated above, that the discriminant of the quadratic polynomial $x^2 + \mu x + \nu$ is $\mu^2 - 4\nu$.

We now resume our discussion of the general theory of determinants.

Theorem 1.4.2. *With the same notation as in Theorem* 1.4.1 *we have for* $r \neq s$,

$$\sum_{k=1}^{n} a_{rk} A_{sk} = 0,$$

$$\sum_{k=1}^{n} a_{kr} A_{ks} = 0.$$

In other words, if each element of a row (or column) is multiplied by the cofactor of the corresponding element of *another* fixed row (or column), then the sum of the n products thus formed is equal to zero. This result is an easy consequence of Theorem 1.4.1. We need, of course, prove only the first of the two stated identities.

If $D' = |a'_{ij}|_n$ denotes the determinant obtained from $D = |a_{ij}|_n$ when the sth row is replaced by the rth row, then

$$a'_{ij} = \begin{cases} a_{ij} & (i \neq s) \\ a_{rj} & (i = s). \end{cases}$$

Denoting by A'_{ij} the cofactor of the element a'_{ij} in D', we clearly have

$$A'_{sk} = A_{sk} \qquad (k = 1, ..., n).$$

† See Burnside and Panton, *The Theory of Equations* (10th edition), i. 165–7, or Perron, **18**, i. 150–1.

Hence, by (1.4.1) (p. 15),

$$D' = \sum_{k=1}^{n} a'_{sk} A'_{sk} = \sum_{k=1}^{n} a_{rk} A_{sk}.$$

But the rth row and sth row of D' are identical, and so $D' = 0$. This completes the proof.

It is often convenient to combine Theorems 1.4.1 and 1.4.2 into a single statement. For this purpose we need a new and most useful notation.

DEFINITION 1.4.2. *The symbol* δ_{rs}, *known as the* KRONECKER DELTA, *is defined as*

$$\delta_{rs} = \begin{cases} 1 & (r = s) \\ 0 & (r \neq s). \end{cases}$$

With the aid of the Kronecker delta Theorems 1.4.1 and 1.4.2 can be combined in the following single theorem.

THEOREM 1.4.3. *If* A_{pq} *denotes the cofactor of* a_{pq} *in the determinant* $D = |a_{ij}|_n$, *then*

$$\left.\begin{aligned} \sum_{k=1}^{n} a_{rk} A_{sk} &= \delta_{rs} D \\ \sum_{k=1}^{n} a_{kr} A_{ks} &= \delta_{rs} D \end{aligned}\right\} \quad (r, s = 1,..., n).$$

1.4.2. Our next object is to obtain a generalization of the Expansion Theorem 1.4.1. We require some preliminary definitions.

DEFINITION 1.4.3. *A k-rowed* MINOR *of an n-rowed determinant D is any k-rowed determinant obtained when $n-k$ rows and $n-k$ columns are deleted from D.*

Alternatively, we may say that a k-rowed minor of D is obtained by retaining, with their relative order unchanged, only the elements common to k specified rows and k specified columns.

For instance, the determinant D_{ij}, obtained from the n-rowed determinant D by deletion of the ith row and jth column, is an $(n-1)$-rowed minor of D. Each element of D is, of course, a 1-rowed minor of D.

EXERCISE 1.4.4. Let $1 \leqslant k < n$, and suppose that all k-rowed minors of a given n-rowed determinant D vanish. Show that all $(k+1)$-rowed minors of D vanish also.

The k-rowed minor obtained from D by retaining only the

elements belonging to rows with suffixes $r_1,..., r_k$ and columns with suffixes $s_1,..., s_k$ will be denoted by

$$D(r_1,..., r_k \mid s_1,..., s_k).$$

Thus, for example, if

$$D = \begin{vmatrix} a_{11} & a_{12} & a_{13} \\ a_{21} & a_{22} & a_{23} \\ a_{31} & a_{32} & a_{33} \end{vmatrix},$$

then
$$D(1, 3 \mid 2, 3) = \begin{vmatrix} a_{12} & a_{13} \\ a_{32} & a_{33} \end{vmatrix}.$$

DEFINITION 1.4.4. *The* COFACTOR (*or* ALGEBRAIC COMPLEMENT) $\tilde{D}(r_1,..., r_k \mid s_1,..., s_k)$ *of the minor* $D(r_1,..., r_k \mid s_1,..., s_k)$ *in a determinant* D *is defined as*

$$\tilde{D}(r_1,..., r_k \mid s_1,..., s_k) = (-1)^{r_1+...+r_k+s_1+...+s_k} D(r_{k+1},..., r_n \mid s_{k+1},..., s_n),$$

where $r_{k+1},..., r_n$ *are the* $n-k$ *numbers among* $1,..., n$ *other than* $r_1,..., r_k$, *and* $s_{k+1},..., s_n$ *are the* $n-k$ *numbers among* $1,..., n$ *other than* $s_1,..., s_k$.

We note that for $k = 1$ this definition reduces to that of a cofactor of an element (Definition 1.4.1, p. 14). If $k = n$, i.e. if a minor coincides with the entire determinant, it is convenient to define its cofactor as 1.

Consider, by way of illustration, the 4-rowed determinant $D = |a_{ij}|_4$. Here

$$D(2, 3 \mid 2, 4) = \begin{vmatrix} a_{22} & a_{24} \\ a_{32} & a_{34} \end{vmatrix},$$

and

$$\tilde{D}(2, 3 \mid 2, 4) = (-1)^{2+3+2+4} D(1, 4 \mid 1, 3) = -\begin{vmatrix} a_{11} & a_{13} \\ a_{41} & a_{43} \end{vmatrix}.$$

Theorem 1.4.4. (Laplace's expansion theorem)

Let D be an n-rowed determinant, and let $r_1,..., r_k$ be integers such that $1 \leqslant k < n$ and $1 \leqslant r_1 < ... < r_k \leqslant n$. Then

$$D = \sum_{1 \leqslant u_1 < ... < u_k \leqslant n} D(r_1,..., r_k \mid u_1,..., u_k) \tilde{D}(r_1,..., r_k \mid u_1,..., u_k).$$

This theorem (which was obtained, in essence, by Laplace in 1772) furnishes us with an expansion of the determinant D in terms of k specified rows, namely, the rows with suffixes $r_1,..., r_k$. We form all possible k-rowed minors of D involving all these rows and multiply each of them by its cofactor; the sum of the $\binom{n}{k}$ products is then equal to D. An analogous expansion applies, of course, to columns. It should be noted that for $k = 1$ Theorem 1.4.4 reduces to the identity (1.4.1) on p. 15.

To prove the theorem, let the numbers $r_{k+1},\ldots, r_n$ be defined by the requirements

$$1 \leqslant r_{k+1} < \ldots < r_n \leqslant n, \qquad (r_1,\ldots, r_n) = \mathscr{A}(1,\ldots, n).$$

Then, by Theorems 1.2.1 (i) (p. 7) and 1.1.4 (p. 4) we have

$$\begin{aligned} D &= \sum_{(s_1,\ldots,s_n)=\mathscr{A}(1,\ldots,n)} \epsilon\begin{pmatrix} r_1,\ldots, r_n \\ s_1,\ldots, s_n \end{pmatrix} a_{r_1 s_1}\ldots a_{r_n s_n} \\ &= \sum_{(s_1,\ldots,s_n)=\mathscr{A}(1,\ldots,n)} (-1)^{r_1+\ldots+r_k+s_1+\ldots+s_k} \epsilon\begin{pmatrix} r_1,\ldots, r_k \\ s_1,\ldots, s_k \end{pmatrix} \times \\ &\qquad \times \epsilon\begin{pmatrix} r_{k+1},\ldots, r_n \\ s_{k+1},\ldots, s_n \end{pmatrix} a_{r_1 s_1}\ldots a_{r_n s_n}. \end{aligned} \tag{1.4.7}$$

Now we can clearly obtain all arrangements $(s_1,\ldots, s_n)$ of $(1,\ldots, n)$ —and each arrangement exactly once—by separating the numbers $1,\ldots, n$ in all possible ways into a set of k and a set of $n-k$ numbers, and letting $(s_1,\ldots, s_k)$ vary over all arrangements of the first and $(s_{k+1},\ldots, s_n)$ over all arrangements of the second set. Thus the condition $(s_1,\ldots, s_n) = \mathscr{A}(1,\ldots, n)$ below the summation sign in (1.4.7) can be replaced by the conditions

$$(u_1,\ldots, u_n) = \mathscr{A}(1,\ldots, n); \tag{1.4.8}$$

$$u_1 < \ldots < u_k; \qquad u_{k+1} < \ldots < u_n; \tag{1.4.9}$$

$$(s_1,\ldots, s_k) = \mathscr{A}(u_1,\ldots, u_k); \qquad (s_{k+1},\ldots, s_n) = \mathscr{A}(u_{k+1},\ldots, u_n).$$

Indicating by an accent that the summation is to be taken over the integers $u_1,\ldots, u_n$ satisfying (1.4.8) and (1.4.9), we therefore have

$$\begin{aligned} D &= \sum{}' (-1)^{r_1+\ldots+r_k+u_1+\ldots+u_k} \sum_{(s_1,\ldots,s_k)=\mathscr{A}(u_1,\ldots,u_k)} \epsilon\begin{pmatrix} r_1,\ldots, r_k \\ s_1,\ldots, s_k \end{pmatrix} a_{r_1 s_1}\ldots a_{r_k s_k} \\ &\qquad \times \sum_{(s_{k+1},\ldots,s_n)=\mathscr{A}(u_{k+1},\ldots,u_n)} \epsilon\begin{pmatrix} r_{k+1},\ldots, r_n \\ s_{k+1},\ldots, s_n \end{pmatrix} a_{r_{k+1} s_{k+1}}\ldots a_{r_n s_n} \\ &= \sum{}'(-1)^{r_1+\ldots+r_k+u_1+\ldots+u_k} \begin{vmatrix} a_{r_1 u_1} & . & . & . & a_{r_1 u_k} \\ . & . & . & . & . \\ a_{r_k u_1} & . & . & . & a_{r_k u_k} \end{vmatrix} \times \\ &\qquad \times \begin{vmatrix} a_{r_{k+1} u_{k+1}} & . & . & . & a_{r_{k+1} u_n} \\ . & . & . & . & . \\ a_{r_n u_{k+1}} & . & . & . & a_{r_n u_n} \end{vmatrix} \end{aligned}$$

$$= \sum{}'(-1)^{r_1+\dots+r_k+u_1+\dots+u_k} D(r_1,\dots, r_k \mid u_1,\dots, u_k) \times$$
$$\times D(r_{k+1},\dots, r_n \mid u_{k+1},\dots, u_n)$$
$$= \sum{}' D(r_1,\dots, r_k \mid u_1,\dots, u_k) \tilde{D}(r_1,\dots, r_k \mid u_1,\dots, u_k)$$
$$= \sum_{1 \leqslant u_1 < \dots < u_k \leqslant n} D(r_1,\dots, r_k \mid u_1,\dots, u_k) \times$$
$$\times \tilde{D}(r_1,\dots, r_k \mid u_1,\dots, u_k) \sum_{u_{k+1},\dots,u_n} 1,$$

where the inner sum is extended over all integers $u_{k+1},\dots, u_n$ satisfying (1.4.8) and (1.4.9). Now the integers $u_{k+1},\dots, u_n$ are clearly determined uniquely for each set of $u_1,\dots, u_k$. Hence the value of the inner sum is equal to 1, and the theorem is proved.

The natural way in which products of minors and their cofactors occur in the expansion of a determinant can be made intuitively clear as follows. To expand an n-rowed determinant in terms of the rows with suffixes $r_1,\dots, r_k$, we write every element a_{ij} in each of these rows in the form $a_{ij}+0$ and every element a_{pq} in each of the remaining rows in the form $0+a_{pq}$. Using the corollary to Theorem 1.2.5 (p. 11) we then obtain the given determinant as a sum of 2^n determinants. Each of these either vanishes or else may be expressed (by virtue of Exercise 1.3.1, p. 13, and after a preliminary rearrangement of rows and columns) as a product of a k-rowed minor and its cofactor. The reader will find it helpful actually to carry out the procedure described here, say for the case $n = 4$, $k = 2$, $r_1 = 1$, $r_2 = 3$.

As an illustration of the use of Laplace's expansion we shall evaluate the determinant

$$D = \begin{vmatrix} 0 & 0 & a_{13} & a_{14} & a_{15} \\ 0 & 0 & a_{23} & a_{24} & 0 \\ 0 & 0 & a_{33} & 0 & 0 \\ 0 & a_{42} & a_{43} & a_{44} & a_{45} \\ a_{51} & a_{52} & a_{53} & a_{54} & a_{55} \end{vmatrix}$$

by expanding it in terms of the first three rows. The only 3-rowed minor which involves these three rows and does not necessarily vanish is

$$D(1, 2, 3 \mid 3, 4, 5) = \begin{vmatrix} a_{13} & a_{14} & a_{15} \\ a_{23} & a_{24} & 0 \\ a_{33} & 0 & 0 \end{vmatrix}.$$

Expanding this minor in terms of the last column, we obtain

$$D(1,2,3\,|\,3,4,5) = a_{15}\begin{vmatrix} a_{23} & a_{24} \\ a_{33} & 0 \end{vmatrix} = -a_{15}a_{24}a_{33}.$$

Furthermore

$$\tilde{D}(1,2,3\,|\,3,4,5) = (-1)^{1+2+3+3+4+5}D(4,5\,|\,1,2)$$

$$= \begin{vmatrix} 0 & a_{42} \\ a_{51} & a_{52} \end{vmatrix} = -a_{42}a_{51},$$

and so, by Theorem 1.4.4,

$$D = D(1,2,3\,|\,3,4,5)\tilde{D}(1,2,3\,|\,3,4,5) = a_{15}a_{24}a_{33}a_{42}a_{51}.$$

1.5. Jacobi's theorem

With every determinant may be associated a second determinant of the same order whose elements are the cofactors of the elements of the first. We propose now to investigate the relation between two such determinants.

DEFINITION 1.5.1. *If A_{rs} denotes the cofactor of a_{rs} in $D = |a_{ij}|_n$, then $D^* = |A_{ij}|_n$ is known as the* ADJUGATE (DETERMINANT) *of D.*

Our object is to express D^* in terms of D and, more generally, to establish the relation between corresponding minors in D and D^*. In discussing these questions we shall require an important general principle concerning polynomials in several variables. We recall that two polynomials, say $f(x_1,..., x_m)$ and $g(x_1,..., x_m)$, are said to be *identically equal* if $f(x_1,..., x_m) = g(x_1,..., x_m)$ for all values of $x_1,..., x_m$. Again, the two polynomials are said to be *formally equal* if the corresponding coefficients in f and g are equal. It is well known that identity and formal equality imply each other. We shall express this relation between the polynomials f and g by writing $f = g$.

THEOREM 1.5.1. *Let f, g, h be polynomials in m variables. If $fg = fh$ and $f \neq 0$, then $g = h$.*

When $m = 1$ this is a well known elementary result. For the proof of the theorem for $m > 1$ we must refer the reader elsewhere.†

THEOREM 1.5.2. *If D is an n-rowed determinant and D^* its adjugate, then*

$$D^* = D^{n-1}.$$

This formula was discovered by Cauchy in 1812. To prove it, we write $D = |a_{ij}|_n$, $D^* = |A_{ij}|_n$, and form the product DD^* rows by

† See, for example, van der Waerden, *Modern Algebra* (English edition), i. 47.

rows. Thus

$$DD^* = \left|\sum_{k=1}^{n} a_{ik} A_{jk}\right|_n = |\delta_{ij} D|_n = \begin{vmatrix} D & 0 & . & . & . & 0 \\ 0 & D & . & . & . & 0 \\ . & . & . & . & . & . \\ 0 & 0 & . & . & . & D \end{vmatrix},$$

and therefore

$$DD^* = D^n. \tag{1.5.1}$$

If now $D \neq 0$, then, dividing both sides of (1.5.1) by D, we obtain the required result. If, however, $D = 0$ this obvious device fails, and we have recourse to Theorem 1.5.1.

Let us regard D as a polynomial in its n^2 elements. The adjugate determinant D^* is then a polynomial in the same n^2 elements, and (1.5.1) is a polynomial identity. But D is not an identically vanishing polynomial and so, by (1.5.1) and Theorem 1.5.1 (with $f = D$, $g = D^*$, $h = D^{n-1}$) we obtain the required result.†

Our next result—the main result of the present section—was discovered by Jacobi in 1833.

THEOREM 1.5.3. (Jacobi's theorem)

If M is a k-rowed minor of a determinant D, M^* the corresponding minor of the adjugate determinant D^*, and $\tilde{M}$ the cofactor of M in D, then

$$M^* = D^{k-1}\tilde{M}. \tag{1.5.2}$$

Before proving this formula we point out a few special cases. The order of D is, as usual, denoted by n. (i) If $k = 1$, then (1.5.2) simply reduces to the definition of cofactors of elements of a determinant. (ii) If $k = n$, then (1.5.2) reduces to Theorem 1.5.2. (iii) For $k = n-1$ the formula (1.5.2) states that if $D = |a_{ij}|_n$, $D^* = |A_{ij}|_n$, then the cofactor of A_{rs} in D^* is equal to $D^{n-2}a_{rs}$. (iv) For $k = 2$ (1.5.2) implies that if $D = 0$, then every 2-rowed minor of D^* vanishes.

To prove (1.5.2) we first consider the special case when M is situated in the top left-hand corner of D, so that

$$M = \begin{vmatrix} a_{11} & . & . & . & a_{1k} \\ . & . & . & . & . \\ a_{k1} & . & . & . & a_{kk} \end{vmatrix}, \qquad \tilde{M} = \begin{vmatrix} a_{k+1,k+1} & . & . & . & a_{k+1,n} \\ . & . & . & . & . \\ a_{n,k+1} & . & . & . & a_{nn} \end{vmatrix},$$

$$M^* = \begin{vmatrix} A_{11} & . & . & . & A_{1k} \\ . & . & . & . & . \\ A_{k1} & . & . & . & A_{kk} \end{vmatrix}.$$

† Alternative proofs which do not depend on Theorem 1.5.1 will be found in § 1.6.3 and § 3.5.

Multiplying determinants rows by rows and using Theorem 1.4.3 (p. 20), we obtain

$$\begin{vmatrix} a_{11} & \cdot & \cdot & \cdot & a_{1n} \\ \cdot & \cdot & \cdot & \cdot & \cdot \\ \cdot & \cdot & \cdot & \cdot & \cdot \\ \cdot & \cdot & \cdot & \cdot & \cdot \\ \cdot & \cdot & \cdot & \cdot & \cdot \\ a_{n1} & \cdot & \cdot & \cdot & a_{nn} \end{vmatrix} \begin{vmatrix} A_{11} & \cdot & \cdot & \cdot & A_{1k} & A_{1,k+1} & \cdot & \cdot & \cdot & A_{1n} \\ \cdot & \cdot & \cdot & \cdot & \cdot & \cdot & \cdot & \cdot & \cdot & \cdot \\ A_{k1} & \cdot & \cdot & \cdot & A_{kk} & A_{k,k+1} & \cdot & \cdot & \cdot & A_{kn} \\ 0 & \cdot & \cdot & \cdot & 0 & 1 & \cdot & \cdot & \cdot & 0 \\ \cdot & \cdot & \cdot & \cdot & \cdot & \cdot & \ddots & & & \\ 0 & \cdot & \cdot & \cdot & 0 & 0 & & & & 1 \end{vmatrix}$$

$$= \begin{vmatrix} D & \cdot & \cdot & \cdot & 0 & a_{1,k+1} & \cdot & \cdot & \cdot & a_{1n} \\ & \ddots & & & \cdot & \cdot & \cdot & \cdot & \cdot & \cdot \\ 0 & \cdot & \cdot & \cdot & D & a_{k,k+1} & \cdot & \cdot & \cdot & a_{kn} \\ 0 & \cdot & \cdot & \cdot & 0 & a_{k+1,k+1} & \cdot & \cdot & \cdot & a_{k+1,n} \\ \cdot & \cdot & \cdot & \cdot & \cdot & \cdot & \cdot & \cdot & \cdot & \cdot \\ 0 & \cdot & \cdot & \cdot & 0 & a_{n,k+1} & \cdot & \cdot & \cdot & a_{nn} \end{vmatrix}$$

Now, by Laplace's expansion theorem, the second determinant on the left is equal to M^*, while the determinant on the right is equal to $D^k\tilde{M}$. Thus

$$DM^* = D^k\tilde{M}.$$

Since this is a polynomial identity in the n^2 elements of D, and since D does not vanish identically, it follows by Theorem 1.5.1 that (1.5.2) is valid for the special case under consideration.

We next turn to the general case, and suppose the minor M to consist of those elements of D which belong to the rows with suffixes $r_1,\ldots,r_k$ and to the columns with suffixes $s_1,\ldots,s_k$ (where $r_1 < \ldots < r_k$ and $s_1 < \ldots < s_k$). We write

$$r_1+\ldots+r_k+s_1+\ldots+s_k = t.$$

Our aim is to reduce the general case to the special case considered above by rearranging the rows and columns of D in such a way that the minor M is moved to the top left-hand corner, while the relative order of the rows and columns not involved in M remains unchanged. We denote the new determinant thus obtained by $\mathscr{D}$, the k-rowed minor in its top left-hand corner by $\mathscr{M}$, the cofactor of $\mathscr{M}$ in $\mathscr{D}$ by $\tilde{\mathscr{M}}$, and the k-rowed minor in the top left-hand corner of the adjugate determinant $\mathscr{D}^*$ by $\mathscr{M}^*$. In view of the special case already discussed we then have

$$\mathscr{M}^* = \mathscr{D}^{k-1}\tilde{\mathscr{M}}. \tag{1.5.3}$$

Now obviously $\mathscr{M} = M$; and, by Exercise 1.2.2 (p. 9),

$$\mathscr{D} = (-1)^t D. \tag{1.5.4}$$

It is, moreover, clear that

$$\tilde{\mathscr{M}} = (-1)^t \tilde{M}. \tag{1.5.5}$$

In view of Exercise 1.4.1 (p. 14) it follows easily that the cofactor of a_{ij} in $\mathscr{D}$ is equal to $(-1)^t A_{ij}$.† Hence

$$\mathscr{M}^* = (-1)^{tk} M^*, \tag{1.5.6}$$

and we complete the proof of the theorem by substituting (1.5.4), (1.5.5), and (1.5.6) in (1.5.3).

EXERCISE 1.5.1. Let $A, H, G,...$ be the cofactors of the elements $a, h, g,...$ in the determinant

$$\Delta = \begin{vmatrix} a & h & g \\ h & b & f \\ g & f & c \end{vmatrix}.$$

Show that $aA+hH+gG = \Delta$, $aH+hB+gF = 0$, and also that the cofactors of the elements $A, H, G,...$ in the determinant

$$\begin{vmatrix} A & H & G \\ H & B & F \\ G & F & C \end{vmatrix}$$

are equal to $a\Delta$, $h\Delta$, $g\Delta$,... respectively.

1.6. Two special theorems on linear equations

We shall next prove two special theorems on linear equations and derive some of their consequences. The second theorem is needed for establishing the basis theorems (Theorems 2.3.2 and 2.3.3) in the next chapter. In touching on the subject of linear equations we do not at present seek to develop a general theory—a task which we defer till Chapter V.

1.6.1. THEOREM 1.6.1. *Let $n \geqslant 1$, and let $D = |a_{ij}|_n$ be a given determinant. Then a necessary and sufficient condition for the existence of numbers $t_1,..., t_n$, not all zero, satisfying the equations*

$$\left.\begin{array}{c} a_{11}t_1+a_{12}t_2+...+a_{1n}t_n = 0 \\ \cdot \quad \cdot \quad \cdot \quad \cdot \quad \cdot \quad \cdot \quad \cdot \quad \cdot \\ a_{n1}t_1+a_{n2}t_2+...+a_{nn}t_n = 0 \end{array}\right\} \tag{1.6.1}$$

is

$$D = 0. \tag{1.6.2}$$

The *sufficiency* of the stated condition is established by induction with respect to n. For $n = 1$ the assertion is true trivially. Suppose it holds for $n-1$, where $n \geqslant 2$; we shall then show that it also holds

† It must, of course, be remembered that a_{ij} does not necessarily stand in the ith row and jth column of $\mathscr{D}$.

for n. Let (1.6.2) be satisfied. If $a_{11} = ... = a_{n1} = 0$, then (1.6.1) is satisfied by

$$t_1 = 1, \quad t_2 = ... = t_n = 0,$$

and the required assertion is seen to hold. If, on the other hand, the numbers $a_{11},..., a_{n1}$ do not all vanish we may assume, without loss of generality, that $a_{11} \neq 0$. In that case we subtract, for $i = 2,..., n$, a_{i1}/a_{11} times the first row from the ith row in D and obtain

$$D = \begin{vmatrix} a_{11} & a_{12} & . & . & . & a_{1n} \\ 0 & b_{22} & . & . & . & b_{2n} \\ . & . & . & . & . & . \\ 0 & b_{n2} & . & . & . & b_{nn} \end{vmatrix} = a_{11} \begin{vmatrix} b_{22} & . & . & . & b_{2n} \\ . & . & . & . & . \\ b_{n2} & . & . & . & b_{nn} \end{vmatrix},$$

where $$b_{ij} = a_{ij} - \frac{a_{i1}a_{1j}}{a_{11}} \qquad (i,j = 2,..., n).$$

Hence $$\begin{vmatrix} b_{22} & . & . & . & b_{2n} \\ . & . & . & . & . \\ b_{n2} & . & . & . & b_{nn} \end{vmatrix} = 0,$$

and so, by the induction hypothesis, there exist numbers $t_2,..., t_n$, not all zero, such that

$$\sum_{j=2}^{n} \left(a_{ij} - \frac{a_{i1}a_{1j}}{a_{11}}\right) t_j = 0 \qquad (i = 2,..., n). \tag{1.6.3}$$

Let t_1 now be defined by the equation

$$t_1 = -\frac{1}{a_{11}} \sum_{j=2}^{n} a_{1j} t_j, \tag{1.6.4}$$

so that $$\sum_{j=1}^{n} a_{1j} t_j = 0. \tag{1.6.5}$$

By (1.6.3) and (1.6.4) we have

$$\sum_{j=1}^{n} a_{ij} t_j = 0 \qquad (i = 2,..., n), \tag{1.6.6}$$

and (1.6.5) and (1.6.6) are together equivalent to (1.6.1). The sufficiency of (1.6.2) is therefore established.

To prove the *necessity* of (1.6.2) we again argue by induction. We have to show that if $D \neq 0$ and the numbers $t_1,..., t_n$ satisfy (1.6.1), then $t_1 = ... = t_n = 0$. For $n = 1$ this assertion is true trivially. Suppose, next, that it holds for $n-1$, where $n \geqslant 2$. The numbers $a_{11},..., a_{n1}$ are not all zero (since $D \neq 0$), and we may, therefore, assume that $a_{11} \neq 0$. If $t_1,..., t_n$ satisfy (1.6.1), then (1.6.4) holds and therefore so does (1.6.3). But

$$\begin{vmatrix} b_{22} & . & . & . & b_{2n} \\ . & . & . & . & . \\ b_{n2} & . & . & . & b_{nn} \end{vmatrix} \neq 0.$$

Hence, by (1.6.3) and the induction hypothesis, $t_2 = ... = t_n = 0$. It follows, by (1.6.4), that $t_1 = 0$; and the proof is therefore complete.†

An alternative proof of the necessity of condition (1.6.2) can be based on Theorem 1.4.3 (p. 20). Suppose that there exist numbers $t_1,..., t_n$, not all zero, satisfying (1.6.1), i.e.

$$\sum_{j=1}^{n} a_{ij} t_j = 0 \qquad (i = 1,..., n).$$

Denoting by A_{ik} the cofactor of a_{ik} in D we therefore have

$$\sum_{i=1}^{n} A_{ik} \sum_{j=1}^{n} a_{ij} t_j = 0 \qquad (k = 1,..., n),$$

i.e.

$$\sum_{j=1}^{n} t_j \sum_{i=1}^{n} a_{ij} A_{ik} = 0 \qquad (k = 1,..., n).$$

Hence, by Theorem 1.4.3,

$$\sum_{j=1}^{n} t_j . \delta_{jk} D = 0 \qquad (k = 1,... n)$$

i.e. $t_k D = 0$ $(k = 1,..., n)$. But, by hypothesis, $t_1,..., t_n$ are not all equal to zero; and therefore $D = 0$.

An obvious but useful consequence of Theorem 1.6.1 is as follows:

THEOREM 1.6.2. *Let a_{ij} $(i = 1,..., n-1;\ j = 1,..., n)$ be given numbers, where $n \geqslant 2$. Then there exists at least one set of numbers $t_1,..., t_n$, not all zero, such that*

$$\left.\begin{array}{r} a_{11} t_1 + ... + a_{1n} t_n = 0 \\ . \quad . \quad . \quad . \quad . \quad . \quad . \quad . \\ a_{n-1,1} t_1 + ... + a_{n-1,n} t_n = 0 \end{array}\right\}. \tag{1.6.7}$$

To the $n-1$ equations comprising (1.6.7) we add the equation

$$0t_1 + ... + 0t_n = 0,$$

† The reader should note that the proof just given depends essentially on the elementary device of reducing the number of 'unknowns' from n to $n-1$ by elimination of t_1.

which does not, of course, affect the choice of permissible sets of the numbers $t_1,..., t_n$. Since

$$\begin{vmatrix} a_{11} & . & . & . & a_{1n} \\ . & . & . & . & . \\ a_{n-1,1} & . & . & . & a_{n-1,n} \\ 0 & . & . & . & 0 \end{vmatrix} = 0,$$

it follows by the previous theorem that there exist values of $t_1,..., t_n$, not all zero, which satisfy (1.6.7).

It is interesting to observe that we can easily give a direct proof of Theorem 1.6.2, without appealing to the theory of determinants, by using essentially the same argument as in the proof of Theorem 1.6.1. For $n = 2$ the assertion is obviously true. Assume that it holds for $n-1$, where $n \geqslant 3$. If now $a_{11} = ... = a_{n-1,1} = 0$, then the equations (1.6.7) are satisfied by $t_1 = 1$, $t_2 = ... = t_n = 0$. If, however, $a_{11},..., a_{n-1,1}$ do not all vanish, we may assume that $a_{11} \neq 0$. In that case we consider the equations

$$\left.\begin{aligned} a_{11}t_1+a_{12}t_2+...+\ a_{1n}t_n &= 0 \\ b_{22}t_2+...+\ b_{2n}t_n &= 0 \\ . \quad . \quad . \quad . \quad . \quad . \quad . \quad & \\ b_{n-1,2}t_2+...+b_{n-1,n}t_n &= 0 \end{aligned}\right\}, \tag{1.6.8}$$

where the b_{ij} are defined as in the proof of Theorem 1.6.1. By the induction hypothesis there exist values of $t_2,..., t_n$, not all 0, satisfying the last $n-2$ equations in (1.6.8); and, with a suitable choice of t_1, the first equation can be satisfied, too. But the values of $t_1,..., t_n$ which satisfy (1.6.8) also satisfy (1.6.7), and the theorem is therefore proved.

Exercise 1.6.1. Let $1 \leqslant m < n$ and let a_{ij} $(i = 1,..., m;\ j = 1,..., n)$ be given numbers. Show that there exist numbers $t_1,... t_n$, not all 0, such that

$$a_{11}t_1+...+a_{1n}t_n = 0,$$
$$. \quad . \quad . \quad . \quad . \quad . \quad .$$
$$a_{m1}t_1+...+a_{mn}t_n = 0.$$

1.6.2. As a first application of Theorem 1.6.1 we shall prove a well-known result on polynomials, which will be useful in later chapters.

Theorem 1.6.3. *If the polynomial*

$$f(x) = c_0 x^n+c_1 x^{n-1}+...+c_{n-1}x+c_n$$

vanishes for $n+1$ distinct values of x, then it vanishes identically.

Let $x_1,\ldots,x_{n+1}$ be distinct numbers, and suppose that $f(x_1) = \ldots = f(x_{n+1}) = 0$, i.e.

$$c_0 x_1^n + c_1 x_1^{n-1} + \ldots + c_{n-1} x_1 + c_n = 0$$
$$\cdot \quad \cdot \quad \cdot \quad \cdot \quad \cdot \quad \cdot \quad \cdot \quad \cdot \quad \cdot \quad \cdot \quad \cdot$$
$$c_0 x_{n+1}^n + c_1 x_{n+1}^{n-1} + \ldots + c_{n-1} x_{n+1} + c_n = 0.$$

Since, by (1.4.5), p. 17, the Vandermonde determinant

$$\begin{vmatrix} x_1^n & x_1^{n-1} & \cdot & \cdot & \cdot & x_1 & 1 \\ \cdot & \cdot & \cdot & \cdot & \cdot & \cdot & \cdot \\ x_{n+1}^n & x_{n+1}^{n-1} & \cdot & \cdot & \cdot & x_{n+1} & 1 \end{vmatrix}$$

is equal to
$$\prod_{1 \leqslant i < j \leqslant n+1} (x_i - x_j)$$

and therefore not equal to zero, it follows by Theorem 1.6.1 that $c_0 = c_1 = \ldots = c_n = 0$, i.e. that $f(x)$ vanishes identically.

COROLLARY. *If $f(x)$, $g(x)$ are polynomials, and there exists a constant x_0 such that*
$$f(x) = g(x)$$
whenever $x > x_0$, then the equality holds for ALL *values of x.*

Let n be the greater of the degrees of f and g. Now $f(x) - g(x)$ vanishes for any $n+1$ distinct values of x which exceed x_0, and the assertion follows, therefore, by Theorem 1.6.3.

1.6.3. Theorem 1.6.1 enables us to dispense with the comparatively deep Theorem 1.5.1 (p. 24) in the proof of Theorem 1.5.2. As we recall, there is only a difficulty when $D = 0$, and in that case we have to show that $D^* = 0$. We write, as before, $D = |a_{ij}|_n$, $D^* = |A_{ij}|_n$, and assume (as we may clearly do) that at least one element in D, say a_{kl}, does not vanish. In view of Theorem 1.4.3 (p. 20) and the assumption $D = 0$ we infer that the relations

$$\sum_{j=1}^{n} A_{ij} t_j = 0 \qquad (i = 1,\ldots,n)$$

are satisfied for
$$t_1 = a_{k1}, \quad \ldots, \quad t_n = a_{kn}.$$

But here $t_l = a_{kl} \neq 0$ and so, by Theorem 1.6.1,

$$D^* = \begin{vmatrix} A_{11} & \cdot & \cdot & \cdot & A_{1n} \\ \cdot & \cdot & \cdot & \cdot & \cdot \\ A_{n1} & \cdot & \cdot & \cdot & A_{nn} \end{vmatrix} = 0.$$

1.6.4. It is useful to possess some easily applicable criteria for deciding whether a determinant does or does not vanish. Below we shall deduce one such criterion due to Minkowski (1900).

Definition 1.6.1. *A determinant* $|a_{ij}|_n$ *is* dominated *by its diagonal elements if*

$$|a_{rr}| > \sum_{\substack{s=1\\s\neq r}}^{n} |a_{rs}| \qquad (r = 1,..., n). \tag{1.6.9}$$

Theorem 1.6.4. *A determinant which is dominated by its diagonal elements does not vanish.*

Let $D = |a_{ij}|_n$ be the determinant in question and suppose that $D = 0$. Then, by Theorem 1.6.1, there exist numbers $t_1,..., t_n$, not all 0, such that

$$\sum_{s=1}^{n} a_{rs} t_s = 0 \qquad (r = 1,..., n). \tag{1.6.10}$$

We have, for some k,

$$\max_{1\leqslant s\leqslant n} |t_s| = |t_k| \neq 0.$$

Then, by (1.6.10),
$$a_{kk} t_k = -\sum_{\substack{s=1\\s\neq k}}^{n} a_{ks} t_s,$$

and so
$$|a_{kk}|\,|t_k| \leqslant \sum_{\substack{s=1\\s\neq k}}^{n} |a_{ks}|\,|t_s| \leqslant \sum_{\substack{s=1\\s\neq k}}^{n} |a_{ks}|\,|t_k|.$$

Therefore
$$|a_{kk}| \leqslant \sum_{\substack{s=1\\s\neq k}}^{n} |a_{ks}|,$$

and this contradicts the hypothesis (1.6.9).

Exercise 1.6.2. Show that if even a single one of the n signs ' $>$ ' in (1.6.9) is replaced by ' $\geqslant$ ', then the conclusion of Theorem 1.6.4 need no longer be true.

A result related to Theorem 1.6.4 is as follows.

Theorem 1.6.5. *If D is a determinant, with real elements and positive diagonal elements, which is dominated by its diagonal elements, then its value is positive.*

We give a proof due to Furtwängler (1936). The theorem is obviously true for $n = 2$. Assume that it holds for $n-1$, where $n \geqslant 3$. We know that $a_{11} \neq 0$; and subtracting, for $r = 2,..., n$, a_{r1}/a_{11} times the first row from the rth row, we obtain $D = a_{11} D'$, where

$$D' = \begin{vmatrix} b_{22} & . & . & . & b_{2n} \\ . & . & . & . & . \\ b_{n2} & . & . & . & b_{nn} \end{vmatrix}$$

and
$$b_{rs} = a_{rs} - \frac{a_{r1} a_{1s}}{a_{11}} \qquad (r, s = 2,..., n).$$

Now, by hypothesis, $\sum_{s=2}^{n} |a_{1s}| < |a_{11}|.$

Therefore $$\sum_{s=2}^{n} \left|\frac{a_{r1}a_{1s}}{a_{11}}\right| \leqslant |a_{r1}| \quad (r = 2,..., n),$$

and hence, for $r = 2,..., n$,

$$\frac{a_{r1}a_{1r}}{a_{11}} + \sum_{\substack{s=2\\s\neq r}}^{n} |b_{rs}| \leqslant \frac{a_{r1}a_{1r}}{a_{11}} + \sum_{\substack{s=2\\s\neq r}}^{n} |a_{rs}| + \sum_{\substack{s=2\\s\neq r}}^{n} \left|\frac{a_{r1}a_{1s}}{a_{11}}\right|$$

$$\leqslant \sum_{\substack{s=2\\s\neq r}}^{n} |a_{rs}| + \sum_{s=2}^{n} \left|\frac{a_{r1}a_{1s}}{a_{11}}\right| \leqslant \sum_{\substack{s=2\\s\neq r}}^{n} |a_{rs}| + |a_{r1}| = \sum_{\substack{s=1\\s\neq r}}^{n} |a_{rs}| < a_{rr}.$$

Thus $$b_{rr} > \sum_{\substack{s=2\\s\neq r}}^{n} |b_{rs}| \qquad (r = 2,..., n).$$

Hence, by the induction hypothesis, $D' > 0$, and so $D > 0$. The proof is therefore complete.

It is possible to obtain further refinements of Theorems 1.6.4 and 1.6.5, and we conclude this chapter by stating two of these without proof. (i) If $D = |a_{ij}|_n$ is a determinant satisfying the conditions of Theorem 1.6.4, then

$$\prod_{r=1}^{n} (|a_{rr}| - a_r^*) \leqslant |D| \leqslant \prod_{r=1}^{n} (|a_{rr}| + a_r^*),$$

where $$a_r^* = \begin{cases} \sum_{s=r+1}^{n} |a_{rs}| & (1 \leqslant r < n) \\ 0 & (r = n). \end{cases}$$

(ii) If $D = |a_{ij}|_n$ is a determinant satisfying the conditions of Theorem 1.6.5, then

$$\prod_{r=1}^{n} (a_{rr} - a_r^*) \leqslant D \leqslant \prod_{r=1}^{n} (a_{rr} + a_r^*),$$

where a_r^* is defined as before.†

PROBLEMS ON CHAPTER I

1. Evaluate the determinant

$$\begin{vmatrix} x_1 & y_1 & a & b \\ 2x_2 & x_2 & y_2 & c \\ 6x_3 & 3x_3 & x_3 & y_3 \\ 24x_4 & 12x_4 & 4x_4 & x_4 \end{vmatrix}.$$

† I am indebted to Professor A. Oppenheim for these estimates. An independent proof has been given by G. B. Price, *Proc. Amer. Math. Soc.* 2 (1951), 497–502.

2. Express the determinant

$$\begin{vmatrix} 1 & a^2-bc & a^4 \\ 1 & b^2-ca & b^4 \\ 1 & c^2-ab & c^4 \end{vmatrix}$$

as the product of one quadratic and four linear factors.

3. By squaring the determinant

$$\begin{vmatrix} b & c & 0 \\ a & 0 & c \\ 0 & a & b \end{vmatrix},$$

show that
$$\begin{vmatrix} b^2+c^2 & ab & ca \\ ab & c^2+a^2 & bc \\ ca & bc & a^2+b^2 \end{vmatrix} = 4a^2b^2c^2.$$

4. Use Laplace's expansion to show that

$$\begin{vmatrix} a & -b & -a & b \\ b & a & -b & -a \\ c & -d & c & -d \\ d & c & d & c \end{vmatrix} = 4(a^2+b^2)(c^2+d^2).$$

5. Find the discriminant of the equation $x^4+ax+b = 0$. Hence show that the equation $x^4+4x+3 = 0$ has at least two coincident roots.

6. If $s_r = \alpha^r+\beta^r+\gamma^r$ and

$$\Delta_n = \begin{vmatrix} s_n & s_{n+1} & s_{n+2} \\ s_{n+1} & s_{n+2} & s_{n+3} \\ s_{n+2} & s_{n+3} & s_{n+4} \end{vmatrix},$$

show that, for any positive integer n,

$$\Delta_n = (\alpha\beta\gamma)^n\{(\beta-\gamma)(\gamma-\alpha)(\alpha-\beta)\}^2.$$

Show that, when α, β, γ are the roots of the equation $x^3+bx+c = 0$, then $\Delta_n = (-1)^{n+1}c^n(4b^3+27c^2)$.

7. Show that $|a_{ij}|_n = |b_{ij}|_n$, where $b_{ij} = (-1)^{i+j}a_{ij}$.

8. If $s = a_1+...+a_n$, $A_k = s-a_k$ $(k = 1,..., n)$, prove that

$$\begin{vmatrix} x-A_1 & a_2 & a_3 & . & . & . & a_n \\ a_1 & x-A_2 & a_3 & . & . & . & a_n \\ . & . & . & . & . & . & . \\ a_1 & a_2 & a_3 & . & . & . & x-A_n \end{vmatrix} = x(x-s)^{n-1}.$$

9. Denoting by Δ_n the n-rowed determinant

$$\begin{vmatrix} 1+x^2 & x & 0 & 0 & . & . & . \\ x & 1+x^2 & x & 0 & . & . & . \\ 0 & x & 1+x^2 & x & . & . & . \\ 0 & 0 & x & 1+x^2 & . & . & . \\ . & . & . & . & . & . & . \end{vmatrix},$$

prove that $\Delta_n-\Delta_{n-1} = x^2(\Delta_{n-1}-\Delta_{n-2})$, and hence evaluate Δ_n.

10. A determinant D is of order n, and a determinant E is obtained from it by subtracting from each element of D the sum of the other elements in the same column. Prove that $E = -(n-2)2^{n-1}D$.

11. D_n is the n-rowed determinant in which the elements in the diagonal are a, those immediately above and immediately below the diagonal are b and c respectively, and all the remaining elements are 0. Obtain a relation between D_n, D_{n-1}, and D_{n-2}; and evaluate D_n when $a = 1+bc$.

12. Let D_n be the n-rowed determinant

$$\begin{vmatrix} x & a & 0 & 0 & . & . & . \\ a & x & a & 0 & . & . & . \\ 0 & a & x & a & . & . & . \\ 0 & 0 & a & x & . & . & . \\ . & . & . & . & . & . & . \end{vmatrix}.$$

Show how to find the value of D_n; and prove that, if $a = 1$, $x = 2\cos\theta$ $(0 < \theta < \pi)$, then $D_n = \sin(n+1)\theta/\sin\theta$.

13. Prove that

$$\begin{vmatrix} 1+a_1 & a_2 & a_3 & . & . & . & . & a_n \\ a_1 & 1+a_2 & a_3 & . & . & . & . & a_n \\ a_1 & a_2 & 1+a_3 & . & . & . & . & a_n \\ . & . & . & . & . & . & . & . \\ a_1 & a_2 & a_3 & . & . & . & . & 1+a_n \end{vmatrix} = 1+a_1+a_2+a_3+...+a_n,$$

and deduce that

$$\begin{vmatrix} x & a_1 & a_1 & . & . & . & a_1 \\ a_2 & x & a_2 & . & . & . & a_2 \\ a_3 & a_3 & x & . & . & . & a_3 \\ . & . & . & . & . & . & . \\ a_n & a_n & a_n & . & . & . & x \end{vmatrix} = \left\{1+\sum_{m=1}^{n}\frac{a_m}{x-a_m}\right\}\prod_{m=1}^{n}(x-a_m).$$

14. Prove the identity

$$\left|\frac{1}{a_i+b_j}\right|_n = \prod_{1\leqslant i<j\leqslant n}(a_j-a_i)(b_j-b_i)\Big/\prod_{i,j=1}^{n}(a_i+b_j).$$

15. Evaluate the determinant $|a_{ij}|_n$, where $a_{ij} = |x_i-x_j|$ and $x_1,...,x_n$ are given real numbers.

Evaluate the determinant

$$\begin{vmatrix} 0 & 1 & 2 & 3 & . & . & . & n-1 \\ 1 & 0 & 1 & 2 & . & . & . & n-2 \\ 2 & 1 & 0 & 1 & . & . & . & n-3 \\ . & . & . & . & . & . & . & . \\ n-1 & n-2 & n-3 & n-4 & . & . & . & 0 \end{vmatrix}.$$

16. Show that

$$\begin{vmatrix} a_{11} & . & . & . & a_{1n} & p_1 \\ . & . & . & . & . & . \\ a_{n1} & . & . & . & a_{nn} & p_n \\ q_1 & . & . & . & q_n & r \end{vmatrix} = r\begin{vmatrix} a_{11} & . & . & . & a_{1n} \\ . & . & . & . & . \\ a_{n1} & . & . & . & a_{nn} \end{vmatrix} - \sum_{i,j=1}^{n} A_{ij}p_i q_j,$$

where A_{ij} is the cofactor of a_{ij} in $|a_{rs}|_n$.

17. Let A_{rs} denote the cofactor of a_{rs} in the determinant $\Delta = |a_{ij}|_n$, which is such that $a_{ij} = a_{ji}$ $(i,j = 1,\ldots,n)$. Show that

$$\begin{vmatrix} A_{11} & . & . & . & A_{1n} & u_1 \\ . & . & . & . & . & . \\ A_{n1} & . & . & . & A_{nn} & u_n \\ u_1 & . & . & . & u_n & 0 \end{vmatrix} = -\Delta^{n-2} \sum_{i,j=1}^{n} a_{ij} u_i u_j.$$

Deduce, (i) by using Theorem 1.4.3 (p. 20), (ii) by using Theorem 1.5.2 (p. 24), that, for $k = 1,\ldots,n$,

$$\begin{vmatrix} A_{11} & . & . & . & A_{1n} & A_{1k} \\ . & . & . & . & . & . \\ A_{n1} & . & . & . & A_{nn} & A_{nk} \\ A_{k1} & . & . & . & A_{kn} & 0 \end{vmatrix} = -\Delta^{n-1} A_{kk}.$$

18. Prove that, if $F(t) = |a_{ij}|_n$, where $a_{ij} = (t+x_i)^j$ $(i,j = 1,\ldots,n)$, then

$$F(t) \prod_{r=1}^{n} x_r = F(0) \prod_{r=1}^{n} (t+x_r).$$

Deduce that

$$\begin{vmatrix} 1 & x_1 & . & . & . & x_1^{k-1} & x_1^{k+1} & . & . & . & x_1^n \\ . & . & . & . & . & . & . & . & . & . & . \\ 1 & x_n & . & . & . & x_n^{k-1} & x_n^{k+1} & . & . & . & x_n^n \end{vmatrix} = s_k \begin{vmatrix} 1 & x_1 & . & . & . & x_1^{n-1} \\ . & . & . & . & . & . \\ 1 & x_n & . & . & . & x_n^{n-1} \end{vmatrix},$$

where s_k is the coefficient of t^k in $\prod_{r=1}^{n} (t+x_r)$.

19. Let $a_{rs}(x)$ $(r,s = 1,\ldots,n)$ be differentiable functions of x and denote by $D(x)$ the determinant

$$\begin{vmatrix} a_{11}(x) & . & . & . & a_{1n}(x) \\ . & . & . & . & . \\ a_{n1}(x) & . & . & . & a_{nn}(x) \end{vmatrix}.$$

Show that $D'(x) = D_1(x)+\ldots+D_n(x)$, where $D_k(x)$ denotes the determinant obtained from $D(x)$ by differentiating the elements of the kth column only.

20. Show that the determinant

$$\begin{vmatrix} f_1(x_1) & . & . & . & f_1(x_n) \\ . & . & . & . & . \\ f_n(x_1) & . & . & . & f_n(x_n) \end{vmatrix},$$

where $f_1,\ldots,f_n$ are polynomials and $x_1,\ldots,x_n$ are variables, is divisible by

$$\prod_{1 \leqslant i < j \leqslant n} (x_i - x_j).$$

21. Show that the value of the 'circulant'

$$\begin{vmatrix} a_1 & a_2 & . & . & . & a_{n-1} & a_n \\ a_n & a_1 & . & . & . & a_{n-2} & a_{n-1} \\ . & . & . & . & . & . & . \\ a_2 & a_3 & . & . & . & a_n & a_1 \end{vmatrix}$$

is equal to

$$\prod_{k=1}^{n} (a_1 + a_2 \omega_k + a_3 \omega_k^2 + \ldots + a_n \omega_k^{n-1}),$$

where $\omega_1,..., \omega_n$ are the nth roots of unity. *Deduce* that

$$x^3+y^3+z^3-3xyz = (x+y+z)(x+y\rho+z\rho^2)(x+y\rho^2+z\rho),$$

where $\rho = e^{2\pi i/3}$.

22. Let $|a_{ij}|_n$ be a determinant whose value is zero. Show that, if A_{ij} denotes the cofactor of a_{ij}, then

$$A_{ir}A_{js} = A_{is}A_{jr} \qquad (i \neq j;\ r \neq s).$$

If no cofactor vanishes, show that the cofactors of elements in any one row (column) are proportional to the cofactors of elements in any other row (column).

23. The determinant $|a_{ij}|_n$ is said to be *skew-symmetric* if $a_{ij} = -a_{ji}$ $(i,j = 1,..., n)$. Show that every skew-symmetric determinant of odd order is equal to zero, and that every skew-symmetric determinant of even order is equal to the square of a polynomial in the elements of the determinant.

24. Let $c_1, c_2,..., c_n$ be given integers whose highest common factor is 1. Show (say by induction with respect to $|c_1|+|c_2|+...+|c_n|$) that there exists an n-rowed determinant with integral elements whose first row is $c_1, c_2,..., c_n$ and whose value is 1.

25. Let $D = |a_{rs}|_n$ be a determinant with complex elements, and write

$$A_r = \sum_{\substack{s=1 \\ s\neq r}}^{n} |a_{rs}|.$$

(i) Show that, if $|a_{11}| = A_1 \neq 0$ and $|a_{rr}| > A_r$ $(r = 2,..., n)$, then $D \neq 0$.
(ii) Show that, if $|a_{rr}|.|a_{ss}| > A_r A_s$ $(r, s = 1,..., n;\ r \neq s)$, then $D \neq 0$.

26. Let $n \geqslant 2$ and let a_{rs} $(r,s = 1,..., n)$ and b_r $(r = 1,..., n-1)$ be given numbers. Suppose that, for every value of r in the range $1 \leqslant r < n$,

$$\begin{vmatrix} b_r & a_{r,r+1} & a_{r,r+2} & \cdots & a_{rn} \\ a_{r+1,r} & a_{r+1,r+1} & a_{r+1,r+2} & \cdots & a_{r+1,n} \\ \cdot & \cdot & \cdot & \cdots & \cdot \\ a_{nr} & a_{n,r+1} & a_{n,r+2} & \cdots & a_{nn} \end{vmatrix} = 0.$$

Prove, by induction with respect to n, that

$$\begin{vmatrix} a_{11} & \cdots & a_{1n} \\ \cdot & \cdots & \cdot \\ a_{n1} & \cdots & a_{nn} \end{vmatrix} = (a_{11}-b_1)(a_{22}-b_2)\ldots(a_{n-1,n-1}-b_{n-1})a_{nn}.$$

27. Let $D = |a_{ij}|_n$ and denote by A_{11} the cofactor of a_{11}. By considering the determinant

$$\begin{vmatrix} a_{11}-D/A_{11} & a_{12} & \cdots & a_{1n} \\ a_{21} & a_{22} & \cdots & a_{2n} \\ \cdot & \cdot & \cdots & \cdot \\ a_{n1} & a_{n2} & \cdots & a_{nn} \end{vmatrix}$$

and making use of Theorem 1.6.4, prove (by induction with respect to n) the two results stated at the end of § 1.6.4.

28. Let $1 \leqslant n < N$ and suppose that the rectangular array of complex numbers

$$\begin{matrix} a_{11} & a_{12} & \cdot & \cdot & \cdot & a_{1N} \\ a_{21} & a_{22} & \cdot & \cdot & \cdot & a_{2N} \\ \cdot & \cdot & \cdot & \cdot & \cdot & \cdot \\ a_{n1} & a_{n2} & \cdot & \cdot & \cdot & a_{nN} \end{matrix}$$

is such that
$$|a_{rr}| > \sum_{\substack{s=1 \\ s \neq r}}^{N} |a_{rs}| \qquad (r = 1,\ldots,n).$$

By using the idea of the proof of Theorem 1.6.5 (p. 32), show that the modulus of the determinant

$$\begin{vmatrix} a_{11} & \cdot & \cdot & \cdot & a_{1n} \\ \cdot & \cdot & \cdot & \cdot & \cdot \\ a_{n1} & \cdot & \cdot & \cdot & a_{nn} \end{vmatrix}$$

is greater than the modulus of any other n-rowed minor selected from the array. Deduce Theorem 1.6.4.

29. Give a proof of Theorem 1.6.5 by considering the determinant

$$D(x) = \begin{vmatrix} a_{11}+x & a_{12} & \cdot & \cdot & \cdot & a_{1n} \\ a_{21} & a_{22}+x & \cdot & \cdot & \cdot & a_{2n} \\ \cdot & \cdot & \cdot & \cdot & \cdot & \cdot \\ a_{n1} & a_{n2} & \cdot & \cdot & \cdot & a_{nn}+x \end{vmatrix}$$

and making use of Theorem 1.6.4 and of properties of continuous functions.

30. Show that

$$\begin{vmatrix} 1 & \alpha_1 & \alpha_1^2 & \cdot & \cdot & \cdot & \alpha_1^{n-2} & \alpha_1^n \\ 1 & \alpha_2 & \alpha_2^2 & \cdot & \cdot & \cdot & \alpha_2^{n-2} & \alpha_2^n \\ \cdot & \cdot & \cdot & \cdot & \cdot & \cdot & \cdot & \cdot \\ 1 & \alpha_n & \alpha_n^2 & \cdot & \cdot & \cdot & \alpha_n^{n-2} & \alpha_n^n \end{vmatrix} = (\alpha_1+\alpha_2+\ldots+\alpha_n) \prod_{1 \leqslant r < s \leqslant n} (\alpha_s - \alpha_r).$$

31. Use the result of Problem 21 to evaluate the n-rowed determinant

$$\begin{vmatrix} a & -1 & -1 & \cdot & \cdot & \cdot & -1 \\ -1 & a & -1 & \cdot & \cdot & \cdot & -1 \\ \cdot & \cdot & \cdot & \cdot & \cdot & \cdot & \cdot \\ -1 & -1 & -1 & \cdot & \cdot & \cdot & a \end{vmatrix}.$$

II

VECTOR SPACES AND LINEAR MANIFOLDS

WE shall find it convenient to use the customary symbolism to denote class membership, writing $x \in X$ to indicate that an object x is a member of a class X and $x \notin X$ to indicate that it is not a member of X. We also write $x, y,... \in X$ to indicate that $x, y,...$ all belong to X. If X and Y are two classes such that, whenever $x \in X$ then $x \in Y$, we say that X is *contained in* Y, and we write $X \subset Y$ (or $Y \supset X$). The relation $X \subset Y$ does not, of course, preclude the possibility that X and Y are the same class.

2.1. The algebra of vectors

In this chapter we shall develop the algebra of vectors and examine the axiomatic foundations of the theory. Many of the ideas we shall meet in the course of our discussion originated in the work of Grassmann.†

2.1.1. In the present section we propose to deal with the simplest properties of operations defined for vectors. We begin by introducing the concept of a field, which plays a fundamental part throughout algebra.

DEFINITION 2.1.1. *A* FIELD *is a set of not fewer than two numbers, which is closed with respect to the four rational operations of addition, subtraction, multiplication, and division by any non-zero number.*‡

This definition may be stated more fully as follows. A set $\mathfrak{F}$ of not fewer than two numbers is a field if, whenever we have $a, b \in \mathfrak{F}$, we also have $a+b$, $a-b$, $ab \in \mathfrak{F}$ and, for $b \neq 0$, $a/b \in \mathfrak{F}$.

Obvious instances of fields are the set of all rational numbers, the set of all real numbers, and the set of all complex numbers. On the other hand, the set of integers is not a field since it is not closed with respect to division, and the set of positive numbers is not a field since it is not closed with respect to subtraction. In what follows we shall normally be concerned with the set of all real

† For bibliography and comment see Hamburger and Grimshaw, **30**, 171.

‡ The system we have defined should, strictly speaking, be called a *number field* since the normal usage of the word 'field' embraces systems more general than that with which we are concerned.

numbers (the *real field*) and the set of all complex numbers (the *complex field*).

It may be of interest to note that every field contains the field of rational numbers. For, since a field $\mathfrak{F}$ contains a number other than 0, suppose that $a \in \mathfrak{F}$, $a \neq 0$. Then $0 = a - a \in \mathfrak{F}$. Also $1 = a/a \in \mathfrak{F}$, and hence, for every positive integer k we have $k = 1+1+...+1 \in \mathfrak{F}$. This, in turn, implies that $-k = 0-k \in \mathfrak{F}$, and so $l/m \in \mathfrak{F}$, where l, m are any integers and $m \neq 0$. Thus every rational number belongs to $\mathfrak{F}$.

We next introduce vectors. In elementary mathematics a vector is described as an entity having both magnitude and direction. We can handle such entities analytically by representing them as pairs or triads of numbers. An immediate extension of this idea leads to the definition of vectors now to be adopted.

DEFINITION 2.1.2. *Let $\mathfrak{F}$ be a field. A* VECTOR $\mathbf{x}$ OF ORDER $n > 0$ OVER $\mathfrak{F}$ *is an ordered set $(x_1, x_2,..., x_n)$ of n numbers which belong to $\mathfrak{F}$.*

Vectors will normally be denoted by small letters in bold-face type. If

$$\mathbf{x} = (x_1, x_2,..., x_n),$$

then $x_1, x_2,..., x_n$ are called the first, second,..., nth components of $\mathbf{x}$.

Further, we shall say that two vectors (of the same order) are equal if and only if all their corresponding components are equal. Thus, if $\mathbf{x} = (x_1, x_2,..., x_n)$ and $\mathbf{y} = (y_1, y_2,..., y_n)$, then $\mathbf{x} = \mathbf{y}$ means that $x_1 = y_1$, $x_2 = y_2$,..., $x_n = y_n$.

The field $\mathfrak{F}$ from which the components are chosen is often called the *reference field*. Vectors over the real field are known as real vectors and those over the complex field as complex vectors. In contrast to vectors over $\mathfrak{F}$, the numbers of $\mathfrak{F}$ are called *scalars*.

DEFINITION 2.1.3. *The* ZERO VECTOR *of order n is the vector $(0, 0,..., 0)$ with n components. It is denoted by* $\mathbf{0}$.

Throughout this chapter all numbers are supposed to belong to a field $\mathfrak{F}$ which is assumed to be specified once and for all.

The introduction of vectors only becomes interesting when we define and investigate operations performed on them.

DEFINITION 2.1.4. *The* MULTIPLICATION OF A VECTOR BY A SCALAR *is defined by the formula*

$$\alpha\mathbf{x} = \mathbf{x}\alpha = (\alpha x_1,..., \alpha x_n),$$

where $\mathbf{x} = (x_1,..., x_n)$.

We note the following obvious results.

$$1\mathbf{x} = \mathbf{x},$$
$$(-1)\mathbf{x} = (-x_1,..., -x_n),$$
$$0\mathbf{x} = \mathbf{0},$$
$$\alpha\mathbf{0} = \mathbf{0},$$
$$(\alpha\beta)\mathbf{x} = \alpha(\beta\mathbf{x}).$$

DEFINITION 2.1.5. VECTOR ADDITION, *i.e. addition of vectors of the same order is defined by the formula*

$$(x_1,..., x_n)+(y_1,..., y_n) = (x_1+y_1,..., x_n+y_n).$$

From the last two definitions a number of important consequences can be inferred immediately. Thus we have the following results.

THEOREM 2.1.1. (i) *Vector addition is commutative and associative, i.e.*

$$\mathbf{x}+\mathbf{y} = \mathbf{y}+\mathbf{x},$$
$$\mathbf{x}+(\mathbf{y}+\mathbf{z}) = (\mathbf{x}+\mathbf{y})+\mathbf{z}.\dagger$$

(ii) *Vector addition and multiplication by scalars are connected by the following distributive laws.*

$$\alpha(\mathbf{x}+\mathbf{y}) = \alpha\mathbf{x}+\alpha\mathbf{y},$$
$$(\alpha+\beta)\mathbf{x} = \alpha\mathbf{x}+\beta\mathbf{x}.$$

EXERCISE 2.1.1. Write out the proof of Theorem 2.1.1.

The next theorem allows us to define the operation of subtraction.

THEOREM 2.1.2. *If* $\mathbf{x}$, $\mathbf{y}$ *are any vectors* (*of the same order*), *then there exists a unique vector* $\mathbf{z}$ *satisfying the equation*

$$\mathbf{y}+\mathbf{z} = \mathbf{x}.$$

For let $\mathbf{x} = (x_1,..., x_n)$, $\mathbf{y} = (y_1,..., y_n)$. Then

$$\mathbf{z} = (x_1-y_1,..., x_n-y_n)$$

satisfies the above equation and is clearly the only vector to do so.

DEFINITION 2.1.6. *The vector* $\mathbf{z}$ *of the previous theorem is called the difference of* $\mathbf{x}$ *and* $\mathbf{y}$ *and is denoted by* $\mathbf{x}-\mathbf{y}$. *Moreover, we write* $-\mathbf{x}$ *for* $\mathbf{0}-\mathbf{x}$.

† The associativity of vector addition enables us to use expressions such as $\mathbf{x}+\mathbf{y}+\mathbf{z}$ without ambiguity.

A number of results relating to subtraction now follow at once. Thus we have

$$\mathbf{x}-\mathbf{x} = \mathbf{0},$$
$$(-1)\mathbf{x} = -\mathbf{x},$$
$$-(-\mathbf{x}) = \mathbf{x},$$
$$\mathbf{x}-\mathbf{y} = -(\mathbf{y}-\mathbf{x}),$$
$$\mathbf{x}-(\mathbf{y}-\mathbf{z}) = (\mathbf{x}-\mathbf{y})+\mathbf{z},$$
$$\alpha(\mathbf{x}-\mathbf{y}) = \alpha\mathbf{x}-\alpha\mathbf{y},$$
$$(\alpha-\beta)\mathbf{x} = \alpha\mathbf{x}-\beta\mathbf{x},$$
$$\mathbf{x}-\mathbf{y} = \mathbf{z} \text{ implies } \mathbf{x} = \mathbf{y}+\mathbf{z}.$$

2.1.2. DEFINITION 2.1.7. *An expression of the form*

$$\alpha_1 \mathbf{x}_1+...+\alpha_k \mathbf{x}_k,$$

where $\alpha_1,..., \alpha_k$ *are scalars, is called a* LINEAR COMBINATION *of the vectors* $\mathbf{x}_1,..., \mathbf{x}_k$. *If a vector* $\mathbf{x}$ *is equal to some linear combination of* $\mathbf{x}_1,..., \mathbf{x}_k$, *it is said to be expressible linearly in terms of, or to depend linearly on,* $\mathbf{x}_1,..., \mathbf{x}_k$.

It is obvious that a linear combination of vectors is again a vector. We shall be particularly interested in sets of vectors having the property that any linear combination of vectors of such a set is again a vector of the set. Sets of this type are introduced in the next definition.

DEFINITION 2.1.8. *A* VECTOR SPACE *of order n over* $\mathfrak{F}$ *is a set* $\mathfrak{V}$ *of vectors of order n over* $\mathfrak{F}$ *with the property that, whenever* $\mathbf{x}, \mathbf{y} \in \mathfrak{V}$, $\alpha \in \mathfrak{F}$, *we have* $\alpha\mathbf{x}, \mathbf{x}+\mathbf{y} \in \mathfrak{V}$.

This definition states, in fact, that a vector space is a set of vectors closed with respect to vector addition and multiplication by scalars. The most obvious instance of a vector space is, of course, the set of *all* vectors of a given order.

DEFINITION 2.1.9. *The set of all vectors of order n over* $\mathfrak{F}$ *is called the* TOTAL VECTOR SPACE OF ORDER n OVER $\mathfrak{F}$, *and is denoted by* $\mathfrak{V}_n(\mathfrak{F})$ *or, more briefly, by* $\mathfrak{V}_n$.

It is not difficult to give other instances of vector spaces. We mention the following.

(i) The set consisting of the zero vector only. This vector space is called the *null space*.

(ii) The set of all vectors of the form $(0, x_2,..., x_n)$.

(iii) The set of all vectors of the form

$$\alpha_1 \mathbf{x}_1 + ... + \alpha_k \mathbf{x}_k, \tag{2.1.1}$$

where $\mathbf{x}_1,..., \mathbf{x}_k$ are fixed vectors.

The last example is particularly important.

DEFINITION 2.1.10. *Let* $\mathbf{x}_1,..., \mathbf{x}_k$ *be given vectors over* $\mathfrak{F}$. *The vector space* $\mathfrak{V}$ *of all vectors of the form* (2.1.1), *where* $\alpha_1,..., \alpha_k$ *are variable scalars in* $\mathfrak{F}$, *is said to be* SPANNED (*or* GENERATED) *by* $\mathbf{x}_1,..., \mathbf{x}_k$; *and these vectors are called* GENERATORS *of* $\mathfrak{V}$.

It is, for instance, obvious that $\mathfrak{V}_n$ (over the real or the complex field) is spanned by the vectors

$$\mathbf{e}_1 = (1, 0,..., 0), \quad \mathbf{e}_2 = (0, 1,..., 0),..., \mathbf{e}_n = (0, 0, ..., 1). \tag{2.1.2}$$

EXERCISE 2.1.2. Can a vector space possess more than one set of generators?

EXERCISE 2.1.3. If $\mathbf{x}_1,..., \mathbf{x}_k, \mathbf{x}_{k+1}$ span a vector space $\mathfrak{V}$ and $\mathbf{x}_{k+1}$ depends linearly on $\mathbf{x}_1,..., \mathbf{x}_k$, show that $\mathbf{x}_1,..., \mathbf{x}_k$ span $\mathfrak{V}$.

2.2. Linear manifolds

2.2.1. It frequently happens that in studying the properties of a mathematical system S we come to recognize that the results obtained do not depend essentially upon the precise definition of S but only upon certain formal relations holding between its elements. In that case it may be useful to select a suitable set R of such relations and then consider a system Σ of entities of which nothing is presupposed except that they, too, satisfy all relations in R. The system Σ constructed in this way is said to be derived from S by a *process of abstraction*. If we now investigate the properties of Σ, the conclusions at which we arrive will, of course, hold for S, since S is a special case of Σ. In this way we obtain results more general than those relating to the system S only; and we are able, moreover, to distinguish between the basic structure of S, and its more superficial features.

This procedure will be adopted in the present section and the following one, the principal object being to study those properties of vector spaces which are concerned with generators. As we shall presently see, however, our investigation depends not so much upon the actual definitions of vectors and vector spaces as upon the fact that vector addition and multiplication by scalars obey certain formal laws of algebra. We shall accordingly introduce the notion of a 'linear manifold', derived by abstraction from that of a vector space, and shall study the properties of this abstract system, returning from time to time to the particular instance of a vector space.

DEFINITION 2.2.1. *Let* $\mathfrak{F}$ *be a field and* $\mathfrak{M}$ *a set of elements* $X, Y, Z,\ldots$. *Suppose that with each* $\alpha \in \mathfrak{F}$ *and each* $X \in \mathfrak{M}$ *is associated a definite element of* $\mathfrak{M}$, *denoted by* αX (*or* $X\alpha$), *and that with each pair* $X, Y \in \mathfrak{M}$ *is associated a definite element of* $\mathfrak{M}$, *denoted by* $X+Y$. *Suppose further that the operations of 'multiplication by scalars' and 'addition' thus defined satisfy the following conditions for all* $X, Y, Z \in \mathfrak{M}$ *and all* $\alpha, \beta \in \mathfrak{F}$.

(i) $$X+Y = Y+X;$$

(ii) $$X+(Y+Z) = (X+Y)+Z;$$

(iii) *the equation* $Y+U = X$ *is soluble for* U;

(iv) $$(\alpha+\beta)X = \alpha X+\beta X;$$

(v) $$\alpha(X+Y) = \alpha X+\alpha Y;$$

(vi) $$(\alpha\beta)X = \alpha(\beta X);$$

(vii) $$1X = X.$$

Then $\mathfrak{M}$ *is called a* LINEAR MANIFOLD OVER $\mathfrak{F}$ (*with respect to multiplication by scalars and addition as defined*).

The terminology used in algebra is, as yet, far from standardized, and the structure we describe as a linear manifold is also referred to in the literature as a vector space, an abstract vector space, a linear space, or a linear system.

It will normally be assumed without explicit mention that a reference field $\mathfrak{F}$ has been chosen and that statements about linear manifolds relate to that field.

It is clear that a vector space is a particular instance of a linear manifold; and in this case the elements of the linear manifold are vectors. There are, of course, many other types of linear manifolds. Consider, for example, the set of all continuous (or all differentiable) functions defined in the interval $0 \leqslant x \leqslant 1$. This set is evidently a linear manifold with respect to the addition of functions and their multiplication by scalars.

DEFINITION 2.2.2. *Let $\mathfrak{M}$ be a linear manifold and let $\mathfrak{M}'$ be a set of elements of $\mathfrak{M}$. If $\mathfrak{M}'$ is also a linear manifold with respect to the same operations of multiplication by scalars and addition as $\mathfrak{M}$, then $\mathfrak{M}'$ is called a* (LINEAR) SUBMANIFOLD *of $\mathfrak{M}$. In particular, if $\mathfrak{M}$ is a vector space, then $\mathfrak{M}'$ is called a* (VECTOR) SUBSPACE of $\mathfrak{M}$.

Suppose that $X_1,..., X_k$ are any given elements of the linear manifold $\mathfrak{M}$, and consider the set $\mathfrak{M}'$ consisting of all linear combinations of $X_1,..., X_k$, i.e. the set of all elements of $\mathfrak{M}$ having the form

$$\alpha_1 X_1 + ... + \alpha_k X_k,$$

where $\alpha_1,..., \alpha_k$ are arbitrary scalars of $\mathfrak{F}$. It is then obvious that $\mathfrak{M}'$ is a submanifold of $\mathfrak{M}$. By analogy with Definition 2.1.10 we shall say that $\mathfrak{M}'$ is *spanned* (or *generated*) by $X_1,..., X_k$, and we shall call these elements *generators* of $\mathfrak{M}'$. Our terminology is then clearly in agreement with that used in the special case of vector spaces.

2.2.2. We next derive a series of results showing that the operations of addition and multiplication by scalars in linear manifolds obey the same formal laws as those defined for vectors.

It is assumed throughout that $\mathfrak{M}$ is a given linear manifold over $\mathfrak{F}$, that $X, Y,...$ are elements of $\mathfrak{M}$, and that α, β are scalars in $\mathfrak{F}$.

THEOREM 2.2.1. *There exists a unique element Θ, called the zero element of $\mathfrak{M}$, such that, for all $X \in \mathfrak{M}$,*

$$X + \Theta = X. \tag{2.2.1}$$

To prove the existence of such an element, let X_0 be any fixed element in $\mathfrak{M}$, and let Θ be a solution—known to exist by condition (iii) in Definition 2.2.1— of the equation

$$X_0 + \Theta = X_0.$$

Let X be an arbitrary element of $\mathfrak{M}$, and let Y be a solution of

$$X_0 + Y = X.$$

Then, using conditions (i) and (ii) of Definition 2.2.1, we obtain

$$X + \Theta = (X_0 + Y) + \Theta = (Y + X_0) + \Theta = Y + (X_0 + \Theta)$$
$$= Y + X_0 = X_0 + Y = X.$$

Thus $\mathfrak{M}$ contains at least one element Θ satisfying (2.2.1) for all $X \in \mathfrak{M}$. If Θ' is another such element, then

$$\Theta' + \Theta = \Theta',$$
$$\Theta + \Theta' = \Theta.$$

Hence, in view of condition (i), $\Theta' = \Theta$; and the uniqueness of Θ is therefore established.

THEOREM 2.2.2. *If $X+U = Y+U$, then $X = Y$.*

This result may be described as a cancellation law for addition. To prove it we denote by U' an element of $\mathfrak{M}$ satisfying

$$U + U' = \Theta.$$

Then $X+U = Y+U$ implies

$$(X+U)+U' = (Y+U)+U'.$$

Hence, by (ii),

$$X+(U+U') = Y+(U+U'),$$

i.e.
$$X+\Theta = Y+\Theta,$$

and so, by (2.2.1), $X = Y$, as asserted.

One of the postulates in Definition 2.2.1 was the solubility of the equation $Y+U = X$ for U. We shall now show that this equation has precisely one solution.

THEOREM 2.2.3. *The solution U of the equation $Y+U = X$ is unique.*

Let U, U' be elements of $\mathfrak{M}$ such that

$$Y+U = X, \qquad Y+U' = X.$$

Then
$$Y+U = Y+U';$$
$$U+Y = U'+Y,$$

and therefore, by Theorem 2.2.2, $U = U'$.

DEFINITION 2.2.3. *The unique element U satisfying the equation $Y+U = X$ is denoted by $X-Y$. In particular, $\Theta - X$ is denoted by $-X$.*

This definition implies the identity

$$(X-Y)+Y = X.$$

THEOREM 2.2.4. *For all* $X \in \mathfrak{M}$

$$X+(-X) = \Theta; \qquad X-X = \Theta.$$

Denote by X' the unique solution of $X+X' = \Theta$. Then $X' = \Theta - X = -X$, and so $X+(-X) = \Theta$. Again, denote by X'' the unique solution of $X+X'' = X$. Then $X'' = X-X$. But, in view of (2.2.1), $X'' = \Theta$ is also a solution of the equation $X+X'' = X$. Hence $X-X = \Theta$.

THEOREM 2.2.5. *For all* $X, Y \in \mathfrak{M}$

$$X-Y = X+(-Y).$$

By condition (i) of Definition 2.2.1 and Theorem 2.2.4 we have

$$(-Y)+Y = \Theta.$$

Hence
$$X+\{(-Y)+Y\} = X+\Theta,$$

and so, by condition (ii) and (2.2.1),

$$\{X+(-Y)\}+Y = X.$$

Thus the element $U = X+(-Y)$ satisfies the equation $Y+U = X$. But the unique solution of this equation is $U = X-Y$, and the theorem therefore follows.

THEOREM 2.2.6. *For all* $X \in \mathfrak{M}$ *and all* $\alpha \in \mathfrak{F}$

$$0X = \Theta,$$

$$\alpha\Theta = \Theta.$$

Using conditions (iv) and (v) of Definition 2.2.1 we have

$$\alpha X+0X = (\alpha+0)X = \alpha X = \alpha X+\Theta,$$

$$\alpha X+\alpha\Theta = \alpha(X+\Theta) = \alpha X = \alpha X+\Theta,$$

and the assertion now follows at once by Theorem 2.2.2.

THEOREM 2.2.7. *For all* $X \in \mathfrak{M}$

$$(-1)X = -X; \qquad -(-X) = X.$$

Using Theorem 2.2.6 and conditions (iv) and (vii) we have

$$\Theta = 0X = \{1+(-1)\}X = 1X+(-1)X = X+(-1)X.$$

Hence
$$(-1)X = \Theta - X = -X.$$

This, together with conditions (vi) and (vii), implies

$$-(-X) = (-1)\{(-1)X\} = \{(-1)(-1)\}X = 1X = X.$$

EXERCISE 2.2.1. Show that, for all $X \in \mathfrak{M}$ and all $\alpha \in \mathfrak{F}$,

$$(-\alpha)X = -\alpha X.$$

EXERCISE 2.2.2. Show that, for all $X, Y \in \mathfrak{M}$ and all $\alpha, \beta \in \mathfrak{F}$,

$$\alpha(X-Y) = \alpha X - \alpha Y,$$
$$(\alpha-\beta)X = \alpha X - \beta X.$$

EXERCISE 2.2.3. Show that, if $\alpha \neq 0$, $X \neq \Theta$, then $\alpha X \neq \Theta$.

EXERCISE 2.2.4. Let $X_1,..., X_k$ be generators of a linear manifold $\mathfrak{M}$, and let $Y_1,..., Y_l$ be any elements of $\mathfrak{M}$. Show that $X_1,..., X_k, Y_1,..., Y_l$ are generators of $\mathfrak{M}$.

EXERCISE 2.2.5. If a linear manifold $\mathfrak{M}$ is spanned by the elements $X_1,..., X_r, Y_1,..., Y_s$, and if each Y is a linear combination of the X's, show that $\mathfrak{M}$ is spanned by $X_1,..., X_r$.

2.3. Linear dependence and bases

2.3.1. Let $\mathfrak{M}$ again denote a linear manifold over a reference field $\mathfrak{F}$. Since every vector space is a linear manifold, the definitions and results developed below apply (with appropriate verbal changes) to the special case of vector spaces.

DEFINITION 2.3.1. *The elements $X_1,..., X_k$ of $\mathfrak{M}$ are* LINEARLY DEPENDENT *if there exist scalars $\alpha_1,..., \alpha_k$, not all zero, such that*

$$\alpha_1 X_1 + ... + \alpha_k X_k = \Theta. \tag{2.3.1}$$

In the contrary case, i.e. when (2.3.1) *implies $\alpha_1 = ... = \alpha_k = 0$, the elements $X_1,..., X_k$ are* LINEARLY INDEPENDENT.

It should be observed that linear dependence or independence is a property of the *unordered* set of elements $X_1,..., X_k$. An analogous remark applies to a number of statements below.

For vectors over $\mathfrak{F}$ Definition 2.3.1 states that $\mathbf{x}_1,..., \mathbf{x}_k$ are linearly dependent or linearly independent according as there exist or do not exist scalars $\alpha_1,..., \alpha_k$, not all zero, such that

$$\alpha_1 \mathbf{x}_1 + ... + \alpha_k \mathbf{x}_k = \mathbf{0}.$$

The concept of linear dependence will be seen to be of crucial importance for almost all topics treated in this book.†

EXERCISE 2.3.1. Using Exercise 2.2.3 interpret the terms 'linear dependence' and 'linear independence' for the case of a set consisting of a single element only.

EXERCISE 2.3.2. Show that a set of linearly independent elements of a linear manifold cannot contain the zero element.

† For geometrical applications of the concept of linear dependence see Bôcher, **16**, Chapter iii.

EXERCISE 2.3.3. (i) Show that if $X_1,..., X_k, Y_1,... Y_l$ are linearly independent, then so are $X_1,..., X_k$. (ii) Show that if $X_1,..., X_k$ are linearly dependent and $Y_1,..., Y_l$ are any elements, then $X_1,..., X_k, Y_1,..., Y_l$ are linearly dependent.

DEFINITION 2.3.2. *A* BASIS *of a linear manifold $\mathfrak{M}$ is a finite ordered set of elements $X_1,..., X_k \in \mathfrak{M}$ having the property that every $X \in \mathfrak{M}$ is expressible as a unique linear combination of $X_1,..., X_k$.*†

Thus the elements $X_1,..., X_k$ constitute a basis if every $X \in \mathfrak{M}$ is expressible in the form

$$X = \alpha_1 X_1 + ... + \alpha_k X_k,$$

where $\alpha_1,..., \alpha_k$ are uniquely determined.‡

If a basis exists in $\mathfrak{M}$, then it furnishes $\mathfrak{M}$ with a 'coordinate system', since we may regard the scalars $\alpha_1,..., \alpha_k$ specifying X as the 'coordinates' of X. This idea will be discussed at greater length in § 2.4.

We shall frequently denote a basis by a single symbol, say $\mathfrak{B}$, and shall indicate the fact that it consists of the elements $X_1,..., X_k$ (in that order) by writing

$$\mathfrak{B} = \{X_1,..., X_k\}.$$

It is, of course, obvious that if $\mathfrak{B}$ is a basis of a linear manifold $\mathfrak{M}$, then so is any (ordered) set $\mathfrak{B}'$ obtained by rearranging in any way the elements in $\mathfrak{B}$. This fact suggests that it might be more convenient to regard a basis as an *unordered* set of elements. However, the order of elements in a basis is relevant in any discussion of 'representations',§ and we prefer, therefore, to retain Definition 2.3.2 as it stands.

The total vector space $\mathfrak{B}_n$ has a basis consisting of the vectors $\mathbf{e}_1,..., \mathbf{e}_n$ defined by (2.1.2) (p. 43); for the vector $(x_1,..., x_n)$ can be written in the form

$$(x_1,..., x_n) = x_1 \mathbf{e}_1 + ... + x_n \mathbf{e}_n,$$

and this representation is clearly unique. We shall always denote the basis $\{\mathbf{e}_1,..., \mathbf{e}_n\}$ of $\mathfrak{B}_n$ by the symbol $\mathfrak{E}$.

Whereas the vector space $\mathfrak{B}_n$ possesses a basis, the linear manifold of functions continuous in the interval $0 \leqslant x \leqslant 1$ has none. Thus we do not know in advance whether a given linear manifold possesses a basis, and the aim of the present section is to clarify the

† What is here described as a basis is more commonly known as a *finite basis*.

‡ In other words, if $X = \alpha_1 X_1 + ... + \alpha_k X_k$ and $X = \alpha'_1 X_1 + ... + \alpha'_k X_k$, then $\alpha_1 = \alpha'_1,..., \alpha_k = \alpha'_k$.

§ See § 2.4 and Chapter IV.

situation. In particular we shall demonstrate that every vector space possesses a basis.

2.3.2. THEOREM 2.3.1. *Every basis of a linear manifold is a set of linearly independent generators, and conversely.*

Let $\{X_1,..., X_k\}$ be a basis of a linear manifold $\mathfrak{M}$. Then $X_1,..., X_k$ are obviously generators of $\mathfrak{M}$. Moreover, suppose scalars $\alpha_1,..., \alpha_k$ are chosen such that

$$\alpha_1 X_1 + ... + \alpha_k X_k = \Theta. \tag{2.3.2}$$

In view of Theorems 2.2.6 (p. 47) and 2.2.1 (p. 45) we also have

$$0X_1 + ... + 0X_k = \Theta.$$

But the representation (2.3.2) of Θ in terms of $X_1,..., X_k$ is unique, and therefore $\alpha_1 = ... = \alpha_k = 0$, i.e. $X_1,..., X_k$ are linearly independent. This establishes the first part of the theorem.

To prove the converse, let $X_1,..., X_k$ be a set of linearly independent generators of $\mathfrak{M}$. If $X \in \mathfrak{M}$, then there exist scalars $\beta_1,..., \beta_k$ such that

$$X = \beta_1 X_1 + ... + \beta_k X_k. \tag{2.3.3}$$

Suppose that we also have

$$X = \gamma_1 X_1 + ... + \gamma_k X_k.$$

In view of Theorem 2.2.4 (p. 47) and Exercise 2.2.2 (p. 48) it follows easily that

$$\Theta = (\beta_1 - \gamma_1)X_1 + ... + (\beta_k - \gamma_k)X_k.$$

But $X_1,..., X_k$ are linearly independent and therefore

$$\beta_1 = \gamma_1, \quad ..., \quad \beta_k = \gamma_k.$$

The representation (2.3.3) of X is therefore unique, and so $\{X_1,..., X_k\}$ is a basis.

Some linear manifolds—for example that consisting of all functions continuous in the interval $0 \leqslant x \leqslant 1$—possess arbitrarily large sets of linearly independent elements. However, our main concern is with those linear manifolds for which that is not the case.

DEFINITION 2.3.3. *The* DIMENSIONALITY $d(\mathfrak{M})$ *of a linear manifold* $\mathfrak{M}$ *is the maximum value of* r *for which* $\mathfrak{M}$ *contains* r *linearly independent elements.*† *If no such maximum value exists, then* $d(\mathfrak{M}) = \infty$.

† If $\mathfrak{M}$ is the *null manifold*, i.e. the linear manifold consisting of the zero element only, we define $d(\mathfrak{M})$ as 0.

The statement $d(\mathfrak{M}) = r$ (where $r > 0$) means, then, that $\mathfrak{M}$ contains at least one set of r linearly independent elements, but no set of $r+1$ (and therefore no set of more than $r+1$) linearly independent elements. In that case we say that $\mathfrak{M}$ is *r-dimensional*.

A linear manifold $\mathfrak{M}$ having finite dimensionality is said to be *finite-dimensional*. In contrast, if $d(\mathfrak{M}) = \infty$ (i.e. if $\mathfrak{M}$ contains arbitrarily large sets of linearly independent elements), then $\mathfrak{M}$ is said to be *infinite-dimensional*.

EXERCISE 2.3.4. Show that $d(\mathfrak{V}_1) = 1$ and $d(\mathfrak{V}_2) = 2$.

EXERCISE 2.3.5. Let the elements $X_1,..., X_n$ of a linear manifold $\mathfrak{M}$ span a submanifold $\mathfrak{M}'$. Show that $d(\mathfrak{M}')$ is equal to the maximum number of linearly independent elements among $X_1,..., X_n$.

We are now able to establish the connexion between the dimensionality of a linear manifold and the nature of its bases by showing that if a linear manifold has a basis consisting of r elements, then the dimensionality of that manifold is r, and conversely.

Theorem 2.3.2. (Basis theorem for linear manifolds)

Let $\mathfrak{M}$ be a linear manifold.

(i) *If $\mathfrak{M}$ has a basis of r elements, then $d(\mathfrak{M}) = r$ and every basis of $\mathfrak{M}$ consists of r (linearly independent) elements.*

(ii) *If $d(\mathfrak{M}) = r$, then $\mathfrak{M}$ has a basis of r elements and, indeed, every set of r linearly independent elements is a basis of $\mathfrak{M}$.*

It should be noted that this theorem asserts, in particular, the existence of at least one basis in every finite-dimensional linear manifold.

Let $\{E_1,..., E_r\}$ be a basis of $\mathfrak{M}$ and let $X_1,..., X_{r+1}$ be any $r+1$ elements. Then there exist scalars α_{ij} $(i = 1,..., r;\ j = 1,..., r+1)$ such that

$$X_j = \sum_{i=1}^{r} a_{ij} E_i \qquad (j = 1,..., r+1).$$

Now, by Theorem 1.6.2 (p. 29), we know that there exist scalars $t_1,..., t_{r+1}$, not all 0, such that

$$\sum_{j=1}^{r+1} \alpha_{ij} t_j = 0 \qquad (i = 1,..., r).$$

Hence

$$\sum_{=1}^{r+1} t_j X_j = \sum_{j=1}^{r+1} t_j \sum_{i=1}^{r} \alpha_{ij} E_i = \sum_{i=1}^{r} E_i \sum_{j=1}^{r+1} \alpha_{ij} t_j$$

$$= \sum_{i=1}^{r} 0 E_i = \Theta,$$

and so $X_1,..., X_{r+1}$ are linearly dependent. In view of Exercise 2.3.3 (ii) (p. 49) it follows that any s elements in $\mathfrak{M}$ are linearly dependent if $s > r$. On the other hand, there exists a set of r linearly independent elements, namely $E_1,..., E_r$. Hence $d(\mathfrak{M}) = r$. Moreover, if $\mathfrak{M}$ has another basis of r' elements, then, by what has just been proved, $d(\mathfrak{M}) = r'$ and so $r' = r$. The proof of (i) is therefore complete.

To establish (ii) suppose that $d(\mathfrak{M}) = r$ and denote by $E_1,..., E_r$ any set of r linearly independent elements in $\mathfrak{M}$. If $X \in \mathfrak{M}$, then, by the definition of r, the $r+1$ elements $X, E_1,..., E_r$ are linearly dependent, i.e. there exist scalars $\alpha, \alpha_1,..., \alpha_r$, not all 0, such

$$\alpha X + \alpha_1 E_1 + ... + \alpha_r E_r = \Theta. \tag{2.3.4}$$

Now if α were 0, then (2.3.4) would imply the linear dependence of $E_1,..., E_r$; hence $\alpha \neq 0$, and

$$X = \left(-\frac{\alpha_1}{\alpha}\right)E_1 + ... + \left(-\frac{\alpha_r}{\alpha}\right)E_r.$$

Thus $E_1,..., E_r$ are generators of $\mathfrak{M}$, and since they are linearly independent they constitute, by Theorem 2.3.1, a basis of $\mathfrak{M}$. This completes the proof.

Theorem 2.3.2 applies, in particular, to vector spaces; but in that special case we can obtain a stronger result, since the finiteness of dimensionality now follows as a consequence of definitions and need not be taken as a separate postulate.

Theorem 2.3.3. (Basis theorem for vector spaces)

Let $\mathfrak{B}$ be a non-null vector space of order n. Then $\mathfrak{B}$ has finite dimensionality $r > 0$, and (i) $r \leqslant n$; (ii) *any r linearly independent vectors of $\mathfrak{B}$ constitute a basis*; (iii) *every basis of $\mathfrak{B}$ consists of r (linearly independent) vectors.*

In view of Theorem 2.3.2 it is sufficient to prove that the dimensionality r of $\mathfrak{B}$ is finite and satisfies the inequality $r \leqslant n$. We have, in fact, to show that any $n+1$ vectors in $\mathfrak{B}$, say $\mathbf{x}_1,..., \mathbf{x}_{n+1}$, are linearly dependent. Write

$$\mathbf{x}_i = (x_{1i},..., x_{ni}) \qquad (i = 1,..., n+1).$$

We know by Theorem 1.6.2 (p. 29) that there exist scalars $t_1,..., t_{n+1}$, not all 0, such that

$$\sum_{i=1}^{n+1} x_{ki} t_i = 0 \qquad (k = 1,..., n).$$

Now these relations are equivalent to the single vector equation

$$\sum_{i=1}^{n+1} t_i \mathbf{x}_i = \mathbf{0},$$

and the proof is therefore complete.†

EXERCISE 2.3.6. Write out a proof of Theorem 2.3.3 without introducing the notion of a linear manifold or quoting the result of Theorem 2.3.2.

EXERCISE 2.3.7. Show that the vectors (a, b), (c, d) constitute a basis of $\mathfrak{V}_2$ if and only if $ad \neq bc$.

COROLLARY 1. *The dimensionality of the total vector space* $\mathfrak{V}_n$ *is* n, *and a basis of* $\mathfrak{V}_n$ *is simply a set of* n *linearly independent vectors of order* n.

Since $\mathfrak{V}_n$ possesses a basis $\mathfrak{E}$ consisting of the n vectors $\mathbf{e}_1,..., \mathbf{e}_n$, it follows that $d(\mathfrak{V}_n) = n$. The remainder of the assertion is then an immediate consequence of Theorem 2.3.3, (ii) and (iii).

COROLLARY 2. *If a vector space is of order* n *and has dimensionality* n, *then it is the total vector space* $\mathfrak{V}_n$.

In view of Theorem 2.3.3 (iii), any basis of the given vector space $\mathfrak{V}$ consists of n linearly independent vectors. But, by Corollary 1, these n vectors span $\mathfrak{V}_n$. Hence $\mathfrak{V}_n \subset \mathfrak{V}$, and since, trivially, $\mathfrak{V} \subset \mathfrak{V}_n$, we obtain $\mathfrak{V} = \mathfrak{V}_n$.

THEOREM 2.3.4. *If* $\mathfrak{V}$, $\mathfrak{V}'$ *are two vector spaces of dimensionalities* r, r' *respectively and* $\mathfrak{V} \subset \mathfrak{V}'$, *then* $r \leqslant r'$. *Moreover, if* $r = r'$, *then* $\mathfrak{V} = \mathfrak{V}'$.

The vector space $\mathfrak{V}$ contains a set of r linearly independent vectors, and these also belong to $\mathfrak{V}'$. Now $\mathfrak{V}'$ contains no set of $r'+1$ linearly independent vectors, and so $r < r'+1$, i.e. $r \leqslant r'$.
If $\mathfrak{V} \subset \mathfrak{V}'$ and $r = r'$, then $\mathfrak{V}$ and $\mathfrak{V}'$ are both spanned by the same set of r vectors, and so are identical.

2.3.3. We know by Theorem 2.3.2 that every finite-dimensional linear manifold possesses a basis. We shall now prove that a basis can be constructed if we start from an arbitrary set of linearly independent elements in the linear manifold.

† For an alternative argument establishing the existence of bases in vector spaces and the invariance of the number of basis elements see MacDuffee, **32**, 205–7, or Lichnerowicz, **25**, 9–12. The former treatment also includes a proof of Steinitz's 'replacement theorem'.

Theorem 2.3.5. (Completion of basis)

Any set of linearly independent elements in a finite-dimensional linear manifold $\mathfrak{M}$ *is part of a basis of* $\mathfrak{M}$.

Let $d(\mathfrak{M}) = r$ and let $X_1,..., X_k$ be a set of linearly independent elements in $\mathfrak{M}$, so that $k \leqslant r$. If $k = r$, then, by Theorem 2.3.2 (ii), $X_1,..., X_k$ constitute a basis of $\mathfrak{M}$, and the conclusion of the theorem therefore holds. If, on the other hand, $k < r$, then let us denote by $\mathfrak{N}$ the set of all linear combinations of $X_1,..., X_k$. Now, in view of Theorem 2.3.2 (i), $X_1,..., X_k$ cannot generate $\mathfrak{M}$ and so there exists some element $X_{k+1} \in \mathfrak{M}$ such that $X_{k+1} \notin \mathfrak{N}$. Now suppose that

$$\alpha_1 X_1 + ... + \alpha_k X_k + \alpha_{k+1} X_{k+1} = \Theta.$$

Then $\alpha_{k+1} = 0$, for otherwise we should have $X_{k+1} \in \mathfrak{N}$. Hence

$$\alpha_1 X_1 + ... + \alpha_k X_k = \Theta,$$

and, since $X_1,..., X_k$ are linearly independent, it follows that $\alpha_1 = ... = \alpha_k = 0$. The elements $X_1,..., X_k, X_{k+1}$ are therefore linearly independent.

We have thus shown that if $k < r$ and $X_1,..., X_k$ are linearly independent elements in $\mathfrak{M}$, then a further element $X_{k+1} \in \mathfrak{M}$ can be found such that $X_1,..., X_k, X_{k+1}$ are linearly independent. Repeating this process as often as is necessary, we ultimately obtain a set of r linearly independent elements $X_1,..., X_k, X_{k+1},..., X_r$. By Theorem 2.3.2 (ii) these elements constitute a basis of $\mathfrak{M}$, and the assertion is therefore proved.

Corollary. *Any set of linearly independent vectors in a vector space* $\mathfrak{V}$ *is part of a basis of* $\mathfrak{V}$.

2.3.4. We shall conclude the present section with a brief discussion of *complements*. It will be understood that $\mathfrak{U}, \mathfrak{U}'$ denote throughout subspaces of $\mathfrak{V}_n$.

Definition 2.3.4. *The vector spaces* $\mathfrak{U}$ *and* $\mathfrak{U}'$ *are* COMPLEMENTS *(of each other) if every* $\mathbf{x} \in \mathfrak{V}_n$ *has a unique representation of the form*

$$\mathbf{x} = \mathbf{u} + \mathbf{u}' \qquad (\mathbf{u} \in \mathfrak{U}, \mathbf{u}' \in \mathfrak{U}'). \tag{2.3.5}$$

We first require a preliminary result.

Theorem 2.3.6. *Suppose that* $\mathfrak{U}, \mathfrak{U}'$ *are subspaces of* $\mathfrak{V}_n$ *whose only common vector is* $\mathbf{0}$. *If* $\mathbf{y}_1,..., \mathbf{y}_r$ *are linearly independent vectors in* $\mathfrak{U}$ *and* $\mathbf{z}_1,..., \mathbf{z}_s$ *are linearly independent vectors in* $\mathfrak{U}'$, *then the* $r+s$ *vectors* $\mathbf{y}_1,..., \mathbf{y}_r, \mathbf{z}_1,..., \mathbf{z}_s$ *are linearly independent.*

Suppose that

$$\alpha_1 \mathbf{y}_1 + \ldots + \alpha_r \mathbf{y}_r + \beta_1 \mathbf{z}_1 + \ldots + \beta_s \mathbf{z}_s = \mathbf{0},$$

i.e.

$$\alpha_1 \mathbf{y}_1 + \ldots + \alpha_r \mathbf{y}_r = -(\beta_1 \mathbf{z}_1 + \ldots + \beta_s \mathbf{z}_s).$$

Here the left-hand side is a vector in $\mathfrak{U}$ and the right-hand side is a vector in $\mathfrak{U}'$. But, by hypothesis, $\mathbf{0}$ is the only common vector of $\mathfrak{U}$ and $\mathfrak{U}'$, and therefore

$$\alpha_1 \mathbf{y}_1 + \ldots + \alpha_r \mathbf{y}_r = \mathbf{0}, \qquad \beta_1 \mathbf{z}_1 + \ldots + \beta_s \mathbf{z}_s = \mathbf{0}.$$

This implies that $\alpha_1 = \ldots = \alpha_r = \beta_1 = \ldots = \beta_s = 0$, and the assertion therefore follows.

THEOREM 2.3.7. *The subspaces $\mathfrak{U}$, $\mathfrak{U}'$ of $\mathfrak{V}_n$ are complements if and only if $\mathbf{0}$ is the only common vector of $\mathfrak{U}$ and $\mathfrak{U}'$, and*

$$d(\mathfrak{U}) + d(\mathfrak{U}') = n.$$

First, let $\mathfrak{U}$ and $\mathfrak{U}'$ be complements. If $\mathbf{x}$ belongs to both $\mathfrak{U}$ and $\mathfrak{U}'$ we may write

$$\mathbf{x} = \mathbf{x} + \mathbf{0} \qquad (\mathbf{x} \in \mathfrak{U}, \ \mathbf{0} \in \mathfrak{U}'),$$

$$\mathbf{x} = \mathbf{0} + \mathbf{x} \qquad (\mathbf{0} \in \mathfrak{U}, \ \mathbf{x} \in \mathfrak{U}').$$

But the representation of $\mathbf{x}$ in the form (2.3.5) is unique, and so $\mathbf{x} = \mathbf{0}$. Thus $\mathbf{0}$ is the only common vector of $\mathfrak{U}$ and $\mathfrak{U}'$.

Now write $d(\mathfrak{U}) = r$, $d(\mathfrak{U}') = s$. If either $\mathfrak{U}$ or $\mathfrak{U}'$ is the null space, then obviously $r+s = n$. We may, therefore, assume that $r \geqslant 1$, $s \geqslant 1$. Let $\{\mathbf{y}_1,\ldots,\mathbf{y}_r\}$, $\{\mathbf{z}_1,\ldots,\mathbf{z}_s\}$ be any bases of $\mathfrak{U}$, $\mathfrak{U}'$ respectively. Then, by Theorem 2.3.6, $\mathbf{y}_1,\ldots,\mathbf{y}_r, \mathbf{z}_1,\ldots,\mathbf{z}_s$ are linearly independent vectors. Since, moreover, every $\mathbf{x} \in \mathfrak{V}_n$ is representable in the form (2.3.5), it follows that it is representable in the form

$$\mathbf{x} = \alpha_1 \mathbf{y}_1 + \ldots + \alpha_r \mathbf{y}_r + \beta_1 \mathbf{z}_1 + \ldots + \beta_s \mathbf{z}_s. \tag{2.3.6}$$

Hence $\mathbf{y}_1,\ldots,\mathbf{y}_r, \mathbf{z}_1,\ldots,\mathbf{z}_s$ are linearly independent generators of $\mathfrak{V}_n$, and so, by Theorems 2.3.1 (p. 50) and 2.3.3 (iii) (p. 52), $r+s = d(\mathfrak{V}_n) = n$.

Next, suppose that $\mathbf{0}$ is the only common vector of $\mathfrak{U}$ and $\mathfrak{U}'$, and that $d(\mathfrak{U})+d(\mathfrak{U}') = n$. Write, as before, $d(\mathfrak{U}) = r$, $d(\mathfrak{U}') = s$, and denote by $\{\mathbf{y}_1,\ldots,\mathbf{y}_r\}$, $\{\mathbf{z}_1,\ldots,\mathbf{z}_s\}$ bases of $\mathfrak{U}$, $\mathfrak{U}'$ respectively. Then, in view of Theorem 2.3.6, the vectors $\mathbf{y}_1,\ldots,\mathbf{y}_r, \mathbf{z}_1,\ldots,\mathbf{z}_s$ constitute a basis of $\mathfrak{V}_n$; every $\mathbf{x} \in \mathfrak{V}_n$ can therefore be represented in the form (2.3.6), and so in the form (2.3.5). The representation (2.3.5) is, moreover, unique; for if we have

$$\mathbf{x} = \mathbf{u}+\mathbf{u}', \quad \mathbf{x} = \mathbf{v}+\mathbf{v}' \qquad (\mathbf{u},\mathbf{v} \in \mathfrak{U};\, \mathbf{u}',\mathbf{v}' \in \mathfrak{U}'),$$

then

$$\mathbf{u}-\mathbf{v} = \mathbf{v}'-\mathbf{u}'.$$

Now $\mathbf{u}-\mathbf{v} \in \mathfrak{U}$, $\mathbf{v}'-\mathbf{u}' \in \mathfrak{U}'$, and since $\mathbf{0}$ is the only common vector of $\mathfrak{U}$ and $\mathfrak{U}'$ we have $\mathbf{u} = \mathbf{v}$, $\mathbf{u}' = \mathbf{v}'$. The theorem is therefore proved.

If $\mathfrak{B}$, $\mathfrak{B}'$ are two sets of objects we shall denote by $\mathfrak{B} \cup \mathfrak{B}'$ the set consisting of all objects belonging to at least one of $\mathfrak{B}$, $\mathfrak{B}'$. Two sets of objects are said to be *disjoint* if they possess no common object.

THEOREM 2.3.8. *Let $\mathfrak{B}$, $\mathfrak{B}'$ be any bases of $\mathfrak{U}$, $\mathfrak{U}'$ respectively. Then $\mathfrak{U}$, $\mathfrak{U}'$ are complements if and only if $\mathfrak{B}$, $\mathfrak{B}'$ are disjoint and $\mathfrak{B} \cup \mathfrak{B}'$† is a basis of $\mathfrak{V}_n$.*

Let $d(\mathfrak{U}) = r$, $d(\mathfrak{U}') = s$, and write

$$\mathfrak{B} = \{\mathbf{y}_1,\ldots,\mathbf{y}_r\}, \qquad \mathfrak{B}' = \{\mathbf{z}_1,\ldots,\mathbf{z}_s\}.$$

If $\mathfrak{U}$, $\mathfrak{U}'$ are complements, then, by Theorem 2.3.7, $r+s = n$ and $\mathbf{0}$ is the only vector common to $\mathfrak{U}$ and $\mathfrak{U}'$; so that, in particular, $\mathfrak{B}$, $\mathfrak{B}'$ are disjoint. It follows by Theorem 2.3.6 that the n vectors $\mathbf{y}_1,\ldots,\mathbf{y}_r$, $\mathbf{z}_1,\ldots,\mathbf{z}_s$ are linearly independent and so constitute a basis of $\mathfrak{V}_n$. In other words, $\mathfrak{B} \cup \mathfrak{B}'$ is a basis of $\mathfrak{V}_n$.

Assume, next, that $\mathfrak{B}$, $\mathfrak{B}'$ are disjoint and that $\mathfrak{B} \cup \mathfrak{B}'$ is a basis of $\mathfrak{V}_n$. Then $r+s = n$. Moreover, if $\mathbf{x} \in \mathfrak{U}$, $\mathbf{x} \in \mathfrak{U}'$, then there exist scalars $\alpha_1,\ldots,\alpha_r$, $\beta_1,\ldots,\beta_s$ such that

$$\mathbf{x} = \alpha_1\mathbf{y}_1+\ldots+\alpha_r\mathbf{y}_r = \beta_1\mathbf{z}_1+\ldots+\beta_s\mathbf{z}_s.$$

Hence $$\alpha_1\mathbf{y}_1+\ldots+\alpha_r\mathbf{y}_r+(-\beta_1)\mathbf{z}_1+\ldots+(-\beta_s)\mathbf{z}_s = \mathbf{0}.$$

But $\mathbf{y}_1,\ldots,\mathbf{y}_r$, $\mathbf{z}_1,\ldots,\mathbf{z}_s$ constitute a basis of $\mathfrak{V}_n$ and so are linearly independent. Thus

$$\alpha_1 = \ldots = \alpha_r = \beta_1 = \ldots = \beta_s = 0.$$

Hence $\mathbf{x} = \mathbf{0}$, i.e. $\mathbf{0}$ is the only vector common to $\mathfrak{U}$ and $\mathfrak{U}'$. It follows, by Theorem 2.3.7, that $\mathfrak{U}$, $\mathfrak{U}'$ are complements. The proof is therefore complete.

EXERCISE 2.3.8. Let $\mathfrak{B}$, $\mathfrak{C}$ be bases of $\mathfrak{U}$, and $\mathfrak{B}'$, $\mathfrak{C}'$ bases of $\mathfrak{U}'$. Show that if $\mathfrak{B}$, $\mathfrak{B}'$ are disjoint and $\mathfrak{B} \cup \mathfrak{B}'$ is a basis of $\mathfrak{V}_n$, then $\mathfrak{C} \cup \mathfrak{C}'$ is also a basis of $\mathfrak{V}_n$.

† The vectors in $\mathfrak{B} \cup \mathfrak{B}'$ may, of course, be ordered in any way.

We have, so far, derived a number of properties of complements without having yet considered the question of their existence. However, the theorem just proved enables us to settle this question.

THEOREM 2.3.9. *Every subspace of $\mathfrak{V}_n$ possesses a complement.*

Let $\mathfrak{U}$ be a subspace of $\mathfrak{V}_n$ and let $\{\mathbf{x}_1,\ldots,\mathbf{x}_r\}$ be a basis of $\mathfrak{U}$. By the corollary to Theorem 2.3.5 there exist vectors $\mathbf{x}_{r+1},\ldots,\mathbf{x}_n$ such that $\{\mathbf{x}_1,\ldots,\mathbf{x}_r,\mathbf{x}_{r+1},\ldots,\mathbf{x}_n\}$ is a basis of $\mathfrak{V}_n$. By Theorem 2.3.8 the vector space $\mathfrak{U}'$, defined as the space spanned by $\mathbf{x}_{r+1},\ldots,\mathbf{x}_n$, is a complement of $\mathfrak{U}$.

The completion of the basis $\{\mathbf{x}_1,\ldots,\mathbf{x}_r\}$ of $\mathfrak{U}$, made use of in the above proof, does not lead to a uniquely determined basis of $\mathfrak{V}_n$. There is, therefore, no reason to suppose that the complement of $\mathfrak{U}$ is unique; and, in fact, if $0 < d(\mathfrak{U}) < n$, it is easy to verify that it is not.

EXERCISE 2.3.9. Extend the theory of complements to more than two subspaces of $\mathfrak{V}_n$.

2.4. Vector representation of linear manifolds

2.4.1. So far we have been careful to distinguish between the general notion of a linear manifold and the special instance of a vector space. We shall now see, however, that for many purposes this distinction need not be maintained.

It is necessary, first of all, to introduce the concept of *isomorphism*. Roughly speaking we call two mathematical systems isomorphic if, however they may differ in the nature of their elements, they possess the same *structure*. This means that it is possible to set up a biunique correspondence between the elements of the two systems in such a way that there is also an exact correspondence between the operations involved in the definitions of the two systems. In the case of linear manifolds the operations in question are, of course, those of addition and multiplication by scalars.

We shall use the symbol $\leftrightarrow$ to denote biunique correspondence. Thus $X \leftrightarrow X'$ ($X \in \mathfrak{M}$, $X' \in \mathfrak{M}'$) means that a biunique correspondence has been set up between $\mathfrak{M}$ and $\mathfrak{M}'$ and that in this correspondence the element X of $\mathfrak{M}$ is associated with the element X' of $\mathfrak{M}'$.

The definition of isomorphism for linear manifolds may now be framed as follows.

DEFINITION 2.4.1. *Two linear manifolds* $\mathfrak{M}$, $\mathfrak{M}'$ *over a reference field* $\mathfrak{F}$ *are* ISOMORPHIC (*in symbols*: $\mathfrak{M} \simeq \mathfrak{M}'$) *if a biunique correspondence* $X \leftrightarrow X'$ ($X \in \mathfrak{M}$, $X' \in \mathfrak{M}'$) *can be set up between* $\mathfrak{M}$ *and* $\mathfrak{M}'$ *in such a way that, whenever* $X \leftrightarrow X'$, $Y \leftrightarrow Y'$, $\alpha \in \mathfrak{F}$, *we have* $\alpha X \leftrightarrow \alpha X'$ *and* $X+Y \leftrightarrow X'+Y'$. *The correspondence itself is called an* ISOMORPHISM *between* $\mathfrak{M}$ and $\mathfrak{M}'$.

In other words, $\mathfrak{M} \simeq \mathfrak{M}'$ if a biunique correspondence can be set up between $\mathfrak{M}$ and $\mathfrak{M}'$ such that, for all $X, Y \in \mathfrak{M}$ and all $\alpha \in \mathfrak{F}$, $(\alpha X)' = \alpha X'$ and $(X+Y)' = X'+Y'$, where Z' denotes the (unique) element of $\mathfrak{M}'$ which corresponds to Z in $\mathfrak{M}$.

It is instructive to express the same idea in yet another way by making use of the functional notation. An isomorphism between $\mathfrak{M}$ and $\mathfrak{M}'$ is a 'function' $\phi(X)$ defined uniquely for all $X \in \mathfrak{M}$ and having the following properties. (i) For every X, $\phi(X)$ is an element of $\mathfrak{M}'$. (ii) Given any element $X' \in \mathfrak{M}'$, there exists precisely one element $X \in \mathfrak{M}$ such that $\phi(X) = X'$. (iii) For all $X, Y \in \mathfrak{M}$ and all $\alpha \in \mathfrak{F}$ we have $\phi(\alpha X) = \alpha\phi(X)$ and $\phi(X+Y) = \phi(X)+\phi(Y)$.

Two linear manifolds are, of course, said to be isomorphic if there exists an isomorphism between them.

EXERCISE 2.4.1. Show that the correspondence $X \leftrightarrow \sigma X$, where σ is any fixed non-zero scalar, is an isomorphism of $\mathfrak{M}$ with itself.

EXERCISE 2.4.2. Show that the correspondence $(x_1, x_2) \leftrightarrow (x_2, -x_1)$ is an isomorphism of $\mathfrak{V}_2$ with itself.

It follows at once from Definition 2.4.1 that if $\mathfrak{M} \simeq \mathfrak{M}'$ and $X_i \leftrightarrow X_i'$ ($X_i \in \mathfrak{M}$, $X_i' \in \mathfrak{M}'$, $i = 1,..., k$) and if $\alpha_1,..., \alpha_k \in \mathfrak{F}$, then

$$\alpha_1 X_1+...+\alpha_k X_k \leftrightarrow \alpha_1 X_1'+...+\alpha_k X_k'.$$

Thus, if $\mathfrak{M}$ and $\mathfrak{M}'$ are isomorphic and we form a linear combination of some elements in $\mathfrak{M}$ and the same linear combination of the corresponding elements in $\mathfrak{M}'$, then the resulting elements in $\mathfrak{M}$ and $\mathfrak{M}'$ again correspond to each other.

Some further properties of isomorphisms between linear manifolds may easily be inferred.

THEOREM 2.4.1. *If* $\mathfrak{M} \simeq \mathfrak{M}'$, *then the zero elements in* $\mathfrak{M}$ *and* $\mathfrak{M}'$ *correspond to each other.*

For let $X \in \mathfrak{M}$, $X' \in \mathfrak{M}'$ be any elements such that $X \leftrightarrow X'$. Then $OX \leftrightarrow OX'$ i.e. $\Theta \leftrightarrow \Theta'$, where Θ, Θ' are the zero elements of $\mathfrak{M}$, $\mathfrak{M}'$ respectively.

THEOREM 2.4.2. *Isomorphism preserves linear dependence and independence. More precisely, if* $\mathfrak{M} \simeq \mathfrak{M}'$ *and* $X_i \leftrightarrow X'_i$ ($X_i \in \mathfrak{M}$, $X'_i \in \mathfrak{M}'$, $i = 1,..., k$) *and if* $X_1,..., X_k$ *are linearly dependent* (*linearly independent*), *then* $X'_1,..., X'_k$ *are linearly dependent* (*linearly independent*).†

Let Θ, Θ' denote the zero elements of $\mathfrak{M}$, $\mathfrak{M}'$ respectively. Suppose that $X_1,..., X_k$ are linearly dependent, i.e. that there exist scalars $\alpha_1,..., \alpha_k$, not all 0, such that

$$\alpha_1 X_1 + ... + \alpha_k X_k = \Theta.$$

Then, in view of Theorem 2.4.1, we have

$$\alpha_1 X'_1 + ... + \alpha_k X'_k = \Theta'.$$

Hence $X'_1,..., X'_k$ are linearly dependent. Again, since isomorphism is a symmetrical relation, it follows that linear dependence of $X'_1,..., X'_k$ implies linear dependence of $X_1,..., X_k$. The proof of the assertion is therefore complete.

As an immediate consequence of Theorem 2.4.2 we have the following result.

COROLLARY. *If two linear manifolds are isomorphic they have the same dimensionality.*

We are now in a position to show in what sense the study of linear manifolds can be replaced by that of vector spaces.

Theorem 2.4.3. (Isomorphism theorem for linear manifolds)

Every linear manifold $\mathfrak{M}$ *over* $\mathfrak{F}$, *which has finite dimensionality r, is isomorphic to the total vector space* $\mathfrak{V}_r(\mathfrak{F})$.

Let $\mathfrak{B} = \{E_1,..., E_r\}$ be any basis of $\mathfrak{M}$. Then, corresponding to each $X \in \mathfrak{M}$, there exist unique scalars $x_1,..., x_r$ such that

$$X = x_1 E_1 + ... + x_r E_r.$$

A biunique correspondence between the elements of $\mathfrak{M}$ and those of $\mathfrak{V}_r$ may now be specified by the scheme

$$X = x_1 E_1 + ... + x_r E_r \leftrightarrow (x_1,..., x_r) = \mathbf{x} \qquad (\mathbf{x} \in \mathfrak{V}_r).$$

It is then clear that, for every $\alpha \in \mathfrak{F}$, αX corresponds to $\alpha \mathbf{x}$. Again, if $Y \in \mathfrak{M}$ corresponds to $\mathbf{y} \in \mathfrak{V}_r$, then $X + Y$ corresponds to $\mathbf{x} + \mathbf{y}$. This proves the theorem.

† The form of statement 'if $P(Q)$ then $R(S)$' means that P implies R, and Q implies S.

DEFINITION 2.4.2. *If* $\mathfrak{B} = \{E_1,..., E_r\}$ *is a basis of* $\mathfrak{M}$ *and the element* $X \in \mathfrak{M}$ *has the form* $X = x_1 E_1 + ... + x_r E_r$, *then the vector* $\mathbf{x} = (x_1,..., x_r)$ *is said to* REPRESENT X *with respect to* $\mathfrak{B}$; *and we write*

$$\mathbf{x} = \mathscr{R}(X; \mathfrak{B}).$$

The scalars $x_1,..., x_r$ may be referred to as the coordinates of X with respect to $\mathfrak{B}$.

We have thus arrived at a representation of $\mathfrak{M}$ by $\mathfrak{B}_r$ which preserves the structure of $\mathfrak{M}$. This representation is not, however, unique since it depends on the arbitrary choice of a basis in $\mathfrak{M}$. By varying this choice of basis we obtain different representations of $\mathfrak{M}$ by $\mathfrak{B}_r$.

EXERCISE 2.4.3. Show that if $\mathfrak{B}$ is any basis of $\mathfrak{M}$ and Θ is the zero element, then

$$\mathbf{0} = \mathscr{R}(\Theta; \mathfrak{B}).$$

EXERCISE 2.4.4. Prove that if two linear manifolds have the same finite dimensionality then they are isomorphic.

For the special case of vector spaces, Theorem 2.4.3 reduces to the following result.

COROLLARY. *A vector space of order n and dimensionality r is isomorphic to the total vector space of order r.*

EXERCISE 2.4.5. Illustrate this corollary by considering the vector space consisting of all vectors of the form $(0, x_2,..., x_n)$.

EXERCISE 2.4.6. Let $\mathfrak{B}$ be a basis of $\mathfrak{B}_n$. If $\mathbf{x}, \mathbf{y} \in \mathfrak{B}_n$ explain what is meant by the statement '$\mathbf{x}$ represents $\mathbf{y}$ with respect to $\mathfrak{B}$'. Show that every vector in $\mathfrak{B}_n$ represents itself with respect to $\mathfrak{E}$.

2.4.2. A vector space which is isomorphic to a linear manifold is said to *represent* it, and the importance of such representations lies in the fact that they furnish us with a simple and effective technique for studying the properties of linear manifolds. In view of this we shall from now on confine our attention for the most part to vector spaces and return only occasionally to the more general idea of a linear manifold. Nevertheless, it is sometimes more illuminating to think in terms of linear manifolds rather than in terms of vector spaces representing them. For in the n-dimensional vector space $\mathfrak{B}_n$ we have a favoured basis $\{\mathbf{e}_1,..., \mathbf{e}_n\}$, and any vector $(x_1,..., x_n)$ is expressible in the simple form

$$(x_1,..., x_n) = x_1 \mathbf{e}_1 + ... + x_n \mathbf{e}_n.$$

In a linear manifold, on the other hand, all bases are of exactly the same standing; and while the existence of a favoured basis in a vector space makes for convenience of calculation, it may obscure some of the essential features of the problem.

Representation of a given linear manifold $\mathfrak{M}$ by a vector space is possible, as we have seen, in more than one way. Indeed, it is clear from the proof of Theorem 2.4.3 that each choice of a basis in $\mathfrak{M}$ leads to a different isomorphism between $\mathfrak{M}$ and $\mathfrak{V}_r$. All such representations of $\mathfrak{M}$ by $\mathfrak{V}_r$ are of equal status, none having precedence over any other. We should think of $\mathfrak{M}$ as an entity possessing infinitely many representations in terms of $\mathfrak{V}_r$, but existing independently of them all.

The idea of representing a linear manifold by a vector space is, of course, already familiar to the reader since it is virtually the same as that of representing points by means of cartesian coordinates. This representation of points is arrived at in the following manner. Let $\mathfrak{S}$ be ordinary three-dimensional space. If an origin O is fixed, then with every point X of $\mathfrak{S}$ is associated a unique directed segment $\overrightarrow{OX}$.† By obvious geometrical constructions the directed segments can be added together and multiplied by real scalars, and it is easy to verify that, with respect to these operations, they constitute a linear manifold $\mathfrak{M}$.

Every choice of coordinate axes, rectangular or oblique, through O is equivalent to a choice of basis in $\mathfrak{M}$. For let a system C of coordinate axes through O be given and let A_1, A_2, A_3 be points on the axes such that the segments $\overrightarrow{OA}_1$, $\overrightarrow{OA}_2$, $\overrightarrow{OA}_3$, measured positively, are all of unit length. Then $\overrightarrow{OA}_1$, $\overrightarrow{OA}_2$, $\overrightarrow{OA}_3$ are linearly independent elements of $\mathfrak{M}$; and if the coordinates, with respect to C, of a point X are (x_1, x_2, x_3), then

$$\overrightarrow{OX} = x_1\,\overrightarrow{OA}_1+x_2\,\overrightarrow{OA}_2+x_3\,\overrightarrow{OA}_3.$$

Thus $\overrightarrow{OA}_1$, $\overrightarrow{OA}_2$, $\overrightarrow{OA}_3$ constitute a basis of $\mathfrak{M}$; and conversely, every choice of basis in $\mathfrak{M}$ (together with a choice of scale) determines a unique coordinate system in $\mathfrak{S}$.

If now the coordinates, with respect to C, of points $X, Y \in \mathfrak{S}$ are (x_1, x_2, x_3), (y_1, y_2, y_3) respectively, then the coordinates of points corresponding to the segments $\alpha\overrightarrow{OX}$, $\overrightarrow{OX}+\overrightarrow{OY}$ are $(\alpha x_1, \alpha x_2, \alpha x_3)$, $(x_1+y_1,\ x_2+y_2,\ x_3+y_3)$. Hence $\mathfrak{M}$ is isomorphic to

† The directed segments are often referred to as 'position vectors' of points with respect to O.

$\mathfrak{B}_3$, and each choice of axes in $\mathfrak{S}$ (i.e. each choice of basis in $\mathfrak{M}$) leads to a different mode of representation of $\mathfrak{M}$ by $\mathfrak{B}_3$.

Situations of the above type where an entity (such as, for instance, a linear manifold) admits of infinitely many representations but exists independently of them are quite common in algebra, and we shall meet further instances when we come to consider linear, bilinear, and quadratic operators in Chapters IV and XII.

2.5. Inner products and orthonormal bases

2.5.1. Let a system of rectangular coordinates be introduced in three-dimensional space, and let (x_1, x_2, x_3), (y_1, y_2, y_3) be the coordinates of two points. We may think of these triads as vectors $\mathbf{x}, \mathbf{y}$. The expression

$$x_1 y_1 + x_2 y_2 + x_3 y_3$$

(which is familiar from elementary geometry) is called the *scalar product* or *inner product* of $\mathbf{x}$ and $\mathbf{y}$. This notion is obviously capable of generalization to any number of dimensions. If $\mathbf{x} = (x_1,..., x_n)$ $\mathbf{y} = (y_1,..., y_n)$ are two *real* vectors we shall define their inner product $(\mathbf{x}, \mathbf{y})$ as the expression

$$(\mathbf{x}, \mathbf{y}) = x_1 y_1 + ... + x_n y_n. \tag{2.5.1}$$

We shall also define the *length* $|\mathbf{x}|$ of the vector $\mathbf{x}$ as

$$|\mathbf{x}| = +(\mathbf{x}, \mathbf{x})^{\frac{1}{2}} = +(x_1^2 + ... + x_n^2)^{\frac{1}{2}}. \tag{2.5.2}$$

When $n = 2$ or $n = 3$, and a system of rectangular cartesian coordinates is given, then $\mathbf{x}$ can be identified (in the obvious way) with a position vector. In that case $|\mathbf{x}|$ is simply the length of the position vector.

The definitions just given are no longer appropriate when we deal with complex vectors. We shall, therefore, introduce modified definitions of inner product and length for the general case of complex vectors; these definitions will naturally reduce to (2.5.1) and (2.5.2) when the vectors are real. It is to be understood that all vectors and scalars which occur below are complex unless the contrary is stated. As usual $\bar{t}$ denotes the complex conjugate of t.

2.5.2. DEFINITION 2.5.1. *The* INNER PRODUCT $(\mathbf{x}, \mathbf{y})$ *of two complex vectors* $\mathbf{x} = (x_1,..., x_n)$, $\mathbf{y} = (y_1,..., y_n)$ *is defined as*

$$(\mathbf{x}, \mathbf{y}) = \bar{x}_1 y_1 + ... + \bar{x}_n y_n.$$

A number of identities follow trivially from this definition.

THEOREM 2.5.1.

(i) $\overline{(\mathbf{x},\mathbf{y})} = (\mathbf{y},\mathbf{x})$;

(ii) $\alpha(\mathbf{x},\mathbf{y}) = (\bar{\alpha}\mathbf{x},\mathbf{y}) = (\mathbf{x},\alpha\mathbf{y})$;

(iii) $(\mathbf{x}+\mathbf{y},\mathbf{z}) = (\mathbf{x},\mathbf{z})+(\mathbf{y},\mathbf{z})$,

$(\mathbf{x},\mathbf{y}+\mathbf{z}) = (\mathbf{x},\mathbf{y})+(\mathbf{x},\mathbf{z})$.

COROLLARY. *If* $\mathbf{x}$, $\mathbf{y}$ *are real vectors and* α *a real scalar, then*

$$(\mathbf{x}, \mathbf{y}) = (\mathbf{y}, \mathbf{x}),$$

$$\alpha(\mathbf{x}, \mathbf{y}) = (\alpha\mathbf{x}, \mathbf{y}) = (\mathbf{x}, \alpha\mathbf{y}).$$

DEFINITION 2.5.2. (i) *The* LENGTH (*or* NORM) $|\mathbf{x}|$ *of the complex vector* $\mathbf{x} = (x_1,..., x_n)$ *is defined as*

$$|\mathbf{x}| = +(\mathbf{x},\mathbf{x})^{\frac{1}{2}} = +(|x_1|^2+...+|x_n|^2)^{\frac{1}{2}}.$$

(ii) *The expression* $|\mathbf{x}-\mathbf{y}|$ *will be called the* SEPARATION *of* $\mathbf{x}$ *and* $\mathbf{y}$.

When $n = 2$ or 3 and the vectors are real, these ideas have a simple geometrical interpretation. Consider the case $n = 3$. If a system of rectangular coordinates is introduced in ordinary geometrical space $\mathfrak{S}$, every point P becomes associated with a definite vector $\overrightarrow{OP} = \mathbf{x}$ with components (x, y, z); and we thus have a biunique correspondence between the points of $\mathfrak{S}$ and the vectors of $\mathfrak{B}_3$. If P, Q are any two points and $\mathbf{x}$, $\mathbf{y}$ are the corresponding vectors, the separation $|\mathbf{x}-\mathbf{y}|$ of the vectors is simply the distance PQ. The length of the vector $\mathbf{x}$, on the other hand, is equal to the distance of the point P from the origin O.

We return now to the consideration of complex vectors and note that $|\mathbf{x}|$ is always real. Moreover, $|\mathbf{x}| \geqslant 0$ for all $\mathbf{x}$, and $|\mathbf{x}| = 0$ if and only if $\mathbf{x} = \mathbf{0}$. Again it is clear that

$$|\alpha\mathbf{x}| = |\alpha| . |\mathbf{x}|.\dagger$$

EXERCISE 2.5.1. Show that $|\mathbf{x}-\mathbf{y}| = |\mathbf{y}-\mathbf{x}|$.

We next consider some inequalities connected with inner products.

THEOREM 2.5.2. $|(\mathbf{x}, \mathbf{y})| \leqslant |\mathbf{x}|\ |\mathbf{y}|$.

To prove this result we write $\mathbf{x} = (x_1,..., x_n)$, $\mathbf{y} = (y_1,..., y_n)$. The assertion then becomes

$$\left|\sum_{r=1}^{n} \bar{x}_r y_r\right|^2 \leqslant \left(\sum_{r=1}^{n} |x_r|^2\right)\left(\sum_{r=1}^{n} |y_r|^2\right).$$

† The reader should be careful to distinguish between the two uses of vertical bars; $|\alpha|$ denotes the modulus of the *number* α and $|\mathbf{x}|$ the length of the *vector* $\mathbf{x}$.

This is the inequality of Cauchy and Schwarz for complex numbers. We may establish it by observing that

$$0 \leqslant \sum_{r,s=1}^{n} |x_r y_s - x_s y_r|^2 = \sum_{r,s=1}^{n} (x_r y_s - x_s y_r)(\bar{x}_r \bar{y}_s - \bar{x}_s \bar{y}_r)$$

$$= \sum_{r,s=1}^{n} (x_r \bar{x}_r y_s \bar{y}_s + x_s \bar{x}_s y_r \bar{y}_r - x_r \bar{y}_r \bar{x}_s y_s - \bar{x}_r y_r x_s \bar{y}_s)$$

$$= 2\Big(\sum_{r=1}^{n} x_r \bar{x}_r\Big)\Big(\sum_{s=1}^{n} y_s \bar{y}_s\Big) - 2\Big(\sum_{r=1}^{n} \bar{x}_r y_r\Big)\Big(\sum_{s=1}^{n} x_s \bar{y}_s\Big)$$

$$= 2\Big(\sum_{r=1}^{n} |x_r|^2\Big)\Big(\sum_{r=1}^{n} |y_r|^2\Big) - 2\Big|\sum_{r=1}^{n} \bar{x}_r y_r\Big|^2.$$

THEOREM 2.5.3. (Triangle inequality)

$$|\mathbf{x}+\mathbf{y}| \leqslant |\mathbf{x}|+|\mathbf{y}|. \tag{2.5.3}$$

Using Theorem 2.5.1, (iii) and (i), we obtain

$$|\mathbf{x}+\mathbf{y}|^2 = (\mathbf{x}+\mathbf{y},\mathbf{x}+\mathbf{y}) = (\mathbf{x},\mathbf{x})+(\mathbf{y},\mathbf{y})+(\mathbf{x},\mathbf{y})+(\mathbf{y},\mathbf{x})$$

$$= |\mathbf{x}|^2+|\mathbf{y}|^2+(\mathbf{x},\mathbf{y})+\overline{(\mathbf{x},\mathbf{y})}$$

$$\leqslant |\mathbf{x}|^2+|\mathbf{y}|^2+2|(\mathbf{x},\mathbf{y})|.$$

The assertion now follows by Theorem 2.5.2.

Theorem 2.5.3 may be restated in the form

$$|\mathbf{x}-\mathbf{y}| \leqslant |\mathbf{x}-\mathbf{z}|+|\mathbf{y}-\mathbf{z}|.$$

If this inequality is interpreted for the case of real vectors of order 2, it is seen to state that the sum of two sides of a triangle is not smaller than the third side. It is for this reason that (2.5.3) is called the 'triangle inequality'.

2.5.3. DEFINITION 2.5.3. (i) *A vector* $\mathbf{x}$ *is a* UNIT VECTOR *if* $|\mathbf{x}| = 1$. (ii) *The process of replacing a non-zero vector* $\mathbf{x}$ *by the unit vector* $|\mathbf{x}|^{-1}\mathbf{x}$ *is known as* NORMALIZATION *of* $\mathbf{x}$.

The next definition is suggested once again by analogy with elementary geometry. If the inner product of two real vectors of order 2 or 3 vanishes, then the corresponding directed segments are orthogonal to each other. We now extend the notion of orthogonality to complex vectors of arbitrary order.

DEFINITION 2.5.4. *The vectors* $\mathbf{x}, \mathbf{y}$ *are* ORTHOGONAL (*to each other*) *if* $(\mathbf{x}, \mathbf{y}) = 0$.

The wording of the definition suggests that orthogonality is a symmetrical relation. This is, in fact, the case since $(\mathbf{x}, \mathbf{y}) = 0$ implies $(\mathbf{y}, \mathbf{x}) = 0$.

EXERCISE 2.5.2. Show that if $\mathbf{x}$ is orthogonal to every vector of a basis of a vector space $\mathfrak{V}$, then it is orthogonal to every vector of $\mathfrak{V}$.

EXERCISE 2.5.3. Show that orthogonality is not preserved by isomorphism, i.e. if the vector spaces $\mathfrak{U}, \mathfrak{U}'$ are isomorphic and $\mathbf{x}, \mathbf{y} \in \mathfrak{U}$ correspond to $\mathbf{x}', \mathbf{y}' \in \mathfrak{U}'$ respectively, then the relation $(\mathbf{x}, \mathbf{y}) = 0$ does not imply $(\mathbf{x}', \mathbf{y}') = 0$.

DEFINITION 2.5.5. (i) *The vectors* $\mathbf{x}_1,..., \mathbf{x}_k$ *form an* ORTHOGONAL SET *if they are orthogonal in pairs.* (ii) *The vectors* $\mathbf{x}_1,..., \mathbf{x}_k$ *form an* ORTHONORMAL SET *if they are orthogonal in pairs and each vector is a unit vector.*†

If a basis of a vector space is an orthogonal (orthonormal) set, we call it an *orthogonal* (*orthonormal*) *basis*.

From an orthogonal set of non-zero vectors we can at once obtain an orthonormal set by normalizing each vector.

THEOREM 2.5.4. *If a number of non-zero vectors form an orthogonal set, then they are linearly independent.*

For let $\mathbf{x}_1,..., \mathbf{x}_m$ be non-zero vectors such that

$$(\mathbf{x}_r, \mathbf{x}_s) = 0 \qquad (r, s = 1,..., m;\ r \neq s).$$

If $\alpha_1,..., \alpha_m$ are scalars such that

$$\alpha_1 \mathbf{x}_1 + ... + \alpha_m \mathbf{x}_m = \mathbf{0},$$

then, forming inner products with $\mathbf{x}_r$, we obtain

$$\alpha_1(\mathbf{x}_r, \mathbf{x}_1) + ... + \alpha_m(\mathbf{x}_r, \mathbf{x}_m) = 0.$$

Hence $\alpha_r(\mathbf{x}_r, \mathbf{x}_r) = 0$ and so $\alpha_r = 0$. Since this result holds for $r = 1,..., m$, the theorem is proved.

COROLLARY 1. *The vectors of an orthonormal set are linearly independent.*

COROLLARY 2. *If n non-zero vectors of order n form an orthogonal set, then they constitute a basis of the total vector space $\mathfrak{V}_n$.*

† It is sometimes more convenient to speak of 'orthogonal vectors' or 'orthonormal vectors' when we refer to an orthogonal *set* or an orthonormal *set* of vectors. An orthonormal set of vectors consisting of a single vector is, of course, a unit vector.

For, in view of the theorem just proved, the n vectors are linearly independent. Hence, by Corollary 1 to Theorem 2.3.3 (p. 53), they constitute a basis of $\mathfrak{V}_n$.

EXERCISE 2.5.4. Suppose that the vectors $\mathbf{u}_1,\ldots,\mathbf{u}_n \in \mathfrak{V}_n$ form an orthonormal set. Show that any vector $\mathbf{v} \in \mathfrak{V}_n$ can be expressed in the form

$$\mathbf{v} = \sum_{i=1}^{n} (\mathbf{u}_i, \mathbf{v})\mathbf{u}_i.$$

We shall next turn to the question of orthonormal bases. The total vector space $\mathfrak{V}_n$ clearly possesses at least one orthonormal basis, namely, $\mathfrak{E}$. It is not, however, obvious whether $\mathfrak{V}_n$ possesses an orthonormal basis containing a given set of orthonormal vectors, though geometrical intuition suggests that this is the case. Again, in dealing with a vector space other than a total vector space we do not as yet know whether an orthonormal basis need exist at all. These questions will now be settled.

Theorem 2.5.5. (Theorem on orthonormal bases)

Every non-null vector space $\mathfrak{V}$ possesses an orthonormal basis. Moreover, every orthonormal set of vectors in $\mathfrak{V}$ is part of an orthonormal basis of $\mathfrak{V}$.

It is instructive to note the resemblance between this result and the corollary to Theorem 2.3.5 (p. 54).

We need only prove the second assertion since the first then follows as a trivial consequence. Write $d(\mathfrak{V}) = r$ $(\geqslant 1)$ and let $\mathbf{x}_1,\ldots,\mathbf{x}_k$ be an orthonormal set of vectors in $\mathfrak{V}$, where $1 \leqslant k < r$. By Corollary 1 to Theorem 2.5.4 and the corollary to Theorem 2.3.5 there exists a vector $\mathbf{x} \in \mathfrak{V}$ such that $\mathbf{x}_1,\ldots,\mathbf{x}_k, \mathbf{x}$ are linearly independent. The vector $\mathbf{y} \in \mathfrak{V}$, given by

$$\mathbf{y} = \mathbf{x} - (\mathbf{x}_1, \mathbf{x})\mathbf{x}_1 - \ldots - (\mathbf{x}_k, \mathbf{x})\mathbf{x}_k,$$

is then non-zero and is evidently orthogonal to each of $\mathbf{x}_1,\ldots,\mathbf{x}_k$. Since $\mathbf{y}$ can be normalized we see that to every orthonormal set of k $(< r)$ vectors in $\mathfrak{V}$ we can add a further vector such that the augmented set is again orthonormal.

The proof of the theorem is completed by a repeated application of this result.† For if $\xi_1,\ldots,\xi_m$ is an orthonormal set in $\mathfrak{V}$ and $m = r$, then the assertion is trivial, while if $m < r$, then we can successively find unit vectors $\xi_{m+1},\ldots,\xi_r$ such that $\xi_1,\ldots,\xi_m, \xi_{m+1},\ldots,\xi_r$ is an orthonormal set, and therefore an orthonormal basis of $\mathfrak{V}$.

† Compare the proof of Theorem 2.3.5.

EXERCISE 2.5.5. Give a geometrical interpretation for the construction of orthonormal bases in real spaces $\mathfrak{B}_2$ and $\mathfrak{B}_3$.

The principle underlying the construction carried out in the proof above can be put even more explicitly. We are concerned with the problem of obtaining an orthogonal set of n vectors (in $\mathfrak{B}_n$) from any given set of n linearly independent vectors. The procedure by means of which this object is achieved—known as *Schmidt's orthogonalization process*—will be used in the proof of the next theorem.

THEOREM 2.5.6. *If* $\mathbf{x}_1,..., \mathbf{x}_n$ *are linearly independent vectors in* $\mathfrak{B}_n$, *then scalars* c_{ij} $(1 \leqslant j < i \leqslant n)$ *can be found such that the vectors* $\mathbf{y}_1,..., \mathbf{y}_n$, *given by the scheme*

$$\left.\begin{aligned} \mathbf{y}_1 &= \mathbf{x}_1 \\ \mathbf{y}_2 &= c_{21}\mathbf{x}_1+\mathbf{x}_2 \\ \mathbf{y}_3 &= c_{31}\mathbf{x}_1+c_{32}\mathbf{x}_2+\mathbf{x}_3 \\ &\cdots\cdots\cdots\cdots \\ \mathbf{y}_n &= c_{n1}\mathbf{x}_1+c_{n2}\mathbf{x}_2+...+c_{n,n-1}\mathbf{x}_{n-1}+\mathbf{x}_n \end{aligned}\right\}, \qquad (2.5.4)$$

form an orthogonal set of non-zero vectors.

Consider the vectors $\mathbf{y}_1,..., \mathbf{y}_n$ defined by the formulae

$$\left.\begin{aligned} \mathbf{y}_1 &= \mathbf{x}_1 \\ \mathbf{y}_2 &= \mathbf{x}_2-\frac{(\mathbf{y}_1,\mathbf{x}_2)}{(\mathbf{y}_1,\mathbf{y}_1)}\mathbf{y}_1 \\ \mathbf{y}_3 &= \mathbf{x}_3-\frac{(\mathbf{y}_1,\mathbf{x}_3)}{(\mathbf{y}_1,\mathbf{y}_1)}\mathbf{y}_1-\frac{(\mathbf{y}_2,\mathbf{x}_3)}{(\mathbf{y}_2,\mathbf{y}_2)}\mathbf{y}_2 \\ &\cdots\cdots\cdots\cdots \\ \mathbf{y}_n &= \mathbf{x}_n-\frac{(\mathbf{y}_1,\mathbf{x}_n)}{(\mathbf{y}_1,\mathbf{y}_1)}\mathbf{y}_1-...-\frac{(\mathbf{y}_{n-1},\mathbf{x}_n)}{(\mathbf{y}_{n-1},\mathbf{y}_{n-1})}\mathbf{y}_{n-1} \end{aligned}\right\}. \qquad (2.5.5)$$

To show that these definitions have a meaning we must verify that each $\mathbf{y}_k \neq \mathbf{0}$. This is done by induction. We clearly have $\mathbf{y}_1 \neq \mathbf{0}$. Assume now that, for some $k \geqslant 2$, $\mathbf{y}_1,..., \mathbf{y}_{k-1}$ are all non-zero. Then the definition of $\mathbf{y}_k$ in (2.5.5) is significant and we recognize, furthermore, that $\mathbf{y}_k$ is a linear combination of $\mathbf{x}_1,..., \mathbf{x}_k$ in which $\mathbf{x}_k$ has the coefficient 1. Since $\mathbf{x}_1,..., \mathbf{x}_k$ are linearly independent it follows that $\mathbf{y}_k \neq \mathbf{0}$. Thus $\mathbf{y}_1 \neq \mathbf{0},..., \mathbf{y}_n \neq \mathbf{0}$.

We next show, once again by induction, that $\mathbf{y}_1,..., \mathbf{y}_n$ form an orthogonal set. Assume that, for some $k \geqslant 2$, $\mathbf{y}_1,\ldots,\mathbf{y}_{k-1}$ form an orthogonal set. Now

$$\mathbf{y}_k = \mathbf{x}_k - \sum_{i=1}^{k-1} \frac{(\mathbf{y}_i, \mathbf{x}_k)}{(\mathbf{y}_i, \mathbf{y}_i)} \mathbf{y}_i,$$

and therefore, forming the inner product $(\mathbf{y}_j, \mathbf{y}_k)$ where $1 \leqslant j < k$, we obtain

$$(\mathbf{y}_j, \mathbf{y}_k) = (\mathbf{y}_j, \mathbf{x}_k) - \sum_{i=1}^{k-1} \frac{(\mathbf{y}_i, \mathbf{x}_k)}{(\mathbf{y}_i, \mathbf{y}_i)} (\mathbf{y}_j, \mathbf{y}_i)$$

$$= (\mathbf{y}_j, \mathbf{x}_k) - \frac{(\mathbf{y}_j, \mathbf{x}_k)}{(\mathbf{y}_j, \mathbf{y}_j)} (\mathbf{y}_j, \mathbf{y}_j) = 0.$$

Thus $\mathbf{y}_1,\ldots,\mathbf{y}_n$ form an orthogonal set. To complete the proof of the theorem we need merely to note that $\mathbf{y}_1,\ldots,\mathbf{y}_n$, as defined by (2.5.5), can, in fact, be written in the form (2.5.4).†

EXERCISE 2.5.6. Let $\mathbf{x}_1,\ldots,\mathbf{x}_n$ be linearly independent vectors in $\mathfrak{V}_n$, and suppose that the first r of them form an orthogonal set, If $\mathbf{y}_1,\ldots,\mathbf{y}_n$ are the vectors defined by (2.5.5), show that $\mathbf{y}_k = \mathbf{x}_k$ $(k = 1,\ldots,r)$.

2.5.4. In the discussion below $\mathfrak{U}$ denotes a subspace of the total vector space $\mathfrak{V}_n$.

DEFINITION 2.5.6. *The* ORTHOGONAL COMPLEMENT *of* $\mathfrak{U}$ *is the set of all vectors in* $\mathfrak{V}_n$ *which are orthogonal to every vector in* $\mathfrak{U}$.‡

EXERCISE 2.5.7. Show that the orthogonal complement of $\mathfrak{V}_n$ is the null space, and that the orthogonal complement of the null space is $\mathfrak{V}_n$.

THEOREM 2.5.7. *If* $\mathfrak{U}'$ *is the orthogonal complement of* $\mathfrak{U}$, *then* (i) $\mathfrak{U}'$ *is a vector space*; (ii) *the orthogonal complement of* $\mathfrak{U}'$ *is* $\mathfrak{U}$; (iii) $d(\mathfrak{U})+d(\mathfrak{U}') = n$.

The verification of (i) is immediate. To prove (ii) and (iii) write $d(\mathfrak{U}) = r$. The cases $r = 0$ and $r = n$ have already been dealt with in Exercise 2.5.7 and we may, therefore, assume that $0 < r < n$. Let $\{\mathbf{x}_1,\ldots,\mathbf{x}_r\}$ denote an orthonormal basis of $\mathfrak{U}$. By Theorem 2.5.5 there exist vectors $\mathbf{x}_{r+1},\ldots,\mathbf{x}_n \in \mathfrak{V}_n$ such that $\{\mathbf{x}_1,\ldots,\mathbf{x}_r, \mathbf{x}_{r+1},\ldots,\mathbf{x}_n\}$ is an orthonormal basis of $\mathfrak{V}_n$. Every $\mathbf{x} \in \mathfrak{V}_n$ is then expressible in the form

$$\mathbf{x} = \alpha_1 \mathbf{x}_1 + \ldots + \alpha_r \mathbf{x}_r + \alpha_{r+1} \mathbf{x}_{r+1} + \ldots + \alpha_n \mathbf{x}_n.$$

† For some interesting applications of Schmidt's orthogonalization process to certain infinitely-dimensional linear manifolds whose elements are functions see Jackson *Fourier Series and Orthogonal Polynomials*, 151–4, and Courant and Hilbert, *Methods of Mathematical Physics*, vol. i, chap. ii, §§ 1, 8, 9.

‡ We do not know at this stage whether the orthogonal complement of $\mathfrak{U}$ is a complement of $\mathfrak{U}$. This question will be settled below.

By Exercise 2.5.2 (p. 65) $\mathbf{x} \in \mathfrak{U}'$ if and only if

$$(\mathbf{x}_1, \mathbf{x}) = \dots = (\mathbf{x}_r, \mathbf{x}) = 0,$$

i.e. if and only if $\alpha_1 = \dots = \alpha_r = 0$. Thus $\mathbf{x} \in \mathfrak{U}'$ if and only if it is of the form

$$\mathbf{x} = \alpha_{r+1}\mathbf{x}_{r+1} + \dots + \alpha_n \mathbf{x}_n.$$

Therefore $\mathfrak{U}'$ is the vector space spanned by $\mathbf{x}_{r+1},\dots,\mathbf{x}_n$. It follows by symmetry that the orthogonal complement of $\mathfrak{U}'$ is spanned by $\mathbf{x}_1,\dots,\mathbf{x}_r$ and so is identical with $\mathfrak{U}$. Moreover, $d(\mathfrak{U}') = n-r$, and therefore $d(\mathfrak{U})+d(\mathfrak{U}') = n$.

THEOREM 2.5.8. *The orthogonal complement of* $\mathfrak{U}$ *is a complement of* $\mathfrak{U}$.

Let $\mathfrak{U}'$ denote the orthogonal complement of $\mathfrak{U}$. If $\mathbf{x} \in \mathfrak{U}, \mathbf{x} \in \mathfrak{U}'$, then $(\mathbf{x},\mathbf{x}) = 0$ and so $\mathbf{x} = \mathbf{0}$. Thus $\mathbf{0}$ is the only vector common to $\mathfrak{U}$ and $\mathfrak{U}'$. Moreover, by Theorem 2.5.7 (iii), $d(\mathfrak{U})+d(\mathfrak{U}') = n$. Hence, by Theorem 2.3.7 (p. 55), $\mathfrak{U}$ and $\mathfrak{U}'$ are complements.

EXERCISE 2.5.8. Let $\mathfrak{U}$ be a subspace of $\mathfrak{V}_n$. Show that, given any vector $\mathbf{x} \in \mathfrak{V}_n$, there exist unique vectors $\mathbf{y}, \mathbf{z} \in \mathfrak{V}_n$ such that

$$\mathbf{x} = \mathbf{y}+\mathbf{z}, \qquad \mathbf{y} \in \mathfrak{U}, \qquad (\mathbf{y},\mathbf{z}) = 0.$$

In our discussion of inner products we confined ourselves to vector spaces and did not attempt to consider linear manifolds. It should be pointed out, however, that the notion of inner product can be based on a set of abstract postulates and axiomatized in much the same way as the notions of vector addition and multiplication by scalars. Since it is not essential for our purpose to pursue this topic, we refer the reader elsewhere for details.†

PROBLEMS ON CHAPTER II

1. Show that the set $\mathfrak{S}$ of numbers of the form $a+b\sqrt{(-5)}$, where a and b are arbitrary rational numbers, is a field. Show also that $\mathfrak{S}$ is not a field if a and b are allowed to take integral values only.

2. Show that if a number of vectors span a vector space $\mathfrak{U}$, then they contain a basis of $\mathfrak{U}$.

3. Let $\mathfrak{U}$ be the set of all vectors (x, y) whose components satisfy the relations $ax+by = 0$, $cx+dy = 0$, where a, b, c, d are given numbers. Show that $\mathfrak{U}$ is a vector space and determine its dimensionality in terms of a, b, c, d.

4. Let $X_1,\dots,X_m$, $Y_1,\dots,Y_n$ be elements of a linear manifold. Let μ be the maximum number of linearly independent elements among $X_1,\dots,X_m$ and ν the maximum number of linearly independent elements among $Y_1,\dots,Y_n$.

† See Birkhoff and MacLane, 7, 183–9, and Stoll, 11, 214–25.

Show that the maximum number of linearly independent elements among $X_1,\ldots,X_m, Y_1,\ldots,Y_n$ does not exceed $\mu+\nu$.

5. Let $X_1,\ldots,X_m, Y_1,\ldots,Y_n$ be elements of a linear manifold and suppose that each Y is a linear combination of the X's. If μ denotes the maximum number of linearly independent elements among the X's and ν the maximum number of linearly independent elements among the Y's, show that $\nu \leqslant \mu$.

6. Let $m \geqslant 2$, and let $X_1,\ldots,X_m$ be any non-zero elements of a linear manifold. Show that these elements are linearly dependent if and only if at least one of them is a linear combination of the elements preceding it. Hence deduce Theorem 2.3.5 (p. 54).

7. Show that a subset $\mathfrak{N}$ of a linear manifold $\mathfrak{M}$ is a submanifold if and only if it is closed with respect to multiplication by scalars and addition.

8. A subspace $\mathfrak{U}$ of $\mathfrak{B}_4$ is spanned by the vectors $(2,-1,0,1)$, $(6,1,4,-5)$, $(4,1,3,-4)$. Find an orthonormal basis of $\mathfrak{U}$.

9. Show that the vectors $(2,-3,1)$, $(0,1,2)$, $(1,1,-2)$ constitute a basis of $\mathfrak{B}_3$, and find the vector representing (α,β,γ) with respect to this basis.

10. Prove that, with respect to suitable definitions of multiplication by scalars and addition, the set of all polynomials in t, of degree $\leqslant 2$, is a linear manifold $\mathfrak{M}$ of dimensionality 3. When do three given quadratic polynomials $f_1(t), f_2(t), f_3(t)$ constitute a basis of $\mathfrak{M}$?

Show that $\{t^2+t,\ t^2-t,\ t+1\}$ is a basis of $\mathfrak{M}$, and obtain the vector representing $2t^2-7t+3$ with respect to this basis.

11. Prove that, for any two complex vectors $\mathbf{x}$ and $\mathbf{y}$ (of the same order),

$$|\mathbf{x}+\mathbf{y}|^2+|\mathbf{x}-\mathbf{y}|^2 = 2|\mathbf{x}|^2+2|\mathbf{y}|^2.$$

Give a geometric interpretation of this result for the case of real vectors of order 2.

12. Show that, for any orthogonal set of complex vectors $\mathbf{x}_1,\ldots,\mathbf{x}_k$,

$$\left|\sum_{\nu=1}^{k} \mathbf{x}_\nu\right|^2 = \sum_{\nu=1}^{k} |\mathbf{x}_\nu|^2.$$

Give a geometric interpretation of this result for the case of two real vectors of order 2.

13. Let $\mathbf{x}_1,\ldots,\mathbf{x}_k$ be an orthonormal set of complex vectors. Show that, for any vector $\mathbf{x}$,

$$\sum_{\nu=1}^{k} |(\mathbf{x},\mathbf{x}_\nu)|^2 \leqslant |\mathbf{x}|^2,$$

with equality if and only if $\mathbf{x}$ is a linear combination of $\mathbf{x}_1,\ldots,\mathbf{x}_k$. (This is known as *Bessel's inequality*.) Hence deduce Theorem 2.5.2 (p. 63).

Also deduce that, if $\{\mathbf{x}_1,\ldots,\mathbf{x}_n\}$ is an orthonormal basis of $\mathfrak{B}_n$, then, for any $\mathbf{x} \in \mathfrak{B}_n$,

$$\sum_{\nu=1}^{n} |(\mathbf{x},\mathbf{x}_\nu)|^2 = |\mathbf{x}|^2.$$

14. By considering the expression $|\lambda\mathbf{x}+\mathbf{y}|^2$ as a quadratic in λ, prove that, for all real vectors $\mathbf{x},\mathbf{y}$,

$$|(\mathbf{x},\mathbf{y})| \leqslant |\mathbf{x}|\,|\mathbf{y}|.$$

Extend the method of proof to the case of complex vectors.

15. Let $\{\mathbf{x}_n\}$ be a sequence of vectors and $\mathbf{x}$ some vector in a vector space. If, as $n \to \infty$, $|\mathbf{x}_n - \mathbf{x}| \to 0$, we say that $\mathbf{x}_n \to \mathbf{x}$. Prove the following results.

(i) If $\mathbf{x}_n \to \mathbf{x}$, $\mathbf{x}_n \to \mathbf{x}'$, then $\mathbf{x} = \mathbf{x}'$.

(ii) If $\mathbf{x}_n \to \mathbf{x}$, $\mathbf{y}_n \to \mathbf{y}$, then $\alpha\mathbf{x}_n + \beta\mathbf{y}_n \to \alpha\mathbf{x} + \beta\mathbf{y}$.

(iii) If $\mathbf{x}_n \to \mathbf{x}$, $\mathbf{y}_n \to \mathbf{y}$, then $(\mathbf{x}_n, \mathbf{y}_n) \to (\mathbf{x}, \mathbf{y})$.

(iv) If $|\mathbf{x}_m - \mathbf{x}_n| \to 0$ as $m, n \to \infty$, then there exists a vector $\mathbf{x}$ such that $\mathbf{x}_n \to \mathbf{x}$.

16. Show that if the subspaces $\mathfrak{U}$, $\mathfrak{U}'$ of $\mathfrak{V}_n$ are orthogonal complements, then $\mathbf{0}$ is the only vector common to $\mathfrak{U}$ and $\mathfrak{U}'$.

17. Show that two subspaces $\mathfrak{U}$, $\mathfrak{U}'$ of $\mathfrak{V}_n$ are complements if and only if $\mathbf{0}$ is the only vector common to $\mathfrak{U}$ and $\mathfrak{U}'$, and every $\mathbf{x} \in \mathfrak{V}_n$ can be expressed in the form $\mathbf{x} = \mathbf{y} + \mathbf{y}'$, where $\mathbf{y} \in \mathfrak{U}$, $\mathbf{y}' \in \mathfrak{U}'$.

18. Let $\mathfrak{U}$, $\mathfrak{V}$ be two subspaces of $\mathfrak{V}_n$ and let $\mathfrak{V}'$ be a complement of $\mathfrak{V}$. Show that, if $\mathfrak{V} \subset \mathfrak{U}$ and the only vector common to $\mathfrak{U}$ and $\mathfrak{V}'$ is $\mathbf{0}$, then $\mathfrak{U} = \mathfrak{V}$.

19. Let $\mathfrak{U}$, $\mathfrak{V}$ be subspaces of $\mathfrak{V}_n$ and let $\mathfrak{U}'$, $\mathfrak{V}'$ be their orthogonal complements. Show that $\mathfrak{U} \subset \mathfrak{V}$ implies $\mathfrak{V}' \subset \mathfrak{U}'$ and that $\mathfrak{U} \subset \mathfrak{V}'$ implies $\mathfrak{V} \subset \mathfrak{U}'$.

20. A subspace $\mathfrak{U}$ of $\mathfrak{V}_3$ consists of all vectors (x_1, x_2, x_3) which satisfy the relations $3x_1 + x_2 - x_3 = 0$, $x_1 - 5x_2 + x_3 = 0$. Determine a basis of the orthogonal complement of $\mathfrak{U}$.

21. The subspace $\mathfrak{U}$ of $\mathfrak{V}_4$ is spanned by the vectors $(1, 0, -1, 2)$ and $(-1, 1, 1, 0)$. Show that the orthogonal complement $\mathfrak{U}'$ of $\mathfrak{U}$ is the set of vectors of the form $(\alpha - 2\beta, -2\beta, \alpha, \beta)$, where α, β are arbitrary numbers. Also find an orthonormal basis of $\mathfrak{U}'$.

22. Let $\mathfrak{U}_1, \ldots, \mathfrak{U}_k$ be subspaces of the total vector space $\mathfrak{V}_n$, and suppose that every $\mathbf{x} \in \mathfrak{V}_n$ can be expressed in the form $\mathbf{x} = \mathbf{y}_1 + \ldots + \mathbf{y}_k$, where $\mathbf{y}_1 \in \mathfrak{U}_1, \ldots, \mathbf{y}_k \in \mathfrak{U}_k$. Show that there exists a basis of $\mathfrak{V}_n$ every vector of which belongs to some $\mathfrak{U}_i$. Deduce that $d(\mathfrak{U}_1) + \ldots + d(\mathfrak{U}_k) \geqslant n$.

23. Let $\mathfrak{U}, \mathfrak{V}$ be two subspaces of $\mathfrak{V}_n$. Let $\mathfrak{U} + \mathfrak{V}$ denote the set of all vectors in $\mathfrak{V}_n$ expressible in the form $\mathbf{x} + \mathbf{y}$, where $\mathbf{x} \in \mathfrak{U}$, $\mathbf{y} \in \mathfrak{V}$; and let $\mathfrak{U} \cap \mathfrak{V}$ denote the set of all vectors common to $\mathfrak{U}$ and $\mathfrak{V}$. Show that $\mathfrak{U} + \mathfrak{V}$ and $\mathfrak{U} \cap \mathfrak{V}$ are vector spaces, and that

$$d(\mathfrak{U} + \mathfrak{V}) + d(\mathfrak{U} \cap \mathfrak{V}) = d(\mathfrak{U}) + d(\mathfrak{V}).$$

What becomes of this result when $\mathfrak{U}, \mathfrak{V}$ are complements?

24. Show that equality occurs in Theorem 2.5.2 if and only if $\mathbf{x}$ and $\mathbf{y}$ are linearly dependent.

III

THE ALGEBRA OF MATRICES

THE algebra of matrices was first developed systematically by Cayley in a series of papers which began to appear in 1857, and most of the results derived in the present book were discovered during the second half of the last century. Our aim in this chapter is to consider in some detail the elementary notions in the theory of matrices.

3.1. Elementary algebra

Mathematicians often find themselves compelled to introduce new types of entities and to define operations applicable to these entities. It then becomes a matter of considerable importance to determine to what extent the operations so defined resemble the familiar operations of elementary algebra. Thus, in the previous chapter we introduced certain objects called 'vectors' and, having defined two operations (addition and multiplication by scalars) to which vectors could be subjected, we proceeded to study the formal nature of these operations. A similar programme is to be carried out in the present chapter with respect to *matrices*, whose definition we defer till the next section. There we shall also define operations applicable to matrices and shall investigate to what extent the resulting matrix laws reflect the structure of the number system. It will, therefore, be useful at the present stage to enumerate briefly the basic algebraic laws valid for real and complex numbers.

We recall, in the first place, that any two numbers a and b are either *equal* ($a = b$) or *unequal* ($a \neq b$).

The fundamental operations in number algebra are those of *addition* and *multiplication*, both operations being applicable to pairs of numbers. The result of adding a and b is a number which is called their *sum* and is denoted by $a+b$; the result of multiplying a and b is a number which is called their *product* and is denoted by $a.b$ or ab. Addition and multiplication are both *commutative*:

$$a+b = b+a, \qquad ab = ba.$$

In other words, it is immaterial in which order the terms of a sum or the factors of product are taken. Addition and multiplication are, furthermore, both *associative*:

$$a+(b+c) = (a+b)+c, \tag{3.1.1}$$

$$a(bc) = (ab)c. \tag{3.1.2}$$

These results imply that we may, without danger of ambiguity, remove the brackets and write simply $a+b+c$ for either side of (3.1.1) and abc for either side of (3.1.2). Hence we may also remove brackets from expressions such as

$$\{a+(b+c)\}+d \quad \text{or} \quad \{(ab)(cd)\}e.$$

The operations of addition and multiplication are involved together in the *distributive law*:

$$a(b+c) = ab+ac.$$

The number system contains two numbers, 1 (unity) and 0 (zero), which have very special properties. Thus, for all a,

$$a+0 = a, \qquad 0a = 0, \qquad 1a = a.$$

Again, if $ab = 0$, then at least one of a, b is equal to 0. This result is known as the *division law*; and we often express it by saying that the set of real (or complex) numbers has no *divisors of zero*.† A consequence of this fact is the *cancellation law* which states that if $ax = ay$ and $a \neq 0$, then $x = y$.

The equation $a+x = b$ possesses a unique solution for x; this solution is denoted by $x = b-a$; in particular we write $-a$ for $0-a$. The number $b-a$ is called the *difference* of b and a, and the operation of forming differences is known as *subtraction*. Of the algebraic laws involving subtraction we may mention the following:

$$a-a = 0, \qquad (-1)a = -a,$$

$$a(-b) = -ab, \qquad a(b-c) = ab-ac.$$

If $a \neq 0$, the equation $ax = 1$ has a unique solution for x, which is denoted by $x = \dfrac{1}{a}$ or $1/a$ or a^{-1}. The number a^{-1} is called the *inverse* or *reciprocal* of a. For $a \neq 0$ the equation $ax = b$ has the unique solution $x = a^{-1}b$ which is usually denoted by $\dfrac{b}{a}$ or b/a. The number b/a is called the *quotient* of b by a, and the operation of

† Cf. Definition 3.6.3 (p. 95).

forming quotients is known as *division*. The following identities are typical of those involving division:

$$\frac{a}{b}\cdot\frac{c}{d}=\frac{ac}{bd} \qquad (b\neq 0,\, d\neq 0),$$

$$\frac{a}{b}\Big/\frac{c}{d}=\frac{ad}{bc} \qquad (b\neq 0,\, c\neq 0,\, d\neq 0),$$

$$\frac{a}{b}+\frac{c}{d}=\frac{ad+bc}{bd} \qquad (b\neq 0,\, d\neq 0).$$

When r is a positive integer, a^r is defined as $a.a\,...\,a$ (r factors). When $a\neq 0$ and r is a negative integer, a^r is defined as $(a^{-1})^{-r}$. Moreover, for every a, we put $a^0=1$. With these conventions we have the following *index laws*, valid for all integers r, s and all numbers $a\neq 0$:

$$a^r a^s = a^{r+s}, \qquad (a^r)^s = a^{rs}.$$

3.2. Preliminary notions concerning matrices

In this section we shall collect together for convenience a number of definitions relating to matrices and explain the notation that is to be used subsequently.

The context in which matrices arise most naturally is that of *linear substitutions*. A linear substitution is a system of relations between two sets of variables, say $x_1,..., x_n$ and $y_1,..., y_m$, of the form

$$\left.\begin{array}{l} y_1 = a_{11}x_1+...+a_{1n}x_n \\ \cdot \quad \cdot \quad \cdot \quad \cdot \quad \cdot \quad \cdot \quad \cdot \quad \cdot \\ y_m = a_{m1}x_1+...+a_{mn}x_n \end{array}\right\}. \tag{3.2.1}$$

The reader will readily call to mind situations where such systems of relations make their appearance. We may conveniently characterize the system of relations (3.2.1) by isolating the array of coefficients a_{ij}. This point of view, first adopted by Cayley, leads at once to the notion of a matrix.

DEFINITION 3.2.1. (i) *An* $m\times n$ MATRIX (*or a matrix of type* $m\times n$) $\mathbf{A}$ *over the field* $\mathfrak{F}$ *is a rectangular array of numbers in* $\mathfrak{F}$, *consisting of m rows and n columns, say*

$$\mathbf{A}=\begin{pmatrix} a_{11} & a_{12} & \cdot & \cdot & \cdot & a_{1n} \\ a_{21} & a_{22} & \cdot & \cdot & \cdot & a_{2n} \\ \cdot & \cdot & \cdot & \cdot & \cdot & \cdot \\ a_{m1} & a_{m2} & \cdot & \cdot & \cdot & a_{mn} \end{pmatrix}. \tag{3.2.2}$$

(ii) *The mn numbers* $a_{11}, a_{12},..., a_{mn}$ *are the* ELEMENTS *of the matrix* **A**.

The term 'matrix' is due to Sylvester (1850). Matrices will normally be denoted by capital letters printed in bold-face type.† The element standing in the ith row and jth column—often called the (i,j)th element—of a matrix **A** will be denoted by $\mathbf{A}_{ij}$. If, as in (3.2.2), $\mathbf{A}_{ij} = a_{ij}$, we shall write $\mathbf{A} = (a_{ij})$. It is important to remember that the first suffix of an element indicates the row and the second the column of that element.

Two matrices of the same type will be said to be equal if and only if their corresponding elements are equal.

In future we shall not, as a rule, mention the reference field $\mathfrak{F}$ explicitly but shall assume tacitly that such a field has been given. If $\mathfrak{F}$ is the real (complex) field, a matrix **A** over $\mathfrak{F}$ is called a real (complex) matrix.

DEFINITION 3.2.2. (i) *A* SQUARE MATRIX *of order n is a matrix of type* $n \times n$.

(ii) *The elements* $a_{11}, a_{22},..., a_{nn}$ *of the square matrix*

$$\begin{pmatrix} a_{11} & a_{12} & . & . & . & a_{1n} \\ a_{21} & a_{22} & . & . & . & a_{2n} \\ . & . & . & . & . & . \\ a_{n1} & a_{n2} & . & . & . & a_{nn} \end{pmatrix}$$

constitute its DIAGONAL, *and are called* DIAGONAL ELEMENTS.

The reader is, in effect, already familiar with square matrices from the discussion of determinants.

When we wish to emphasize that we are not necessarily dealing with square matrices we speak of *rectangular matrices.*

DEFINITION 3.2.3. *The* ZERO MATRIX (*of type* $m \times n$) *is the* $m \times n$ *matrix all of whose elements are equal to zero. It is denoted by* $\mathbf{O}_m^n$, *or, when no confusion is likely, by* **O**.

DEFINITION 3.2.4. *The* UNIT MATRIX *of order n (denoted by* $\mathbf{I}_n$, *or simply by* **I**) *is the square matrix of order n whose diagonal elements are equal to* 1 *and whose remaining elements are all equal to* 0.

† The exceptions to this convention are explained after Definition 3.2.8.

Thus

$$\mathbf{I}_n = \begin{pmatrix} 1 & 0 & 0 & . & . & . & 0 \\ 0 & 1 & 0 & . & . & . & 0 \\ . & . & . & . & . & . & . \\ 0 & 0 & 0 & . & . & . & 1 \end{pmatrix} = (\delta_{ij}),$$

where δ_{ij} is the Kronecker delta. We note that the linear substitution characterized by $\mathbf{I}_n$ has the form $y_1 = x_1, y_2 = x_2,..., y_n = x_n$. This is known as the *identical substitution.*

DEFINITION 3.2.5. (i) *A* DIAGONAL MATRIX *is a square matrix all of whose elements outside the diagonal are equal to zero.* (ii) *A* SCALAR MATRIX *is a diagonal matrix all of whose diagonal elements are equal to one another.*

Thus a diagonal matrix of order n has the form

$$\begin{pmatrix} \alpha_1 & 0 & . & . & . & 0 \\ 0 & \alpha_2 & . & . & . & 0 \\ . & . & . & . & . & . \\ 0 & 0 & . & . & . & \alpha_n \end{pmatrix};$$

we shall denote this matrix by $\mathbf{dg}(\alpha_1, \alpha_2,..., \alpha_n)$. A scalar matrix has the form $\mathbf{dg}(\alpha, \alpha,..., \alpha)$. Particular instances of scalar matrices are the unit matrix and the square zero matrix. We shall presently see that scalar matrices behave essentially like scalars.

DEFINITION 3.2.6. *An* UPPER (LOWER) TRIANGULAR MATRIX *is a square matrix all of whose elements below (above) the diagonal are equal to zero. A* TRIANGULAR MATRIX *is one that is either upper triangular or lower triangular.*

If a matrix is both upper triangular and lower triangular, it is clearly diagonal.

DEFINITION 3.2.7. *The* TRANSPOSE $\mathbf{A}^T$ *of a matrix* $\mathbf{A}$ *is the matrix obtained from* $\mathbf{A}$ *by the interchange of rows and columns. The operation of deriving* $\mathbf{A}^T$ *from* $\mathbf{A}$ *is known as* TRANSPOSITION.

Thus, if

$$\mathbf{A} = \begin{pmatrix} a_{11} & . & . & . & a_{1n} \\ . & . & . & . & . \\ a_{m1} & . & . & . & a_{mn} \end{pmatrix}, \quad \text{then} \quad \mathbf{A}^T = \begin{pmatrix} a_{11} & . & . & . & a_{m1} \\ . & . & . & . & . \\ a_{1n} & . & . & . & a_{mn} \end{pmatrix},$$

i.e. if $\mathbf{A} = (a_{ij})$, then $\mathbf{A}^T = (b_{ij})$, where $b_{ij} = a_{ji}$ $(i = 1,..., n;$ $j = 1,..., m)$. The transpose of an $m \times n$ matrix is an $n \times m$ matrix. The transpose of an upper triangular matrix is a lower triangular matrix, and conversely.

Clearly $(\mathbf{A}^T)^T = \mathbf{A}$, and so the relation between $\mathbf{A}$ and $\mathbf{A}^T$ is symmetrical, either matrix being the transpose of the other.

DEFINITION 3.2.8. *A* COLUMN MATRIX *is a matrix of type* $n \times 1$, *say*

$$\begin{pmatrix} x_1 \\ x_2 \\ \cdot \\ \cdot \\ \cdot \\ x_n \end{pmatrix}.$$

A ROW MATRIX *is a matrix of type* $1 \times n$, *say*

$$(x_1 \quad x_2 \quad \ldots \quad x_n).$$

Matrices of both these types may be regarded as vectors† and referred to respectively as *column vectors* and *row vectors*. A matrix may thus be thought of as a generalization of a vector.

In the majority of cases it is more convenient to use column vectors in preference to row vectors. We shall, accordingly, adopt the convention that, unless the contrary is stated, all vectors are understood to be column vectors. In particular, the vectors $\mathbf{e}_1,\ldots,\mathbf{e}_n$ defined by (2.1.2) on p. 43 will from now on be regarded as column vectors.

We shall continue to denote vectors by small letters in bold-face type. Suppose, for instance, that

$$\mathbf{x} = \begin{pmatrix} x_1 \\ \cdot \\ \cdot \\ \cdot \\ x_n \end{pmatrix}. \tag{3.2.3}$$

The transpose of a column vector is a row vector, and we shall then have

$$\mathbf{x}^T = (x_1,\ldots,x_n).$$

For typographical reasons it is generally preferable to avoid the form (3.2.3) and to write instead

$$\mathbf{x} = (x_1,\ldots,x_n)^T.$$

The notion of row vectors and column vectors enables us to introduce a very convenient notation connected with matrices.

† Strictly speaking this assertion will have been justified only after the operations defined for vectors have been extended to matrices in § 3.3. It will then also be seen that 1×1 matrices can, in fact, be treated as scalars. We shall make use of this identification whenever it is convenient to do so.

Definition 3.2.9. *The i-th row of a matrix* $\mathbf{A}$ *(regarded as a row vector) is denoted by* $\mathbf{A}_{i*}$; *the j-th column of* $\mathbf{A}$ *(regarded as a column vector) is denoted by* $\mathbf{A}_{*j}$.

Exercise 3.2.1. Establish the identities

$$(\mathbf{A}_{i*})^T = (\mathbf{A}^T)_{*i}; \qquad (\mathbf{A}_{*j})^T = (\mathbf{A}^T)_{j*}.$$

3.3. Addition and multiplication of matrices

3.3.1. Matrix algebra† is based on four operations—multiplication by scalars, addition, matrix multiplication, and transposition. The first three of these will be discussed in the present section. It should be noted that the definitions of multiplication by scalars and of addition given below coincide with the previous definitions (Definitions 2.1.4 and 2.1.5) for the special case when the matrices are vectors.

Definition 3.3.1. *The product* $\alpha\mathbf{A}$ *(or* $\mathbf{A}\alpha$*) of an* $m \times n$ *matrix* $\mathbf{A}$ *by a scalar* α *is a matrix defined by the relations*

$$(\alpha\mathbf{A})_{ij} = (\mathbf{A}\alpha)_{ij} = \alpha\mathbf{A}_{ij} \qquad (i = 1,..., m;\, j = 1,..., n).$$

We obtain at once a number of obvious results, such as

$$1\mathbf{A} = \mathbf{A},$$
$$(-1)\mathbf{A} = (-1)(a_{ij}) = (-a_{ij}),$$
$$0\mathbf{A} = \mathbf{O},$$
$$\alpha\mathbf{O} = \mathbf{O},$$
$$(\alpha\beta)\mathbf{A} = \alpha(\beta\mathbf{A}),$$
$$\alpha\mathbf{I} = \mathbf{dg}(\alpha,..., \alpha).$$

The last result shows that a scalar matrix is a scalar multiple of the unit matrix.

Definitions 3.3.2. *Let* $\mathbf{A}, \mathbf{B}$ *be matrices of the same type, say of type* $m \times n$. *Then their sum* $\mathbf{A}+\mathbf{B}$ *is a matrix defined by the relations*

$$(\mathbf{A}+\mathbf{B})_{ij} = \mathbf{A}_{ij}+\mathbf{B}_{ij} \qquad (i = 1,..., m;\, j = 1,..., n).$$

Thus, if $\mathbf{A} = (a_{ij})$, $\mathbf{B} = (b_{ij})$, then $\mathbf{A}+\mathbf{B} = (a_{ij}+b_{ij})$. For instance

$$\begin{pmatrix} 2 & 3 & -1 \\ 0 & 1 & 2 \end{pmatrix} + \begin{pmatrix} 0 & -1 & 1 \\ 2 & 0 & 3 \end{pmatrix} = \begin{pmatrix} 2 & 2 & 0 \\ 2 & 1 & 5 \end{pmatrix}.$$

When $\mathbf{A}$ and $\mathbf{B}$ are not of the same type, $\mathbf{A}+\mathbf{B}$ is not defined.

† Matrix algebra can be studied at different levels of sophistication. An elementary exposition will be found, for instance, in Durell and Robson, **14**, chap. xvii. For treatment from an advanced point of view, on the other hand, see MacDuffee, **32**, chap. vii, or Albert, **33**, chaps. iii–v.

We note that we can now form linear combinations of matrices, i.e. expressions of the form

$$\alpha_1 \mathbf{A}_1 + \ldots + \alpha_k \mathbf{A}_k,$$

provided only that $\mathbf{A}_1, \ldots, \mathbf{A}_k$ are of the same type.

The usual algebraic laws relating to the two operations so far introduced follow at once from the definitions.† Some of these are incorporated in the next theorem.

THEOREM 3.3.1. *Matrix addition is commutative and associative, i.e.*

$$\mathbf{A}+\mathbf{B} = \mathbf{B}+\mathbf{A},$$

$$\mathbf{A}+(\mathbf{B}+\mathbf{C}) = (\mathbf{A}+\mathbf{B})+\mathbf{C}.$$

Furthermore, matrix addition and multiplication by scalars are connected by the following distributive laws:

$$\alpha(\mathbf{A}+\mathbf{B}) = \alpha\mathbf{A}+\alpha\mathbf{B},$$

$$(\alpha+\beta)\mathbf{A} = \alpha\mathbf{A}+\beta\mathbf{A}.$$

The associative law implies, in particular, that when we have an expression involving only addition of matrices and containing brackets, all such brackets can be removed without danger of ambiguity.

There is no difficulty in introducing *subtraction* of matrices. By Definition 3.3.2 it is obvious that if $\mathbf{A}$, $\mathbf{B}$ are matrices of the same type, the equation $\mathbf{A}+\mathbf{X} = \mathbf{B}$ possesses a unique solution for $\mathbf{X}$; this solution is denoted by $\mathbf{X} = \mathbf{B}-\mathbf{A}$ and is called the difference of $\mathbf{B}$ and $\mathbf{A}$. In fact, if $\mathbf{A} = (a_{ij})$, $\mathbf{B} = (b_{ij})$, then $\mathbf{X} = (b_{ij}-a_{ij})$. For $\mathbf{O}-\mathbf{A}$ we write $-\mathbf{A}$. It is easy to see that subtraction satisfies all the usual rules. We have, for instance, the following identities.

$$\mathbf{A}-\mathbf{A} = \mathbf{O}, \qquad (-1)\mathbf{A} = -\mathbf{A}, \qquad -(-\mathbf{A}) = \mathbf{A},$$

$$\alpha(\mathbf{A}-\mathbf{B}) = \alpha\mathbf{A}-\alpha\mathbf{B}, \qquad (\alpha-\beta)\mathbf{A} = \alpha\mathbf{A}-\beta\mathbf{A}.$$

EXERCISE 3.3.1. Verify that the scalar matrices of order n form a one-dimensional linear manifold $\mathfrak{M}$ and that the scalars form a linear manifold $\mathfrak{M}'$. Prove that $\mathfrak{M} \simeq \mathfrak{M}'$.

† The reader is advised to write out in detail the proofs of the results given below and to observe how the various laws relating to matrices are derived from analogous laws of number algebra.

EXERCISE 3.3.2. Show that the set $\mathfrak{M}_{m,n}$ of all $m \times n$ matrices over $\mathfrak{F}$ is a linear manifold over $\mathfrak{F}$.† By considering the mn matrices of $\mathfrak{M}_{m,n}$ whose elements consist of a single 1 and $mn-1$ zeros, or by using the corollary to Theorem 2.4.2 (p. 59), show that $d(\mathfrak{M}_{m,n}) = mn$.

3.3.2. Matrix addition and multiplication of matrices by scalars are obvious extensions of the corresponding ideas for vectors. In introducing next the idea of matrix multiplication, however, we break fresh ground.

DEFINITION 3.3.3. *Suppose that* $\mathbf{A}$ *is an* $l \times m$ *matrix and* $\mathbf{B}$ *an* $m \times n$ *matrix, so that the number of columns of* $\mathbf{A}$ *is equal to the number of rows of* $\mathbf{B}$. *Then the matrix product* $\mathbf{AB}$ *is the* $l \times n$ *matrix defined by the relations*

$$(\mathbf{AB})_{ij} = \sum_{k=1}^{m} \mathbf{A}_{ik}\,\mathbf{B}_{kj} \qquad (i = 1,\ldots,l;\ j = 1,\ldots,n).$$

In other words, the (i,j)th element of $\mathbf{AB}$ is obtained by multiplying together corresponding elements in the ith row of $\mathbf{A}$ and the jth column of $\mathbf{B}$ and then adding the products.

EXERCISE 3.3.3. Show that, if the product $\mathbf{AB}$ exists, then

$$(\mathbf{AB})_{ij} = \mathbf{A}_{i*}\,\mathbf{B}_{*j}.$$

We note that $\mathbf{AB}$, when it exists, has as many rows as $\mathbf{A}$ and as many columns as $\mathbf{B}$. Thus, for instance, the product of a 4×2 matrix and a 2×3 matrix (in that order) is a 4×3 matrix, e.g.

$$\begin{pmatrix} 2 & 1 \\ -1 & 3 \\ 1 & 0 \\ 5 & 2 \end{pmatrix} \begin{pmatrix} 1 & 1 & -2 \\ 3 & -4 & 3 \end{pmatrix} = \begin{pmatrix} 5 & -2 & -1 \\ 8 & -13 & 11 \\ 1 & 1 & -2 \\ 11 & -3 & -4 \end{pmatrix}.$$

If the number of columns of $\mathbf{A}$ is not equal to the number of rows of $\mathbf{B}$, then $\mathbf{AB}$ is not defined.

The two products $\mathbf{AB}$ and $\mathbf{BA}$ are quite distinct entities and, indeed, one of them may exist whereas the other does not. The condition for both $\mathbf{AB}$ and $\mathbf{BA}$ to exist is that if $\mathbf{A}$ is of type $m \times n$, $\mathbf{B}$ should be of type $n \times m$. In that case $\mathbf{AB}$, $\mathbf{BA}$ are of type $m \times m$, $n \times n$ respectively, and consequently the two products cannot even be compared unless $m = n$. The question whether $\mathbf{AB}$ and $\mathbf{BA}$ are

† We could have begun our discussion of matrices by proving this result as soon as the operations of addition and multiplication by scalars had been defined. It would then have been unnecessary to verify successively all the various laws relating to these two operations since these laws would have been known to hold by virtue of the results derived in § 2.2.

equal only arises, therefore, when **A** and **B** are square matrices of the same order. Analogy with elementary algebra may lead us to expect that in this case the two products are necessarily equal, but this is not in fact so.

THEOREM 3.3.2. *Matrix multiplication is non-commutative, i.e. the equation* **AB** = **BA** *need not be satisfied even when both* **AB** *and* **BA** *exist and are of the same type.*

To establish this negative conclusion we need only construct a single example where **AB** ≠ **BA**. In fact, almost any two square matrices taken at random satisfy this requirement. Thus, taking

$$\mathbf{A} = \begin{pmatrix} 1 & 3 \\ 2 & 1 \end{pmatrix}, \qquad \mathbf{B} = \begin{pmatrix} 1 & -1 \\ 1 & 2 \end{pmatrix},$$

we obtain
$$\mathbf{AB} = \begin{pmatrix} 4 & 5 \\ 3 & 0 \end{pmatrix}, \qquad \mathbf{BA} = \begin{pmatrix} -1 & 2 \\ 5 & 5 \end{pmatrix}.$$

In view of Theorem 3.3.2 it is essential when referring to a matrix multiplication to state unambiguously the order in which the factors are taken.

DEFINITION 3.3.4. *The matrix* **A** *is said to* PREMULTIPLY **B** *and* **B** *is said to* POSTMULTIPLY **A** *in the product* **AB**.

DEFINITION 3.3.5. *Two matrices* **A**, **B** COMMUTE (*with each other*), *or are* COMMUTING MATRICES, *if* **AB** = **BA**.

EXERCISE 3.3.4. Let $\mathbf{\Lambda}$ be a diagonal matrix whose diagonal elements are distinct. Show that if **A** commutes with $\mathbf{\Lambda}$, then **A** is also a diagonal matrix.

Although multiplication of matrices differs from that of numbers in that it does not obey the commutative law, in most other respects the two operations have similar properties.

THEOREM 3.3.3. *Matrix multiplication is associative, i.e.*

$$\mathbf{A}(\mathbf{BC}) = (\mathbf{AB})\mathbf{C}, \tag{3.3.1}$$

provided that either side of (3.3.1) *is defined.*

Let the matrices **A**, **B**, **C** be of type $p \times q$, $r \times s$, $t \times u$ respectively. Then **BC** exists and is of type $r \times u$ provided that $s = t$. In that case **A**(**BC**) exists and is of type $p \times u$ provided that $q = r$. Thus **A**(**BC**) exists if and only if $q = r$, $s = t$, and in that case it is of type $p \times u$. In exactly the same way we see that the same is true of (**AB**)**C**. Hence the existence of either side of (3.3.1) implies the existence of the other side, and the matrices on the two sides are then of the same type. It now remains to prove that they are equal. Writing $\mathbf{A} = (a_{ij})$ and so on, we have, in view of the associative and the distributive laws for numbers,

$$\{\mathbf{A}(\mathbf{BC})\}_{ij} = \sum_k a_{ik}(\mathbf{BC})_{kj} = \sum_k a_{ik}\Big\{\sum_l b_{kl}c_{lj}\Big\} = \sum_{k,l} a_{ik}b_{kl}c_{lj},$$

where k, l run over the appropriate ranges of values.† Similarly

$$\{(\mathbf{AB})\mathbf{C}\}_{ij} = \sum_l (\mathbf{AB})_{il}c_{lj} = \sum_l \Big\{\sum_k a_{ik}b_{kl}\Big\}c_{lj} = \sum_{k,l} a_{ik}b_{kl}c_{lj},$$

and the proof is therefore complete.

The result just proved shows that the matrix product

$$\mathbf{A}(\mathbf{BC}) = (\mathbf{AB})\mathbf{C}$$

can be written unambiguously as **ABC**.

The associative law extends immediately to any number of factors. Thus we have, for instance,

$$(\mathbf{AB})(\mathbf{CD}) = \mathbf{A}(\mathbf{BC})\mathbf{D} = (\mathbf{ABC})\mathbf{D} = \mathbf{A}(\mathbf{BCD}),$$

provided that any one of these expressions is defined. In fact, we can write **ABCD** for the product in question. Quite generally we may say that in forming products of matrices we need only pay attention to the order of the factors but not to the way in which they are bracketed.

We saw above that

$$(\mathbf{ABC})_{ij} = \sum_{k,l} \mathbf{A}_{ik}\mathbf{B}_{kl}\mathbf{C}_{lj}.$$

Analogous formulae hold for products of more than three factors, e.g.

$$(\mathbf{ABCD})_{ij} = \sum_{k,l,m} \mathbf{A}_{ik}\mathbf{B}_{kl}\mathbf{C}_{lm}\mathbf{D}_{mj}.$$

EXERCISE 3.3.5. If $\mathbf{A}_1,\ldots,\mathbf{A}_k$ are matrices of type $m_1 \times n_1,\ldots, m_k \times n_k$ respectively, what is the condition for the product $\mathbf{A}_1\ldots\mathbf{A}_k$ to exist; and, when it exists, what is its type?

A matrix can be multiplied by itself if and only if it is a square matrix, and in that case the index notation can be conveniently employed.

DEFINITION 3.3.6. *If* **A** *is a square matrix of order* n, *then*

$$\mathbf{A}^0 = \mathbf{I}_n, \qquad \mathbf{A}^r = \mathbf{A}^{r-1}\mathbf{A} \quad (r \geqslant 1).$$

† In fact, $k = 1, 2,\ldots, q$; $l = 1, 2,\ldots, s$.

In virtue of the associative law we have, in fact,

$$\mathbf{A}^1 = \mathbf{A}, \quad \mathbf{A}^2 = \mathbf{AA}, \quad \mathbf{A}^3 = \mathbf{AAA}, \quad \mathbf{A}^4 = \mathbf{AAAA},$$

and so on. It is now an easy matter to verify the index laws

$$\mathbf{A}^r\mathbf{A}^s = \mathbf{A}^{r+s}, \qquad (\mathbf{A}^r)^s = \mathbf{A}^{rs}$$

for all non-negative integers, r, s.

The zero matrix and the unit matrix play a particularly interesting part in matrix algebra. If $\mathbf{A}$ is any $m \times n$ matrix, then clearly

$$\mathbf{A} + \mathbf{O}_m^n = \mathbf{A},$$

$$\mathbf{AO}_n^r = \mathbf{O}_m^r, \qquad \mathbf{O}_l^m \mathbf{A} = \mathbf{O}_l^n.$$

Moreover, since $\mathbf{I}_n$ is the $n \times n$ matrix (δ_{ij}), we have

$$(\mathbf{AI}_n)_{ij} = \sum_{k=1}^{n} \mathbf{A}_{ik}\,\delta_{kj} = \mathbf{A}_{ij}.$$

Hence
$$\mathbf{AI}_n = \mathbf{A}, \tag{3.3.2}$$

and similarly
$$\mathbf{I}_m\mathbf{A} = \mathbf{A}. \tag{3.3.3}$$

These results show that in matrix algebra the zero matrix and the unit matrix play roles corresponding to those of the numbers 0 and 1 in elementary algebra.

We may also note that, in view of (3.3.2) and (3.3.3), multiplication by scalars can be interpreted as matrix multiplication, for we have

$$\alpha\mathbf{A} = (\alpha\mathbf{I}_m)\mathbf{A} = \mathbf{A}(\alpha\mathbf{I}_n).$$

THEOREM 3.3.4. *Matrix multiplication is distributive with respect to addition, i.e.*

$$\mathbf{A}(\mathbf{B}+\mathbf{C}) = \mathbf{AB}+\mathbf{AC}, \tag{3.3.4}$$

provided that either side of (3.3.4) *is defined; and*

$$(\mathbf{B}+\mathbf{C})\mathbf{A} = \mathbf{BA}+\mathbf{CA}, \tag{3.3.5}$$

provided that either side of (3.3.5) *is defined.*

We confine ourselves to the proof of the first identity, since the second can be dealt with in exactly the same manner. Let $\mathbf{A}$, $\mathbf{B}$, $\mathbf{C}$ be matrices of type $p \times q$, $r \times s$, $t \times u$ respectively. It is easily seen that either side of (3.3.4) is defined if and only if $q = r = t$, $s = u$, and that in that case both sides are $p \times u$ matrices. It remains, therefore, to show that the corresponding elements in the two matrices are equal. We write $\mathbf{A} = (a_{ij})$, and so on. Making use of

Definitions 3.3.3 and 3.3.2 and the distributive law for numbers, we obtain

$$\{\mathbf{A}(\mathbf{B}+\mathbf{C})\}_{ij} = \sum_k a_{ik}(\mathbf{B}+\mathbf{C})_{kj} = \sum_k a_{ik}(b_{kj}+c_{kj})$$
$$= \sum_k a_{ik}b_{kj}+\sum_k a_{ik}c_{kj}$$
$$= (\mathbf{AB})_{ij}+(\mathbf{AC})_{ij} = (\mathbf{AB}+\mathbf{AC})_{ij}.$$

The proof is therefore complete.

The distributive law can at once be extended to more complicated expressions. For example, we have

$$(\mathbf{A}+\mathbf{B})(\mathbf{C}+\mathbf{D}+\mathbf{E}) = \mathbf{AC}+\mathbf{AD}+\mathbf{AE}+\mathbf{BC}+\mathbf{BD}+\mathbf{BE}.$$

EXERCISE 3.3.6. Show that the set of all matrices which commute with a given matrix is closed with respect to addition and multiplication.

3.3.3. The results contained in the next two theorems are almost trivial. Thus, for instance, Theorem 3.3.5 (i) merely states that the ith row of **AB** can be obtained by premultiplying **B** by the ith row of **A**. Nevertheless identities such as these are useful since they help to reduce the manipulation of matrices to a purely mechanical procedure.

THEOREM 3.3.5.

(i) $(\mathbf{AB})_{i*} = \mathbf{A}_{i*}\mathbf{B}$;

(ii) $(\mathbf{AB})_{*j} = \mathbf{AB}_{*j}$;

(iii) $(\mathbf{ABC})_{ij} = \mathbf{A}_{i*}\mathbf{BC}_{*j}$.

By Exercise 3.3.3 (p. 80) we have

$$(\mathbf{AB})_{ij} = \mathbf{A}_{i*}\mathbf{B}_{*j}. \tag{3.3.6}$$

Hence $\quad (\mathbf{A}_{i*}\mathbf{B})_{1j} = (\mathbf{A}_{i*})_{1*}\mathbf{B}_{*j} = \mathbf{A}_{i*}\mathbf{B}_{*j} = (\mathbf{AB})_{ij}.$

Moreover $\quad \{(\mathbf{AB})_{i*}\}_{1j} = (\mathbf{AB})_{ij},$

and so, for all j, $\quad \{(\mathbf{AB})_{i*}\}_{1j} = (\mathbf{A}_{i*}\mathbf{B})_{1j}.$

Since both sides of (i) are row vectors, this identity is now proved. Identity (ii) is proved similarly. Again, using (3.3.6) and (i), we have

$$(\mathbf{ABC})_{ij} = (\mathbf{AB})_{i*}\mathbf{C}_{*j} = \mathbf{A}_{i*}\mathbf{BC}_{*j}.$$

THEOREM 3.3.6. *If* **A** *is an* $m\times n$ *matrix, and* $\mathbf{x} = (x_1,..., x_n)^T$, $\mathbf{y} = (y_1,..., y_m)^T$, *then*

$$\mathbf{Ax} = x_1\mathbf{A}_{*1}+...+x_n\mathbf{A}_{*n},$$
$$\mathbf{y}^T\mathbf{A} = y_1\mathbf{A}_{1*}+...+y_m\mathbf{A}_{m*}.$$

Let $\mathbf{A} = (a_{ij})$. Then

$$x_1\mathbf{A}_{*1}+...+x_n\mathbf{A}_{*n} = x_1\begin{pmatrix} a_{11} \\ \cdot \\ \cdot \\ \cdot \\ a_{m1} \end{pmatrix} +...+x_n\begin{pmatrix} a_{1n} \\ \cdot \\ \cdot \\ \cdot \\ a_{mn} \end{pmatrix}$$

$$= \begin{pmatrix} a_{11}\,x_1+...+a_{1n}\,x_n \\ \cdot \quad \cdot \quad \cdot \quad \cdot \quad \cdot \\ a_{m1}\,x_1+...+a_{mn}\,x_n \end{pmatrix} = \mathbf{Ax},$$

and the second assertion is established similarly.

EXERCISE 3.3.7. Show that the inner product $(\mathbf{x},\mathbf{y})$ (as given by Definition 2.5.1, p. 62) satisfies the identity $(\mathbf{x},\mathbf{y}) = \bar{\mathbf{x}}^T\mathbf{y} = \mathbf{y}^T\bar{\mathbf{x}}$.†

EXERCISE 3.3.8. Let $\mathbf{A}$ be an $n \times n$ matrix such that $\mathbf{Ax} = \mathbf{0}$ for all vectors $\mathbf{x}$ (of order n). By taking, in turn, $\mathbf{x} = \mathbf{e}_1,..., \mathbf{x} = \mathbf{e}_n$, show that $\mathbf{A} = \mathbf{O}$.

EXERCISE 3.3.9. Show that, if $\mathbf{A}$ is an $n \times n$ matrix, then

$$\mathbf{e}_i^T\mathbf{A}\mathbf{e}_j = \mathbf{A}_{ij} \qquad (i,j = 1,...,n).$$

3.4. Application of matrix technique to linear substitutions

We have now carried the discussion of matrix algebra a fairly long way without, as yet, having shown much need for the introduction and development of the notion of a matrix. We know, indeed, that the $m \times n$ matrix $\mathbf{A} = (a_{ij})$ provides a convenient means for specifying the system of relations

$$\left.\begin{aligned} y_1 &= a_{11}\,x_1+...+a_{1n}\,x_n \\ &\cdot \quad \cdot \quad \cdot \quad \cdot \quad \cdot \quad \cdot \quad \cdot \\ y_m &= a_{m1}\,x_1+...+a_{mn}\,x_n \end{aligned}\right\}. \tag{3.4.1}$$

It is easily seen, moreover, that (3.4.1) can be written as a single matrix equation

$$\mathbf{y} = \mathbf{Ax}, \tag{3.4.2}$$

where $\mathbf{x} = (x_1,..., x_n)^T$, $\mathbf{y} = (y_1,..., y_m)^T$. Nevertheless, it is not in the first place obvious that the introduction of matrices gives us anything more than a contracted notation for systems of relations such as (3.4.1). However, we are now in a position to demonstrate that the advantages arising from the use of matrices are not purely notational and that matrix technique enables us to handle systems of the type of (3.4.1) rapidly and efficiently.

† If $\mathbf{x} = (x_1,..., x_n)$, then $\bar{\mathbf{x}}$ denotes the vector $(\bar{x}_1,..., \bar{x}_n)$; and similarly for column vectors.

Suppose, for example, that

$$\left.\begin{aligned} y_1 &= a_{11}x_1+a_{12}x_2+a_{13}x_3 \\ y_2 &= a_{21}x_1+a_{22}x_2+a_{23}x_3 \end{aligned}\right\} \tag{3.4.3}$$

and

$$\left.\begin{aligned} z_1 &= b_{11}y_1+b_{12}y_2 \\ z_2 &= b_{21}y_1+b_{22}y_2 \\ z_3 &= b_{31}y_1+b_{32}y_2 \\ z_4 &= b_{41}y_1+b_{42}y_2 \end{aligned}\right\}. \tag{3.4.4}$$

Substituting (3.4.3) in (3.4.4) we obtain

$$\left.\begin{aligned} z_1 &= b_{11}(a_{11}x_1+a_{12}x_2+a_{13}x_3)+b_{12}(a_{21}x_1+a_{22}x_2+a_{23}x_3) \\ z_2 &= b_{21}(a_{11}x_1+a_{12}x_2+a_{13}x_3)+b_{22}(a_{21}x_1+a_{22}x_2+a_{23}x_3) \\ z_3 &= b_{31}(a_{11}x_1+a_{12}x_2+a_{13}x_3)+b_{32}(a_{21}x_1+a_{22}x_2+a_{23}x_3) \\ z_4 &= b_{41}(a_{11}x_1+a_{12}x_2+a_{13}x_3)+b_{42}(a_{21}x_1+a_{22}x_2+a_{23}x_3) \end{aligned}\right\}.$$

Thus

$$\left.\begin{aligned} z_1 &= (b_{11}a_{11}+b_{12}a_{21})x_1+(b_{11}a_{12}+b_{12}a_{22})x_2+(b_{11}a_{13}+b_{12}a_{23})x_3 \\ z_2 &= (b_{21}a_{11}+b_{22}a_{21})x_1+(b_{21}a_{12}+b_{22}a_{22})x_2+(b_{21}a_{13}+b_{22}a_{23})x_3 \\ z_3 &= (b_{31}a_{11}+b_{32}a_{21})x_1+(b_{31}a_{12}+b_{32}a_{22})x_2+(b_{31}a_{13}+b_{32}a_{23})x_3 \\ z_4 &= (b_{41}a_{11}+b_{42}a_{21})x_1+(b_{41}a_{12}+b_{42}a_{22})x_2+(b_{41}a_{13}+b_{42}a_{23})x_3 \end{aligned}\right\}. \tag{3.4.5}$$

The relation between the x's and the z's is therefore specified by the matrix

$$\begin{pmatrix} b_{11}a_{11}+b_{12}a_{21} & b_{11}a_{12}+b_{12}a_{22} & b_{11}a_{13}+b_{12}a_{23} \\ b_{21}a_{11}+b_{22}a_{21} & b_{21}a_{12}+b_{22}a_{22} & b_{21}a_{13}+b_{22}a_{23} \\ b_{31}a_{11}+b_{32}a_{21} & b_{31}a_{12}+b_{32}a_{22} & b_{31}a_{13}+b_{32}a_{23} \\ b_{41}a_{11}+b_{42}a_{21} & b_{41}a_{12}+b_{42}a_{22} & b_{41}a_{13}+b_{42}a_{23} \end{pmatrix},$$

which is seen to be equal to the matrix product

$$\begin{pmatrix} b_{11} & b_{12} \\ b_{21} & b_{22} \\ b_{31} & b_{32} \\ b_{41} & b_{42} \end{pmatrix} \begin{pmatrix} a_{11} & a_{12} & a_{13} \\ a_{21} & a_{22} & a_{23} \end{pmatrix} = \mathbf{BA},$$

where $\mathbf{A} = (a_{ij})$, $\mathbf{B} = (b_{ij})$.

The result just obtained has been arrived at after a certain amount of computation, and the work would naturally be even heavier if a larger number of variables were involved. However, the conclusion that **BA** is the matrix specifying the relation between the x's and z's can be reached much more rapidly as follows.

Put $\mathbf{x} = (x_1, x_2, x_3)^T$, $\mathbf{y} = (y_1, y_2)^T$, $\mathbf{z} = (z_1, z_2, z_3, z_4)^T$.

Then (3.4.3) and (3.4.4) can be written in the form

$$\mathbf{y} = \mathbf{A}\mathbf{x}, \qquad \mathbf{z} = \mathbf{B}\mathbf{y}.$$

Hence

$$\mathbf{z} = \mathbf{B}(\mathbf{A}\mathbf{x}),$$

and so, by the associative law, we have

$$\mathbf{z} = (\mathbf{B}\mathbf{A})\mathbf{x},$$

which is equivalent to (3.4.5).

The rapidity of the process demonstrates the superiority of matrix technique over straightforward calculation. Matrices are thus seen to be an effective tool for manipulating linear substitutions. The principal reason for their effectiveness lies in the appropriateness of the definition of a matrix product; we can now see that this has been defined in precisely such a way that the resultant of two successive substitutions specified respectively by the matrices **A** and **B** is itself specified by **BA**.

A matrix was originally defined as a rectangular array of numbers. We see now that it is convenient to think of a matrix as a 'transformation of vectors'. Thus the equation (3.4.2) associates with every vector $\mathbf{x}$ of order n a certain vector $\mathbf{y}$ of order m, and the mode of associating these pairs of vectors is represented by the matrix **A**. The idea of a transformation of vectors points to fresh possibilities inherent in the concept of a matrix. We shall not, for the moment, pursue this topic but return to it in Chapter IV.

Exercise 3.4.1. Let $\mathbf{x}$ range over a vector space of order n and let $\mathbf{A}$ be a fixed $m \times n$ matrix. Show that $\mathbf{A}\mathbf{x}$ ranges over a vector space of order m.

3.5. Adjugate matrices

In this section all matrices are assumed to be square matrices of order n.

Definition 3.5.1. (i) *Let* $\mathbf{A} = (a_{ij})$. *The determinant* $|a_{ij}|_n$ *is known as the* determinant of $\mathbf{A}$, *and is denoted by* $|\mathbf{A}|$ *or* $\det \mathbf{A}$.

(ii) *The matrix* $\mathbf{A}$ *is* singular *or* non-singular *according as* $|\mathbf{A}| = 0$ *or* $|\mathbf{A}| \neq 0$.

Exercise 3.5.1. Show that $|\alpha\mathbf{A}| = \alpha^n|\mathbf{A}|$.

Theorem 3.5.1. $$|\mathbf{AB}| = |\mathbf{A}|.|\mathbf{B}| = |\mathbf{BA}|.$$

In other words, the determinant of a matrix product is equal to the product of the determinants of the matrix factors. We note, in particular, that although **AB** and **BA** are generally distinct, their determinants are equal.

Theorem 3.5.1 is an immediate corollary of the multiplication theorem for determinants (Theorem 1.3.1, p. 12). For, writing $\mathbf{A} = (a_{ij})$, $\mathbf{B} = (b_{ij})$, $\mathbf{AB} = (c_{ij})$, we have

$$\sum_{k=1}^{n} a_{ik} b_{kj} = c_{ij} \qquad (i,j = 1,\ldots, n),$$

and so
$$|a_{ij}|_n \cdot |b_{ij}|_n = |c_{ij}|_n,$$

i.e.
$$|\mathbf{A}|.|\mathbf{B}| = |\mathbf{AB}|.$$

Interchanging **A** and **B** we obtain $|\mathbf{B}|.|\mathbf{A}| = |\mathbf{BA}|$, and the theorem is therefore proved.

Theorem 3.5.1 extends at once to any number of factors. In particular we have, for any non-negative integer k,

$$|\mathbf{A}^k| = |\mathbf{A}|^k.$$

Our next definition makes use of cofactors. These are defined as in the theory of determinants (Definition 1.4.1, p. 14).

DEFINITION 3.5.2. *The* ADJUGATE MATRIX $\mathbf{A}^*$ *of* **A** *is the transpose of the matrix of cofactors of the elements of* **A**.

In other words, $\mathbf{A}^* = (A_{ij})^T$, where $\mathbf{A} = (a_{ij})$ and A_{ij} denotes the cofactor of the element a_{ij} in **A**.

EXERCISE 3.5.2. Show that $(\alpha\mathbf{A})^* = \alpha^{n-1}\mathbf{A}^*$.

THEOREM 3.5.2. $\mathbf{AA}^* = \mathbf{A}^*\mathbf{A} = |\mathbf{A}|\mathbf{I}.$

Let $\mathbf{A} = (a_{ij})$, $\mathbf{A}^* = (b_{ij})$, so that $b_{ij} = A_{ji}$. Then, using Theorem 1.4.3 (p. 20), we obtain, for $i,j = 1,\ldots, n$,

$$(\mathbf{AA}^*)_{ij} = \sum_{k=1}^{n} a_{ik} b_{kj} = \sum_{k=1}^{n} a_{ik} A_{jk} = |\mathbf{A}|\delta_{ij},$$

$$(\mathbf{A}^*\mathbf{A})_{ij} = \sum_{k=1}^{n} b_{ik} a_{kj} = \sum_{k=1}^{n} a_{kj} A_{ki} = |\mathbf{A}|\delta_{ji}.$$

These relations are equivalent to the assertion.

It is interesting to recall at this stage the result of Theorem 1.5.2 (p. 24). In the language of matrices this states that

$$|\mathbf{A}^*| = |\mathbf{A}|^{n-1}. \tag{3.5.1}$$

The proof given in § 1.5 may now be formulated as follows. By Theorem 3.5.2 we have

$$|\mathbf{AA}^*| = |(\det \mathbf{A})\mathbf{I}|.$$

Hence, by Theorem 3.5.1 and Exercise 3.5.1,

$$\begin{aligned} |\mathbf{A}|.|\mathbf{A}^*| &= |\mathbf{A}|^n|\mathbf{I}| \\ &= |\mathbf{A}|^n, \end{aligned} \tag{3.5.2}$$

and (3.5.1) now follows by virtue of Theorem 1.5.1.

In § 1.6.3 we gave an alternative proof of (3.5.1) which was independent of Theorem 1.5.1 We shall now sketch yet another proof, again independent of Theorem 1.5.1, and employing a useful new device. Let $\mathbf{A} = (a_{ij})$, and consider the matrix

$$\mathbf{A}-x\mathbf{I} = \begin{pmatrix} a_{11}-x & a_{12} & \cdot & \cdot & \cdot & a_{1n} \\ a_{21} & a_{22}-x & \cdot & \cdot & \cdot & a_{2n} \\ \cdot & \cdot & \cdot & \cdot & \cdot & \cdot \\ a_{n1} & a_{n2} & \cdot & \cdot & \cdot & a_{nn}-x \end{pmatrix}.$$

The determinant $|\mathbf{A}-x\mathbf{I}|$ is obviously a polynomial in x with leading term $(-1)^n x^n$. Hence, for all sufficiently large values of x (say for $x > x_0$) $|\mathbf{A}-x\mathbf{I}| \neq 0$. Again, each element of $(\mathbf{A}-x\mathbf{I})^*$ is a polynomial in x, and so $|(\mathbf{A}-x\mathbf{I})^*|$ is a polynomial in x.

Rewriting (3.5.2) with $\mathbf{A}-x\mathbf{I}$ in place of $\mathbf{A}$ we have

$$|\mathbf{A}-x\mathbf{I}|.|(\mathbf{A}-x\mathbf{I})^*| = |\mathbf{A}-x\mathbf{I}|^n.$$

Hence, for $x > x_0$,

$$|(\mathbf{A}-x\mathbf{I})^*| = |\mathbf{A}-x\mathbf{I}|^{n-1}.$$

Here both sides are polynomials in x and since they are equal for $x > x_0$ they are equal for all values of x, by the corollary to Theorem 1.6.3 (p. 31). In particular, the two sides are equal for $x = 0$, and the identity (3.5.1) is therefore proved.

THEOREM 3.5.3. $\quad (\mathbf{A}^*)^* = |\mathbf{A}|^{n-2}\mathbf{A}.$

For $n = 2$ the factor $|\mathbf{A}|^{n-2}$ is interpreted as 1, no matter whether $\mathbf{A}$ is singular or not. In that case the proof is trivial and we shall at once assume that $n > 2$. We write, for simplicity, $\mathbf{A}^{**}$ for $(\mathbf{A}^*)^*$. Applying Theorem 3.5.2 with $\mathbf{A}^*$ in place of $\mathbf{A}$ and using (3.5.1) we obtain

$$\mathbf{A}^*\mathbf{A}^{**} = |\mathbf{A}^*|\mathbf{I} = |\mathbf{A}|^{n-1}\mathbf{I}.$$

Hence

$$(\mathbf{A}\mathbf{A}^*)\mathbf{A}^{**} = |\mathbf{A}|^{n-1}\mathbf{A},$$

and so, again by Theorem 3.5.2,

$$|\mathbf{A}|\mathbf{A}^{**} = |\mathbf{A}|^{n-1}\mathbf{A}.$$

Thus

$$|\mathbf{A}|(\mathbf{A}^{**})_{ij} = |\mathbf{A}|^{n-1}\mathbf{A}_{ij} \quad (i,j = 1,\ldots,n).$$

Both sides in this relation are polynomials in the n^2 elements of $\mathbf{A}$, and since $|\mathbf{A}|$ does not vanish identically it follows, by Theorem 1.5.1, that

$$(\mathbf{A}^{**})_{ij} = |\mathbf{A}|^{n-2}\mathbf{A}_{ij} \qquad (i,j = 1,\ldots,n).$$

This is equivalent to the theorem. Just as in the case of the identity (3.5.1) so here, too, an alternative proof depending on the consideration of the matrix $\mathbf{A}-x\mathbf{I}$ can easily be constructed.

EXERCISE 3.5.3. Write out such a proof in detail.

THEOREM 3.5.4. $\qquad (\mathbf{AB})^* = \mathbf{B}^*\mathbf{A}^*.$

Making use of Theorems 3.5.1 and 3.5.2 we obtain

$$\begin{aligned} |\mathbf{A}|\,.\,|\mathbf{B}|\mathbf{B}^*\mathbf{A}^* &= |\mathbf{AB}|\mathbf{I}(\mathbf{B}^*\mathbf{A}^*) = (\mathbf{AB})^*(\mathbf{AB})(\mathbf{B}^*\mathbf{A}^*) \\ &= (\mathbf{AB})^*\mathbf{A}(\mathbf{BB}^*)\mathbf{A}^* = (\mathbf{AB})^*\mathbf{A}|\mathbf{B}|\mathbf{A}^* \\ &= |\mathbf{A}|\,.\,|\mathbf{B}|(\mathbf{AB})^*. \end{aligned} \tag{3.5.3}$$

Hence $\quad |\mathbf{A}|\,.\,|\mathbf{B}|(\mathbf{B}^*\mathbf{A}^*)_{ij} = |\mathbf{A}|\,.\,|\mathbf{B}|\{(\mathbf{AB})^*\}_{ij} \quad (i,j = 1,\ldots,n).$

All the expressions involved here are polynomials in the $2n^2$ elements of $\mathbf{A}$ and $\mathbf{B}$. Since, moreover, $|\mathbf{A}|.|\mathbf{B}|$ does not vanish identically, it follows by Theorem 1.5.1 that

$$(\mathbf{B}^*\mathbf{A}^*)_{ij} = \{(\mathbf{AB})^*\}_{ij} \qquad (i,j = 1,\ldots,n),$$

and this is, in fact, our assertion.

An alternative argument of the now familiar kind runs as follows. Rewriting (3.5.3) with $\mathbf{A}-x\mathbf{I}$, $\mathbf{B}-x\mathbf{I}$ in place of $\mathbf{A}$, $\mathbf{B}$ respectively, we have

$$\begin{aligned} &|\mathbf{A}-x\mathbf{I}|\,.\,|\mathbf{B}-x\mathbf{I}|(\mathbf{B}-x\mathbf{I})^*(\mathbf{A}-x\mathbf{I})^* \\ &\qquad = |\mathbf{A}-x\mathbf{I}|\,.\,|\mathbf{B}-x\mathbf{I}|\{(\mathbf{A}-x\mathbf{I})(\mathbf{B}-x\mathbf{I})\}^*. \end{aligned}$$

Now there exists a number x_1 such that, for all $x > x_1$,

$$|\mathbf{A}-x\mathbf{I}| \neq 0, \qquad |\mathbf{B}-x\mathbf{I}| \neq 0.$$

It follows that

$$(\mathbf{B}-x\mathbf{I})^*(\mathbf{A}-x\mathbf{I})^* = \{(\mathbf{A}-x\mathbf{I})(\mathbf{B}-x\mathbf{I})\}^* \qquad (x > x_1).$$

Comparing the matrices on the two sides element by element and using the corollary to Theorem 1.6.3 we complete the proof of Theorem 3.5.4.

3.6. Inverse matrices

3.6.1. In § 3.3 we dealt with addition, subtraction, and multiplication of matrices. We shall next investigate to what extent the laws of division carry over from number algebra to matrix algebra. It will be recalled that we introduced division in number algebra by considering the equation $ax = 1$. An analogous procedure is now to be followed for matrices. Unless the contrary is stated it is again assumed that all matrices are of type $n \times n$.

THEOREM 3.6.1. (i) *If* $|\mathbf{A}| \neq 0$, *then the matrix equations*

$$\mathbf{AX} = \mathbf{I}, \tag{3.6.1}$$

$$\mathbf{XA} = \mathbf{I} \tag{3.6.2}$$

are both uniquely soluble for $\mathbf{X}$ *and possess the common solution*

$$\mathbf{X} = \frac{1}{|\mathbf{A}|}\mathbf{A}^*. \tag{3.6.3}$$

(ii) *If at least one of the equations* (3.6.1), (3.6.2) *is soluble for* $\mathbf{X}$, *then* $|\mathbf{A}| \neq 0$, *and so both equations are soluble and possess the common unique solution* (3.6.3).

If $|\mathbf{A}| \neq 0$, then, by virtue of Theorem 3.5.2, the matrix $\mathbf{X}$ given by (3.6.3) satisfies $\mathbf{AX} = \mathbf{XA} = \mathbf{I}$, and is therefore a solution of both (3.6.1) and (3.6.2). To show that this solution is unique, suppose that $\mathbf{Y}$ also satisfies (3.6.1), i.e. $\mathbf{AY} = \mathbf{I}$. Then

$$\mathbf{X} = \mathbf{X}(\mathbf{AY}) = (\mathbf{XA})\mathbf{Y} = \mathbf{IY} = \mathbf{Y}.$$

Hence the solution of (3.6.1) is unique; and a similar argument applies to (3.6.2).

To prove (ii) suppose, for instance, that (3.6.1) is soluble. Then, by Theorem 3.5.1, $|\mathbf{A}| . |\mathbf{X}| = |\mathbf{I}| = 1$, and so $|\mathbf{A}| \neq 0$.

DEFINITION 3.6.1. *If* $|\mathbf{A}| \neq 0$, *then the common unique solution of equations* (3.6.1) *and* (3.6.2) *of the previous theorem is called the* INVERSE (MATRIX) *of* $\mathbf{A}$, *and is denoted by* $\mathbf{A}^{-1}$. *The operation of obtaining* $\mathbf{A}^{-1}$ *from* $\mathbf{A}$ *is known as* INVERSION.

Consider, for instance, the matrix

$$\mathbf{A} = \begin{pmatrix} a & b \\ c & d \end{pmatrix},$$

where $|\mathbf{A}| = ad - bc \neq 0$. Clearly

$$\mathbf{A}^* = \begin{pmatrix} d & -b \\ -c & a \end{pmatrix},$$

and so $$\mathbf{A}^{-1} = \begin{pmatrix} d/(ad-bc) & -b/(ad-bc) \\ -c/(ad-bc) & a/(ad-bc) \end{pmatrix}.$$

If $\mathbf{A}$ is not a square matrix, or if it is a singular square matrix, then $\mathbf{A}^{-1}$ is not defined.

Both terminology and notation relating to inverse matrices are chosen by analogy with elementary algebra, where with any number $a \neq 0$ is associated a unique inverse number a^{-1} having the property that $aa^{-1} = a^{-1}a = 1$. As we have seen, every non-singular square matrix $\mathbf{A}$ possesses a unique inverse such that

$$\mathbf{AA}^{-1} = \mathbf{A}^{-1}\mathbf{A} = \mathbf{I}.$$

Exercise 3.6.1. Show that $\mathbf{I}^{-1} = \mathbf{I}$. Also find the inverse of $\mathbf{dg}(\alpha_1,..., \alpha_n)$, assuming that $\alpha_1 \neq 0,..., \alpha_n \neq 0$.

The introduction of inverse matrices enables us to recognize a result anticipated a little earlier, namely, the analogy between the behaviour of scalar matrices and that of scalars. We have, in fact,

$$\alpha\mathbf{I} \pm \beta\mathbf{I} = (\alpha \pm \beta)\mathbf{I},$$
$$\alpha\mathbf{I}.\beta\mathbf{I} = (\alpha\beta)\mathbf{I},$$
$$(\alpha\mathbf{I})^{-1} = \alpha^{-1}\mathbf{I} \qquad (\alpha \neq 0).$$

Thus there is a complete correspondence between the scalars and the scalar matrices, each scalar α being associated with the matrix $\alpha\mathbf{I}$. It is this correspondence which motivates the use of the term 'scalar matrix'.

In elementary algebra one of the first problems we deal with is the solution of the equation $ax = b$. The corresponding problem in matrix algebra will now be considered.

Theorem 3.6.2. (i) *If* $\mathbf{A}$ *is a non-singular* $m \times m$ *matrix and* $\mathbf{B}$ *is an* $m \times n$ *matrix, then the matrix equation*

$$\mathbf{AX} = \mathbf{B} \tag{3.6.4}$$

possesses the unique solution

$$\mathbf{X} = \mathbf{A}^{-1}\mathbf{B}. \tag{3.6.5}$$

(ii) *If* $\mathbf{C}$ *is a non-singular* $n \times n$ *matrix and* $\mathbf{D}$ *is an* $m \times n$ *matrix, then the matrix equation*

$$\mathbf{YC} = \mathbf{D}$$

possesses the unique solution

$$\mathbf{Y} = \mathbf{DC}^{-1}.$$

It is sufficient to consider (i) since (ii) is proved similarly. If the equation (3.6.4) possesses a solution, then, premultiplying both sides by $\mathbf{A}^{-1}$, we obtain (3.6.5). Thus a solution, if it exists, is unique and is given by (3.6.5). Moreover, $\mathbf{X}$ as given by (3.6.5), is clearly a solution of (3.6.4).

THEOREM 3.6.3. (i) *If* $\mathbf{A}$ *is a non-singular matrix, then so is* $\mathbf{A}^{-1}$ *and* $(\mathbf{A}^{-1})^{-1} = \mathbf{A}$. *Moreover* $|\mathbf{A}^{-1}| = |\mathbf{A}|^{-1}$.

(ii) *If* $\mathbf{A}, \mathbf{B}$ *are non-singular matrices, then so is* $\mathbf{AB}$; *and*

$$(\mathbf{AB})^{-1} = \mathbf{B}^{-1}\mathbf{A}^{-1}. \tag{3.6.6}$$

If $\mathbf{A}$ is non-singular, then $\mathbf{A}^{-1}$ exists and

$$\mathbf{A}^{-1}\mathbf{A} = \mathbf{I} \tag{3.6.7}$$

Hence the equation $\mathbf{A}^{-1}\mathbf{X} = \mathbf{I}$ is soluble and has $\mathbf{X} = \mathbf{A}$ as a solution. In view of Theorem 3.6.1 it therefore follows that $\mathbf{A}^{-1}$ is non-singular and that the above equation has the unique solution $\mathbf{X} = (\mathbf{A}^{-1})^{-1}$. Hence $(\mathbf{A}^{-1})^{-1} = \mathbf{A}$. Moreover, by (3.6.7) and Theorem 3.5.1, $|\mathbf{A}^{-1}|.|\mathbf{A}| = 1$, so that $|\mathbf{A}^{-1}| = |\mathbf{A}|^{-1}$.

Again, if $\mathbf{A}, \mathbf{B}$ are non-singular, then, by Theorem 3.5.1, $\mathbf{AB}$ is non-singular; and we have

$$(\mathbf{AB})(\mathbf{B}^{-1}\mathbf{A}^{-1}) = \mathbf{A}(\mathbf{BB}^{-1})\mathbf{A}^{-1} = \mathbf{AA}^{-1} = \mathbf{I}.$$

Hence the equation $(\mathbf{AB})\mathbf{X} = \mathbf{I}$ has a solution $\mathbf{X} = \mathbf{B}^{-1}\mathbf{A}^{-1}$. But, by Theorem 3.6.1, the unique solution of this equation is given by $\mathbf{X} = (\mathbf{AB})^{-1}$, and (3.6.6) therefore follows. We may note in passing the analogy between (3.6.6) and Theorem 3.5.4.

EXERCISE 3.6.2. Show that, if $\mathbf{A}, \mathbf{B}, \ldots, \mathbf{K}$ are non-singular, then so is their product $\mathbf{AB}\ldots\mathbf{K}$; and $(\mathbf{AB}\ldots\mathbf{K})^{-1} = \mathbf{K}^{-1}\ldots\mathbf{B}^{-1}\mathbf{A}^{-1}$.

DEFINITION 3.6.2. *If* $\mathbf{A}$ *is non-singular and* r *is a positive integer, then*

$$\mathbf{A}^{-r} = (\mathbf{A}^{-1})^r.$$

We can now extend the index laws, established in § 3.3 for non-negative indices, to all integral indices.

THEOREM 3.6.4. *For all integral values of* r, s

$$\mathbf{A}^r\mathbf{A}^s = \mathbf{A}^{r+s}, \tag{3.6.8}$$

$$(\mathbf{A}^r)^s = \mathbf{A}^{rs}, \tag{3.6.9}$$

provided only that the matrices in question are defined.

If r is a positive integer, then, by Exercise 3.6.2,

$$(\mathbf{AA}\ldots\mathbf{A})^{-1} = \mathbf{A}^{-1}\ldots\mathbf{A}^{-1}\mathbf{A}^{-1},$$

where each product comprises r factors. Thus

$$(\mathbf{A}^r)^{-1} = (\mathbf{A}^{-1})^r,$$

and so, by Definition 3.6.2,

$$\mathbf{A}^{-r} = (\mathbf{A}^{-1})^r = (\mathbf{A}^r)^{-1} \qquad (r \geqslant 0). \tag{3.6.10}$$

We first prove (3.6.8) . This is already known to hold for $r \geqslant 0$, $s \geqslant 0$. If both r and s are negative, put $r = -\rho$, $s = -\sigma$. Then, by (3.6.10),

$$\mathbf{A}^r\mathbf{A}^s = \mathbf{A}^{-\rho}\mathbf{A}^{-\sigma} = (\mathbf{A}^{-1})^\rho(\mathbf{A}^{-1})^\sigma = (\mathbf{A}^{-1})^{\rho+\sigma} = \mathbf{A}^{-\rho-\sigma} = \mathbf{A}^{r+s}.$$

If just one of r, s is negative, let, for instance, $r \geqslant 0$, $s < 0$ and write $s = -\sigma$. Then, by (3.6.10),

$$\mathbf{A}^r\mathbf{A}^s = \mathbf{A}^r\mathbf{A}^{-\sigma} = \mathbf{A}^r(\mathbf{A}^{-1})^\sigma. \tag{3.6.11}$$

Now, if $r \geqslant \sigma$, then, since $\mathbf{A}\mathbf{A}^{-1} = \mathbf{I}$, the right-hand side of (3.6.11) is equal to $\mathbf{A}^{r-\sigma} = \mathbf{A}^{r+s}$. If, on the other hand, $r < \sigma$, then the right-hand side of (3.6.11) is equal to

$$(\mathbf{A}^{-1})^{\sigma-r} = \mathbf{A}^{r-\sigma} = \mathbf{A}^{r+s}.$$

Next, consider (3.6.9). This is already known to hold for $r \geqslant 0$, $s \geqslant 0$. When $r \geqslant 0$, $s < 0$, write $s = -\sigma$. Then, by (3.6.10),

$$(\mathbf{A}^r)^s = (\mathbf{A}^r)^{-\sigma} = \{(\mathbf{A}^r)^\sigma\}^{-1} = (\mathbf{A}^{r\sigma})^{-1} = \mathbf{A}^{-r\sigma} = \mathbf{A}^{rs}.$$

Again, let $r < 0$, $s \geqslant 0$. Writing $r = -\rho$ and using (3.6.10) we obtain

$$(\mathbf{A}^r)^s = (\mathbf{A}^{-\rho})^s = \{(\mathbf{A}^\rho)^{-1}\}^s = (\mathbf{A}^\rho)^{-s};$$

hence, by the previous case,

$$(\mathbf{A}^r)^s = \mathbf{A}^{-\rho s} = \mathbf{A}^{rs}.$$

Finally, let $r < 0$, $s < 0$ and write $r = -\rho$, $s = -\sigma$. Then, by (3.6.10),

$$(\mathbf{A}^r)^{-1} = (\mathbf{A}^{-\rho})^{-1} = \{(\mathbf{A}^\rho)^{-1}\}^{-1} = \mathbf{A}^\rho = \mathbf{A}^{-r},$$

and therefore

$$(\mathbf{A}^r)^s = (\mathbf{A}^r)^{-\sigma} = \{(\mathbf{A}^r)^{-1}\}^\sigma = (\mathbf{A}^{-r})^\sigma.$$

But, by the previous case, the expression on the right is equal to $\mathbf{A}^{-r\sigma} = \mathbf{A}^{rs}$, and the proof is therefore complete.

3.6.2. One of the most far-reaching results in elementary algebra is the division law, which states that if $ab = 0$, then $a = 0$ or $b = 0$. It is easy to see that this does not apply to matrices.

THEOREM 3.6.5. *The division law is not valid in matrix algebra, i.e. the equation* $\mathbf{AB} = \mathbf{O}$, *where* $\mathbf{A}$, $\mathbf{B}$ *are rectangular matrices, does not imply that* $\mathbf{A} = \mathbf{O}$ *or* $\mathbf{B} = \mathbf{O}$.†

† Each symbol $\mathbf{O}$ must, of course, be interpreted as a zero matrix of the appropriate type.

Take, for instance,

$$\mathbf{A} = \begin{pmatrix} 2 & -1 \\ 8 & -4 \\ -2 & 1 \end{pmatrix}, \qquad \mathbf{B} = \begin{pmatrix} 3 & 1 \\ 6 & 2 \end{pmatrix}.$$

Then $\mathbf{A} \neq \mathbf{O}$, $\mathbf{B} \neq \mathbf{O}$, but $\mathbf{AB} = \mathbf{O}_3^2$.

Though the division law does not hold for matrices, there is a modified form which is still valid.

THEOREM 3.6.6. *If* $\mathbf{A}$, $\mathbf{B}$ *are square matrices and* $\mathbf{AB} = \mathbf{O}$, *then* $\mathbf{A} = \mathbf{O}$, *or* $\mathbf{B} = \mathbf{O}$, *or* $\mathbf{A}$ *and* $\mathbf{B}$ *are both singular.*

For suppose that $\mathbf{A} \neq \mathbf{O}$, $\mathbf{B} \neq \mathbf{O}$. If at least one of $\mathbf{A}$, $\mathbf{B}$ is non-singular, let, for instance, $|\mathbf{A}| \neq 0$. Then $\mathbf{AB} = \mathbf{O}$ implies $\mathbf{A}^{-1}\mathbf{AB} = \mathbf{O}$, i.e. $\mathbf{B} = \mathbf{O}$, and we have a contradiction.

DEFINITION 3.6.3. *A rectangular matrix* $\mathbf{A}$ *is a* DIVISOR OF ZERO *if* $\mathbf{A} \neq \mathbf{O}$ *and if there exists a matrix* $\mathbf{B} \neq \mathbf{O}$ *such that* $\mathbf{AB} = \mathbf{O}$ *or a matrix* $\mathbf{C} \neq \mathbf{O}$ *such that* $\mathbf{CA} = \mathbf{O}$.

Theorem 3.6.5 asserts the existence of divisors of zero in matrix algebra. For instance, the matrices $\mathbf{A}$ and $\mathbf{B}$ which appear in the proof of that theorem are both divisors of zero.

In elementary algebra an important consequence of the division law is the cancellation law which is deduced as follows. Let $a \neq 0$ and suppose that $ax = ay$. Then $a(x-y) = 0$ and hence, by the division law, $x-y = 0$, i.e. $x = y$. An analagous argument cannot be applied to matrices, for if $\mathbf{A} \neq \mathbf{O}$ and $\mathbf{AX} = \mathbf{AY}$, then $\mathbf{A}(\mathbf{X}-\mathbf{Y}) = \mathbf{O}$, but we cannot infer from this that $\mathbf{X}-\mathbf{Y} = \mathbf{O}$, since $\mathbf{A}$ may be a divisor of zero. In fact, it is easy to write down examples which show that the relations $\mathbf{AX} = \mathbf{AY}$, $\mathbf{A} \neq \mathbf{O}$, may be compatible with $\mathbf{X} \neq \mathbf{Y}$. One such example is provided by the matrices

$$\mathbf{A} = \begin{pmatrix} 1 & -3 \\ 0 & 0 \end{pmatrix}, \qquad \mathbf{X} = \begin{pmatrix} 5 & 8 \\ 2 & 5 \end{pmatrix}, \qquad \mathbf{Y} = \begin{pmatrix} 2 & 2 \\ 1 & 3 \end{pmatrix}.$$

Similarly, we can find matrices which satisfy the relations $\mathbf{XB} = \mathbf{YB}$, $\mathbf{B} \neq \mathbf{O}$, $\mathbf{X} \neq \mathbf{Y}$. We thus arrive at the following result.

THEOREM 3.6.7. *The relations* $\mathbf{AX} = \mathbf{AY}$, $\mathbf{A} \neq \mathbf{O}$, *where* $\mathbf{A}$, $\mathbf{X}$, $\mathbf{Y}$ *are rectangular matrices, do not imply that* $\mathbf{X} = \mathbf{Y}$. *Similarly, the relations* $\mathbf{XB} = \mathbf{YB}$, $\mathbf{B} \neq \mathbf{O}$, *where* $\mathbf{B}$ *is also rectangular, do not imply that* $\mathbf{X} = \mathbf{Y}$.

EXERCISE 3.6.3. Let $\mathbf{A}$, $\mathbf{B}$ be non-singular square matrices. Show that either of the relations $\mathbf{AX} = \mathbf{AY}$, $\mathbf{XB} = \mathbf{YB}$ implies $\mathbf{X} = \mathbf{Y}$.

The problem of determining which matrices are divisors of zero now naturally presents itself. We shall solve it here for the case of square matrices and deal with the general case in § 5.5.2.

THEOREM 3.6.8. *A (non-zero) square matrix is a divisor of zero if and only if it is singular.*

This result is almost obvious. For if $\mathbf{A}$ is a non-singular matrix and $\mathbf{AX} = \mathbf{O}$ $(\mathbf{YA} = \mathbf{O})$, then, premultiplying (postmultiplying) by $\mathbf{A}^{-1}$ we obtain $\mathbf{X} = \mathbf{O}$ $(\mathbf{Y} = \mathbf{O})$. Hence a non-singular matrix is not a divisor of zero. On the other hand, if $\mathbf{A}$ is singular, there exists (by Theorem 1.6.1, p. 27) a vector $\mathbf{x} \neq \mathbf{0}$ such that $\mathbf{Ax} = \mathbf{0}$. Hence $\mathbf{A}$ is then a divisor of zero.

It is, in fact easy to obtain the stronger result that if $\mathbf{A}$ is singular, then there exist *square* matrices $\mathbf{X}$, $\mathbf{Y}$ such that $\mathbf{AX} = \mathbf{O}$, $\mathbf{YA} = \mathbf{O}$. For suppose $\mathbf{A}$ is of type $n \times n$ and denote by $\mathbf{x}_1,..., \mathbf{x}_n$ any vectors (of order n), not all zero, such that $\mathbf{Ax}_1 = ... = \mathbf{Ax}_n = \mathbf{0}$. Let the $n \times n$ matrix $\mathbf{X}$ be defined by the relations $\mathbf{X}_{*i} = \mathbf{x}_i$ $(i = 1,..., n)$. Then $\mathbf{X} \neq \mathbf{O}$ and, by Theorem 3.3.5 (ii), p. 84,

$$(\mathbf{AX})_{*i} = \mathbf{AX}_{*i} = \mathbf{Ax}_i = \mathbf{0} \qquad (i = 1,..., n).$$

Hence $\mathbf{AX} = \mathbf{O}$, as required. Again, in view of Theorem 1.6.1, there exist scalars $t_1,..., t_n$, not all zero, such that $(t_1,..., t_n)\mathbf{A} = \mathbf{0}$. Denote by $\mathbf{y}_1,..., \mathbf{y}_n$ any n row vectors, not all zero, satisfying $\mathbf{y}_1\mathbf{A} = ... = \mathbf{y}_n\mathbf{A} = \mathbf{0}$, and let $\mathbf{Y}$ by the $n \times n$ matrix defined by the relations $\mathbf{Y}_{i*} = \mathbf{y}_i$ $(i = 1,..., n)$. Then $\mathbf{Y} \neq \mathbf{O}$ and, by Theorem 3.3.5 (i),

$$(\mathbf{YA})_{i*} = \mathbf{Y}_{i*}\mathbf{A} = \mathbf{y}_i\mathbf{A} = \mathbf{0} \qquad (i = 1,..., n),$$

so that $\mathbf{YA} = \mathbf{O}$.

We may summarize the conclusions reached so far about matrix algebra by saying that in most respects it resembles the algebra of numbers, with the unit matrix and the zero matrix playing the parts of the numbers 1 and 0 respectively. The vital points on which the two algebras differ is that matrix multiplication is non-commutative and that the division law (and consequently the cancellation law) is not valid for matrices.

3.6.3. We conclude the present section by deriving some identities relating to transposition.

It is, in the first place, obvious that

$$(\alpha\mathbf{A})^T = \alpha\mathbf{A}^T, \qquad (\mathbf{A}+\mathbf{B})^T = \mathbf{A}^T+\mathbf{B}^T.$$

A more interesting identity concerns the transpose of a product.

THEOREM 3.6.9. *If* **A**, **B** *are rectangular matrices, then*

$$(\mathbf{AB})^T = \mathbf{B}^T\mathbf{A}^T, \tag{3.6.12}$$

provided that either side of the equation is defined.

We note the analogy between this result, Theorem 3.5.4, and equation (3.6.6).

Let **A**, **B** be of type $p\times q$, $r\times s$ respectively. Then $(\mathbf{AB})^T$ exists if and only if $q = r$, and is then of type $s\times p$. The same is true of $\mathbf{B}^T\mathbf{A}^T$; and therefore, if either side of (3.6.12) exists, so does the other side and both sides are of the same type. To prove the actual equality it now suffices to note that

$$\{(\mathbf{AB})^T\}_{ij} = (\mathbf{AB})_{ji} = \sum_k \mathbf{A}_{jk}\mathbf{B}_{ki} = \sum_k (\mathbf{B}^T)_{ik}(\mathbf{A}^T)_{kj} = (\mathbf{B}^T\mathbf{A}^T)_{ij}.$$

EXERCISE 3.6.4. Show that $(\mathbf{AB}...\mathbf{K})^T = \mathbf{K}^T...\mathbf{B}^T\mathbf{A}^T$, provided that either side is defined.

A useful deduction from Theorem 3.6.9 is as follows.

THEOREM 3.6.10. *If* **A** *is a non-singular square matrix, then*

$$(\mathbf{A}^T)^{-1} = (\mathbf{A}^{-1})^T.$$

In other words, the operations of transposition and inversion can be carried out in either order.

To prove the theorem we put $\mathbf{B} = \mathbf{A}^{-1}$ in (3.6.12) and obtain

$$(\mathbf{AA}^{-1})^T = (\mathbf{A}^{-1})^T\mathbf{A}^T,$$

i.e.

$$(\mathbf{A}^{-1})^T\mathbf{A}^T = \mathbf{I}.$$

The assertion now follows immediately.

3.7. Rational functions of a square matrix

We have already considered positive and negative powers of matrices. We shall now extend this idea still further.

DEFINITION 3.7.1. *Let $f(x)$ be a polynomial in the scalar variable x, say*

$$f(x) = c_0+c_1x+c_2x^2+...+c_kx^k.$$

If **A** *is a square matrix, of order n, then $f(\mathbf{A})$ is defined as the (square) matrix given by the relation*

$$f(\mathbf{A}) = c_0\mathbf{I}_n+c_1\mathbf{A}+c_2\mathbf{A}^2+...+c_k\mathbf{A}^k. \tag{3.7.1}$$

An expression of the type (3.7.1) is called a *matrix polynomial* or polynomial in a matrix. By contrast, a polynomial in a scalar variable is known as a scalar polynomial.

We recall that two scalar polynomials $f(x)$, $g(x)$ are said to be identically equal if $f(x) = g(x)$ for all values of x.

THEOREM 3.7.1. *If the polynomials $f(x)$, $g(x)$ are identically equal, then, for any square matrix $\mathbf{A}$, $f(\mathbf{A}) = g(\mathbf{A})$.*

This result is obvious since $f(x) = g(x)$ implies, as we know, that the coefficients of like powers of x in f and g are equal. The conclusion therefore follows.

It is not difficult to extend the scope of Theorem 3.7.1.

THEOREM 3.7.2. *Any polynomial identity between scalar polynomials remains valid for the corresponding matrix polynomials.*

Thus, for instance, if $f_1,\ldots,f_6$ are polynomials, and

$$\{f_1(x)f_2(x)+f_3(x)\}f_4(x) = f_5(x)-f_6(x),$$

then, for every square matrix $\mathbf{A}$,

$$\{f_1(\mathbf{A})f_2(\mathbf{A})+f_3(\mathbf{A})\}f_4(\mathbf{A}) = f_5(\mathbf{A})-f_6(\mathbf{A}).$$

In view of Theorem 3.7.1 the proof of Theorem 3.7.2 will be complete if we can show that

$$f(x)+g(x) = \phi(x) \quad \text{implies} \quad f(\mathbf{A})+g(\mathbf{A}) = \phi(\mathbf{A}), \qquad (3.7.2)$$

$$f(x)g(x) = \phi(x) \quad \text{implies} \quad f(\mathbf{A})g(\mathbf{A}) = \phi(\mathbf{A}). \qquad (3.7.3)$$

The proof of (3.7.2) is trivial and we leave it to the reader. To establish (3.7.3) we write

$$f(x) = \sum_{i=0}^{r} a_i x^i, \qquad g(x) = \sum_{j=0}^{s} b_j x^j, \qquad \phi(x) = \sum_{k=0}^{t} c_k x^k.$$

Since $f(x)g(x) = \phi(x)$, the coefficients of like powers of x may be equated; hence $r+s = t$, and

$$c_k = \sum_{i,j} a_i b_j \quad (k = 0, 1,\ldots, t), \qquad (3.7.4)$$

where the summation extends over all pairs of integers i, j satisfying the conditions

$$i+j = k, \qquad 0 \leqslant i \leqslant r, \qquad 0 \leqslant j \leqslant s. \qquad (3.7.5)$$

Next, using the distributive law for matrices and identity (3.6.8) (p. 93), we obtain

$$f(\mathbf{A})g(\mathbf{A}) = \Big(\sum_{i=0}^{r} a_i \mathbf{A}^i\Big)\Big(\sum_{j=0}^{s} b_j \mathbf{A}^j\Big) = \sum_{i=0}^{r} \sum_{j=0}^{s} a_i \mathbf{A}^i . b_j \mathbf{A}^j$$

$$= \sum_{i=0}^{r} \sum_{j=0}^{s} a_i b_j \mathbf{A}^{i+j} = \sum_{k=0}^{r+s} d_k \mathbf{A}^k,$$

where $$d_k = \sum_{i,j} a_i b_j \quad (k = 0, 1, \ldots, r+s),$$

with the summation extending over all pairs of integers i, j satisfying (3.7.5). Hence, by virtue of (3.7.4), $d_k = c_k$, and so

$$f(\mathbf{A})g(\mathbf{A}) = \phi(\mathbf{A}).$$

Since, for any two polynomials $f(x)$, $g(x)$, we have

$$f(x)g(x) = g(x)f(x)$$

it now follows that $f(\mathbf{A})g(\mathbf{A}) = g(\mathbf{A})f(\mathbf{A})$. We have therefore the following corollary.

COROLLARY. *Any two polynomials in* **A** *commute with each other.*

Again, let $f(x)$, $g(x)$ be two scalar polynomials and suppose that the square matrix **A** satisfies the relation $|g(\mathbf{A})| \neq 0$. We have just seen that

$$f(\mathbf{A})g(\mathbf{A}) = g(\mathbf{A})f(\mathbf{A}). \tag{3.7.6}$$

Hence, premultiplying and postmultiplying by $\{g(\mathbf{A})\}^{-1}$, we obtain

$$\{g(\mathbf{A})\}^{-1}f(\mathbf{A}) = f(\mathbf{A})\{g(\mathbf{A})\}^{-1}. \tag{3.7.7}$$

This identity enables us to define rational functions of **A**.

DEFINITION 3.7.2. *Let $f(x)$, $g(x)$ be scalar polynomials and let* **A** *be any square matrix such that $|g(\mathbf{A})| \neq 0$. Then the matrix appearing on either side of* (3.7.7) *is known as the* QUOTIENT *of $f(\mathbf{A})$ by $g(\mathbf{A})$, and is denoted by $\dfrac{f(\mathbf{A})}{g(\mathbf{A})}$ or $f(\mathbf{A})/g(\mathbf{A})$.*

A quotient of two polynomials in **A** is called a *rational function* of **A**. If $g(x)$ is identically equal to 1, then $g(\mathbf{A}) = \mathbf{I}$, and it follows that any polynomial in **A** is a special rational function of **A**.

EXERCISE 3.7.1. If f, g, ϕ, ψ are polynomials, show that

$$f(x)/g(x) = \phi(x)/\psi(x) \quad \text{implies} \quad f(\mathbf{A})/g(\mathbf{A}) = \phi(\mathbf{A})/\psi(\mathbf{A})$$

for any square matrix **A** such that $|g(\mathbf{A})| \neq 0$, $|\psi(\mathbf{A})| \neq 0$.

Theorem 3.7.2 can now be extended to rational functions.

THEOREM 3.7.3. *Any identity between scalar rational functions remains valid for the corresponding rational functions of a square matrix, provided that all the latter functions are defined.*

It is clearly sufficient to establish the following results.

$$\frac{f_1(x)}{f_2(x)}+\frac{f_3(x)}{f_4(x)}=\frac{f_5(x)}{f_6(x)} \quad \text{implies} \quad \frac{f_1(\mathbf{A})}{f_2(\mathbf{A})}+\frac{f_3(\mathbf{A})}{f_4(\mathbf{A})}=\frac{f_5(\mathbf{A})}{f_6(\mathbf{A})}, \tag{3.7.8}$$

$$\frac{f_1(x)}{f_2(x)}\cdot\frac{f_3(x)}{f_4(x)}=\frac{f_5(x)}{f_6(x)} \quad \text{implies} \quad \frac{f_1(\mathbf{A})}{f_2(\mathbf{A})}\cdot\frac{f_3(\mathbf{A})}{f_4(\mathbf{A})}=\frac{f_5(\mathbf{A})}{f_6(\mathbf{A})},$$

$$\frac{f_1(x)}{f_2(x)}\Big/\frac{f_3(x)}{f_4(x)}=\frac{f_5(x)}{f_6(x)} \quad \text{implies} \quad \frac{f_1(\mathbf{A})}{f_2(\mathbf{A})}\Big/\frac{f_3(\mathbf{A})}{f_4(\mathbf{A})}=\frac{f_5(\mathbf{A})}{f_6(\mathbf{A})}.$$

Here all the f_i are polynomials and it is assumed that the rational functions of **A** involved in each case are defined. Thus, in (3.7.8) it is assumed that $|f_2(\mathbf{A})| \neq 0$, $|f_4(\mathbf{A})| \neq 0$, $|f_6(\mathbf{A})| \neq 0$. We shall give a proof of (3.7.8) and leave the proofs of the remaining two statements to the reader.

The identity on the left-hand side of (3.7.8) implies

$$f_1(x)f_4(x)f_6(x)+f_3(x)f_2(x)f_6(x) = f_5(x)f_2(x)f_4(x).$$

Hence, by Theorem 3.7.2,

$$f_1(\mathbf{A})f_4(\mathbf{A})f_6(\mathbf{A})+f_3(\mathbf{A})f_2(\mathbf{A})f_6(\mathbf{A}) = f_5(\mathbf{A})f_2(\mathbf{A})f_4(\mathbf{A}).$$

Premultiplying both sides by $\{f_2(\mathbf{A})\}^{-1}\{f_4(\mathbf{A})\}^{-1}\{f_6(\mathbf{A})\}^{-1}$ and making use of (3.7.6) and (3.7.7), we obtain

$$f_1(\mathbf{A})\{f_2(\mathbf{A})\}^{-1}+f_3(\mathbf{A})\{f_4(\mathbf{A})\}^{-1} = f_5(\mathbf{A})\{f_6(\mathbf{A})\}^{-1}.$$

The proof of (3.7.8) is therefore complete.

Theorem 3.7.3 shows that rational functions of matrices can be manipulated in accordance with the same formal rules that are used for the rational functions of a scalar variable. It might now be concluded that the theory of rational functions of matrices is virtually the same as that of scalar rational functions. In one important respect, however, the analogy breaks down completely for, as we shall see in § 7.4, every rational function of **A** is equal to a polynomial in **A**—a result which has, of course, no analogue in the theory of scalar rational functions.

3.8. Partitioned matrices

In the present section we shall introduce a very useful technical device which will frequently facilitate the manipulation of matrices.

The results to be established below are almost obvious intuitively, even though the details of the formal argument may be found rather tedious.

Consider a rectangular matrix **A**, and let a number of lines be drawn between its rows, or columns, or both. These lines will then *partition* **A** into a number of smaller arrays. Thus, if **A** is the 4×3 matrix

$$\mathbf{A} = \begin{pmatrix} a_{11} & a_{12} & a_{13} \\ a_{21} & a_{22} & a_{23} \\ a_{31} & a_{32} & a_{33} \\ a_{41} & a_{42} & a_{43} \end{pmatrix},$$

three of the possible ways of partitioning it are as follows:

$$\text{(i)} \quad \left(\begin{array}{c|cc} a_{11} & a_{12} & a_{13} \\ a_{21} & a_{22} & a_{23} \\ \hline a_{31} & a_{32} & a_{33} \\ a_{41} & a_{42} & a_{43} \end{array}\right); \qquad \text{(ii)} \quad \left(\begin{array}{cc|c} a_{11} & a_{12} & a_{13} \\ \hline a_{21} & a_{22} & a_{23} \\ \hline a_{31} & a_{32} & a_{33} \\ a_{41} & a_{42} & a_{43} \end{array}\right);$$

$$\text{(iii)} \quad \left(\begin{array}{ccc} a_{11} & a_{12} & a_{13} \\ a_{21} & a_{22} & a_{23} \\ a_{31} & a_{32} & a_{33} \\ \hline a_{41} & a_{42} & a_{43} \end{array}\right).$$

A matrix devided in some such way by horizontal or vertical lines is called a *partitioned matrix*. If $\mathfrak{P}$ stands for the mode of partitioning of the original matrix **A**, then the resulting partitioned matrix will be denoted by $\mathbf{A}_{\mathfrak{P}}$. If $\mathfrak{P}$ consists of all lines drawn between every pair of consecutive rows and every pair of consecutive columns of **A**, then $\mathbf{A}_{\mathfrak{P}}$ is, of course, indistinguishable from **A**.

We can represent a partitioned matrix most economically by denoting each constituent array by a single matrix symbol. Thus in the case (i) above we can write

$$\mathbf{A}_{\mathfrak{P}} = \begin{pmatrix} \mathbf{A}^{(11)} & \mathbf{A}^{(12)} \\ \mathbf{A}^{(21)} & \mathbf{A}^{(22)} \end{pmatrix},$$

where

$$\mathbf{A}^{(11)} = \begin{pmatrix} a_{11} \\ a_{21} \end{pmatrix}, \qquad \mathbf{A}^{(12)} = \begin{pmatrix} a_{12} & a_{13} \\ a_{22} & a_{23} \end{pmatrix}, \qquad \mathbf{A}^{(21)} = \begin{pmatrix} a_{31} \\ a_{41} \end{pmatrix},$$

$$\mathbf{A}^{(22)} = \begin{pmatrix} a_{32} & a_{33} \\ a_{42} & a_{43} \end{pmatrix}.$$

In the case (ii) we have

$$\mathbf{A}_{\mathfrak{P}} = \begin{pmatrix} \mathbf{A}^{(11)} & \mathbf{A}^{(12)} \\ \mathbf{A}^{(21)} & \mathbf{A}^{(22)} \\ \mathbf{A}^{(31)} & \mathbf{A}^{(32)} \end{pmatrix},$$

where

$$\mathbf{A}^{(11)} = (a_{11} \quad a_{12}), \qquad \mathbf{A}^{(21)} = (a_{21} \quad a_{22}), \qquad \mathbf{A}^{(31)} = \begin{pmatrix} a_{31} & a_{32} \\ a_{41} & a_{42} \end{pmatrix},$$

$$\mathbf{A}^{(12)} = (a_{13}), \qquad \mathbf{A}^{(22)} = (a_{23}), \qquad \mathbf{A}^{(32)} = \begin{pmatrix} a_{33} \\ a_{43} \end{pmatrix}.$$

Again, in (iii), $$\mathbf{A}_{\mathfrak{P}} = \begin{pmatrix} \mathbf{A}^{(11)} \\ \mathbf{A}^{(21)} \end{pmatrix},$$

where

$$\mathbf{A}^{(11)} = \begin{pmatrix} a_{11} & a_{12} & a_{13} \\ a_{21} & a_{22} & a_{23} \\ a_{31} & a_{32} & a_{33} \end{pmatrix}, \qquad \mathbf{A}^{(21)} = (a_{41} \quad a_{42} \quad a_{43}).$$

The constituent matrices of a partitioned matrix will be called its elements. It must, however, be borne in mind that an arbitrary rectangular array of matrices is not, in general, a partitioned matrix since when the constituent matrices are written out in full the resulting array of numbers need not, of course, be a rectangular array.

If $\mathbf{\Gamma}$ is any partitioned matrix we shall denote by $\{\mathbf{\Gamma}\}$ the ordinary matrix obtained from $\mathbf{\Gamma}$ by removing its partitions. Obviously $\{\mathbf{A}_{\mathfrak{P}}\} = \mathbf{A}$.

We can at once extend the definitions of addition and multiplication to partitioned matrices simply by replacing the operations carried out on numerical elements by the corresponding operations carried out on the matrix elements. Addition and multiplication so defined will be found to obey the same formal laws as those established for ordinary matrices. What is equally important is that in forming sums or products of partitioned matrices we arrive at precisely the same results we should have obtained if the partitions had been removed initially.† It is this last fact that makes partitioning so convenient a device.

Let us first consider addition. Here the discussion is almost trivial.

† To be more precise, the two results are the same apart from the partitions. This qualification must be added wherever the context requires it.

DEFINITION 3.8.1. *Let* **A, B** *be matrices of the same type and let the same partitioning* $\mathfrak{P}$ *be applied to them, so that*

$$\mathbf{A}_{\mathfrak{P}} = \begin{pmatrix} \mathbf{A}^{(11)} & . & . & . & \mathbf{A}^{(1q)} \\ . & . & . & . & . \\ \mathbf{A}^{(p1)} & . & . & . & \mathbf{A}^{(pq)} \end{pmatrix}, \quad \mathbf{B}_{\mathfrak{P}} = \begin{pmatrix} \mathbf{B}^{(11)} & . & . & . & \mathbf{B}^{(1q)} \\ . & . & . & . & . \\ \mathbf{B}^{(p1)} & . & . & . & \mathbf{B}^{(pq)} \end{pmatrix}.$$

Then addition is defined by the equation†

$$\mathbf{A}_{\mathfrak{P}}+\mathbf{B}_{\mathfrak{P}} = \begin{pmatrix} \mathbf{A}^{(11)}+\mathbf{B}^{(11)} & . & . & . & \mathbf{A}^{(1q)}+\mathbf{B}^{(1q)} \\ . & . & . & . & . \\ \mathbf{A}^{(p1)}+\mathbf{B}^{(p1)} & . & . & . & \mathbf{A}^{(pq)}+\mathbf{B}^{(pq)} \end{pmatrix}.$$

We note that this definition reduces to Definition 3.3.2 (p. 78) for the case of ordinary matrices.

It is at once clear that

$$\mathbf{A}_{\mathfrak{P}}+\mathbf{B}_{\mathfrak{P}} = (\mathbf{A}+\mathbf{B})_{\mathfrak{P}};$$

in other words $$\{\mathbf{A}_{\mathfrak{P}}+\mathbf{B}_{\mathfrak{P}}\} = \mathbf{A}+\mathbf{B}.$$

Thus, no matter whether we add two matrices or their partitioned forms, we obtain the same result.

It is equally clear that addition of partitioned matrices is commutative and associative, i.e.

$$\mathbf{A}_{\mathfrak{P}}+\mathbf{B}_{\mathfrak{P}} = \mathbf{B}_{\mathfrak{P}}+\mathbf{A}_{\mathfrak{P}},$$

$$\mathbf{A}_{\mathfrak{P}}+(\mathbf{B}_{\mathfrak{P}}+\mathbf{C}_{\mathfrak{P}}) = (\mathbf{A}_{\mathfrak{P}}+\mathbf{B}_{\mathfrak{P}})+\mathbf{C}_{\mathfrak{P}}.$$

The discussion of multiplication requires a little more care.

DEFINITION 3.8.2. *Let* **A, B** *be two matrices for which* **AB** *is defined, and suppose that partitions* $\mathfrak{P}$, $\mathfrak{Q}$ *are applied to* **A, B** *respectively, so that*

$$\mathbf{A}_{\mathfrak{P}} = \begin{pmatrix} \mathbf{A}^{(11)} & . & . & . & \mathbf{A}^{(1q)} \\ . & . & . & . & . \\ \mathbf{A}^{(p1)} & . & . & . & \mathbf{A}^{(pq)} \end{pmatrix}, \quad \mathbf{B}_{\mathfrak{Q}} = \begin{pmatrix} \mathbf{B}^{(11)} & . & . & . & \mathbf{B}^{(1s)} \\ . & . & . & . & . \\ \mathbf{B}^{(r1)} & . & . & . & \mathbf{B}^{(rs)} \end{pmatrix}.$$

If $q = r$ *and if every matrix product*

$$\mathbf{A}^{(ik)}\mathbf{B}^{(kj)} \quad (i = 1,..., p;\ j = 1,..., s;\ k = 1,..., q)$$

is defined, then the product $\mathbf{A}_{\mathfrak{P}}\,\mathbf{B}_{\mathfrak{Q}}$ *is defined as the partitioned matrix of* p *rows and* s *columns whose* (i,j)*th element is the matrix*

$$\sum_{k=1}^{q} \mathbf{A}^{(ik)}\mathbf{B}^{(kj)} \quad (i = 1,..., p;\ j = 1,..., s).\ddagger$$

† It is obvious that $\mathbf{A}^{(ij)}$ and $\mathbf{B}^{(ij)}$ are of the same type.

‡ It should, in the first place, be noted that the q matrices $\mathbf{A}^{(ik)}\mathbf{B}^{(kj)}$ $(k = 1,..., q)$ are all of the same type, so that they can be added together. Furthermore, any two elements of the matrix $\mathbf{A}_{\mathfrak{P}}\,\mathbf{B}_{\mathfrak{Q}}$ are matrices with the same number of rows (columns) if they stand in the same column (row) of $\mathbf{A}_{\mathfrak{P}}\,\mathbf{B}_{\mathfrak{Q}}$. Hence $\mathbf{A}_{\mathfrak{P}}\,\mathbf{B}_{\mathfrak{Q}}$ is a partitioned matrix in the sense in which we use this term.

This definition reduces to Definition 3.3.3 (p. 80) for the case of ordinary matrices.

THEOREM 3.8.1. *If the product* $\mathbf{A}_{\mathfrak{P}}\,\mathbf{B}_{\mathfrak{Q}}$ *is defined, then*

$$\{\mathbf{A}_{\mathfrak{P}}\,\mathbf{B}_{\mathfrak{Q}}\} = \mathbf{AB}.$$

This result states, in effect, that partitioning of matrices does not affect their multiplication. It is extremely useful in certain cases where it is found to be more economical to multiply matrices in their partitioned form.†

Theorem 3.8.1. is almost obvious intuitively, and before we give a formal proof we shall illustrate it by a simple example.

Let $\mathbf{A}$, $\mathbf{B}$ be two matrices of type 4×4, 4×2 respectively, and let them be partitioned as follows:

$$\mathbf{A}_{\mathfrak{P}} = \left(\begin{array}{ccc|c} a_{11} & a_{12} & a_{13} & a_{14} \\ a_{21} & a_{22} & a_{23} & a_{24} \\ \hline a_{31} & a_{32} & a_{33} & a_{34} \\ a_{41} & a_{42} & a_{43} & a_{44} \end{array}\right) = \begin{pmatrix} \mathbf{A}^{(11)} & \mathbf{A}^{(12)} \\ \mathbf{A}^{(21)} & \mathbf{A}^{(22)} \end{pmatrix},$$

$$\mathbf{B}_{\mathfrak{Q}} = \left(\begin{array}{cc} b_{11} & b_{12} \\ b_{21} & b_{22} \\ b_{31} & b_{32} \\ \hline b_{41} & b_{42} \end{array}\right) = \begin{pmatrix} \mathbf{B}^{(11)} \\ \mathbf{B}^{(21)} \end{pmatrix}.$$

Then, by Definition 3.8.2, $\mathbf{A}_{\mathfrak{P}}\,\mathbf{B}_{\mathfrak{Q}}$ is the 2×1 matrix

$$\mathbf{A}_{\mathfrak{P}}\,\mathbf{B}_{\mathfrak{Q}} = \begin{pmatrix} \mathbf{A}^{(11)}\mathbf{B}^{(11)}+\mathbf{A}^{(12)}\mathbf{B}^{(21)} \\ \mathbf{A}^{(21)}\mathbf{B}^{(11)}+\mathbf{A}^{(22)}\mathbf{B}^{(21)} \end{pmatrix} = \begin{pmatrix} \mathbf{C}^{(11)} \\ \mathbf{C}^{(21)} \end{pmatrix}, \text{ say.}$$

Here we have

$$\mathbf{C}^{(11)} = \mathbf{A}^{(11)}\mathbf{B}^{(11)}+\mathbf{A}^{(12)}\mathbf{B}^{(21)} = \begin{pmatrix} a_{11} & a_{12} & a_{13} \\ a_{21} & a_{22} & a_{23} \end{pmatrix}\begin{pmatrix} b_{11} & b_{12} \\ b_{21} & b_{22} \\ b_{31} & b_{32} \end{pmatrix} + \begin{pmatrix} a_{14} \\ a_{24} \end{pmatrix}(b_{41}\;\; b_{42})$$

$$= \begin{pmatrix} a_{11}b_{11}+a_{12}b_{21}+a_{13}b_{31}+a_{14}b_{41} & a_{11}b_{12}+a_{12}b_{22}+a_{13}b_{32}+a_{14}b_{42} \\ a_{21}b_{11}+a_{22}b_{21}+a_{23}b_{31}+a_{24}b_{41} & a_{21}b_{12}+a_{22}b_{22}+a_{23}b_{32}+a_{24}b_{42} \end{pmatrix};$$

$$\mathbf{C}^{(21)} = \mathbf{A}^{(21)}\mathbf{B}^{(11)}+\mathbf{A}^{(22)}\mathbf{B}^{(21)} = \begin{pmatrix} a_{31} & a_{32} & a_{33} \\ a_{41} & a_{42} & a_{43} \end{pmatrix}\begin{pmatrix} b_{11} & b_{12} \\ b_{21} & b_{22} \\ b_{31} & b_{32} \end{pmatrix} + \begin{pmatrix} a_{34} \\ a_{44} \end{pmatrix}(b_{41}\;\; b_{42})$$

$$= \begin{pmatrix} a_{31}b_{11}+a_{32}b_{21}+a_{33}b_{31}+a_{34}b_{41} & a_{31}b_{12}+a_{32}b_{22}+a_{33}b_{32}+a_{34}b_{42} \\ a_{41}b_{11}+a_{42}b_{21}+a_{43}b_{31}+a_{44}b_{41} & a_{41}b_{12}+a_{42}b_{22}+a_{43}b_{32}+a_{44}b_{42} \end{pmatrix}.$$

† See, for instance, the proofs of Theorems 5.6.2., 6.4.1, and 10.3.4.

Therefore

$$\mathbf{A}_{\mathfrak{P}}\mathbf{B}_{\mathfrak{Q}}$$

$$= \left(\begin{array}{cc} a_{11}b_{11}+a_{12}b_{21}+a_{13}b_{31}+a_{14}b_{41} & a_{11}b_{12}+a_{12}b_{22}+a_{13}b_{32}+a_{14}b_{42} \\ a_{21}b_{11}+a_{22}b_{21}+a_{23}b_{31}+a_{24}b_{41} & a_{21}b_{12}+a_{22}b_{22}+a_{23}b_{32}+a_{24}b_{42} \\ \hline a_{31}b_{11}+a_{32}b_{21}+a_{33}b_{31}+a_{34}b_{41} & a_{31}b_{12}+a_{32}b_{22}+a_{33}b_{32}+a_{34}b_{42} \\ a_{41}b_{11}+a_{42}b_{21}+a_{43}b_{31}+a_{44}b_{41} & a_{41}b_{12}+a_{42}b_{22}+a_{43}b_{32}+a_{44}b_{42} \end{array}\right),$$

and so $\{\mathbf{A}_{\mathfrak{P}}\mathbf{B}_{\mathfrak{Q}}\} = \mathbf{AB}$, as stated by the theorem.

To prove the assertion in general we shall, for the sake of brevity, write $\mathbf{\Gamma} = \mathbf{A}_{\mathfrak{P}}\mathbf{B}_{\mathfrak{Q}}$. We note, in the first place, that $\{\mathbf{\Gamma}\}$ and $\mathbf{AB}$ are matrices of the same type.

Next, consider a typical element $(\mathbf{AB})_{ij}$ of $\mathbf{AB}$. Suppose that the ith row of $\mathbf{A}$ is made up of the λth rows of the matrices

$$\mathbf{A}^{(u1)},\ldots,\mathbf{A}^{(uq)},$$

and that the jth column of $\mathbf{B}$ is made up of the μth columns of the matrices

$$\mathbf{B}^{(1v)},\ldots,\mathbf{B}^{(qv)}.$$

Let $\mathbf{\Gamma}^{(uv)}$ denote the (u, v)th element of $\mathbf{\Gamma}$. Then, by Definition 3.8.2,

$$\mathbf{\Gamma}^{(uv)} = \sum_{k=1}^{q} \mathbf{A}^{(uk)}\mathbf{B}^{(kv)},$$

and so†

$$\begin{aligned}(\mathbf{\Gamma}^{(uv)})_{\lambda\mu} &= \sum_{k=1}^{q} (\mathbf{A}^{(uk)}\mathbf{B}^{(kv)})_{\lambda\mu} \\ &= \sum_{k=1}^{q} (\mathbf{A}^{(uk)})_{\lambda *}(\mathbf{B}^{(kv)})_{*\mu} \\ &= (\mathbf{AB})_{ij}. \qquad (3.8.1)\end{aligned}$$

Hence $(\mathbf{AB})_{ij}$ is an element of $\{\mathbf{\Gamma}\}$.

The matrices $\{\mathbf{\Gamma}\}$ and $\mathbf{AB}$ are thus of the same type and have the same elements. It remains now only to prove that the elements in the two matrices occur in the same order. Suppose that, in addition to (3.8.1), we also have

$$(\mathbf{\Gamma}^{(u'v')})_{\lambda'\mu'} = (\mathbf{AB})_{i'j'}.$$

It is clear that if $i' \geqslant i$, then either $u' > u$ or $u' = u$, $\lambda' \geqslant \lambda$. Similarly, if $j' \geqslant j$, then either $v' > v$ or $v' = v$, $\mu' \geqslant \mu$. This means, in fact, that

$$(\{\mathbf{\Gamma}\})_{ij} = (\mathbf{AB})_{ij}.$$

The proof of the theorem is therefore complete.

† It is essential to be quite clear about the precise significance of each symbol. We repeat, then, that $\mathbf{\Gamma}$ is the partitioned matrix $\mathbf{A}_{\mathfrak{P}}\mathbf{B}_{\mathfrak{Q}}$; its (u, v)th element $\mathbf{\Gamma}^{(uv)}$ is, of course, itself a matrix. Again $(\mathbf{\Gamma}^{(uv)})_{\lambda\mu}$ denotes the (λ, μ)th element of $\mathbf{\Gamma}^{(uv)}$; this is therefore an element of $\{\mathbf{\Gamma}\}$ and so is a scalar.

In conclusion it should be mentioned that multiplication of partitioned matrices is associative and also distributive with respect to addition, i.e.

$$\mathbf{A}_{\mathfrak{P}}(\mathbf{B}_{\mathfrak{Q}}\,\mathbf{C}_{\mathfrak{R}}) = (\mathbf{A}_{\mathfrak{P}}\,\mathbf{B}_{\mathfrak{Q}})\mathbf{C}_{\mathfrak{R}},$$
$$\mathbf{A}_{\mathfrak{P}}(\mathbf{B}_{\mathfrak{Q}}+\mathbf{C}_{\mathfrak{R}}) = \mathbf{A}_{\mathfrak{P}}\,\mathbf{B}_{\mathfrak{Q}}+\mathbf{A}_{\mathfrak{P}}\,\mathbf{C}_{\mathfrak{R}},$$
$$(\mathbf{B}_{\mathfrak{Q}}+\mathbf{C}_{\mathfrak{R}})\mathbf{A}_{\mathfrak{P}} = \mathbf{B}_{\mathfrak{Q}}\,\mathbf{A}_{\mathfrak{P}}+\mathbf{C}_{\mathfrak{R}}\,\mathbf{A}_{\mathfrak{P}},$$

provided that the matrices in question are defined. Multiplication of partitioned matrices is obviously non-commutative.

PROBLEMS ON CHAPTER III

1. Compute the matrix products

$$\text{(i)}\quad \begin{pmatrix} -1 & 5 \\ 2 & 0 \\ 3 & 1 \end{pmatrix}\begin{pmatrix} 7 \\ 2 \end{pmatrix}; \qquad \text{(ii)}\quad \begin{pmatrix} 0 & 2 & 1 \\ -1 & 0 & 3 \end{pmatrix}\begin{pmatrix} 1 & 1 & -1 & 1 \\ 2 & 0 & 0 & 2 \\ 1 & -2 & -1 & 0 \end{pmatrix};$$

$$\text{(iii)}\quad (1,-1)\begin{pmatrix} 1/\sqrt{2} & 1/\sqrt{2} \\ -1/\sqrt{2} & 1/\sqrt{2} \end{pmatrix}.$$

2. Find **AB** and **BA** when

$$\text{(i)}\quad \mathbf{A} = \begin{pmatrix} 1 & 2 & 1 \\ 1 & 3 & 2 \\ 1 & 4 & 6 \end{pmatrix}, \qquad \mathbf{B} = \begin{pmatrix} 0 & 0 & 1 \\ 0 & 1 & 0 \\ 1 & 0 & 0 \end{pmatrix};$$

$$\text{(ii)}\quad \mathbf{A} = \begin{pmatrix} 1 & i \\ -i & 1 \end{pmatrix}, \qquad \mathbf{B} = \begin{pmatrix} i & -1 \\ -1 & -i \end{pmatrix}.$$

In case (ii) find a non-zero matrix **C** such that $\mathbf{BC} = \mathbf{O}$.

3. If

$$\mathbf{U} = \begin{pmatrix} 0 & 1 & 0 \\ 0 & 0 & 1 \\ 0 & 0 & 0 \end{pmatrix},$$

find $\mathbf{U}\mathbf{U}^T$, $\mathbf{U}^T\mathbf{U}$, $\mathbf{U}^2$, $\mathbf{U}^3$.

4. Find $\mathbf{A}^2$, $\mathbf{A}^3$, and $\mathbf{A}^4$ when

$$\text{(i)}\quad \mathbf{A} = \begin{pmatrix} 1 & 1 & 1 \\ 1 & \omega & \omega^2 \\ 1 & \omega^2 & \omega \end{pmatrix} \quad (\omega = e^{2\pi i/3}); \quad \text{(ii)}\quad \mathbf{A} = \begin{pmatrix} 1 & 1 & 1 & 1 \\ 1 & i & -1 & -i \\ 1 & -1 & 1 & -1 \\ 1 & -i & -1 & i \end{pmatrix}.$$

5. Find $\mathbf{A}^k$, given that every element of the $n \times n$ matrix **A** is equal to a.

6. Show that, if $\mathbf{A}^T = \mathbf{A}$, then $(\mathbf{A}^*)^T = \mathbf{A}^*$.

7. If

$$\mathbf{P} = \begin{pmatrix} 1 & a & a^2 \\ 1 & b & b^2 \\ 1 & c & c^2 \end{pmatrix},$$

show that

$$\mathbf{P}^* = \begin{pmatrix} -bc & -ca & -ab \\ b+c & c+a & a+b \\ -1 & -1 & -1 \end{pmatrix}\begin{pmatrix} b-c & 0 & 0 \\ 0 & c-a & 0 \\ 0 & 0 & a-b \end{pmatrix}.$$

8. Show that the product of two lower triangular matrices is again a lower triangular matrix. Also show that the inverse of a non-singular lower triangular matrix is a lower triangular matrix.

Find the inverse matrix of

$$\begin{pmatrix} 1 & 0 & 0 & 0 \\ 1 & 2 & 0 & 0 \\ 2 & 1 & 3 & 0 \\ 1 & 2 & 1 & 4 \end{pmatrix}.$$

9. Find the inverse matrix of

$$\begin{pmatrix} 1 & 1 & 1 \\ 1 & \omega & \omega^2 \\ 1 & \omega^2 & \omega \end{pmatrix},$$

where $\omega = e^{2\pi i/3}$.

10. Let

$$\mathbf{A}(\theta) = \begin{pmatrix} \cos\theta & -\sin\theta \\ \sin\theta & \cos\theta \end{pmatrix}.$$

Show that $\mathbf{A}(\theta)\mathbf{A}(\theta') = \mathbf{A}(\theta+\theta')$, and give a geometrical interpretation of this result.

11. $\mathbf{A}$ is a given square matrix. Writing $\mathbf{A}_t = (t\mathbf{I}-\mathbf{A})^{-1}$ whenever $t\mathbf{I}-\mathbf{A}$ is non-singular, prove that

$$(t-u)\mathbf{A}_t\mathbf{A}_u = \mathbf{A}_u-\mathbf{A}_t.$$

12. Determine the condition for the matrix

$$\begin{pmatrix} 1 & -n & m \\ n & 1 & -l \\ -m & l & 1 \end{pmatrix}$$

to be non-singular; and in that case finds its inverse.

13. Show that if $\mathbf{B}$ commutes with $\mathbf{A}$, then it commutes with every polynomial in $\mathbf{A}$.

14. Show that if $\mathbf{A}$ and $\mathbf{B}$ commute, so do $\mathbf{A}^k$ and $\mathbf{B}^l$, where k, l are positive integers. Discuss the case when k, l are not necessarily positive.

15. Show that, if $\mathbf{P}^{-1}\mathbf{AP} = \mathbf{Q}^{-1}\mathbf{BQ}$, then there exist matrices $\mathbf{R}, \mathbf{S}$ such that $\mathbf{A} = \mathbf{RS}$, $\mathbf{B} = \mathbf{SR}$.

16. Let $\mathbf{A} = (a_{rs})$ be an $n\times n$ matrix and write $\bar{\mathbf{A}} = (\bar{a}_{rs})$. Show that, if $\mathbf{x} = (x_1,..., x_n)^T$, then

$$\bar{\mathbf{x}}^T\bar{\mathbf{A}}^T\mathbf{A}\mathbf{x} = \sum_{k=1}^{n}\left|\sum_{r=1}^{n} a_{kr}x_r\right|^2.$$

17. A (square) matrix $\mathbf{A}$ is said to be *idempotent* if $\mathbf{A}^2 = \mathbf{A}$. Show that the sum of two idempotent matrices $\mathbf{A}$ and $\mathbf{B}$ is idempotent if and only if $\mathbf{AB} = \mathbf{BA} = \mathbf{O}$.

18. Show that if the product of two matrices is equal to a non-zero scalar matrix, then the two matrices commute. Can the term 'non-zero' be omitted in this statement?

19. Let $\mathbf{A}$ be a fixed $m\times m$ matrix and $\mathbf{B}$ a variable $m\times n$ matrix. Show that the equation $\mathbf{AX} = \mathbf{B}$ in $\mathbf{X}$ is soluble uniquely for every choice of $\mathbf{B}$ if and only if $|\mathbf{A}| \neq 0$.

20. Find all matrices which commute with the matrix

$$\begin{pmatrix} \lambda & 1 & 0 \\ 0 & \lambda & 1 \\ 0 & 0 & \lambda \end{pmatrix}.$$

21. Show that any two matrices which commute with $\begin{pmatrix} 0 & 1 \\ -1 & 0 \end{pmatrix}$, commute with each other.

22. Show that a necessary and sufficient condition for an $n \times n$ matrix to commute with *every* $n \times n$ matrix is that it should be scalar.

23. Express the matrix

$$\mathbf{A} = \begin{pmatrix} 13 & 7 & 3 \\ 7 & 5 & 4 \\ 3 & 4 & 6 \end{pmatrix}$$

in the form $\mathbf{A} = \mathbf{B}\mathbf{B}^T$, where $\mathbf{B}$ is an upper triangular matrix with positive elements.

24. Let the vectors $\mathbf{x}_1,\ldots,\mathbf{x}_n$, $\mathbf{y}_1,\ldots,\mathbf{y}_n$, of order n, be connected by the equations

$$\mathbf{y}_r = \sum_{s=1}^{n} a_{rs}\mathbf{x}_s \quad (r = 1,\ldots,n).$$

Show that $\mathbf{Y} = \mathbf{A}\mathbf{X}$, where $\mathbf{A} = (a_{rs})$, and $\mathbf{X}$ and $\mathbf{Y}$ are the matrices having $\mathbf{x}_1,\ldots,\mathbf{x}_n$ and $\mathbf{y}_1,\ldots,\mathbf{y}_n$ respectively as their rows.

25. Let $\mathbf{A} = \begin{pmatrix} a & b \\ 0 & 1 \end{pmatrix}$, where $a \neq 1$. Show that, for $n \geqslant 0$,

$$\mathbf{A}^n = \begin{pmatrix} a_n & b_n \\ 0 & 1 \end{pmatrix},$$

where $a_n = a^n$, $b_n = b(a^n-1)/(a-1)$, and discuss the validity of this formula for $n < 0$. Also consider the case $a = 1$.

26. Let $\mathbf{A} = (a_{rs})$ be a rectangular matrix, and write

$$N(\mathbf{A}) = \left\{ \sum_{r,s} |a_{rs}|^2 \right\}^{\frac{1}{2}}.$$

Show that $N(\mathbf{AB}) \leqslant N(\mathbf{A})N(\mathbf{B})$, and deduce Theorem 2.5.2 (p. 63).

27. Defining $N(\mathbf{A})$ as in the preceding question, prove, by means of Minkowski's inequality,† that $N(\mathbf{A}+\mathbf{B}) \leqslant N(\mathbf{A})+N(\mathbf{B})$. Also show that this relation is equivalent to the triangle inequality for vectors (Theorem 2.5.3, p. 64).

28. By considering powers of the $n \times n$ matrix

$$\begin{pmatrix} 0 & 1 & 0 & 0 & . & . & . & 0 \\ 0 & 0 & 1 & 0 & . & . & . & 0 \\ 0 & 0 & 0 & 1 & . & . & . & 0 \\ . & . & . & . & . & . & . & . \\ 0 & 0 & 0 & 0 & . & . & . & 1 \\ 0 & 0 & 0 & 0 & . & . & . & 0 \end{pmatrix},$$

† See Hardy, Littlewood, and Pólya, *Inequalities*, 31.

show that

$$\begin{pmatrix} \lambda & 1 & 0 & 0 & . & . & . & 0 \\ 0 & \lambda & 1 & 0 & . & . & . & 0 \\ 0 & 0 & \lambda & 1 & . & . & . & 0 \\ 0 & 0 & 0 & \lambda & . & . & . & 0 \\ . & . & . & . & . & . & . & . \\ 0 & 0 & 0 & 0 & . & . & . & \lambda \end{pmatrix}^N$$

$$= \begin{pmatrix} \lambda^N & \binom{N}{1}\lambda^{N-1} & \binom{N}{2}\lambda^{N-2} & \binom{N}{3}\lambda^{N-3} & . & . & . & \binom{N}{n-1}\lambda^{N-n+1} \\ 0 & \lambda^N & \binom{N}{1}\lambda^{N-1} & \binom{N}{2}\lambda^{N-2} & . & . & . & \binom{N}{n-2}\lambda^{N-n+2} \\ 0 & 0 & \lambda^N & \binom{N}{1}\lambda^{N-1} & . & . & . & \binom{N}{n-3}\lambda^{N-n+3} \\ . & . & . & . & . & . & . & . \\ 0 & 0 & 0 & 0 & . & . & . & \lambda^N \end{pmatrix},$$

where $\binom{N}{k}$ is interpreted as 0 when $k > N$.

29. Let $f(t)$, $g(t)$ be polynomials whose highest common factor is 1; let $\mathbf{M}$ be an $n \times n$ matrix; and write $\mathbf{A} = f(\mathbf{M})$, $\mathbf{B} = g(\mathbf{M})$. Show that every solution of the equation $\mathbf{ABx} = \mathbf{0}$ can be written in the form $\mathbf{x} = \mathbf{y}+\mathbf{z}$, where $\mathbf{By} = \mathbf{Az} = \mathbf{0}$.

30. Show that the sum and the product of two matrices of the type $\begin{pmatrix} a & b \\ -b & a \end{pmatrix}$ are again matrices of the same type. Deduce that the product of two matrices of the type

$$\begin{pmatrix} p & q & r & s \\ -q & p & -s & r \\ -r & s & p & -q \\ -s & -r & q & p \end{pmatrix}$$

is again a matrix of this type.

31. $\mathfrak{U}$ is a subspace of $\mathfrak{B}_n$ $(n > 2)$, which consists of all complex vectors $\mathbf{x}$ whose last $n-2$ components all vanish; and $\mathbf{A}$ is a complex $n \times n$ matrix such that $|\mathbf{Ax}| = k|\mathbf{x}|$ for all $\mathbf{x} \in \mathfrak{U}$, where k is a constant. Prove that the first two columns of $\mathbf{A}$ are orthogonal to each other and have length k.

32. Referred to a system of rectangular cartesian coordinates OX_1, OX_2, OX_3, the coordinates of a point X are given by the column vector $\mathbf{x}$, where $\mathbf{x}^T = (x_1, x_2, x_3)$. The direction cosines of a given line l through O are given by the column vector $\mathbf{l}$, where $\mathbf{l}^T = (l_1, l_2, l_3)$. Prove that the foot of the perpendicular from X to l is the point $\mathbf{Ax}$, where $\mathbf{A} = \mathbf{l}\mathbf{l}^T$; and show that

$$\mathbf{A}^2 = \mathbf{A}, \qquad |\mathbf{A}| = 0.$$

33. Find all 2×2 matrices which satisfy the equation $\mathbf{X}^2 = \mathbf{X}$, and show that all but one of these matrices are singular.

34. Let $a_1,\dots,a_n$ be numbers such that $a_1^2+\dots+a_n^2 = 1$, and put

$$\mathbf{A}_x = \begin{pmatrix} a_1^2+x & a_1 a_2 & . & . & . & a_1 a_n \\ a_2 a_1 & a_2^2+x & . & . & . & a_2 a_n \\ . & . & . & . & . & . \\ a_n a_1 & a_n a_2 & . & . & . & a_n^2+x \end{pmatrix}.$$

Show that, for every value of x, $\mathbf{A}_x^* = -x^{n-2}\mathbf{A}_{-x-1}$.

35. Let $\omega_1,\dots,\omega_k$ be distinct numbers and $t_1,\dots,t_k$ arbitrary numbers. Prove that there exists one and only one polynomial f of degree $\leqslant k-1$ such that $f(\omega_i) = t_i$ $(i = 1,\dots,k)$.

Show that, for any square matrix $\mathbf{A}$ and any distinct numbers $\omega_1,\dots,\omega_k$,

$$\sum_{i=1}^{k} \prod_{\substack{\kappa=1\\ \kappa\neq i}}^{k} \left(\frac{\mathbf{A}-\omega_\kappa \mathbf{I}}{\omega_i-\omega_\kappa}\right) = \mathbf{I}.$$

36. **A, B, C, D** are $n\times n$ matrices, **A** being non-singular and **AC** = **CA**. Prove that the determinant of the $2n\times 2n$ matrix

$$\begin{pmatrix} \mathbf{A} & \mathbf{B} \\ \mathbf{C} & \mathbf{D} \end{pmatrix}$$

is equal to that of the $n\times n$ matrix $\mathbf{AD}-\mathbf{CB}$. By replacing $\mathbf{A}$ by $\mathbf{A}+x\mathbf{I}$, prove that the restriction that **A** is non-singular can be dropped.

37. If $\mathbf{E}_{ij}$ is the $n\times n$ matrix whose elements all vanish except for a 1 in the ith row and jth column, show that $\mathbf{E}_{ij}\mathbf{E}_{kl} = \delta_{jk}\mathbf{E}_{il}$.

If

$$\mathbf{A}_1 = \begin{pmatrix} 1 & 0 \\ 0 & 0 \end{pmatrix}, \quad \mathbf{A}_2 = \begin{pmatrix} 0 & 1 \\ 0 & 0 \end{pmatrix}, \quad \mathbf{A}_3 = \begin{pmatrix} 0 & 0 \\ 1 & 0 \end{pmatrix}, \quad \mathbf{A}_4 = \begin{pmatrix} 0 & 0 \\ 0 & 1 \end{pmatrix},$$

show that

$$\sum_{i,j,k,l=1}^{4} \epsilon_{ijkl}\mathbf{A}_i\mathbf{A}_j\mathbf{A}_k\mathbf{A}_l = \mathbf{O},$$

where $\epsilon_{ijkl} = \epsilon\begin{pmatrix} i\,j\,k\,l \\ 1\,2\,3\,4 \end{pmatrix}$. Deduce that this result still holds when $\mathbf{A}_1, \mathbf{A}_2, \mathbf{A}_3, \mathbf{A}_4$ are replaced by arbitrary 2×2 matrices.

38. Let $\phi(n)$ denote the number of integers not exceeding n and prime to n, and write

$$D = \begin{vmatrix} (1,1) & (1,2) & . & . & . & (1,n) \\ (2,1) & (2,2) & . & . & . & (2,n) \\ . & . & . & . & . & . \\ (n,1) & (n,2) & . & . & . & (n,n) \end{vmatrix},$$

where (r,s) denotes the highest common factor of r and s. By considering the matrix product $\mathbf{P}^T\mathbf{\Phi}\mathbf{P}$, where $\mathbf{\Phi} = \mathbf{dg}\{\phi(1),\phi(2),\dots,\phi(n)\}$ and $\mathbf{P}_{rs} = 1$ or 0 according as r divides or does not divide s, show that

$$D = \phi(1)\phi(2)\dots\phi(n).$$

39. Let $N(n)$ denote the number of non-singular $n\times n$ matrices each of whose elements is 0 or 1. Show that

$$N(n) = (2^2-1)(2^3-1)\dots(2^n-1)2^{\frac{1}{2}n(n-1)}.$$

IV

LINEAR OPERATORS

THE object of the present chapter (which may be omitted at the first reading) is to elaborate the remark about 'transformation of vectors' made at the end of § 3.4. The discussion in that section already pointed to a close relation between matrices and linear substitutions or transformations. We now propose to study this relation in greater detail and to show how, when we attempt to construct an analytical apparatus for handling linear transformations, we are naturally led to develop the calculus of matrices. In particular, we shall discover that, contrary to the impression that may have been derived from § 3.4, matrices and linear transformations cannot be identified. What, in fact, we shall find is that a linear transformation can be 'represented' by any matrix of an infinite class of matrices and that all such 'representations' have equal status. As it is desirable to frame the argument with the greatest possible degree of generality, we shall base our discussion on the idea of a linear manifold.

4.1. Change of basis in a linear manifold

Throughout the present section we denote by $\mathfrak{M}$ a given linear manifold, of finite dimensionality $n > 0$, over some specified field $\mathfrak{F}$. The zero element of $\mathfrak{M}$ is denoted by Θ.

THEOREM 4.1.1. *Let* $\mathfrak{B} = \{E_1,..., E_n\}$ *be a basis of* $\mathfrak{M}$, *and suppose that the elements* $\tilde{E}_1,..., \tilde{E}_n$ *are defined by the equations*

$$\tilde{E}_j = \sum_{i=1}^{n} p_{ij} E_i \quad (j = 1,..., n). \tag{4.1.1}$$

Then $\tilde{\mathfrak{B}} = \{\tilde{E}_1,..., \tilde{E}_n\}$ *is a basis of* $\mathfrak{M}$ *if and only if the matrix* $\mathbf{P} = (p_{ij})$ *is non-singular.*

If $t_1,..., t_n$ are any scalars, then, by (4.1.1),

$$\sum_{j=1}^{n} t_j \tilde{E}_j = \sum_{j=1}^{n} t_j \sum_{i=1}^{n} p_{ij} E_i$$

$$= \sum_{i=1}^{n} E_i \sum_{j=1}^{n} p_{ij} t_j. \tag{4.1.2}$$

Suppose, in the first place, that $|\mathbf{P}| \neq 0$. If $t_1,\ldots,t_n$ are scalars such that

$$\sum_{j=1}^{n} t_j \tilde{E}_j = \Theta, \tag{4.1.3}$$

then, since $E_1,\ldots,E_n$ are linearly independent, we have by (4.1.2)

$$\sum_{j=1}^{n} p_{ij} t_j = 0 \quad (i = 1,\ldots,n). \tag{4.1.4}$$

Hence, by Theorem 1.6.1 (p. 27), $t_1 = \ldots = t_n = 0$. The elements $\tilde{E}_1,\ldots,\tilde{E}_n$ are therefore linearly independent, and $\tilde{\mathfrak{B}}$ is a basis.

Next, let $|\mathbf{P}| = 0$. Then there exist scalars $t_1,\ldots,t_n$, not all zero, satisfying (4.1.4). In view of (4.1.2) these scalars also satisfy (4.1.3). Hence $\tilde{E}_1,\ldots,\tilde{E}_n$ are linearly dependent and so do not constitute a basis.

We recall from § 2.4 that once a basis in a linear manifold has been chosen, the elements of that manifold can be represented by vectors. We shall now investigate the relation between any two such representations arising from two different choices of basis.

THEOREM 4.1.2. (i) *Let $\mathfrak{B}$, $\tilde{\mathfrak{B}}$ be any two bases in $\mathfrak{M}$. If, for any $X \in \mathfrak{M}$,*

$$\mathbf{x} = \mathscr{R}(X; \mathfrak{B}), \qquad \tilde{\mathbf{x}} = \mathscr{R}(X; \tilde{\mathfrak{B}}), \tag{4.1.5}$$

then there exists a unique non-singular $n \times n$ matrix $\mathbf{P}$, *independent of X, such that*

$$\tilde{\mathbf{x}} = \mathbf{P}\mathbf{x}. \tag{4.1.6}$$

(ii) *If* $\mathbf{P}$ *is a given non-singular $n \times n$ matrix and $\mathfrak{B}$ is any given basis in $\mathfrak{M}$, then a second basis $\tilde{\mathfrak{B}}$ can be found such that, whenever* $\mathbf{x}$, $\tilde{\mathbf{x}}$ *satisfy* (4.1.5), *they also satisfy* (4.1.6).

This result states, loosely speaking, that when the elements of a linear manifold are represented in terms of vectors, a change of basis in $\mathfrak{M}$ is appropriately described by a matrix multiplication. Moreover, each such multiplication corresponds to a change of basis.

To prove (i), let $\mathfrak{B} = \{E_1,\ldots,E_n\}$, $\tilde{\mathfrak{B}} = \{\tilde{E}_1,\ldots,\tilde{E}_n\}$, and write

$$\mathbf{x} = (x_1,\ldots,x_n)^T, \qquad \tilde{\mathbf{x}} = (\tilde{x}_1,\ldots,\tilde{x}_n)^T. \tag{4.1.7}$$

By (4.1.5) we have

$$X = \sum_{j=1}^{n} x_j E_j = \sum_{i=1}^{n} \tilde{x}_i \tilde{E}_i. \tag{4.1.8}$$

Let

$$E_j = \sum_{i=1}^{n} p_{ij} \tilde{E}_i \quad (j = 1,\ldots,n). \tag{4.1.9}$$

Then, by Theorem 4.1.1, the matrix $\mathbf{P} = (p_{ij})$ is non-singular. By (4.1.8) and (4.1.9) we have

$$X = \sum_{j=1}^{n} x_j E_j = \sum_{j=1}^{n} x_j \sum_{i=1}^{n} p_{ij} \tilde{E}_i = \sum_{i=1}^{n} \tilde{E}_i \sum_{j=1}^{n} p_{ij} x_j,$$

and so, since the coordinates of an element with respect to a basis are uniquely determined,

$$\tilde{x}_i = \sum_{j=1}^{n} p_{ij} x_j \quad (i = 1,..., n).$$

These relations are, of course, equivalent to (4.1.6). Furthermore, if in addition to (4.1.6) we also have $\tilde{\mathbf{x}} = \mathbf{Q}\mathbf{x}$, then, for all $\mathbf{x}$, $(\mathbf{P}-\mathbf{Q})\mathbf{x} = \mathbf{0}$; and it follows by Exercise 3.3.8 (p. 85) that $\mathbf{Q} = \mathbf{P}$.

To prove (ii), let $\mathbf{P}$ be a given non-singular $n \times n$ matrix and write $\mathbf{P}^{-1} = (p^*_{ij})$. Let $\mathfrak{B} = \{E_1,..., E_n\}$ be any given basis in $\mathfrak{M}$, and put

$$\tilde{E}_j = \sum_{i=1}^{n} p^*_{ij} E_i \quad (j = 1,..., n).$$

Then, by Theorem 4.1.1, $\tilde{\mathfrak{B}} = \{\tilde{E}_1,..., \tilde{E}_n\}$ is also a basis of $\mathfrak{M}$. If $\mathbf{x}$ and $\tilde{\mathbf{x}}$ are defined by the relations (4.1.5) and if the notation (4.1.7) is used again, then

$$X = \sum_{j=1}^{n} \tilde{x}_j \tilde{E}_j = \sum_{j=1}^{n} \tilde{x}_j \sum_{i=1}^{n} p^*_{ij} E_i = \sum_{i=1}^{n} E_i \sum_{j=1}^{n} p^*_{ij} \tilde{x}_j,$$

and so

$$x_i = \sum_{j=1}^{n} p^*_{ij} \tilde{x}_j \quad (i = 1,..., n);$$

in other words, $\mathbf{x} = \mathbf{P}^{-1}\tilde{\mathbf{x}}$, and the proof is complete.

4.2. Linear operators and their representations

4.2.1. Let us consider two sets of objects $\mathfrak{M}$ and $\mathfrak{M}^*$.

DEFINITION 4.2.1. *If with each object X in $\mathfrak{M}$ is associated a unique object $L(X)$ in $\mathfrak{M}^*$, then L is called a* MAPPING *or a* TRANSFORMATION OF $\mathfrak{M}$ INTO $\mathfrak{M}^*$, *and $\mathfrak{M}$ is said to be mapped into $\mathfrak{M}^*$ by L. Furthermore, for each $X \in \mathfrak{M}$, $L(X)$ is called the* IMAGE *of X.*

Other terms such as operator, operation, or function are also used to describe L, and the idea they express is an exceedingly common one. The term 'function' is particularly apt; for $L(X)$, as defined above, is in fact an obvious extension of the elementary notion of a function. Here the argument X is taken from the set of objects $\mathfrak{M}$ and the functional value $L(X)$ lies in the set $\mathfrak{M}^*$.

For example, let $\mathfrak{M}$ be the set of real and $\mathfrak{M}^*$ the set of complex numbers. The function

$$L(X) = e^{2\pi i X}$$

is then clearly a mapping of $\mathfrak{M}$ into $\mathfrak{M}^*$. Or, again, let $\mathfrak{M}$ be the set of points of a plane and $\mathfrak{M}^*$ the set of non-negative real numbers. If, for any $X \in \mathfrak{M}$, $L(X)$ is defined as the distance of X from some fixed point in $\mathfrak{M}$ (the unit of measurement having been chosen), then L is a mapping of $\mathfrak{M}$ into $\mathfrak{M}^*$.

Our main concern is, however, with linear manifolds and we shall throughout this chapter use the symbols $\mathfrak{M}$, $\mathfrak{M}^*$ to denote two linear manifolds, of finite non-zero dimensionality, over the same reference field $\mathfrak{F}$. Until the contrary is stated we shall also write $d(\mathfrak{M}) = m$, $d(\mathfrak{M}^*) = n$. We propose to study a particularly simple and interesting class of transformations of one linear manifold into another.

DEFINITION 4.2.2. *Let L be a mapping of $\mathfrak{M}$ into $\mathfrak{M}^*$ and suppose that, for all X, $Y \in \mathfrak{M}$ and all $\alpha \in \mathfrak{F}$,*

$$L(\alpha X) = \alpha L(X), \qquad L(X+Y) = L(X)+L(Y). \tag{4.2.1}$$

Then L is said to be LINEAR.

Similarly we speak of *linear transformations*, or *linear operations*, or *linear operators*, and we say that $\mathfrak{M}$ is mapped linearly into $\mathfrak{M}^*$ by L.

The requirements (4.2.1) are referred to as the conditions of linearity, and an immediate consequence of them is that, for all $X_1,..., X_k \in \mathfrak{M}$ and all $\alpha_1,..., \alpha_k \in \mathfrak{F}$,

$$L(\alpha_1 X_1+...+\alpha_k X_k) = \alpha_1 L(X_1)+...+\alpha_k L(X_k). \tag{4.2.2}$$

EXERCISE 4.2.1. Let $\mathfrak{N}_{m,n}$ denote the linear manifold of $m \times n$ matrices. Show that transposition is a linear mapping of $\mathfrak{N}_{m,n}$ into $\mathfrak{N}_{n,m}$.

EXERCISE 4.2.2. Let L be a linear mapping of $\mathfrak{M}$ into $\mathfrak{M}^*$ and denote by Θ, Θ^* the zero elements of $\mathfrak{M}$, $\mathfrak{M}^*$ respectively. Show that $L(\Theta) = \Theta^*$.

It may be recalled that we have already encountered one important instance of a linear mapping. If $\mathbf{A}$ is a given $n \times m$ matrix and $\mathbf{x}$ is any vector of order m, then the function L, specified by the equation

$$L(\mathbf{x}) = \mathbf{A}\mathbf{x}, \tag{4.2.3}$$

is a linear mapping of $\mathfrak{V}_m$ into $\mathfrak{V}_n$. This fact can, of course, be verified immediately.

We know (by Theorem 2.4.3, p. 59) that every linear manifold of finite dimensionality can be represented by a total vector space. Equation (4.2.3) therefore suggests that it may be possible to represent every linear mapping of $\mathfrak{M}$ into $\mathfrak{M}^*$ by means of a suitable matrix. This is, indeed, the case as is shown by the next result.†

THEOREM 4.2.1. *Let L be a linear mapping of $\mathfrak{M}$ into $\mathfrak{M}^*$ and let $\mathfrak{B}$, $\mathfrak{B}^*$ be any bases of $\mathfrak{M}$, $\mathfrak{M}^*$ respectively. If, for any $X \in \mathfrak{M}$, we write*

$$\mathbf{x} = \mathscr{R}(X;\, \mathfrak{B}), \qquad \mathbf{x}^* = \mathscr{R}(L(X);\, \mathfrak{B}^*),$$

then there exists a unique $n \times m$ matrix $\mathbf{A}$, independent of X, such that

$$\mathbf{x}^* = \mathbf{A}\mathbf{x}. \tag{4.2.4}$$

Let $\mathfrak{B} = \{E_1,\ldots, E_m\}$, $\mathfrak{B}^* = \{E_1^*,\ldots, E_n^*\}$. Each of $L(E_1),\ldots, L(E_m)$ is an element of $\mathfrak{M}^*$ and so is expressible as a linear combination of $E_1^*,\ldots, E_n^*$, say

$$L(E_j) = \sum_{i=1}^{n} a_{ij}\, E_i^* \quad (j = 1,\ldots, m). \tag{4.2.5}$$

If $\mathbf{x} = (x_1,\ldots, x_m)^T$, $\mathbf{x}^* = (x_1^*,\ldots, x_n^*)^T$, then, by hypothesis,

$$X = \sum_{j=1}^{m} x_j\, E_j, \qquad L(X) = \sum_{i=1}^{n} x_i^*\, E_i^*.$$

Hence, using (4.2.2) and (4.2.5), we obtain

$$L(X) = L\Big\{\sum_{j=1}^{m} x_j\, E_j\Big\} = \sum_{j=1}^{m} x_j\, L(E_j)$$
$$= \sum_{j=1}^{m} x_j \sum_{i=1}^{n} a_{ij}\, E_i^* = \sum_{i=1}^{n} E_i^* \sum_{j=1}^{m} a_{ij}\, x_j.$$

But the expression for $L(X)$ as a linear combination of $E_1^*,\ldots, E_n^*$ is unique; therefore

$$x_i^* = \sum_{j=1}^{m} a_{ij}\, x_j \quad (i = 1,\ldots, n),$$

and this is equivalent to (4.2.4) with $\mathbf{A} = (a_{ij})$, The uniqueness of $\mathbf{A}$ is proved by the same argument as that used in the proof of Theorem 4.1.2 (i).

The theorem just proved shows that every linear operator can be represented in terms of a matrix multiplication. The representation

† Throughout the discussion below we make use of properties of matrices to deduce properties of linear operators. For a treatment of linear algebra in which the study of operators precedes that of matrices see, for example, Halmos, **20**, and Lichnerowicz, **25**.

is not, of course, uniquely determined since it depends on the arbitrary choice of bases in $\mathfrak{M}$ and $\mathfrak{M}^*$.

DEFINITION 4.2.3. *The matrix* $\mathbf{A}$ *in* (4.2.4) *is said to* REPRESENT *the linear operator* L *with respect to the bases* $\mathfrak{B}$, $\mathfrak{B}^*$ *of* $\mathfrak{M}$, $\mathfrak{M}^*$. *In symbols*:

$$\mathbf{A} = \mathscr{R}(L;\, \mathfrak{B},\, \mathfrak{B}^*).$$

Suppose now that we take new bases in $\mathfrak{M}$ and $\mathfrak{M}^*$. With respect to these bases the linear operator L will be represented by a new matrix; and we see, in fact, that L possesses an infinity of matrix representations corresponding to the variable choice of bases in $\mathfrak{M}$ and $\mathfrak{M}^*$. We are therefore led to consider what relations exist between different matrices representing the same operator.

THEOREM 4.2.2. *Let* L *be a linear mapping of* $\mathfrak{M}$ *into* $\mathfrak{M}^*$, *and let* $\mathfrak{B}$, $\tilde{\mathfrak{B}}$ *be any bases in* $\mathfrak{M}$ *and* $\mathfrak{B}^*$, $\tilde{\mathfrak{B}}^*$ *any bases in* $\mathfrak{M}^*$. *If*

$$\mathbf{A} = \mathscr{R}(L;\, \mathfrak{B},\, \mathfrak{B}^*), \qquad \tilde{\mathbf{A}} = \mathscr{R}(L;\, \tilde{\mathfrak{B}},\, \tilde{\mathfrak{B}}^*),$$

then there exist non-singular matrices $\mathbf{P}$, $\mathbf{Q}$, *of order* m, n *respectively, such that*

$$\tilde{\mathbf{A}} = \mathbf{QAP}^{-1}. \tag{4.2.6}$$

The proof is almost immediate. If X is any element in $\mathfrak{M}$, write

$$\mathbf{x} = \mathscr{R}(X;\, \mathfrak{B}), \qquad \tilde{\mathbf{x}} = \mathscr{R}(X;\, \tilde{\mathfrak{B}}),$$
$$\mathbf{x}^* = \mathscr{R}(L(X);\, \mathfrak{B}^*), \quad \tilde{\mathbf{x}}^* = \mathscr{R}(L(X);\, \tilde{\mathfrak{B}}^*).$$

Then, by hypothesis,

$$\mathbf{x}^* = \mathbf{Ax}, \qquad \tilde{\mathbf{x}}^* = \tilde{\mathbf{A}}\tilde{\mathbf{x}}.$$

Moreover, by Theorem 4.1.2 (i), there exist non-singular matrices $\mathbf{P}$, $\mathbf{Q}$, of order m, n respectively, such that

$$\tilde{\mathbf{x}} = \mathbf{Px}, \qquad \tilde{\mathbf{x}}^* = \mathbf{Qx}^*.$$

Hence
$$\tilde{\mathbf{x}}^* = \mathbf{Qx}^* = \mathbf{QAx} = \mathbf{QAP}^{-1}\tilde{\mathbf{x}},$$

and (4.2.6) follows by virtue of the uniqueness of $\tilde{\mathbf{A}}$.

4.2.2. The situation described by Theorem 4.2.2 becomes even more interesting if we take $\mathfrak{M}$ and $\mathfrak{M}^*$ as identical linear manifolds. In that case both the variable X and the functional value $L(X)$ are elements of $\mathfrak{M}$, and we speak now of a linear mapping, or linear transformation, of $\mathfrak{M}$ *into itself*.† The mapping $L(X) = X$ (for all $X \in \mathfrak{M}$) is called the *identical transformation* of $\mathfrak{M}$.

† It is important to stress that if L is a linear mapping of $\mathfrak{M}$ into itself, then to each element $X \in \mathfrak{M}$ corresponds a unique element $L(X) \in \mathfrak{M}$. On the other hand, given $X' \in \mathfrak{M}$, there may be no $X \in \mathfrak{M}$ such that $L(X) = X'$; or there may possibly be more than one X satisfying this relation.

EXERCISE 4.2.3. Show that the operator which transforms every element of a linear manifold $\mathfrak{M}$ into the zero element is a linear mapping of $\mathfrak{M}$ into itself.

By specializing Theorems 4.2.1 and 4.2.2 we could obtain a number of results about transformations of $\mathfrak{M}$ into itself. But it is essential that the reader should become thoroughly familiar with the idea of representation of operators by matrices and we prefer, therefore, to give an independent derivation of the theorems relating to linear transformations of a linear manifold into itself. Throughout the remainder of the present secton we shall write $d(\mathfrak{M}) = n$.

Theorem 4.2.3. *Let L be a linear transformation of $\mathfrak{M}$ into itself, and let $\mathfrak{B}$ be any basis of $\mathfrak{M}$. If, for any $X \in \mathfrak{M}$, we write*

$$\mathbf{x} = \mathscr{R}(X;\, \mathfrak{B}), \qquad \mathbf{x}' = \mathscr{R}(L(X);\, \mathfrak{B}),$$

then there exists a unique $n \times n$ matrix $\mathbf{A}$, independent of X, such that

$$\mathbf{x}' = \mathbf{A}\mathbf{x}. \tag{4.2.7}$$

Let $\mathfrak{B} = \{E_1,..., E_n\}$ and write

$$L(E_j) = \sum_{i=1}^{n} a_{ij} E_i \qquad (j = 1,..., n), \tag{4.2.8}$$

$$\mathbf{x} = (x_1,..., x_n)^T, \qquad \mathbf{x}' = (x_1',..., x_n')^T.$$

Then
$$L(X) = L\Big\{\sum_{j=1}^{n} x_j E_j\Big\} = \sum_{j=1}^{n} x_j L(E_j)$$

$$= \sum_{j=1}^{n} x_j \sum_{i=1}^{n} a_{ij} E_i = \sum_{i=1}^{n} E_i \sum_{j=1}^{n} a_{ij} x_j,$$

so that
$$x_i' = \sum_{j=1}^{n} a_{ij} x_j \qquad (i = 1,..., n).$$

This is, in fact, (4.2.7) with $\mathbf{A} = (a_{ij})$; and the uniqueness of $\mathbf{A}$ follows in the usual manner.

Theorem 4.2.3 shows, in particular, that a linear transformation of $\mathfrak{M}$ into itself can be represented by a linear transformation of $\mathfrak{B}_n$ into itself.

We may also remark that the matrix equation $\mathbf{x}' = \mathbf{A}\mathbf{x}$, where $\mathbf{A}$ is non-singular, can be interpreted in two different ways. On the one hand, we may think of $\mathfrak{M}$ as undergoing a transformation which changes the element X, represented by $\mathbf{x}$, into the element $L(X)$, represented by $\mathbf{A}\mathbf{x}$. On the other hand, we can consider the effect produced by a change of basis in $\mathfrak{M}$. In view of Theorem 4.1.2 (ii) (p. 112) we know that if $\mathbf{x}$ represents $X \in \mathfrak{M}$ with respect to $\mathfrak{B}$, then a second basis $\mathfrak{B}$ can be found such that $\mathbf{Ax}$ represents the same element X with respect to $\mathfrak{B}$. The transition from $\mathbf{x}$ to $\mathbf{Ax}$ can thus be regarded as resulting either from a linear transformation of $\mathfrak{M}$ or from a change of basis in $\mathfrak{M}$. The same twofold aspect of matrix transformations appears in projective geometry when we consider collineations and changes of the coordinate system.

The terminology and notation of Definition 4.2.3 can, of course, be suitably modified to meet our present requirements.

DEFINITION 4.2.4. *The matrix* $\mathbf{A}$ *in* (4.2.7) *is said to* REPRESENT *the linear transformation* L *with respect to the basis* $\mathfrak{B}$ *of* $\mathfrak{M}$. *In symbols*:

$$\mathbf{A} = \mathscr{R}(L;\, \mathfrak{B}).$$

It is easy to obtain a converse of Theorem 4.2.3. This shows, in particular, that, for a fixed choice of basis, there is a biunique correspondence between the linear transformations of $\mathfrak{M}$ into itself and the matrices representing them.

THEOREM 4.2.4. *Let* $\mathbf{A}$ *be an* $n \times n$ *matrix and* $\mathfrak{B}$ *a basis of* $\mathfrak{M}$. *Then there exists a unique linear transformation* L *of* $\mathfrak{M}$ *into itself which is represented by* $\mathbf{A}$ *with respect to* $\mathfrak{B}$.

Let $\mathbf{A} = (a_{ij})$ and $\mathfrak{B} = \{E_1,..., E_n\}$, and let the elements $L(E_j)$ of $\mathfrak{M}$ be defined by (4.2.8). Then, in view of the linearity of L, $L(X)$ is uniquely determined for every $X \in \mathfrak{M}$. Using now the equations appearing in the proof of Theorem 4.2.3 we can easily show that $\mathbf{A} = \mathscr{R}(L;\, \mathfrak{B})$.

Suppose now that we also have $\mathbf{A} = \mathscr{R}(M;\, \mathfrak{B})$, where M is some linear transformation of $\mathfrak{M}$ into itself. Then, if $X \in \mathfrak{M}$ and $\mathbf{x} = \mathscr{R}(X;\, \mathfrak{B})$,

$$\mathscr{R}(L(X);\, \mathfrak{B}) = \mathbf{Ax} = \mathscr{R}(M(X);\, \mathfrak{B}).$$

Hence $L(X) = M(X)$ for all $X \in \mathfrak{M}$, i.e. $L = M$ as required.

EXERCISE 4.2.4. Write out the details of the proof sketched above.

EXERCISE 4.2.5. Let L_1, L_2 be linear transformations of $\mathfrak{M}$ into itself, and let the transformations αL_1, $L_1 + L_2$, $L_2 L_1$ be defined respectively by the formulae

$$(\alpha L_1)(X) = \alpha L_1(X); \qquad (L_1 + L_2)(X) = L_1(X) + L_2(X);$$
$$(L_2 L_1)(X) = L_2\{L_1(X)\} \qquad (X \in \mathfrak{M}).$$

Show that αL_1, $L_1 + L_2$, $L_2 L_1$ are all linear.

If L_1, L_2 are represented, with respect to a basis $\mathfrak{B}$, by the matrices $\mathbf{A}_1$, $\mathbf{A}_2$ respectively, show that αL_1, $L_1 + L_2$, $L_2 L_1$ are represented by $\alpha \mathbf{A}_1$, $\mathbf{A}_1 + \mathbf{A}_2$, $\mathbf{A}_2 \mathbf{A}_1$ respectively.

In comparing matrix representations of a linear operator with respect to different bases we encounter a type of relation between matrices which will play a prominent part in our subsequent discussion.

DEFINITION 4.2.5. *If* **A** *and* **B** *are square matrices, and*

$$\mathbf{B} = \mathbf{S}^{-1}\mathbf{A}\mathbf{S}, \tag{4.2.9}$$

where **S** *is some non-singular matrix, then* **B** *is said to be obtained from* **A** *by a* SIMILARITY TRANSFORMATION, *in which* **S** *is the* TRANSFORMING MATRIX. *Furthermore,* **B** *is said to be* SIMILAR *to* **A**.

Since (4.2.9) may be rewritten in the form $\mathbf{A} = (\mathbf{S}^{-1})^{-1}\mathbf{B}\mathbf{S}^{-1}$, it follows that if **B** is similar to **A**, then **A** is similar to **B**. There is, therefore, no ambiguity in speaking simply of two matrices as being similar.

The next theorem states, roughly speaking, that two matrices are similar if and only if they represent the same linear transformation of $\mathfrak{M}$ into itself.

Theorem 4.2.5. (Representation theorem for linear transformations)

(i) *Let L be a linear transformation of $\mathfrak{M}$ into itself. If* $\mathbf{A}$, $\tilde{\mathbf{A}}$ *are the $n \times n$ matrices representing L with respect to the bases $\mathfrak{B}$, $\tilde{\mathfrak{B}}$ of $\mathfrak{M}$ respectively, then* $\mathbf{A}$, $\tilde{\mathbf{A}}$ *are similar.*

(ii) *Let* $\mathbf{A}$, $\tilde{\mathbf{A}}$ *be similar $n \times n$ matrices, and let $\mathfrak{B}$ be any basis of $\mathfrak{M}$. Then a second basis $\tilde{\mathfrak{B}}$ and a linear transformation L of $\mathfrak{M}$ into itself can be found such that*

$$\mathbf{A} = \mathscr{R}(L; \mathfrak{B}), \qquad \tilde{\mathbf{A}} = \mathscr{R}(L; \tilde{\mathfrak{B}}).$$

To prove (i), let $X \in \mathfrak{M}$ and write

$$\mathbf{x} = \mathscr{R}(X; \mathfrak{B}), \qquad \tilde{\mathbf{x}} = \mathscr{R}(X; \tilde{\mathfrak{B}}),$$
$$\mathbf{x}' = \mathscr{R}(L(X); \mathfrak{B}), \qquad \tilde{\mathbf{x}}' = \mathscr{R}(L(X); \tilde{\mathfrak{B}}).$$

By hypothesis we have

$$\mathbf{x}' = \mathbf{A}\mathbf{x}, \qquad \tilde{\mathbf{x}}' = \tilde{\mathbf{A}}\tilde{\mathbf{x}}.$$

Furthermore, by Theorem 4.1.2 (i) (p. 112), there exists a non-singular matrix **P** such that

$$\tilde{\mathbf{x}} = \mathbf{P}\mathbf{x}, \qquad \tilde{\mathbf{x}}' = \mathbf{P}\mathbf{x}'.$$

From these relations we infer that

$$\tilde{\mathbf{x}}' = \mathbf{P}\mathbf{x}' = \mathbf{P}\mathbf{A}\mathbf{x} = \mathbf{P}\mathbf{A}\mathbf{P}^{-1}\tilde{\mathbf{x}},$$

and it follows that
$$\tilde{\mathbf{A}} = \mathbf{P}\mathbf{A}\mathbf{P}^{-1}; \qquad (4.2.10)$$
thus $\mathbf{A}$, $\tilde{\mathbf{A}}$ are similar.

Assume, next, that $\mathbf{A}$, $\tilde{\mathbf{A}}$ are given similar matrices, so that (4.2.10) is satisfied for some matrix $\mathbf{P}$. If $\mathfrak{B}$ is a given basis, then, by Theorem 4.2.4, there exists a linear transformation L of $\mathfrak{M}$ into itself such that $\mathbf{A} = \mathscr{R}(L;\, \mathfrak{B})$. This means, of course, that if $X \in \mathfrak{M}$ and
$$\mathbf{x} = \mathscr{R}(X;\, \mathfrak{B}), \qquad \mathbf{x}' = \mathscr{R}(L(X);\, \mathfrak{B}),$$
then
$$\mathbf{x}' = \mathbf{A}\mathbf{x}.$$
Again, by Theorem 4.1.2 (ii) (p. 112), $\mathfrak{M}$ possesses a second basis $\tilde{\mathfrak{B}}$ such that, if
$$\tilde{\mathbf{x}} = \mathscr{R}(X;\, \tilde{\mathfrak{B}}), \qquad \tilde{\mathbf{x}}' = \mathscr{R}(L(X);\, \tilde{\mathfrak{B}}),$$
then
$$\tilde{\mathbf{x}} = \mathbf{P}\mathbf{x}, \qquad \tilde{\mathbf{x}}' = \mathbf{P}\mathbf{x}'.$$
Hence, in view of (4.2.10), $\tilde{\mathbf{x}}' = \tilde{\mathbf{A}}\tilde{\mathbf{x}}$, i.e. $\tilde{\mathbf{A}} = \mathscr{R}(L;\, \tilde{\mathfrak{B}})$. The proof is therefore complete.

The preceding theorems show that in dealing with linear transformations of a linear manifold into itself we encounter a situation of a type with which we are already familiar from the study of representations of a linear manifold by means of vector spaces.† A linear transformation of $\mathfrak{M}$ into itself exists quite independently of any representation; but it possesses, at the same time, infinitely many different representations (in terms of matrices) none of which has special precedence over the others. Our choice of representation (that is, essentially our choice of basis in $\mathfrak{M}$) is therefore governed in any particular problem solely by considerations of convenience.

4.2.3. Since in the problems that actually arise we deal more often with vector spaces than with linear manifolds, it is useful to reformulate some of the results obtained in § 4.2.2.

We note, in the first place, that if L is a linear transformation of $\mathfrak{B}_n$ into itself and $\mathfrak{X} = \{\mathbf{x}_1,\ldots,\mathbf{x}_n\}$ is a basis of $\mathfrak{B}_n$, then there exists a (unique) $n \times n$ matrix $\mathbf{A}$ such that
$$\mathbf{A} = \mathscr{R}(L;\, \mathfrak{X}). \qquad (4.2.11)$$
This means, in fact, that if
$$\mathbf{x} = \alpha_1\mathbf{x}_1+\ldots+\alpha_n\mathbf{x}_n, \qquad L(\mathbf{x}) = \beta_1\mathbf{x}_1+\ldots+\beta_n\mathbf{x}_n,$$
then
$$(\beta_1,\ldots,\beta_n)^T = \mathbf{A}(\alpha_1,\ldots,\alpha_n)^T. \qquad (4.2.12)$$

† See § 2.4.

A more convenient way of characterizing the representing matrix $\mathbf{A}$ follows from (4.2.8) (p. 117). This set of equations shows that if

$$L(\mathbf{x}_j) = \sum_{i=1}^{n} a_{ij}\,\mathbf{x}_i \qquad (j = 1,\ldots, n), \tag{4.2.13}$$

then $\mathbf{A} = (a_{ij})$.

In the particular case when $\mathfrak{X}$ is taken as $\mathfrak{E} = \{\mathbf{e}_1,\ldots, \mathbf{e}_n\}$ (4.2.11) means, then, that

$$L(\mathbf{e}_j) = \sum_{i=1}^{n} a_{ij}\,\mathbf{e}_i = \mathbf{A}_{*j} \qquad (j = 1,\ldots, n).$$

Writing $\mathbf{x} = (x_1,\ldots, x_n)^T$ and using Theorem 3.3.6 (p. 84) we see that

$$L(\mathbf{x}) = L(x_1\,\mathbf{e}_1 + \ldots + x_n\,\mathbf{e}_n) = \sum_{j=1}^{n} x_j\,L(\mathbf{e}_j) = \sum_{j=1}^{n} x_j\,\mathbf{A}_{*j} = \mathbf{A}\mathbf{x}.$$

We have, therefore, the following result:

THEOREM 4.2.6. *If L is a linear mapping of $\mathfrak{B}_n$ into itself, and $\mathbf{A} = \mathscr{R}(L; \mathfrak{E})$, then, for all $\mathbf{x} \in \mathfrak{B}_n$, $L(\mathbf{x}) = \mathbf{A}\mathbf{x}$.*

EXERCISE 4.2.6. Deduce Theorem 4.2.6 from (4.2.12).

THEOREM 4.2.7. *Let L be a linear mapping of $\mathfrak{B}_n$ into itself. Let $\mathfrak{X} = \{\mathbf{x}_1,\ldots, \mathbf{x}_n\}$ and $\tilde{\mathfrak{X}} = \{\tilde{\mathbf{x}}_1,\ldots, \tilde{\mathbf{x}}_n\}$ be any two bases in $\mathfrak{B}_n$, and suppose that*

$$\tilde{\mathbf{x}}_j = \sum_{i=1}^{n} p_{ij}\,\mathbf{x}_i \qquad (j = 1,\ldots, n).$$

If
$$\mathbf{A} = \mathscr{R}(L; \mathfrak{X}), \qquad \tilde{\mathbf{A}} = \mathscr{R}(L; \tilde{\mathfrak{X}}),$$

then
$$\tilde{\mathbf{A}} = \mathbf{P}^{-1}\mathbf{A}\mathbf{P},$$

where $\mathbf{P} = (p_{ij})$.†

This result follows immediately from Theorem 4.2.5 (i) and the construction of $\mathbf{P}$ in the proof of Theorem 4.1.2 (i), when an obvious change of notation has been made.

We may specialize Theorem 4.2.7 still further by taking $\mathfrak{E}$ to be one of the given bases.

Theorem 4.2.8. *Let $\mathfrak{X} = \{\mathbf{x}_1,\ldots, \mathbf{x}_n\}$ be a basis of $\mathfrak{B}_n$ and let $\mathbf{X}$ be the $n \times n$ matrix having $\mathbf{x}_1,\ldots,\mathbf{x}_n$ (in that order) as its columns. If the $n \times n$ matrix $\mathbf{A}$ represents the linear transformation L of $\mathfrak{B}_n$ into itself with respect to $\mathfrak{E}$, then $\mathbf{X}^{-1}\mathbf{A}\mathbf{X}$ represents L with respect to $\mathfrak{X}$.*

† The matrix $\mathbf{P}$ is non-singular by virtue of Theorem 4.1.1.

We apply Theorem 4.2.7 with $\mathfrak{E}$, $\mathfrak{X}$ in place of $\mathfrak{X}$, $\tilde{\mathfrak{X}}$ respectively. This shows that if

$$\mathbf{x}_j = \sum_{i=1}^{n} p_{ij}\,\mathbf{e}_i \qquad (j = 1,..., n), \tag{4.2.14}$$

then $\mathscr{R}(L;\, \mathfrak{X}) = \mathbf{P}^{-1}\mathbf{A}\mathbf{P}$, where $\mathbf{P} = (p_{ij})$. But (4.2.14) can be written as

$$\mathbf{x}_j = (p_{1j},..., p_{nj})^T \qquad (j = 1,..., n).$$

Hence

$$\mathbf{X} = \begin{pmatrix} p_{11} & \cdot & \cdot & \cdot & p_{1n} \\ \cdot & \cdot & \cdot & \cdot & \cdot \\ p_{n1} & \cdot & \cdot & \cdot & p_{nn} \end{pmatrix} = \mathbf{P},$$

and the assertion follows.

4.2.4. Now that the relations between matrices and linear operators have been discussed at some length, it becomes clear that the properties a square matrix may possess are of two distinct kinds, which might be termed 'invariant' and 'non-invariant'—an 'invariant' property being one that is shared by all matrices representing (with respect to suitable bases) the same linear operator as the given matrix. In view of Theorem 4.2.5 this simply means that a property is invariant if it is possessed by an entire class of similar matrices. It is, for instance, obvious that similar matrices are either all singular or all non-singular, and singularity is thus seen to be an invariant property. On the other hand, the symmetry of a matrix $\mathbf{A}$, i.e. the property $\mathbf{A}^T = \mathbf{A}$, is clearly non-invariant.

Invariant properties of matrices are fundamental since they express intrinsic characteristics of the underlying linear operators, whereas non-invariant properties depend on the arbitrary choice of a basis in a linear manifold and so are, as it were, accidental. To determine the invariant features of a given matrix $\mathbf{A}$ (that is, to determine the intrinsic characteristics of the operator represented by $\mathbf{A}$) we endeavour to find and then to examine some matrix $\mathbf{\Lambda}$, similar to $\mathbf{A}$, and having at the same time as simple a form as possible. This procedure leads to the discussion of canonical forms which will be undertaken in Chapter X.

In projective geometry, too, similarity transformations play a prominent part. Projective geometry is the study of 'projective space'. Although this space is not a linear manifold, its properties can be made to depend in a simple way on the properties of such manifolds. For example, projective space can be represented by means of systems of 'projective coordinates'; each such system is derived from a basis of a vector space, and the change from one system to another can be described in terms of matrices. In discussing projective geometry we have often to deal with *collineations*, i.e. with linear transformations of projective space into itself. These transformations have properties closely resembling those of linear transformations of a linear manifold; in particular, they can be represented by matrices, and two matrices represent the same collineation with respect to two systems of coordinates if and only if each is similar to a scalar multiple of the other. It follows that in investigating the nature of a given collineation it is advantageous to choose the system of coordinates in such a way that the representing matrix assumes as simple a form as possible; and this amounts to finding a matrix having a simple form and similar to the given matrix. Thus, no matter whether our interest lies primarily in pure matrix theory or in geometry, we are led naturally to the recognition of the importance of similarity transformations and to the systematic study of these transformations.†

4.3. Isomorphisms and automorphisms of linear manifolds

The reader has probably already noticed the close resemblance between the notion of an isomorphism (as specified in Definition 2.4.1, p. 58) and that of a linear transformation. An isomorphism between two linear manifolds $\mathfrak{M}$ and $\mathfrak{M}^*$ is obviously a linear transformation of $\mathfrak{M}$ into $\mathfrak{M}^*$ (and equally of $\mathfrak{M}^*$ into $\mathfrak{M}$). On the other hand, a linear transformation of $\mathfrak{M}$ into $\mathfrak{M}^*$ need not be an isomorphism, as may be seen, for example, by considering the linear transformation which maps every element of $\mathfrak{M}$ into the zero element of $\mathfrak{M}^*$. If, however, in addition to the requirements imposed on $\mathfrak{M}$, $\mathfrak{M}^*$, and L in Definition 4.2.2 (p. 114) we also assume that to every element X' of $\mathfrak{M}^*$ there corresponds one and only one element X of $\mathfrak{M}$ such that $L(X) = X'$, then the linear transformation L of $\mathfrak{M}$ into $\mathfrak{M}^*$ becomes an isomorphism, and $\mathfrak{M}$, $\mathfrak{M}^*$ are isomorphic.

A somewhat weaker requirement is specified in the next definition.

† For a discussion of projective geometry based on the methods of linear algebra the reader may be referred to Semple and Kneebone, *Algebraic Projective Geometry*, or, for a more comprehensive treatment, to Todd, *Projective and Analytic Geometry*.

DEFINITION 4.3.1. *A linear transformation (or linear mapping) of* $\mathfrak{M}$ ONTO $\mathfrak{M}^*$ *is a linear transformation* L, *of* $\mathfrak{M}$ *into* $\mathfrak{M}^*$, *having the additional property that, corresponding to every* $X' \in \mathfrak{M}^*$, *there exists* AT LEAST *one* $X \in \mathfrak{M}$ *which satisfies the equation* $L(X) = X'$.†

In particular, we may speak of a linear transformation of $\mathfrak{M}$ *onto itself*.

We note at once that an isomorphism between $\mathfrak{M}$ and $\mathfrak{M}^*$ is a mapping of $\mathfrak{M}$ onto $\mathfrak{M}^*$. The converse need not be the case. Thus, take $\mathfrak{M} = \mathfrak{V}_3$, $\mathfrak{M}^* = \mathfrak{V}_2$ and define L as the transformation which associates with every vector (x_1, x_2, x_3) of $\mathfrak{V}_3$ the vector (x_1, x_2) of $\mathfrak{V}_2$. It is easily verified that L is a linear mapping of $\mathfrak{V}_3$ onto $\mathfrak{V}_2$, but it is obviously not an isomorphism since it does not set up a biunique correspondence between the vectors of $\mathfrak{V}_3$ and those of $\mathfrak{V}_2$. If, however, the dimensionalities of $\mathfrak{M}$ and $\mathfrak{M}^*$ are equal, then a linear mapping of $\mathfrak{M}$ onto $\mathfrak{M}^*$ and an isomorphism between $\mathfrak{M}$ and $\mathfrak{M}^*$ are equivalent concepts. This is shown by the next theorem.

THEOREM 4.3.1. *If* $\mathfrak{M}$, $\mathfrak{M}^*$ *are linear manifolds having the same (finite) dimensionality and if* L *is a linear transformation of* $\mathfrak{M}$ *onto* $\mathfrak{M}^*$, *then* L *is an isomorphism.*

Let $d(\mathfrak{M}) = d(\mathfrak{M}^*) = n$ and let $\{E_1,..., E_n\}$ be any basis of $\mathfrak{M}$. If X' is any element of $\mathfrak{M}^*$ then there exists, by hypothesis, at least one element X in $\mathfrak{M}$ such that $X' = L(X)$. Writing

$$X = \alpha_1 E_1 + ... + \alpha_n E_n$$

we therefore obtain

$$X' = \alpha_1 L(E_1) + ... + \alpha_n L(E_n).$$

Thus $L(E_1),..., L(E_n)$ are generators of $\mathfrak{M}^*$ and, since $d(\mathfrak{M}^*) = n$, it follows that these elements constitute a basis and so are linearly independent.

Now let $Y' \in \mathfrak{M}^*$ and suppose that $Y_1, Y_2 \in \mathfrak{M}$ and $L(Y_1) = Y'$, $L(Y_2) = Y'$. Then $L(Y_1 - Y_2) = \Theta^*$, where Θ^* denotes the zero element of $\mathfrak{M}^*$. Writing

$$Y_1 - Y_2 = \beta_1 E_1 + ... + \beta_n E_n$$

we obtain $$\beta_1 L(E_1) + ... + \beta_n L(E_n) = \Theta^*,$$

and since $L(E_1),..., L(E_n)$ are linearly independent this implies that $\beta_1 = ... = \beta_n = 0$. Hence $Y_1 = Y_2$, and thus corresponding to any element $Y' \in \mathfrak{M}^*$ there exists precisely one element $Y \in \mathfrak{M}$ such that $L(Y) = Y'$. The mapping L is therefore an isomorphism.

† Thus 'mapping onto' is a special case of 'mapping into'.

We are particularly interested in the case when $\mathfrak{M} = \mathfrak{M}^*$.

DEFINITION 4.3.2. *An* AUTOMORPHISM *of a linear manifold* $\mathfrak{M}$ *is an isomorphism of* $\mathfrak{M}$ *with itself.*

The mapping $L(X) = X$ (for all $X \in \mathfrak{M}$) is called the *identical automorphism* of $\mathfrak{M}$.

EXERCISE 4.3.1. Show that the biunique correspondence between the elements of $\mathfrak{B}_2$, specified by the scheme

$$(x_1, x_2) \leftrightarrow (x_1+x_2, x_1-2x_2),$$

is an automorphism of $\mathfrak{B}_2$.

As a special case of Theorem 4.3.1 we have the following result.

THEOREM 4.3.2. *A linear transformation of a finite-dimensional linear manifold onto itself is an automorphism.*

A criterion for deciding whether a linear transformation L of $\mathfrak{M}$ *into* itself is an automorphism can be given in terms of matrix representations of L. We need to observe in the first place that the (infinitely many) matrices representing L are either all singular or all non-singular. For any two such matrices $\mathbf{A}$, $\tilde{\mathbf{A}}$ are similar by Theorem 4.2.5, i.e. they are connected by a relation of the form $\tilde{\mathbf{A}} = \mathbf{PAP}^{-1}$; and hence, by Theorems 3.5.1 (p. 87) and 3.6.3 (i) (p. 93), $|\tilde{\mathbf{A}}| = |\mathbf{A}|$.

THEOREM 4.3.3. *A linear transformation* L *of a finite-dimensional linear manifold* $\mathfrak{M}$ *into itself is an automorphism if and only if the matrices representing* L *are non-singular.*

Let $\mathfrak{B} = \{E_1,..., E_n\}$ by any basis of $\mathfrak{M}$, and write

$$L(E_j) = \sum_{i=1}^{n} a_{ij} E_i \qquad (j = 1,..., n).$$

Then $\mathbf{A} = (a_{ij}) = \mathscr{R}(L; \mathfrak{B})$. Suppose first that $|\mathbf{A}| \neq 0$. Let X' be any given element of $\mathfrak{M}$ and write $\mathbf{x}' = \mathscr{R}(X'; \mathfrak{B})$. Let $X \in \mathfrak{M}$ be defined by the relation $\mathbf{A}^{-1}\mathbf{x}' = \mathscr{R}(X; \mathfrak{B})$. Then

$$\mathscr{R}(X'; \mathfrak{B}) = \mathbf{A} . \mathscr{R}(X; \mathfrak{B}),$$

and it follows from Theorem 4.2.3 (p. 117) that $L(X) = X'$. Thus L is a linear mapping of $\mathfrak{M}$ onto itself; and so, by Theorem 4.3.2, it is an automorphism.

If, on the other hand, $|\mathbf{A}| = 0$, then $L(E_1),..., L(E_n)$ are linearly dependent by virtue of Theorem 4.1.1, i.e. there exist scalars $\alpha_1,..., \alpha_n$, not all 0, such that

$$L(\alpha_1 E_1+...+\alpha_n E_n) = \alpha_1 L(E_1)+...+\alpha_n L(E_n) = \Theta,$$

where Θ is the zero element of $\mathfrak{M}$. Thus, for some $X \neq \Theta$, $L(X) = \Theta$. But clearly $L(\Theta) = \Theta$, and therefore L is not an automorphism of $\mathfrak{M}$. The proof is now complete.

We may summarize the results of Theorems 4.3.2 and 4.3.3 by saying that the following statements, relating to a linear mapping L of a finite-dimensional linear manifold of $\mathfrak{M}$ into itself, are equivalent.

(i) L is a mapping of $\mathfrak{M}$ onto itself;
(ii) L is an automorphism of $\mathfrak{M}$;
(iii) the matrices representing L are non-singular.

EXERCISE 4.3.2. Let $\mathfrak{M}$, $\mathfrak{M}^*$ be linear manifolds of the same finite dimensionality and let L be a linear mapping of $\mathfrak{M}$ into $\mathfrak{M}^*$. Show that the matrices representing L are either all singular or all non-singular.

EXERCISE 4.3.3. Show that if $\mathfrak{M}$, $\mathfrak{M}^*$ are linear manifolds of the same finite dimensionality and L is a linear mapping of $\mathfrak{M}$ into $\mathfrak{M}^*$, then the following statements are equivalent.

(i) L is a mapping of $\mathfrak{M}$ onto $\mathfrak{M}^*$;
(ii) L is an isomorphism between $\mathfrak{M}$ and $\mathfrak{M}^*$;
(iii) the matrices representing L are non-singular.

4.4. Further instances of linear operators

The idea of a linear operator introduced in Definition 4.2.2 pervades, in one form or another, a large part of mathematics. The operators we have encountered so far have all been associated with matrices, but there also exist linear operators of quite different types, and many of these are important in analysis and especially in the theories of differential and integral equations. The basic fact in this context is that differentiation and integration are linear operators, i.e.

$$\frac{d}{dx}\{\alpha f(x)+\beta g(x)\} = \alpha f'(x)+\beta g'(x);$$

$$\int_a^b \{\alpha f(x)+\beta g(x)\}\,dx = \alpha \int_a^b f(x)\,dx + \beta \int_a^b g(x)\,dx.$$

The linearity of all operators mentioned below is, as we shall see, an almost immediate consequence of this fact.

Consider, for example, the class $\mathfrak{C}$ of all real-valued functions of t and its subclass $\mathfrak{C}^{(n)}$ consisting of those functions which possess derivatives of the nth order. Multiplication by scalars and addition of elements in $\mathfrak{C}$ can be defined in the obvious way, and with respect to these operations $\mathfrak{C}$ and $\mathfrak{C}^{(n)}$ are, of course, linear manifolds (of infinite dimensionality). Let, now, Ω be the operator

$$\Omega = a_0 \frac{d^n}{dt^n}+a_1 \frac{d^{n-1}}{dt^{n-1}}+\ldots+a_{n-1}\frac{d}{dt}+a_n,$$

where $a_0, a_1,..., a_{n-1}, a_n$ are constants. If Ω operates upon a function $x = x(t)$ in $\mathfrak{C}^{(n)}$, the resulting function is denoted by $\Omega(x)$ or Ωx, and is defined by the equation

$$\Omega x = a_0 \frac{d^n x}{dt^n} + a_1 \frac{d^{n-1}x}{dt^{n-1}} + ... + a_{n-1} \frac{dx}{dt} + a_n x.$$

It is clear that, for every real number α and every pair of functions $x, y \in \mathfrak{C}^{(n)}$, we have

$$\Omega(\alpha x) = \alpha\Omega x, \qquad \Omega(x+y) = \Omega x + \Omega y.$$

Hence Ω is a linear transformation of $\mathfrak{C}^{(n)}$ into $\mathfrak{C}$, and it is precisely this fact which underlies the theory of the differential equation

$$\Omega x = 0,$$

since it implies that every linear combination of solutions of the equation is again a solution. The set of all solutions is, in fact, a linear manifold and, as we may recall, its dimensionality is n (provided that $a_0 \neq 0$); thus every solution may be expressed as a linear combination of n linearly independent solutions.

A somewhat similar situation arises with regard to numerous partial differential equations. We may, by way of illustration, consider the three-dimensional equation of heat conduction, namely,

$$\frac{\partial^2 u}{\partial x^2} + \frac{\partial^2 u}{\partial y^2} + \frac{\partial^2 u}{\partial z^2} = \kappa \frac{\partial u}{\partial t}. \tag{4.4.1}$$

If Ω denotes the operator

$$\frac{\partial^2}{\partial x^2} + \frac{\partial^2}{\partial y^2} + \frac{\partial^2}{\partial z^2} - \kappa \frac{\partial}{\partial t},$$

then this equation can be rewritten in the operational form $\Omega u = 0$. Now the functions of the four variables x, y, z, t which possess all the requisite partial derivatives form a linear manifold $\mathfrak{M}$, and it is at once clear that Ω is a linear operator in that manifold. Hence any linear combination of solutions of $\Omega u = 0$ is again a solution, and the method of treatment of the equation (4.4.1) depends essentially upon this fact. Exactly the same remarks apply to most of the other standard equations of mathematical physics, such as the wave equation, and Laplace's equation. Generally speaking, when the operator specifying an (ordinary or partial) differential equation is linear, the equation is far more likely to be tractable than when it is not.

We may also mention certain operators connected with integration. Let $\mathfrak{J}$ be the class of real-valued functions integrable in the range $a \leqslant t \leqslant b$. This class is obviously a linear manifold. The operator Ω defined by the relation

$$\Omega x = \int_a^b x(t)\, dt$$

is clearly a linear transformation of $\mathfrak{J}$ into the manifold of real numbers. A similar, but slightly more complicated linear operator Ω, defined for the linear manifold of functions integrable in $-\infty \leqslant t \leqslant \infty$, is given by the relation

$$\Omega x = \int_a^b K(t)x(u-t)\, dt, \tag{4.4.2}$$

where $K(t)$ is a given integrable function, known as the *kernel* of Ω. Here Ωx is a function of u, and so (when $K(t)$ satisfies suitable conditions) Ω is a linear transformation of the class of integrable functions into itself. Operators such as (4.4.2) occur in the theory of integral equations.

Many other 'integral transforms' involve linear operators. The *Laplace transform* of a function $x(t)$, for example, is defined as the function $\bar{x}(p)$ given by the formula

$$\bar{x}(p) = \int_0^\infty e^{-pt}x(t)\,dt,$$

and the operation of changing $x(t)$ into $\bar{x}(p)$ is obviously linear. The treatment of many types of differential equations depends precisely upon this fact.†

PROBLEMS ON CHAPTER IV

1. L is a linear transformation of a linear manifold $\mathfrak{M}$ into a linear manifold $\mathfrak{M}^*$. Show that the set of all $X \in \mathfrak{M}$ such that $L(X) = \Theta^*$, where Θ^* denotes the zero element of $\mathfrak{M}^*$, is a submanifold of $\mathfrak{M}$.

2. Show that every linear mapping of one linear manifold onto another preserves linear dependence but not necessarily linear independence. Show further that if the dimensionalities of the two linear manifolds are equal, then linear independence is also preserved.

3. L is a linear transformation of a linear manifold $\mathfrak{M}$ into itself; $\mathfrak{B} = \{E_1,..., E_n\}$ is a basis of $\mathfrak{M}$; $\nu(\mathfrak{B})$ is the maximum number of linearly independent elements among $L(E_1),..., L(E_n)$; and $\mathfrak{N}(\mathfrak{B})$ is the submanifold of $\mathfrak{M}$ spanned by $L(E_1),..., L(E_n)$. Show that both $\nu(\mathfrak{B})$ and $\mathfrak{N}(\mathfrak{B})$ are independent of the choice of $\mathfrak{B}$.

4. Let L, L' be linear transformations of a linear manifold $\mathfrak{M}$ into itself. Show that, if α, α' are any complex numbers and $\mathfrak{B}$ any basis of $\mathfrak{M}$, then

$$\mathscr{R}(\alpha L+\alpha' L'; \mathfrak{B}) = \alpha\mathscr{R}(L; \mathfrak{B})+\alpha'\mathscr{R}(L'; \mathfrak{B}).$$

5. L is a linear transformation of a linear manifold $\mathfrak{M}$ into a linear manifold $\mathfrak{M}^*$. Show that $\mathfrak{M}^*$ possesses a submanifold $\mathfrak{M}'$ such that L effects a linear transformation of $\mathfrak{M}$ *onto* $\mathfrak{M}'$.

6. Let L be a linear transformation of $\mathfrak{V}_3$ into itself which is represented, with respect to the basis $(-1, 1, 1)$, $(1, 0, -1)$, $(0, 1, 1)$, by the matrix

$$\begin{pmatrix} 1 & 0 & 1 \\ 1 & 1 & 0 \\ -1 & 2 & 1 \end{pmatrix}.$$

Find the matrix representing L with respect to $\mathfrak{E}$.

7. A linear mapping L of $\mathfrak{V}_3$ into itself transforms the vectors $(-1, 0, 2)$, $(0, 1, 1)$, $(3, -1, 0)$ into $(-5, 0, 3)$, $(0, -1, 6)$, $(-5, -1, 9)$ respectively. Find the matrix representing L with respect to (i) the basis consisting of the first three vectors given above; (ii) $\mathfrak{E}$.

† The reader who wishes to see how the algebraic idea of linearity can be systematically employed in analysis should consult Courant and Hilbert, *Methods of Mathematical Physics*.

8. The matrix $\begin{pmatrix} a & b \\ -b & a \end{pmatrix}$ represents the linear transformation L of the complex space $\mathfrak{V}_2$ with respect to the basis $\mathfrak{E}$. Find the matrix representing L with respect to the basis consisting of the vectors $(1, i)$, $(1, -i)$.

9. L is a linear transformation of $\mathfrak{V}_n$ into itself, $\mathfrak{B}$ is a basis of $\mathfrak{V}_n$, and $\mathfrak{B}'$ is another basis obtained from $\mathfrak{B}$ by rearrangement of the vectors in $\mathfrak{B}$. Determine the relation between $\mathscr{R}(L;\, \mathfrak{B})$ and $\mathscr{R}(L;\, \mathfrak{B}')$.

10. L is a linear transformation of the n-dimensional linear manifold $\mathfrak{M}$ into itself; $\mathfrak{M}_1$ is the submanifold consisting of elements of the form $L(X)$ $(X \in \mathfrak{M})$; and $\mathfrak{M}_2$ is the submanifold of elements $X \in \mathfrak{M}$ such that $L(X) = \Theta$, where Θ is the zero element of $\mathfrak{M}$. Prove that

$$d(\mathfrak{M}_1) + d(\mathfrak{M}_2) = n.$$

11. Verify the result of the preceding question for the case when $\mathfrak{M} = \mathfrak{V}_3$ and L is the matrix operator

$$\begin{pmatrix} 0 & 1 & 0 \\ 0 & 0 & 1 \\ 0 & 0 & 0 \end{pmatrix}.$$

12. Let $\mathfrak{P}_n$ be the set of all polynomials in t of degree $< n$. Show that, with the obvious definitions of addition and multiplication by scalars, $\mathfrak{P}_n$ is a linear manifold and that $\mathfrak{B} = \{1, t, t^2, \ldots, t^{n-1}\}$ is a basis of $\mathfrak{P}_n$. Show further that the operator D, defined by the relation $Dp(t) = p'(t)$, is linear and find $\mathscr{R}(D;\, \mathfrak{B})$.

13. Let $\mathfrak{U}$, $\mathfrak{U}'$ be complements in $\mathfrak{V}_n$, and let L, M be linear transformations of $\mathfrak{V}_n$ into itself such that $L(\mathbf{x}) = M(\mathbf{x})$ whenever $\mathbf{x} \in \mathfrak{U}$ or $\mathbf{x} \in \mathfrak{U}'$. Show that $L = M$.

14. Let $\mathfrak{M}, \mathfrak{M}^*$ be isomorphic linear manifolds, and let L be a linear mapping of $\mathfrak{M}$ into itself. Show that there exists a linear mapping L^* of $\mathfrak{M}^*$ into itself such that $L(X) \leftrightarrow L^*(X^*)$ whenever $X \leftrightarrow X^*$.

15. L is a linear mapping of an n-dimensional linear manifold $\mathfrak{M}$ into itself, and $X_1, \ldots, X_n$ are linearly independent elements of $\mathfrak{M}$. Prove that L is an automorphism of $\mathfrak{M}$ if and only if $L(X_1), \ldots, L(X_n)$ are linearly independent.

16. $\mathfrak{M}$ is an n-dimensional linear manifold over $\mathfrak{F}$, and X, X' are two distinct elements in $\mathfrak{M}$. Show that there exists a linear transformation f of $\mathfrak{M}$ into $\mathfrak{F}$ such that $f(X) \neq f(X')$.

17. $\mathfrak{M}$ is a linear manifold of (finite) dimensionality n, over a field $\mathfrak{F}$, and $\mathfrak{M}^*$ is the set of all linear transformations of $\mathfrak{M}$ into $\mathfrak{F}$. Show that, with suitable definitions of multiplication by scalars and addition, $\mathfrak{M}^*$ is also a linear manifold of dimensionality n.

18. $\mathfrak{M}$ is a linear manifold of dimensionality n. Show that the set of all linear transformations of $\mathfrak{M}$ into itself is a linear manifold of dimensionality n^2. Generalize this result for the case of linear transformations of $\mathfrak{M}$ into a second linear manifold $\mathfrak{M}'$.

19. Let L be a linear transformation of $\mathfrak{V}_n$ into itself and let

$$\mathfrak{B} = \{\mathbf{x}_1, \ldots, \mathbf{x}_n\}, \qquad \mathfrak{B}' = \{\mathbf{x}_1', \ldots, \mathbf{x}_n'\}$$

be bases of $\mathfrak{V}_n$ such that, for $k = 1,\ldots,n$, $\mathbf{x}_k'$ is a linear combination of $\mathbf{x}_1,\ldots,\mathbf{x}_k$. Show that there exists a triangular matrix $\mathbf{\Delta}$ such that

$$\mathbf{\Delta}\mathscr{R}(L;\mathfrak{B}') = \mathscr{R}(L;\mathfrak{B})\mathbf{\Delta}.$$

20. Let $\mathfrak{B} = \{\mathbf{x}_1,\ldots,\mathbf{x}_n\}$ be a basis of $\mathfrak{V}_n$, and suppose that $\mathbf{A}$ is an $n\times n$ matrix such that $\mathbf{Ax}_{r+1} = \ldots = \mathbf{Ax}_n = \mathbf{0}$, while each one of $\mathbf{Ax}_1,\ldots,\mathbf{Ax}_r$ is a linear combination of $\mathbf{x}_1,\ldots,\mathbf{x}_r$. Prove that, if $\mathbf{A}$ represents the linear transformation L with respect to $\mathfrak{E}$, then the matrix representing L with respect to $\mathfrak{B}$ has the form

$$\begin{pmatrix} \mathbf{A}_1 & \mathbf{O} \\ \mathbf{O} & \mathbf{O} \end{pmatrix},$$

where $\mathbf{A}_1$ is an $r\times r$ matrix.

21. Let $1 \leqslant r < n$, and suppose that $\mathfrak{U}$ is an r-dimensional subspace of $\mathfrak{V}_n$. Show that the general formula for the matrix $\mathbf{A}$, which satisfies the conditions

$$\mathbf{Ax} = \mathbf{0} \text{ (for all } \mathbf{x} \in \mathfrak{U}), \qquad \mathbf{Ax} \neq \mathbf{0} \text{ (for all } \mathbf{x} \notin \mathfrak{U}),$$

is

$$\mathbf{A} = \mathbf{S}\begin{pmatrix} \mathbf{O} & \mathbf{P} \\ \mathbf{O} & \mathbf{Q} \end{pmatrix}\mathbf{S}^{-1},$$

where $\mathbf{S}$ is any *given* non-singular matrix whose first r columns constitute a basis of $\mathfrak{U}$, while $\mathbf{P}$ is an arbitrary $r\times(n-r)$ matrix and $\mathbf{Q}$ an arbitrary non-singular $(n-r)\times(n-r)$ matrix.

22. A transformation L of $\mathfrak{V}_n$ into itself is called a *projection* if there exist complements $\mathfrak{U}$, $\mathfrak{U}'$ such that $L(\mathbf{x}) = \mathbf{x}$ for all $\mathbf{x} \in \mathfrak{U}$ and $L(\mathbf{x}) = \mathbf{0}$ for all $\mathbf{x} \in \mathfrak{U}'$. Show that (i) a projection is a linear transformation; (ii) a linear transformation L is a projection if and only if $L^2 = L$; (iii) a linear transformation L is a projection if and only if $I - L$ is a projection, where I denotes the identical transformation of $\mathfrak{V}_n$.

23. Let L_1, L_2 be two projections. Show that (i) $L_1 + L_2$ is a projection if and only if $L_1 L_2 = L_2 L_1 = O$; (ii) $L_1 - L_2$ is a projection if and only if $L_1 L_2 = L_2 L_1 = L_2$.

24. Let $\mathfrak{M}_n$ denote the linear manifold of all $n\times n$ matrices and L a linear transformation of $\mathfrak{M}_n$ into itself. Show that there exist an integer m and matrices $\mathbf{Q}_k$, $\mathbf{R}_k$ $(1 \leqslant k \leqslant m)$ such that, for any $\mathbf{X} \in \mathfrak{M}_n$,

$$L(\mathbf{X}) = \sum_{k=1}^{m} \mathbf{Q}_k \mathbf{X} \mathbf{R}_k.$$

V

SYSTEMS OF LINEAR EQUATIONS AND RANK OF MATRICES

In the present chapter we shall give a complete account of the theory of simultaneous linear equations. The results of Chapter IV will not be needed here and we shall make use only of the simplest properties of determinants, vectors, and matrices. The most important new idea which we shall introduce is that of rank of a matrix.

5.1. Preliminary results

In this section we shall explain the terminology and notation and consider the simplest cases of our problem.

5.1.1. DEFINITION 5.1.1. *A* LINEAR EQUATION *in the unknowns* $x_1,..., x_n$ *is an equation of the form*

$$a_1 x_1+...+a_n x_n = b. \tag{5.1.1}$$

If $b = 0$, *then* (5.1.1) *is known as a* HOMOGENEOUS LINEAR EQUATION.

A *system* of linear equations in the unknowns $x_1,..., x_n$ has the form

$$\left.\begin{matrix} a_{11} x_1+...+a_{1n} x_n = b_1 \\ \cdot \quad \cdot \quad \cdot \quad \cdot \quad \cdot \quad \cdot \quad \cdot \\ a_{m1} x_1+...+a_{mn} x_n = b_m \end{matrix}\right\}, \tag{5.1.2}$$

where the a_{ij} and b_i are given numbers. The *associated system* of homogeneous equations is given by

$$\left.\begin{matrix} a_{11} x_1+...+a_{1n} x_n = 0 \\ \cdot \quad \cdot \quad \cdot \quad \cdot \quad \cdot \quad \cdot \\ a_{m1} x_1+...+a_{mn} x_n = 0 \end{matrix}\right\}. \tag{5.1.3}$$

Our problem is to investigate whether a set or sets of values of $x_1,..., x_n$ satisfying (5.1.2), or (5.1.3), can be found; how all such sets may be determined in the case when they exist; and what are the relations between them.

A certain amount of complication is introduced into the study of linear equations by the possible presence of *redundant* equations. Consider, for instance, the system of equations

$$x+y = 5, \qquad 2x+y = 3, \qquad 3x+2y = 8. \tag{5.1.4}$$

It is clear that the last equation adds nothing to the information provided by the first two, since it can be obtained by adding these equations. Such an equation which arises from a linear combination of other equations of the system is called redundant (with respect to the system in question.) It is easy to see that redundant equations can be discarded.

EXERCISE 5.1.1. Discuss the fallacy involved in the following argument. 'The second and third equations in (5.1.4) are both redundant since the second is equal to the difference of the third and first while the third is equal to the sum of the first and second. Hence the second and third equations may both be discarded, and the solutions of the system (5.1.4) are precisely the solutions of the single equation $x+y = 5$.'

DEFINITION 5.1.2. *A* SOLUTION *of a system of equations in the unknowns* $x_1,..., x_n$ *is a set of numbers* $\xi_1,..., \xi_n$ *such that every equation of the system is satisfied for the values* $x_1 = \xi_1,..., x_n = \xi_n$.

Even quite trivial instances reveal that a variety of different cases may arise with regard to the nature of solutions of a system of linear equations. Thus the system

$$x_1+x_2 = 1, \qquad x_1+x_2 = 0$$

possesses *no solution*; the system

$$x_1+x_2 = 2, \qquad x_1-x_2 = 0$$

possesses *precisely one solution* ($x_1 = x_2 = 1$); and the system

$$x_1+x_2 = 1, \qquad 2x_1+2x_2 = 2$$

possesses an *infinity of solutions* ($x_1 = t$, $x_2 = 1-t$, t arbitrary).

DEFINITION 5.1.3. *A system of equations is* CONSISTENT *if it possesses at least one solution*; *otherwise it is* INCONSISTENT.

A homogeneous system† is necessarily consistent since (5.1.3) always possesses the *zero solution* $x_1 = ... = x_n = 0$. Such a solution is, however, of little interest.

DEFINITION 5.1.4. *The solution* $x_1 = ... = x_n = 0$ *of a homogeneous system in the unknowns* $x_1,..., x_n$ *is called* TRIVIAL; *any other solution is called* NON-TRIVIAL.

A system of equations not all of which are homogeneous may, of course, be consistent or inconsistent; the examples given a little earlier illustrate both alternatives.

† i.e. a system of homogeneous linear equations.

In our discussion below we shall make systematic use of matrix technique. Thus the system of linear equations (5.1.2) will generally be written in the form

$$\mathbf{Ax} = \mathbf{b},$$

where $\mathbf{A} = \begin{pmatrix} a_{11} & . & . & . & a_{1n} \\ . & . & . & . & . \\ a_{m1} & . & . & . & a_{mn} \end{pmatrix}$, $\mathbf{x} = \begin{pmatrix} x_1 \\ \vdots \\ x_n \end{pmatrix}$, $\mathbf{b} = \begin{pmatrix} b_1 \\ \vdots \\ b_m \end{pmatrix}$.

Correspondingly, the homogeneous system (5.1.3) assumes the form

$$\mathbf{Ax} = \mathbf{0}.$$

A solution $x_1 = \xi_1,..., x_n = \xi_n$ (of either system) will, accordingly, be regarded as a column vector $(\xi_1,..., \xi_n)^T$.

5.1.2. It is convenient to determine at once the connexion between a system of linear equations $\mathbf{Ax} = \mathbf{b}$ and the associated homogeneous system $\mathbf{Ax} = \mathbf{0}$.

THEOREM 5.1.1. *Suppose that the system* $\mathbf{Ax} = \mathbf{b}$ *possesses a solution* $\mathbf{x}_0$. *Then* (i) *any solution of this system is expressible as the sum of* $\mathbf{x}_0$ *and a suitable solution of* $\mathbf{Ax} = \mathbf{0}$; (ii) *the sum of* $\mathbf{x}_0$ *and any solution of* $\mathbf{Ax} = \mathbf{0}$ *is a solution of* $\mathbf{Ax} = \mathbf{b}$.

We may express this result informally by saying that the general solution of $\mathbf{Ax} = \mathbf{b}$ is equal to the sum of any particular solution of $\mathbf{Ax} = \mathbf{b}$ and the general solution of $\mathbf{Ax} = \mathbf{0}$.

The proof is trivial For let $\mathbf{x}_1$ be any solution of $\mathbf{Ax} = \mathbf{b}$. Then $\mathbf{x}_1 = \mathbf{x}_0 + (\mathbf{x}_1 - \mathbf{x}_0)$, and

$$\mathbf{A}(\mathbf{x}_1 - \mathbf{x}_0) = \mathbf{Ax}_1 - \mathbf{Ax}_0 = \mathbf{b} - \mathbf{b} = \mathbf{0}.$$

Thus $\mathbf{x}_1 = \mathbf{x}_0 + \mathbf{y}$, where $\mathbf{y}$ is a solution of $\mathbf{Ax} = \mathbf{0}$, and so (i) is proved. Again, let $\mathbf{x}_2$ be any solution of $\mathbf{Ax} = \mathbf{0}$. Then

$$\mathbf{A}(\mathbf{x}_0 + \mathbf{x}_2) = \mathbf{Ax}_0 + \mathbf{Ax}_2 = \mathbf{b} + \mathbf{0} = \mathbf{b},$$

and so $\mathbf{x}_0 + \mathbf{x}_2$ is a solution of $\mathbf{Ax} = \mathbf{b}$. This proves (ii).

The demonstration just given succeeds by virtue of the fact that matrix multiplication is a linear operation.† In this context it is useful to recall that a result analogous to Theorem 5.1.1 holds in the theory of differential equations.‡ Let $Q(t)$ be a given function of t, f a polynomial, and D the operator d/dt. Consider the differential equation

$$f(D)x = Q(t) \tag{5.1.5}$$

and its auxiliary equation

$$f(D)x = 0. \tag{5.1.6}$$

† See Definition 4.2.2 (p. 114). ‡ Compare the remarks in § 4.4.

Then, as we know, the general solution of (5.1.5) is equal to the sum of any particular solution of (5.1.5) and the general solution of (5.1.6). The proof of this result depends on the linearity of the differential operator $f(D)$ in precisely the same way as the proof of Theorem 5.1.1 depends on the linearity of the matrix operator **A**.

5.1.3. The simplest case of a system of linear equations is that in which the number of unknowns is equal to the number of equations. Since the general process of solution depends on reducing any given system to an equivalent system of this special type, we begin by discussing systems of the form (5.1.2) for which $m = n$.

THEOREM 5.1.2. (Cramer's rule)

If the $n \times n$ matrix $\mathbf{A} = (a_{ij})$ is non-singular, then the system of linear equations

$$\left.\begin{array}{c} a_{11}x_1 + \ldots + a_{1n}x_n = b_1 \\ \cdot \quad \cdot \quad \cdot \quad \cdot \quad \cdot \quad \cdot \\ a_{n1}x_1 + \ldots + a_{nn}x_n = b_n \end{array}\right\} \tag{5.1.7}$$

possesses a unique solution given by

$$x_i = \frac{|\mathbf{A}^{(i)}|}{|\mathbf{A}|} \quad (i = 1, \ldots, n), \tag{5.1.8}$$

where $\mathbf{A}^{(i)}$ is the matrix obtained when the i-th column of **A** *is replaced by the vector $\mathbf{b} = (b_1, \ldots, b_n)^T$.*

The solution (5.1.8) was found by Cramer in 1750. Essentially the same result was known, however, to Leibnitz some fifty years earlier.

If we rewrite (5.1.7) in the matrix form $\mathbf{Ax} = \mathbf{b}$, where $\mathbf{x} = (x_1, \ldots, x_n)^T$, we see at once that the existence and uniqueness of the solution are guaranteed by Theorem 3.6.2 (i) (p. 92), which also shows that the solution is given by the formula $\mathbf{x} = \mathbf{A}^{-1}\mathbf{b}$. Denoting the cofactor of the (i, j)th element in **A** by A_{ij} we therefore obtain

$$\mathbf{x} = \frac{1}{|\mathbf{A}|}\mathbf{A}^*\mathbf{b} = \frac{1}{|\mathbf{A}|}\begin{pmatrix} A_{11} & \cdot & \cdot & \cdot & A_{n1} \\ \cdot & \cdot & \cdot & \cdot & \cdot \\ A_{1n} & \cdot & \cdot & \cdot & A_{nn} \end{pmatrix}\begin{pmatrix} b_1 \\ \vdots \\ b_n \end{pmatrix}.$$

Hence, by Theorem 1.4.1 (p. 15), we have for $i = 1, \ldots, n$,

$$x_i = \frac{1}{|\mathbf{A}|}(A_{1i}b_1 + \ldots + A_{ni}b_n) = \frac{|\mathbf{A}^{(i)}|}{|\mathbf{A}|}.$$

As an illustration consider the system of equations

$$ax+by = k, \qquad a'x+b'y = k'.$$

Cramer's rule shows that, if

$$\begin{vmatrix} a & b \\ a' & b' \end{vmatrix} \neq 0,$$

then the system has a unique solution given by

$$x = \begin{vmatrix} k & b \\ k' & b' \end{vmatrix} \Big/ \begin{vmatrix} a & b \\ a' & b' \end{vmatrix}, \qquad y = \begin{vmatrix} a & k \\ a' & k' \end{vmatrix} \Big/ \begin{vmatrix} a & b \\ a' & b' \end{vmatrix}.$$

Again, if

$$D = \begin{vmatrix} a & b & c \\ a' & b' & c' \\ a'' & b'' & c'' \end{vmatrix} \neq 0,$$

then the system

$$\begin{aligned} ax+by+cz &= k \\ a'x+b'y+c'z &= k' \\ a''x+b''y+c''z &= k'' \end{aligned}$$

has a unique solution given by

$$x = D^{-1}\begin{vmatrix} k & b & c \\ k' & b' & c' \\ k'' & b'' & c'' \end{vmatrix}, \qquad y = D^{-1}\begin{vmatrix} a & k & c \\ a' & k' & c' \\ a'' & k'' & c'' \end{vmatrix},$$

$$z = D^{-1}\begin{vmatrix} a & b & k \\ a' & b' & k' \\ a'' & b'' & k'' \end{vmatrix}.$$

COROLLARY. *If* $\mathbf{A}$ *is a non-singular square matrix, then the only solution of the homogeneous system* $\mathbf{Ax} = \mathbf{0}$ *is the trivial solution* $\mathbf{x} = \mathbf{0}$.

An equivalent statement is that, if $\mathbf{A}$ is a square matrix and $\mathbf{Ax} = \mathbf{0}$ possesses a non-trivial solution, then $\mathbf{A}$ is singular. This result is not, of course, new to us as it is contained in Theorem 1.6.1 (p. 27).

EXERCISE 5.1.2. Let a_{ij} $(i,j = 1,\ldots,n)$ be given numbers. Show (say by considering components of vectors) that a necessary and sufficient condition for the existence of vectors $\mathbf{x}_1,\ldots,\mathbf{x}_n$ (of order n), not all zero and satisfying the equations

$$\sum_{j=1}^{n} a_{ij}\mathbf{x}_j = \mathbf{0} \qquad (i = 1,\ldots,n),$$

is the vanishing of the determinant $|a_{ij}|_n$.

5.2. The rank theorem

In this section we shall establish a link between the notion of a determinant and that of linear dependence, and thereby obtain the preliminary results necessary for the solution of the main problem. The key idea in this context is that of rank, first discussed explicitly by Sylvester in 1851.

DEFINITION 5.2.1. (i) *Let* $\mathbf{A}$ *be an* $m\times n$ *matrix. If* $k \leqslant m$, $l \leqslant n$, *then any* k *rows and* l *columns of* $\mathbf{A}$ *determine a* $k\times l$ SUBMATRIX *of* $\mathbf{A}$. (ii) *The determinant of a* $k\times k$ *submatrix of* $\mathbf{A}$ *is called a* k-ROWED MINOR *of* $\mathbf{A}$, *or a* MINOR OF ORDER k.

DEFINITION 5.2.2. (i) *The* RANK (*sometimes called* DETERMINANT RANK) $R(\mathbf{A})$ *of a non-zero matrix* $\mathbf{A}$ *is the maximum value of* r *for which there exists a non-vanishing* r*-rowed minor of* $\mathbf{A}$. (ii) *A* CRITICAL MINOR *of a non-zero matrix is any non-vanishing minor of maximum order.* (iii) *The rank of any zero matrix is equal to zero.*

Thus, for $\mathbf{A} \neq \mathbf{O}$, the statement $R(\mathbf{A}) = r$ means that (i) $\mathbf{A}$ contains at least one non-vanishing minor of order r; (ii) $\mathbf{A}$ contains no non-vanishing minor of order greater than r. In order to show for a particular matrix $\mathbf{A}$ that (ii) is satisfied it is, of course, sufficient to verify that all minors of order $r+1$ vanish.

EXERCISE 5.2.1. Use Exercise 1.4.4 (p. 20) to prove this statement.

It is obvious from Definition 5.2.2 (i) that, for any non-zero matrix $\mathbf{A}$ of type $m\times n$, the rank cannot exceed m or n, i.e.

$$0 < R(\mathbf{A}) \leqslant \min(m, n).$$

It is equally obvious that, for a square matrix $\mathbf{A}$ of order n, $R(\mathbf{A}) < n$ or $R(\mathbf{A}) = n$ according as $\mathbf{A}$ is singular or non-singular.

To illustrate the notion of rank, consider the matrix

$$\mathbf{A} = \begin{pmatrix} -1 & 0 & 2 & 1 \\ 0 & 1 & 1 & -1 \\ 2 & 0 & -4 & -2 \end{pmatrix}.$$

This matrix possesses no minors of order greater than 3; and its four minors of order 3 all vanish, since

$$\begin{vmatrix} -1 & 0 & 2 \\ 0 & 1 & 1 \\ 2 & 0 & -4 \end{vmatrix} = \begin{vmatrix} -1 & 0 & 1 \\ 0 & 1 & -1 \\ 2 & 0 & -2 \end{vmatrix} = \begin{vmatrix} -1 & 2 & 1 \\ 0 & 1 & -1 \\ 2 & -4 & -2 \end{vmatrix}$$

$$= \begin{vmatrix} 0 & 2 & 1 \\ 1 & 1 & -1 \\ 0 & -4 & -2 \end{vmatrix} = 0.$$

On the other hand, $\mathbf{A}$ contains at least one non-vanishing 2-rowed minor, e.g. that situated in the top left-hand corner of $\mathbf{A}$ and indicated in bold-face type. Hence $R(\mathbf{A}) = 2$. The reader should have no difficulty in determining a number of critical minors of $\mathbf{A}$ in addition to the one just mentioned.

EXERCISE 5.2.2. Show that if $\mathbf{A}'$ is a submatrix of $\mathbf{A}$, then $R(\mathbf{A}') \leqslant R(\mathbf{A})$.

EXERCISE 5.2.3. Show that, for every non-zero scalar λ, $R(\lambda\mathbf{A}) = R(\mathbf{A})$.

EXERCISE 5.2.4. Show that the rank of a diagonal matrix with $n-k$ zero elements and k non-zero elements on the diagonal is equal to k.

EXERCISE 5.2.5. Show that the rank of a matrix remains unchanged if the rows, or the columns, are permuted, or if the matrix is transposed.

In what follows we shall have frequent occasion to speak of linear combinations and of linear dependence of the rows (or columns) of a matrix. This terminology introduces no new ideas and simply means that the rows or columns of the matrix are treated as vectors.

DEFINITION 5.2.3. (i) *The* ROW RANK (COLUMN RANK) *of a matrix* $\mathbf{A} \neq \mathbf{O}$ *is the maximum value of* r *for which there exist* r *linearly independent rows* (*columns*) *of* $\mathbf{A}$. (ii) *The row rank and column rank of any zero matrix are both equal to zero.*

In view of Exercise 2.3.5 (p. 51) it is clear that the row rank (column rank) of $\mathbf{A}$ is equal to the dimensionality of the vector space spanned by the rows (columns) of $\mathbf{A}$.

We need the next theorem in order to demonstrate that the rank, row rank, and column rank of a matrix are all equal.

THEOREM 5.2.1. *Let* $R(\mathbf{A}) = r$ $(\geqslant 1)$. *Then any* r *rows* (*columns*) *of* $\mathbf{A}$ *which contain a critical minor are linearly independent, and every row* (*column*) *of* $\mathbf{A}$ *may be expressed linearly in terms of these* r *rows* (*columns*).

It is, of course, sufficient to prove this theorem for rows. Let $\mathbf{A}$ be the $m \times n$ matrix (a_{ij}) and assume, as may be done without loss of generality, that a critical minor Δ is situated in the top left-hand corner of $\mathbf{A}$, i.e.

$$\Delta = \begin{vmatrix} a_{11} & \cdot & \cdot & \cdot & a_{1r} \\ \cdot & \cdot & \cdot & \cdot & \cdot \\ a_{r1} & \cdot & \cdot & \cdot & a_{rr} \end{vmatrix} \neq 0. \tag{5.2.1}$$

We then have to show that (i) the first r rows of $\mathbf{A}$ are linearly independent; (ii) every row of $\mathbf{A}$ is expressible linearly in terms of the first r rows.

Suppose $\gamma_1, \gamma_2,..., \gamma_r$ are numbers such that

$$\gamma_1(a_{11},..., a_{1n})+\gamma_2(a_{21},..., a_{2n})+...+\gamma_r(a_{r1},..., a_{rn}) = \mathbf{0}.$$

Equating to zero the first r components of the vector on the left, we obtain

$$\begin{aligned} \gamma_1 a_{11}+...+\gamma_r a_{r1} &= 0, \\ \cdot \quad \cdot \quad \cdot \quad &\cdot \quad \cdot \quad \cdot \\ \gamma_1 a_{1r}+...+\gamma_r a_{rr} &= 0. \end{aligned}$$

Hence, by (5.2.1) and the corollary to Theorem 5.1.2 it follows that $\gamma_1 = ... = \gamma_r = 0$. Thus the first r rows of $\mathbf{A}$ are, in fact, linearly independent.

We shall show next that every row of $\mathbf{A}$ is expressible linearly in terms of the first r rows. When $r = m$ there is nothing to prove, and we may therefore assume that $r < m$, $r \leqslant n$. Consider the determinant

$$D = \begin{vmatrix} a_{11} & \cdot & \cdot & \cdot & a_{1r} & a_{1j} \\ \cdot & \cdot & \cdot & \cdot & \cdot & \cdot \\ a_{r1} & \cdot & \cdot & \cdot & a_{rr} & a_{rj} \\ a_{i1} & \cdot & \cdot & \cdot & a_{ir} & a_{ij} \end{vmatrix},$$

where $r+1 \leqslant i \leqslant m$, $1 \leqslant j \leqslant n$. If $j \leqslant r$, then D possesses two identical columns and therefore vanishes. If $j > r$, then D is an $(r+1)$-rowed minor of $\mathbf{A}$ and so vanishes in virtue of the assumption $R(\mathbf{A}) = r$. Thus, in every case, $D = 0$.

The cofactor of the element a_{tj} $(t = 1,..., r)$ in D depends on t and i (but not on j), and so may be denoted by λ_{ti}. The cofactor of a_{ij} in D is obviously Δ. Expanding D in terms of the elements of the last column we therefore obtain

$$0 = D = a_{1j}\lambda_{1i}+...+a_{rj}\lambda_{ri}+a_{ij}\Delta.$$

Hence, by (5.2.1), there exist numbers μ_{ki} such that

$$a_{ij} = \mu_{1i}a_{1j}+...+\mu_{ri}a_{rj} \qquad (i = r+1,..., m; j = 1,..., n).$$

These relations may be written as

$$(a_{i1},..., a_{in}) = \mu_{1i}(a_{11},..., a_{1n})+...+\mu_{ri}(a_{r1},..., a_{rn}) \quad (i = r+1,..., m),$$

and the proof is therefore complete.

Theorem 5.2.2. (Rank theorem)

The rank, the row rank, and the column rank of a matrix are all equal.

Let the row rank and the column rank of a matrix **A** be denoted by $R_1(\mathbf{A})$ and $R_2(\mathbf{A})$ respectively. It is clearly sufficient to show that $R(\mathbf{A}) = R_1(\mathbf{A})$, for $R(\mathbf{A}) = R_2(\mathbf{A})$ will then follow by an analogous argument. Alternatively, we may deduce the second relation from the first by observing that

$$R(\mathbf{A}) = R(\mathbf{A}^T) = R_1(\mathbf{A}^T) = R_2(\mathbf{A}).$$

The theorem is trivially true for $\mathbf{A} = \mathbf{O}$, and we may therefore assume that **A** is a non-zero matrix, say of type $m \times n$. If $R(\mathbf{A}) = r$ ($\geqslant 1$), then, by Theorem 5.2.1, **A** possesses r linearly independent rows, say $\xi_1,..., \xi_r$, and every row of **A** is expressible as a linear combination of $\xi_1,..., \xi_r$. Denote by $\mathfrak{B}$ the vector space (of order n) spanned by all rows of **A**. Since (in view of Exercise 2.2.5, p. 48) $\mathfrak{B}$ is also spanned by $\xi_1,..., \xi_r$ and since these vectors are linearly independent it follows, by Theorem 2.3.1 (p. 50), that they constitute a basis of $\mathfrak{B}$. Hence $d(\mathfrak{B}) = r = R(\mathbf{A})$. But, as was pointed out immediately after Definition 5.2.3, $R_1(\mathbf{A}) = d(\mathfrak{B})$, and so $R(\mathbf{A}) = R_1(\mathbf{A})$.

Corollary. *The maximum number of linearly independent rows of a matrix is equal to the maximum number of its linearly independent columns.*

Alternatively, we may say that the rows and the columns of a matrix span vector spaces of the same dimensionality. It is worth noting that although this result involves only the notion of linear dependence, we have found it simplest to establish it indirectly by appealing to the theory of determinants.†

Since rank, row rank, and column rank are now known to be equal, we need no longer to differentiate between them. When in future the term 'rank' is used we shall bear in mind that it denotes a number which possesses all the three properties specified in Definitions 5.2.2 and 5.2.3.

Exercise 5.2.6. Show that a square matrix is singular if and only if its rows (or columns) are linearly dependent.

† For proofs independent of determinant theory see Artin, *Galois Theory* (2nd edition), 7–9; Schreier and Sperner, **8**, 117–19; or Perlis, **10**, 56. See also Hasse, **19**, 68–103, for a systematic development of linear algebra without the use of determinants.

The following simple result will prove very useful in our subsequent discussion.

THEOREM 5.2.3. *Let* $\mathbf{A}$ *be a given* $m \times n$ *matrix. The set of all vectors of the form* $\mathbf{Ax}$, *where* $\mathbf{x}$ *is an arbitrary vector of order* n, *is a vector space of order* m *and dimensionality* $R(\mathbf{A})$.

Let $\mathfrak{B}$ denote the set of all vectors of the form $\mathbf{Ax}$. In view of the linearity of matrix multiplication $\mathfrak{B}$ is obviously a vector space (of order m). By Theorem 3.3.6 (p. 84) we have

$$\mathbf{Ax} = x_1 \mathbf{A}_{*1} + \ldots + x_n \mathbf{A}_{*n},$$

where $\mathbf{x} = (x_1,\ldots,x_n)^T$. Hence $\mathfrak{B}$ is spanned by $\mathbf{A}_{*1},\ldots,\mathbf{A}_{*n}$, and so $d(\mathfrak{B})$ is equal to the maximum number of linearly independent columns of $\mathbf{A}$, i.e. $d(\mathfrak{B}) = R(\mathbf{A})$.

5.3. The general theory of linear equations

5.3.1. The results of the previous section furnish all the necessary preliminary material and enable us to construct a complete theory of linear equations.†

DEFINITION 5.3.1. *In the system of linear equations*

$$\left.\begin{array}{c} a_{11} x_1 + \ldots + a_{1n} x_n = b_1 \\ \cdot \quad \cdot \quad \cdot \quad \cdot \quad \cdot \quad \cdot \quad \cdot \quad \cdot \\ a_{m1} x_1 + \ldots + a_{mn} x_n = b_m \end{array}\right\} \qquad (5.3.1)$$

the matrix

$$\mathbf{A} = \begin{pmatrix} a_{11} & \cdot & \cdot & \cdot & a_{1n} \\ \cdot & \cdot & \cdot & \cdot & \cdot \\ a_{m1} & \cdot & \cdot & \cdot & a_{mn} \end{pmatrix}$$

is known as the MATRIX OF COEFFICIENTS, *while*

$$\mathbf{B} = \begin{pmatrix} a_{11} & \cdot & \cdot & \cdot & a_{1n} & b_1 \\ \cdot & \cdot & \cdot & \cdot & \cdot & \cdot \\ a_{m1} & \cdot & \cdot & \cdot & a_{mn} & b_m \end{pmatrix}$$

is known as the AUGMENTED MATRIX.

Theorem 5.3.1. (Consistency theorem)

A necessary and sufficient condition for a system of linear equations to be consistent is that the matrix of coefficients should have the same rank as the augmented matrix.

† For an alternative treatment see Wade, **1**, 147–55.

The system of equations (5.3.1) is, in view of Theorem 3.3.6, equivalent to

$$x_1 \mathbf{A}_{*1} + \ldots + x_n \mathbf{A}_{*n} = \mathbf{b}, \tag{5.3.2}$$

where $\mathbf{b} = (b_1,\ldots, b_m)^T$. Moreover, the augmented matrix $\mathbf{B}$ has $\mathbf{A}_{*1},\ldots, \mathbf{A}_{*n}$, $\mathbf{b}$ as its columns.

To prove *necessity*, suppose that there exists a solution $x_1,\ldots, x_n$ of (5.3.2). Then $\mathbf{b}$ is a linear combination of the columns of $\mathbf{A}$, and so $\mathbf{B}$ cannot have a greater number of linearly independent columns than $\mathbf{A}$. Hence $R(\mathbf{B}) \leqslant R(\mathbf{A})$. Furthermore, since (by Exercise 5.2.2, p. 137) $R(\mathbf{A}) \leqslant R(\mathbf{B})$, we infer that $R(\mathbf{A}) = R(\mathbf{B})$.

Sufficiency is established as follows. Suppose that

$$R(\mathbf{A}) = R(\mathbf{B}) = r \geqslant 1$$

(the case $r = 0$ being trivial). Consider r columns of $\mathbf{A}$ which contain a critical minor, say Δ, and assume, without loss of generality, that $\mathbf{A}_{*1},\ldots, \mathbf{A}_{*r}$ are such columns. By hypothesis, Δ is also a critical minor of $\mathbf{B}$ and therefore, by Theorem 5.2.1, $\mathbf{b}$ is a linear combination of $\mathbf{A}_{*1},\ldots, \mathbf{A}_{*r}$, say

$$\mathbf{b} = \beta_1 \mathbf{A}_{*1} + \ldots + \beta_r \mathbf{A}_{*r}.$$

It follows that (5.3.2) is satisfied by

$$x_1 = \beta_1, \ \ldots, \ x_r = \beta_r, \ x_{r+1} = 0, \ \ldots, \ x_n = 0;\dagger$$

and the system is therefore consistent.

Since the rank of a matrix can be found by evaluating a number of determinants, it follows that the result just proved furnishes not merely a theoretical criterion but also a practical procedure for testing a system of linear equations for consistency. Consider, for example, the system of equations

$$3x + y - 5z = -1, \quad x - 2y + z = -5, \quad x + 5y - 7z = 2. \tag{5.3.3}$$

Here the matrix of coefficients is

$$\mathbf{A} = \begin{pmatrix} 3 & 1 & -5 \\ 1 & -2 & 1 \\ 1 & 5 & -7 \end{pmatrix}.$$

† For $r = n$ we have simply $x_1 = \beta_1,\ldots, x_n = \beta_n$.

It can be verified at once that the determinant of **A** vanishes, and that therefore $R(\mathbf{A}) < 3$. Again, the augmented matrix is

$$\mathbf{B} = \begin{pmatrix} 3 & 1 & -5 & -1 \\ 1 & -2 & 1 & -5 \\ 1 & 5 & -7 & 2 \end{pmatrix},$$

and the 3-rowed minor of **B** consisting of the first, second, and fourth columns has the value 49. Hence $R(\mathbf{B}) = 3$, and so $R(\mathbf{A}) < R(\mathbf{B})$. The system (5.3.3) is therefore inconsistent.

It should be noted that the method illustrated by this example has an obvious disadvantage. It depends on the evaluation of a number of determinants and so often involves a considerable amount of numerical work. This can be avoided by a more expeditious method of determining rank which will be described in the next chapter.†

Theorem 5.3.1 does not, of course, exhaust the theory of linear equations, for though it enables us to carry out tests for consistency, it provides no method for finding solutions of a system of equations or for determining the nature of the totality of the solutions.

DEFINITION 5.3.2. *Two systems of linear equations (in the same unknowns) are* EQUIVALENT *if they have the same solutions.*

EXERCISE 5.3.1. Show that the systems

$$x+y+z = 0, \quad 2x+y-2z = 1, \quad \text{and} \quad x+y+z = 0, \quad 3x+2y-z = 1$$

are equivalent.

THEOREM 5.3.2. (Complete solution of a system of linear equations)

Let $\mathfrak{S}$ be a consistent system of linear equations in n unknowns, and let Δ be a critical r-rowed minor of the matrix of coefficients. Then

(i) *the r equations of $\mathfrak{S}$ whose coefficients are involved in Δ form a system $\mathfrak{S}'$ equivalent to $\mathfrak{S}$;*

(ii) *if arbitrary values are assigned to the $n-r$ 'disposable' unknowns in $\mathfrak{S}'$, whose coefficients are not elements of Δ, then the remaining r unknowns are uniquely determined;*

(iii) *by assigning all possible sets of arbitrary values to the disposable unknowns in $\mathfrak{S}'$ and determining in each case the remaining unknowns, we obtain all solutions of the original system $\mathfrak{S}$.*

† See § 6.2.2.

For $r = n$ the theorem must be interpreted as meaning that there are no disposable unknowns, and that all unknowns are determined uniquely.

Let (5.3.1) be the given system $\mathfrak{S}$ and assume that

$$\Delta = \begin{vmatrix} a_{11} & . & . & . & a_{1r} \\ . & . & . & . & . \\ a_{r1} & . & . & . & a_{rr} \end{vmatrix} \neq 0.$$

The r equations whose coefficients are involved in Δ, namely,

$$\begin{aligned} a_{11}x_1+...+a_{1n}x_n &= b_1, \\ . \quad . \quad . \quad . \quad . \quad . \quad . & \quad . \\ a_{r1}x_1+...+a_{rn}x_n &= b_r, \end{aligned}$$

then constitute the system $\mathfrak{S}'$. Any solution of $\mathfrak{S}$ is trivially a solution of $\mathfrak{S}'$. To prove the converse we may clearly assume that $r < m$. Since $\mathfrak{S}$ is consistent by hypothesis, Δ is a critical minor of the augmented matrix; hence, by Theorem 5.2.1, every row of the augmented matrix is expressible linearly in terms of the first r rows. Thus there exist numbers λ_{ik} ($i = r+1,..., m$; $k = 1,..., r$) such that

$$a_{ij} = \sum_{k=1}^{r} \lambda_{ik} a_{kj} \qquad (i = r+1,..., m; \, j = 1,..., n),$$

$$b_i = \sum_{k=1}^{r} \lambda_{ik} b_k \qquad (i = r+1,..., m).$$

Let $x_1,..., x_n$ be any solution of $\mathfrak{S}'$. Then, for $r < i \leqslant m$,

$$\begin{aligned} \sum_{=1}^{n} a_{ij}x_j - b_i &= \sum_{j=1}^{n} x_j \sum_{k=1}^{r} \lambda_{ik} a_{kj} - \sum_{k=1}^{r} \lambda_{ik} b_k \\ &= \sum_{k=1}^{r} \lambda_{ik} \left\{ \sum_{j=1}^{n} a_{kj} x_j - b_k \right\} = 0, \end{aligned}$$

and so $x_1,..., x_n$ is a solution of $\mathfrak{S}$. Hence $\mathfrak{S}$ and $\mathfrak{S}'$ are equivalent. The reason why we obtain a solution of the entire system by considering a solution of r selected equations only is that the other equations are, in fact, redundant and do not add anything to the information provided by the r selected ones.

We now consider the system $\mathfrak{S}'$. If $r = n$, then, since $\Delta \neq 0$, it follows by Cramer's rule (Theorem 5.1.2) that all the unknowns are uniquely determined. If, on the other hand, $r < n$, then the coefficients of $x_{r+1},..., x_n$ are not elements of Δ. Assign any arbitrary values to these $n-r$ unknowns (which may be called disposable unknowns) and rewrite $\mathfrak{S}'$ in the form

$$\begin{aligned} a_{11}x_1+\dots+a_{1r}x_r &= b_1-a_{1,r+1}x_{r+1}-\dots-a_{1n}x_n, \\ &\cdots \\ a_{r1}x_1+\dots+a_{rr}x_r &= b_r-a_{r,r+1}x_{r+1}-\dots-a_{rn}x_n. \end{aligned}$$

The remaining r unknowns $x_1,\dots,x_r$ are now seen to be uniquely determined by Cramer's rule.

It remains to show that by giving all possible sets of values to $x_{r+1},\dots,x_n$ and in each case determining the corresponding (unique) values of $x_1,\dots,x_r$ we obtain *all* solutions of $\mathfrak{S}$. Let $y_1,\dots,y_n$ be any solution of $\mathfrak{S}$, and so of $\mathfrak{S}'$. In the procedure described above we may take $x_{r+1}=y_{r+1},\dots,x_n=y_n$. We then obtain a solution $y'_1,\dots,y'_r, y_{r+1},\dots,y_n$. But, in view of Theorem 5.1.2 (p. 134), $y'_1=y_1,\dots,y'_r=y_r$, and so our procedure does, indeed, yield $y_1,\dots,y_n$ as a solution of the system. The theorem is therefore established.

From the proof of Theorem 5.3.2 and from Theorem 5.1.2 it is clear that the general solution of the system $\mathfrak{S}$ has been obtained in the form

$$\left.\begin{aligned} x_1 &= p_{1,r+1}\lambda_{r+1}+\dots+p_{1n}\lambda_n+q_1 \\ &\cdots \\ x_r &= p_{r,r+1}\lambda_{r+1}+\dots+p_{rn}\lambda_n+q_r \\ x_{r+1} &= \lambda_{r+1} \\ &\cdots \\ x_n &= \lambda_n \end{aligned}\right\}, \qquad (5.3.4)$$

where $\lambda_{r+1},\dots,\lambda_n$ are parameters, i.e. numbers to which arbitrary values may be assigned. The scheme (5.3.4) is not symmetrical since in our discussion the position of a critical minor was chosen in such a way that $x_{r+1},\dots,x_n$ became disposable unknowns. However, the following formulation is independent of the position of critical minors.

THEOREM 5.3.3. *The general solution of a consistent system of linear equations in the unknowns $x_1,\dots,x_n$ is given by*

$$\left.\begin{aligned} x_1 &= p_{11}\lambda_1+\dots+p_{1s}\lambda_s+q_1 \\ &\cdots \\ x_n &= p_{n1}\lambda_1+\dots+p_{ns}\lambda_s+q_n \end{aligned}\right\}. \qquad (5.3.5)$$

Here the p_{ij}, q_i are constants;† *$\lambda_1,\dots,\lambda_s$ are parameters; and $s=n-r$, where r is the rank of the matrix of coefficients.*

† These constants are not, of course, uniquely determined since they depend on the choice of disposable unknowns.

Thus all solutions depend linearly on $n-r$ parameters. We note that (5.3.5) can be expressed in the form

$$\mathbf{x} = \mathbf{P}\boldsymbol{\lambda}+\mathbf{q},$$

where $\mathbf{x} = (x_1,..., x_n)^T$, $\mathbf{P}$ is an $n \times s$ matrix, $\mathbf{q}$ a vector of order n, and $\boldsymbol{\lambda}$ a variable vector of order s.

The information derived from Theorems 5.3.1 and 5.3.2 about the number of solutions of a system of linear equations may be summarized as follows.

THEOREM 5.3.4. *Let* $\mathbf{A}$ *be the matrix of coefficients and* $\mathbf{B}$ *the augmented matrix of a system of linear equations in n unknowns. Then the system possesses an infinity of solutions if and only if*

$$R(\mathbf{A}) = R(\mathbf{B}) < n.$$

It possesses a unique solution if and only if

$$R(\mathbf{A}) = R(\mathbf{B}) = n.$$

It possesses no solution if and only if

$$R(\mathbf{A}) < R(\mathbf{B}).$$

EXERCISE 5.3.2. Illustrate the above result by reference to (i) the systems of equations mentioned immediately after Definition 5.1.2 (p. 132); (ii) a single equation $px = q$.

EXERCISE 5.3.3. Suppose that a system of linear equations in n unknowns is consistent and that the rank of the matrix of coefficients is n. Show that the solution of the system is unique.

Since the determination of solutions of systems of linear equations involves only rational operations we obtain the following result which is implicit in the previous work.

THEOREM 5.3.5. *If all the coefficients of a system of linear equations belong to a reference field* $\mathfrak{F}$ *and if all the disposable unknowns (if any exist) are also restricted to values in* $\mathfrak{F}$, *then all solutions of the system are vectors over* $\mathfrak{F}$.

5.3.2. As an illustration of the procedure established by Theorem 5.3.2 let us consider the following system of equations in 4 unknowns:

$$\left.\begin{aligned} 3x-7y+14z-8w &= 24 \\ x-4y+3z-w &= -2 \\ y+z-w &= 6 \\ 2x-15y-z+5w &= -46 \end{aligned}\right\}. \tag{5.3.6}$$

The augmented matrix of this system is

$$\begin{pmatrix} 3 & -7 & 14 & -8 & 24 \\ \mathbf{1} & \mathbf{-4} & 3 & -1 & -2 \\ \mathbf{0} & \mathbf{1} & 1 & -1 & 6 \\ 2 & -15 & -1 & 5 & -46 \end{pmatrix}.$$

It can be verified that all 3-rowed minors of this matrix vanish. There are, however, non-vanishing 2-rowed minors, and the figures in bold-face type indicate one such minor. It is, of course, critical both for the matrix of coefficients and for the augmented matrix. Thus the system (5.3.6) is consistent and we have, in fact, $m = 4$, $n = 4$, $r = 2$. We may now discard the first and fourth equations since none of their coefficients is an element of the critical minor we are considering. In the system consisting of the second and third equations, namely,

$$x-4y+3z-w = -2, \qquad y+z-w = 6,$$

the coefficients of y and z are not elements of the critical minor. We therefore take y and z as the disposable unknowns, and giving them the arbitrary values λ, μ respectively obtain

$$x = 5\lambda-2\mu-8, \qquad w = \lambda+\mu-6.$$

The general solution of (5.3.6) is therefore given by

$$x = 5\lambda-2\mu-8, \qquad y = \lambda, \qquad z = \mu, \qquad w = \lambda+\mu-6,$$

where λ, μ are parameters. In vector notation this may be written as†

$$\begin{aligned}(x,y,z,w) &= (5\lambda-2\mu-8,\ \lambda,\ \mu,\ \lambda+\mu-6) \\ &= \lambda(5,1,0,1)+\mu(-2,0,1,1)+(-8,0,0,-6). \quad (5.3.7)\end{aligned}$$

Here $(-8,0,0,-6)$ is a particular solution of (5.3.6) while

$$\lambda(5,1,0,1)+\mu(-2,0,1,1)$$

is the general solution of the associated homogeneous system.‡

The form in which the general solution of a system of linear equations is written out is not, as a rule, unique, since it depends on the initial choice of a critical minor. In the case of the system (5.3.6) just considered we might, for instance, discard the first and

† We use row vectors for typographical convenience.

‡ Compare Theorem 5.1.1 and also Theorem 5.4.2 below.

second equations and consider the third and fourth only, since these contain the critical minor

$$\begin{vmatrix} 1 & 1 \\ -15 & -1 \end{vmatrix}.$$

In that case x and w became our disposable unknowns, and we obtain

$$y = \tfrac{1}{7}x + \tfrac{2}{7}w + \tfrac{20}{7}, \qquad z = -\tfrac{1}{7}x + \tfrac{5}{7}w + \tfrac{22}{7}.$$

Putting $x = 7\rho$, $w = 7\sigma$, where ρ, σ are arbitrary parameters, we are led to the general solution of (5.3.6) in the form

$$(x, y, z, w) = \rho(7, 1, -1, 0) + \sigma(0, 2, 5, 7) + (0, \tfrac{20}{7}, \tfrac{22}{7}, 0). \quad (5.3.8)$$

This formula must, of course, yield the same set of solutions as (5.3.7), and we may also verify this directly by putting

$$\lambda = \rho + 2\sigma + \tfrac{20}{7}, \qquad \mu = -\rho + 5\sigma + \tfrac{22}{7}$$

in (5.3.7), or

$$\rho = \tfrac{5}{7}\lambda - \tfrac{2}{7}\mu - \tfrac{8}{7}, \qquad \sigma = \tfrac{1}{7}\lambda + \tfrac{1}{7}\mu - \tfrac{6}{7}$$

in (5.3.8).

EXERCISE 5.3.4. Obtain the general solution of the system (5.3.6) by discarding the second and third equations, and taking z, w as the disposable unknowns.

The solution of a system of linear equations by the method just illustrated can result in a great deal of tedious work since it involves the evaluation of numerous determinants. A better practical procedure is to treat the given set of equations according to the rules familiar from elementary mathematics, i.e. by combining equations linearly and obtaining a simpler system equivalent to the original one. The justification of this rule lies in the fact that if some multiple of an equation is added to another equation of the system, the resulting system is equivalent to the original system. It follows, therefore, in particular, that redundant equations can be discarded.

In the case of the system (5.3.6) we may, for instance, proceed as follows. Subtracting the third equation from the second we obtain (5.3.6) in the equivalent form

$$\begin{aligned} 3x - 7y + 14z - 8w &= 24, \\ x - 5y + 2z &= -8, \\ y + z - w &= 6, \\ 2x - 15y - z + 5w &= -46. \end{aligned}$$

Next, we add 5 times the third equation to the fourth, and obtain

$$\begin{aligned} 3x-7y+14z-8w &= 24, \\ x-5y+2z &= -8, \\ y+z-w &= 6, \\ 2x-10y+4z &= -16. \end{aligned}$$

Here the fourth equation is simply equal to twice the second. It may therefore be discarded, and the system is equivalent to

$$\begin{aligned} 3x-7y+14z-8w &= 24, \\ x-5y+2z &= -8, \\ y+z-w &= 6. \end{aligned}$$

Subtracting the second equation from the first and multiplying the result by $\frac{1}{2}$ we have

$$\begin{aligned} x-y+6z-4w &= 16, \\ x-5y+2z &= -8, \\ y+z-w &= 6. \end{aligned}$$

Here the second equation is equal to the difference between the first and 4 times the third. Hence it may be discarded, and we are consequently left with the system

$$x-y+6z-4w = 16, \qquad y+z-w = 6.$$

We may take y, z as the disposable unknowns; and putting $y = \lambda$, $z = \mu$, where λ, μ are parameters, we are at once led to the general solution in the form (5.3.7).

In § 6.3.2 we shall see how the method for solving systems of linear equations illustrated above can be used with even less labour.

5.4. Systems of homogeneous linear equations

5.4.1. All conclusions of § 5.3.1 are, of course, valid for the special case of homogeneous systems. There are, however, certain results that relate to homogeneous systems only, and we propose next to deal with these results.

The question as to the existence of non-trivial solutions is settled at once by the previous theory.

THEOREM 5.4.1. *The homogeneous system* $\mathbf{Ax} = \mathbf{0}$ *in* n *unknowns possesses a non-trivial solution if and only if* $R(\mathbf{A}) < n$.

The matrix of coefficients and the augmented matrix of a homogeneous system have obviously the same rank. Hence, by Theorem

5.3.4, the system possesses an infinity of solutions, and hence at least one non-trivial solution, if and only if $R(\mathbf{A}) < n$.

COROLLARY 1. *A homogeneous system in which the number of equations is equal to the number of unknowns, possesses a non-trivial solution if and only if the matrix of coefficients is singular.*

This result is, of course, identical with Theorem 1.6.1.

COROLLARY 2. *If a homogeneous system has fewer equations than unknowns, then it possesses a non-trivial solution.*

The next result furnishes us with important new information concerning the totality of solutions of a homogeneous system.

Theorem 5.4.2. (Dimensionality theorem for homogeneous systems)

The solutions of the homogeneous system

$$\mathbf{Ax} = \mathbf{0} \tag{5.4.1}$$

in n unknowns constitute a vector space of dimensionality $n - R(\mathbf{A})$.

Let $\mathfrak{B}$ be the set of all solutions of (5.4.1). If $\mathbf{x}, \mathbf{y} \in \mathfrak{B}$, then clearly $\alpha\mathbf{x}, \mathbf{x}+\mathbf{y} \in \mathfrak{B}$, and so $\mathfrak{B}$ is a vector space. We write $R(\mathbf{A}) = r$ and have now to prove

$$d(\mathfrak{B}) = n - r. \tag{5.4.2}$$

We shall give three proofs of this important result.

(i) Write $d(\mathfrak{B}) = k$. It follows by Theorem 5.4.1 that $k = 0$ implies $r = n$, while obviously $k = n$ implies $r = 0$. We may therefore assume that $0 < k < n$.

Let $\{\mathbf{x}_1,..., \mathbf{x}_k\}$ be a basis of $\mathfrak{B}$. By the Corollary to Theorem 2.3.5 (p. 54), there exist vectors $\mathbf{x}_{k+1},..., \mathbf{x}_n$ such that

$$\{\mathbf{x}_1,..., \mathbf{x}_k, \mathbf{x}_{k+1},..., \mathbf{x}_n\}$$

is a basis of $\mathfrak{B}_n$. The vectors $\mathbf{Ax}_1,..., \mathbf{Ax}_n$ therefore span the space $\mathfrak{B}'$ of vectors of the form $\mathbf{Ax}$ ($\mathbf{x} \in \mathfrak{B}_n$). But $\mathbf{Ax}_1 = ... = \mathbf{Ax}_k = \mathbf{0}$, and so the vectors

$$\mathbf{Ax}_{k+1}, \quad ..., \quad \mathbf{Ax}_n \tag{5.4.3}$$

span $\mathfrak{B}'$. Let the scalars $\alpha_{k+1},..., \alpha_n$ be such that

$$\alpha_{k+1}(\mathbf{Ax}_{k+1}) + ... + \alpha_n(\mathbf{Ax}_n) = \mathbf{0},$$

i.e.

$$\mathbf{A}(\alpha_{k+1}\mathbf{x}_{k+1} + ... + \alpha_n \mathbf{x}_n) = \mathbf{0}.$$

This means that $\alpha_{k+1}\mathbf{x}_{k+1} + ... + \alpha_n\mathbf{x}_n \in \mathfrak{B}$. But $\{\mathbf{x}_1,..., \mathbf{x}_k\}$ is a basis of $\mathfrak{B}$, and so, for suitable scalars $\alpha_1,..., \alpha_k$,

$$\alpha_{k+1}\mathbf{x}_{k+1} + ... + \alpha_n\mathbf{x}_n = \alpha_1\mathbf{x}_1 + ... + \alpha_k\mathbf{x}_k.$$

Since $\mathbf{x}_1,\ldots,\mathbf{x}_n$ are linearly independent this implies, in particular, that

$$\alpha_{k+1} = \ldots = \alpha_n = 0.$$

The vectors (5.4.3) are therefore linearly independent and constitute a basis of $\mathfrak{V}'$; hence $d(\mathfrak{V}') = n-k$. But, by Theorem 5.2.3 (p. 140), $d(\mathfrak{V}') = r$, and (5.4.2) is therefore proved.

Though in the proof just given we speak of the matrix $\mathbf{A}$ rather than of a linear transformation, the argument is, in fact, of the 'invariant' type; and of this the reader should have no difficulty in satisfying himself. The 'invariant' restatement of Theorem 5.4.2 is as follows. Let L be a linear mapping† of $\mathfrak{V}_n$ into $\mathfrak{V}_m$. Denote by $\mathfrak{U}$ the vector space of vectors in $\mathfrak{V}_m$ which are images of vectors in $\mathfrak{V}_n$, and by $\mathfrak{U}'$ the vector space of vectors in $\mathfrak{V}_n$ which map into the zero vector of $\mathfrak{V}_m$. Then $d(\mathfrak{U}') = n-d(\mathfrak{U})$.

(ii) The next proof depends on the general theory of § 5.3.1. If $r = 0$, then clearly $\mathfrak{V} = \mathfrak{V}_n$ and if $r = n$, then $\mathfrak{V}$ is the null space. In these two cases (5.4.2) is therefore valid and we may now assume that $0 < r < n$.

In view of the discussion preceding Theorem 5.3.3 (p. 144), we know that the general solution of (5.4.1) is given by

$$\mathbf{x} = \mathbf{P}\boldsymbol{\lambda}, \tag{5.4.4}$$

where $\boldsymbol{\lambda}$ is an arbitrary vector of order $n-r$ and $\mathbf{P}$ is an $n \times (n-r)$ matrix of the form

$$\begin{pmatrix} p_{1,r+1} & . & . & . & p_{1n} \\ . & . & . & . & . \\ p_{r,r+1} & . & . & . & p_{rn} \\ 1 & 0 & . & . & 0 \\ 0 & 1 & . & . & 0 \\ . & . & . & . & . \\ 0 & 0 & . & . & 1 \end{pmatrix}.$$

Since $\mathbf{P}$ contains a non-vanishing $(n-r)$-rowed minor (namely, that consisting of its last $n-r$ rows), it follows that $R(\mathbf{P}) = n-r$. But, by (5.4.4) and Theorem 5.2.3, $d(\mathfrak{V}) = R(\mathbf{P})$. Hence (5.4.2) is valid.

(iii) Finally, we give a short proof depending on orthogonal complements. It is obvious that the vector $\mathbf{x}$ satisfies the equation $\mathbf{Ax} = \mathbf{0}$ if and only if it is orthogonal to every column of the matrix $\bar{\mathbf{A}}^T$.‡ This implies that $\mathbf{x} \in \mathfrak{V}$ if and only if $\mathbf{x}$ is orthogonal to every vector of the space $\mathfrak{W}$ generated by the columns of $\bar{\mathbf{A}}^T$. Thus $\mathfrak{V}$ and $\mathfrak{W}$ are orthogonal complements and so, by Theorem 2.5.7 (iii) (p. 68), $d(\mathfrak{V})+d(\mathfrak{W}) = n$. But $d(\mathfrak{W}) = R(\bar{\mathbf{A}}^T) = r$, and (5.4.2) follows.

Exercise 5.4.1. Let $\mathbf{A}$ be the matrix of coefficients of a consistent system of linear equations in n unknowns. Show that there exists a vector space $\mathfrak{V}$, of order n and dimensionality $n-R(\mathbf{A})$, such that the general

† L corresponds, of course, to $\mathbf{A}$.

‡ If $\mathbf{A} = (a_{ij})$, then $\bar{\mathbf{A}}$ denotes the matrix $(\bar{a}_{ij})$.

solution of the system is of the form $\mathbf{x}_0 + \boldsymbol{\xi}$, where $\mathbf{x}_0$ is any particular solution and $\boldsymbol{\xi}$ is an arbitrary vector in $\mathfrak{V}$.

5.4.2. DEFINITION 5.4.1. *A set of solutions* $\mathbf{x}_1, \ldots, \mathbf{x}_k$ *of a homogeneous system is a* FUNDAMENTAL SET *of solutions if* (i) $\mathbf{x}_1, \ldots, \mathbf{x}_k$ *are linearly independent, and* (ii) *every solution of the system is expressible as a linear combination of* $\mathbf{x}_1, \ldots, \mathbf{x}_k$.

In fact, a set of solutions is a fundamental set if and only if it is a basis of the vector space of all solutions. By Theorem 5.4.2 it follows that, if $R(\mathbf{A}) = r$, then any $n-r$ linearly independent solutions of the system $\mathbf{Ax} = \mathbf{0}$ in n unknowns form a fundamental set.

It is easy to see how a fundamental set may be constructed. Let us suppose, to fix our ideas, that $x_1, \ldots, x_{n-r}$ are taken as the disposable unknowns. Then the $n-r$ uniquely defined solutions corresponding to

$$\begin{array}{llll} x_1 = 1, & x_2 = 0, & \ldots, & x_{n-r} = 0, \\ x_1 = 0, & x_2 = 1, & \ldots, & x_{n-r} = 0, \\ \cdot & \cdot & \cdot & \cdot \\ x_1 = 0, & x_2 = 0, & \ldots, & x_{n-r} = 1, \end{array}$$

constitute a fundamental set.

EXERCISE 5.4.2. Prove the statement just made, and use it as a basis for a proof of Theorem 5.4.2.

Consider, by way of illustration, the homogeneous system

$$6x+y+z+w = 0, \quad 16x+y-z+5w = 0, \quad 7x+2y+3z = 0. \tag{5.4.5}$$

The matrix of coefficients is

$$\begin{pmatrix} 6 & 1 & \mathbf{1} & \mathbf{1} \\ 16 & 1 & -1 & 5 \\ 7 & 2 & \mathbf{3} & \mathbf{0} \end{pmatrix}.$$

This has rank 2 and a critical minor is indicated in bold-face type. Discarding, as we may, the second equation, and treating x, y as the disposable unknowns, we rewrite (5.4.5) in the equivalent form

$$z+w = -6x-y, \qquad 3z = -7x-2y.$$

If $x = 1$, $y = 0$, then $z = -\frac{7}{3}$, $w = -\frac{11}{3}$; if $x = 0$, $y = 1$, then $z = -\frac{2}{3}$, $w = -\frac{1}{3}$. Hence the vectors

$$(1, 0, -\tfrac{7}{3}, -\tfrac{11}{3}), \qquad (0, 1, -\tfrac{2}{3}, -\tfrac{1}{3})$$

constitute a fundamental set of solutions of (5.4.5). The general solution may therefore be written in the form

$$(x, y, z, w) = \lambda(3, 0, -7, -11) + \mu(0, 3, -2, -1),$$

where λ, μ are parameters.

The importance of fundamental sets of solutions derives, of course, from the fact that any solution can be represented as a unique linear combination of the solutions of a fundamental set. We recall that an analogous result is valid for the linear differential equation

$$\frac{d^n x}{dt^n} + c_1 \frac{d^{n-1} x}{dt^{n-1}} + \dots + c_{n-1} \frac{dx}{dt} + c_n x = 0,$$

any solution of which can be represented as a unique linear combination of n fixed linearly independent solutions.

5.5. Miscellaneous applications

5.5.1. The notion of linear dependence of vectors was introduced in § 2.3.1, but at that stage we had no means of testing effectively whether the vectors of a given set were linearly dependent or not. We are now able to deduce simple criteria for deciding these questions.

By a matrix of a set of vectors (all of the same order) we shall mean any matrix having these vectors as its rows (or columns).†

THEOREM 5.5.1. *Let* $\mathbf{A}$ *be a matrix of* m *vectors of order* n. *A necessary and sufficient condition for these vectors to be linearly dependent is that* $R(\mathbf{A}) < m$.

Let $\mathbf{x}_1, \dots, \mathbf{x}_m$ be the given vectors and let $\mathbf{A}$ be the matrix of which they are the rows. Denoting by μ the maximum number of linearly independent vectors among $\mathbf{x}_1, \dots, \mathbf{x}_m$, we have $\mu = R(\mathbf{A})$. Now $\mathbf{x}_1, \dots, \mathbf{x}_m$ are obviously linearly independent or linearly dependent according as $\mu = m$ or $\mu < m$, and the assertion therefore follows.

Theorem 5.5.1 states, in particular, that if $n < m$, then any m vectors of order n are linearly dependent—a result with which we are, of course, already familiar.

If $m = n$, the condition $R(\mathbf{A}) < m$ means that $|\mathbf{A}| = 0$. We therefore have the following consequence of Theorem 5.5.1.

† We speak here of 'a matrix' rather than 'the matrix', since the order in which the vectors are taken is not laid down.

COROLLARY. *A necessary and sufficient condition for n vectors of order n to be linearly dependent is that a determinant formed by their components should vanish.*

A problem closely allied to the problem of linear dependence of vectors is concerned with the linear dependence of linear forms.

DEFINITION 5.5.1. *A* LINEAR FORM *L in the variables $x_1,..., x_n$ is a polynomial of the type*

$$L = L(x_1,..., x_n) = a_1 x_1 + ... + a_n x_n.$$

EXERCISE 5.5.1. Show that, with respect to obvious definitions of multiplication by scalars and addition, the set of linear forms in n variables is a linear manifold of dimensionality n.

It is easy to see that linear forms can be used to 'represent' linear transformations of a linear manifold into its reference field. We leave it to the reader to supply the details.

DEFINITION 5.5.2. *The linear forms $L_1,..., L_m$ in the variables $x_1,..., x_n$ are said to be linearly dependent if there exist numbers $t_1,..., t_m$, not all zero, such that, identically in $x_1,..., x_n$,*

$$t_1 L_1 + ... + t_m L_m = 0.$$

THEOREM 5.5.2. *A necessary and sufficient condition for m linear forms in n variables to be linearly dependent is that $R(\mathbf{A}) < m$, where* **A** *is a matrix of the linear forms.*†

Let $$L_i(x_1,..., x_n) = a_{i1} x_1 + ... + a_{in} x_n \quad (i = 1,..., m)$$

be the given linear forms. It is seen at once that they are linearly dependent if and only if the m vectors $(a_{i1},..., a_{in})$, $i = 1,..., m$, are linearly dependent. The assertion now follows as an immediate consequence of Theorem 5.5.1.

It should be observed that Theorems 5.5.1 and 5.5.2 are not based on the detailed theory of linear equations and could have been proved immediately after § 5.2.

5.5.2. The question concerning the characterization of divisors of zero in matrix algebra, which was first raised in § 3.6.2, can be settled now without difficulty.

THEOREM 5.5.3. *A (non-zero) matrix* **A** *of type $m \times n$ is a divisor of zero if and only if $R(\mathbf{A}) < \max(m, n)$.*

† By a matrix of a set of linear forms (in the same variables) we mean, of course, any matrix formed by the array of their coefficients.

This statement includes Theorem 3.6.8 (p. 96) and shows further that a (non-zero) non-square matrix is necessarily a divisor of zero.

We recall that a matrix $\mathbf{A} \neq \mathbf{O}$ is called a divisor of zero if and only if there exists a matrix $\mathbf{X} \neq \mathbf{O}$ such that $\mathbf{AX} = \mathbf{O}$ or a matrix $\mathbf{Y} \neq \mathbf{O}$ such that $\mathbf{YA} = \mathbf{O}$. In the former case $\mathbf{X}$ possesses a non-zero column, say $\mathbf{X}_{*j} = \mathbf{x}$. Then $(\mathbf{AX})_{*j} = \mathbf{0}$, and so, in view of Theorem 3.3.5 (ii) (p. 84),

$$\mathbf{Ax} = \mathbf{0}, \qquad \mathbf{x} \neq \mathbf{0}. \tag{5.5.1}$$

In the latter case $\mathbf{Y}$ possesses a non-zero row, say $\mathbf{Y}_{i*} = \mathbf{y}^T$. Then $(\mathbf{YA})_{i*} = \mathbf{0}$, and so, in view of Theorems 3.3.5 (i) and 3.6.9 (p. 97),

$$\mathbf{A}^T\mathbf{y} = \mathbf{0}, \qquad \mathbf{y} \neq \mathbf{0}. \tag{5.5.2}$$

It follows that $\mathbf{A}$ is a divisor of zero if and only if there exists a vector $\mathbf{x}$ satisfying (5.5.1) or a vector $\mathbf{y}$ satisfying (5.5.2). But, by Theorem 5.4.1, a vector $\mathbf{x}$ of the required type exists if and only if $R(\mathbf{A}) < n$, and a vector $\mathbf{y}$ of the required type exists if and only if $R(\mathbf{A}) = R(\mathbf{A}^T) < m$. Hence $\mathbf{A}$ is a divisor of zero if and only if $R(\mathbf{A}) < \max(m, n)$.

5.5.3. We shall next give an interesting proof that if n (complex) non-zero vectors of order n form an orthogonal set, then they are linearly independent—a result which is a special case of Theorem 2.5.4 (p. 65).

Let the vectors in question be denoted by

$$\mathbf{x}_i = (x_{i1}, \ldots, x_{in}) \qquad (i = 1, \ldots, n),$$

and consider the n-rowed determinant $D = |x_{ij}|_n$. Then, multiplying $\bar{D}$ and D rows by rows, we obtain

$$|D|^2 = \begin{vmatrix} \bar{x}_{11} & \cdot & \cdot & \cdot & \bar{x}_{1n} \\ \cdot & \cdot & \cdot & \cdot & \cdot \\ \bar{x}_{n1} & \cdot & \cdot & \cdot & \bar{x}_{nn} \end{vmatrix} \begin{vmatrix} x_{11} & \cdot & \cdot & \cdot & x_{1n} \\ \cdot & \cdot & \cdot & \cdot & \cdot \\ x_{n1} & \cdot & \cdot & \cdot & x_{nn} \end{vmatrix}$$

$$= \begin{vmatrix} (\mathbf{x}_1, \mathbf{x}_1) & \cdot & \cdot & \cdot & (\mathbf{x}_1, \mathbf{x}_n) \\ \cdot & \cdot & \cdot & \cdot & \cdot \\ (\mathbf{x}_n, \mathbf{x}_1) & \cdot & \cdot & \cdot & (\mathbf{x}_n, \mathbf{x}_n) \end{vmatrix}$$

$$= \begin{vmatrix} (\mathbf{x}_1, \mathbf{x}_1) & 0 & \cdot & \cdot & \cdot & 0 \\ 0 & (\mathbf{x}_2, \mathbf{x}_2) & \cdot & \cdot & \cdot & 0 \\ \cdot & \cdot & \cdot & \cdot & \cdot & \cdot \\ 0 & 0 & \cdot & \cdot & \cdot & (\mathbf{x}_n, \mathbf{x}_n) \end{vmatrix}$$

$$= (\mathbf{x}_1, \mathbf{x}_1)(\mathbf{x}_2, \mathbf{x}_2) \ldots (\mathbf{x}_n, \mathbf{x}_n).$$

Hence $D \neq 0$ and so, by the corollary to Theorem 5.5.1, $\mathbf{x}_1,\ldots,\mathbf{x}_n$ are linearly independent.

Determinants such as $|(\mathbf{x}_i, \mathbf{x}_j)|_n$ occur frequently in algebra, and are often referred to as Gram determinants, after J. P. Gram (1850–1916).

DEFINITION 5.5.3. *Let* **A** *be a rectangular matrix. Then the (square) matrix* $\mathbf{G} = \bar{\mathbf{A}}^T\mathbf{A}$ *is known as the* GRAM MATRIX *of* **A**, *and* $|\mathbf{G}|$ *as the* GRAM DETERMINANT *of* **A**.

Our procedure in the argument above consisted essentially in constructing the Gram matrix of the matrix having $\mathbf{x}_1,\ldots,\mathbf{x}_n$ as its columns.

EXERCISE 5.5.2. Let **A** be the $m \times n$ matrix having the vectors $\mathbf{x}_1,\ldots, \mathbf{x}_n$, of order m, as its columns and let **G** be the Gram matrix of **A**. Show that

$$G_{ij} = (\mathbf{x}_i, \mathbf{x}_j) \qquad (i, j = 1,\ldots, n).$$

There is a simple connexion between the rank of any matrix and the rank of its Gram matrix.

THEOREM 5.5.4. *If* **G** *is the Gram matrix of* **A**, *then* $R(\mathbf{G}) = R(\mathbf{A})$.

Let **x** be a vector such that $\bar{\mathbf{A}}^T\mathbf{A}\mathbf{x} = \mathbf{0}$. Then $\bar{\mathbf{x}}^T\bar{\mathbf{A}}^T\mathbf{A}\mathbf{x} = 0$, and so $(\overline{\mathbf{A}\mathbf{x}})^T\mathbf{A}\mathbf{x} = 0$, i.e. $|\mathbf{A}\mathbf{x}|^2 = 0$. We therefore have $\mathbf{A}\mathbf{x} = \mathbf{0}$. Conversely, if **x** satisfies the relation $\mathbf{A}\mathbf{x} = \mathbf{0}$, then clearly $\bar{\mathbf{A}}^T\mathbf{A}\mathbf{x} = \mathbf{0}$. It follows that the homogeneous systems $\mathbf{A}\mathbf{x} = \mathbf{0}$ and $\mathbf{G}\mathbf{x} = \mathbf{0}$ are equivalent. Hence, by Theorem 5.4.2,

$$n - R(\mathbf{A}) = n - R(\mathbf{G}),$$

where n is the number of columns of **A**. The theorem is therefore proved.

5.5.4. The theory of homogeneous linear equations enables us to give an alternative proof of the theorem on orthonormal bases (Theorem 2.5.5, p. 66). This proof is based on Theorem 5.4.1 (p. 148) and makes no use of an explicit construction such as that furnished by Schmidt's orthogonalization process.

Let $\mathbf{x}_1,\ldots,\mathbf{x}_k$ be an orthonormal set in a vector space $\mathfrak{V}$ of dimensionality r. We have to show that this set may be augmented in such a way as to become an orthonormal basis of $\mathfrak{V}$. We may assume that $1 \leqslant k < r$.

By the corollary to Theorem 2.3.5 (p. 54) there exist vectors $\mathbf{y}_{k+1},\ldots,\mathbf{y}_r \in \mathfrak{V}$ such that $\{\mathbf{x}_1,\ldots,\mathbf{x}_k, \mathbf{y}_{k+1},\ldots,\mathbf{y}_r\}$ is a basis of $\mathfrak{V}$. Now it is possible to choose a non-zero vector $\mathbf{x}_{k+1} \in \mathfrak{V}$ such that

$$(\mathbf{x}_1, \mathbf{x}_{k+1}) = \ldots = (\mathbf{x}_k, \mathbf{x}_{k+1}) = 0. \tag{5.5.3}$$

For, if $\quad \mathbf{x}_{k+1} = \alpha_1 \mathbf{x}_1 + ... + \alpha_k \mathbf{x}_k + \alpha_{k+1} \mathbf{y}_{k+1} + ... + \alpha_r \mathbf{y}_r,$

then (5.5.3) is equivalent to the system of equations

$$\alpha_1(\mathbf{x}_1, \mathbf{x}_1) + ... + \alpha_k(\mathbf{x}_1, \mathbf{x}_k) + \alpha_{k+1}(\mathbf{x}_1, \mathbf{y}_{k+1}) + ... + \alpha_r(\mathbf{x}_1, \mathbf{y}_r) = 0,$$
$$\cdot \quad \cdot \quad \cdot \quad \cdot \quad \cdot \quad \cdot \quad \cdot \quad \cdot \quad \cdot \quad \cdot$$
$$\alpha_1(\mathbf{x}_k, \mathbf{x}_1) + ... + \alpha_k(\mathbf{x}_k, \mathbf{x}_k) + \alpha_{k+1}(\mathbf{x}_k, \mathbf{y}_{k+1}) + ... + \alpha_r(\mathbf{x}_k, \mathbf{y}_r) = 0,$$

and, since $k < r$, there exist, by Corollary 2 to Theorem 5.4.1, values of $\alpha_1,..., \alpha_r$, not all zero, satisfying these equations.

Since $\mathbf{x}_{k+1} \neq \mathbf{0}$, it may be normalized; and it follows that if $\mathbf{x}_1,..., \mathbf{x}_k$ is an orthonormal set in $\mathfrak{V}$ and $k < d(\mathfrak{V})$, then there exists a vector $\mathbf{x}_{k+1} \in \mathfrak{V}$ such that the augmented set $\mathbf{x}_1,..., \mathbf{x}_k, \mathbf{x}_{k+1}$ is again orthonormal. The proof of the theorem is now completed by a repeated application of this result.

5.5.5. An interesting application of matrix technique can be made in the calculus of observations. Consider the system of linear equations

$$\left.\begin{array}{c} a_{11} x_1 + ... + a_{1n} x_n = b_1 \\ \cdot \quad \cdot \quad \cdot \quad \cdot \quad \cdot \quad \cdot \quad \cdot \\ a_{m1} x_1 + ... + a_{mn} x_n = b_m \end{array}\right\} \qquad (5.5.4)$$

in which $m > n$, and suppose that the coefficients a_{ij}, b_i (which have real values) have been determined as the result of experiments and that, owing to errors of observation, the system (5.5.4) is, in fact, inconsistent. There exist, then, no values of $x_1,..., x_n$ satisfying (5.5.4) and we are consequently faced with the problem of determining the set of values satisfying the system with the 'least degree of inaccuracy'. The *principle of least squares* states that such a set of values makes the expression

$$(a_{11} x_1 + ... + a_{1n} x_n - b_1)^2 + ... + (a_{m1} x_1 + ... + a_{mn} x_n - b_m)^2 \qquad (5.5.5)$$

a minimum.† Taking this as our starting-point we can easily determine the appropriate values of $x_1,..., x_n$. We rewrite (5.5.4) in the

† It will have been noted that whereas the system (5.5.4) remains unaffected if its equations are multiplied by any non-zero constants, the expression (5.5.5) is certainly changed as the result of such an operation; and so are therefore the values of $x_1,..., x_n$ which are obtained by the application of the principle of least squares. The determination of the appropriate multiples by which the equations are to be multiplied is therefore vital; but since it depends on the 'weighting' of observations and is not relevant in the present context we shall ignore it. The reader who wishes to pursue this topic further may consult Whittaker and Robinson, *The Calculus of Observations* (2nd edition), chap. ix.

familiar form $\mathbf{Ax} = \mathbf{b}$ and put $\boldsymbol{\delta} = \mathbf{Ax}-\mathbf{b}$. The principle of least squares just enunciated requires that the expression

$$\boldsymbol{\delta}^T\boldsymbol{\delta} = \mathbf{x}^T\mathbf{A}^T\mathbf{Ax}-\mathbf{x}^T\mathbf{A}^T\mathbf{b}-\mathbf{b}^T\mathbf{Ax}+\mathbf{b}^T\mathbf{b}$$

should be a minimum, so that

$$\frac{\partial}{\partial\mathbf{x}}(\boldsymbol{\delta}^T\boldsymbol{\delta}) = \mathbf{0}, \tag{5.5.6}$$

where the differential operator $\partial/\partial\mathbf{x}$ is defined by the formula

$$\frac{\partial}{\partial\mathbf{x}}\phi(x_1,\ldots, x_n) = \left(\frac{\partial\phi}{\partial x_1},\ldots, \frac{\partial\phi}{\partial x_n}\right)^T.$$

It is not difficult to verify that

$$\frac{\partial}{\partial\mathbf{x}}(\boldsymbol{\delta}^T\boldsymbol{\delta}) = 2\mathbf{A}^T\mathbf{Ax}-2\mathbf{A}^T\mathbf{b},$$

and so (5.5.6) can be written as $\mathbf{A}^T\mathbf{Ax} = \mathbf{A}^T\mathbf{b}$. Hence, if $\mathbf{A}^T\mathbf{A}$ is non-singular, our problem has the unique solution

$$\mathbf{x} = (\mathbf{A}^T\mathbf{A})^{-1}\mathbf{A}^T\mathbf{b}. \tag{5.5.7}$$

We note that the $n\times n$ matrix $\mathbf{A}^T\mathbf{A}$ is non-singular if and only if $R(\mathbf{A}^T\mathbf{A}) = n$ and, in view of Theorem 5.5.4, this is equivalent to the requirement that $R(\mathbf{A}) = n$. This means that among the equations (5.5.4) there is at least one set of n linearly independent equations.

As an example consider the determination of a straight line by means of the measurement of the position of k ($\geqslant 3$) points on that line. Let the coordinates of these points, as observed, be

$$(x_1, y_1),\quad \ldots,\quad (x_k, y_k), \tag{5.5.8}$$

and write the equation of the required line in the form $y = mx+c$. Then the unknowns m, c, should satisfy the equations

$$mx_1+c-y_1 = 0,\quad \ldots,\quad mx_k+c-y_k = 0.$$

If these equations are inconsistent—as, in almost every practical case they are bound to be—we have to determine the values of m, c which will make the line $y = mx+c$ pass 'as nearly as possible' through the points (5.5.8). The formula (5.5.7) shows that these values are given by

$$(m,c)^T = (\mathbf{A}^T\mathbf{A})^{-1}\mathbf{A}^T(y_1,\ldots,y_k)^T,$$

where

$$\mathbf{A} = \begin{pmatrix} x_1 & 1 \\ \cdot & \cdot \\ \cdot & \cdot \\ x_k & 1 \end{pmatrix}.$$

Hence

$$(m,c)^T = \begin{pmatrix} x_1^2+\ldots+x_k^2 & x_1+\ldots+x_k \\ x_1+\ldots+x_k & k \end{pmatrix}^{-1} \begin{pmatrix} x_1y_1+\ldots+x_ky_k \\ y_1+\ldots+y_k \end{pmatrix},$$

and it is easy to show that the first matrix on the right-hand side is non-singular except when $x_1 = \ldots = x_k$. The reader who is familiar with statistical terminology will have now no difficulty in verifying that the gradient m of the required line is given by

$$m = r\frac{\sigma_y}{\sigma_x},$$

where σ_x, σ_y are the standard deviations of the x's and y's respectively, and r is the coefficient of correlation between these two sets of numbers.

EXERCISE 5.5.3. Determine the parabola $y = a+bx+cx^2$ which passes 'as nearly as possible' through the points (x_i, y_i), $i = 1,\ldots, k$, where $k \geqslant 4$.

5.6. Further theorems on rank of matrices

Since the idea of rank is of fundamental importance in linear algebra we proceed now to derive a number of results involving the ranks of sums and products of matrices.

THEOREM 5.6.1. *If* $\mathbf{A}$ *and* $\mathbf{B}$ *are matrices (of the same type), then*

$$R(\mathbf{A}+\mathbf{B}) \leqslant R(\mathbf{A})+R(\mathbf{B}).$$

Write $R(\mathbf{A}) = r$, $R(\mathbf{B}) = s$. Denote by $\mathbf{x}_1,\ldots,\mathbf{x}_r$ a set of r linearly independent columns of $\mathbf{A}$ and by $\mathbf{y}_1,\ldots,\mathbf{y}_s$ a set of s linearly independent columns of $\mathbf{B}$. Then every column of $\mathbf{A}+\mathbf{B}$ is expressible as a linear combination of the $r+s$ vectors

$$\mathbf{x}_1,\ldots,\mathbf{x}_r, \quad \mathbf{y}_1,\ldots,\mathbf{y}_s. \tag{5.6.1}$$

Let $\mathfrak{U}$ denote the vector space spanned by the columns of $\mathbf{A}+\mathbf{B}$. Then $\mathfrak{U}$ is also spanned by the vectors (5.6.1), and we have

$$R(\mathbf{A}+\mathbf{B}) = d(\mathfrak{U}) \leqslant r+s = R(\mathbf{A})+R(\mathbf{B}).$$

COROLLARY. $\quad R(\mathbf{A}-\mathbf{B}) \geqslant |R(\mathbf{A})-R(\mathbf{B})|.$

This inequality follows at once from Theorem 5.6.1 if we replace $\mathbf{A}$ by $\mathbf{A}-\mathbf{B}$ in that result and also make use of the fact that $R(-\mathbf{C}) = R(\mathbf{C})$.

Exercise 5.6.1. Show that Theorem 5.6.1 is *best possible* in the sense that the sign '$\leqslant$' cannot be replaced by '$<$'.

Theorem 5.6.2. *The rank of a product of two matrices is not greater than the rank of either factor, i.e. if* $\mathbf{AB}$ *exists, then*

$$R(\mathbf{AB}) \leqslant \min\{R(\mathbf{A}),\, R(\mathbf{B})\}.$$

We shall give three proofs of this result.

(i) Let $\mathfrak{U}$ be the vector space of vectors $\mathbf{x}$ such that $\mathbf{Bx} = \mathbf{0}$, and let $\mathfrak{V}$ be the vector space of vectors $\mathbf{x}$ such that $\mathbf{ABx} = \mathbf{0}$. If $\mathbf{x} \in \mathfrak{U}$, then $\mathbf{x} \in \mathfrak{V}$, and so $\mathfrak{U} \subset \mathfrak{V}$. Hence, by Theorem 2.3.4 (p. 53), $d(\mathfrak{U}) \leqslant d(\mathfrak{V})$ and hence, by the dimensionality theorem (Theorem 5.4.2),

$$n - R(\mathbf{B}) \leqslant n - R(\mathbf{AB}),$$

where n is the number of columns of $\mathbf{B}$. Hence $R(\mathbf{AB}) \leqslant R(\mathbf{B})$; and from this we also infer that

$$R(\mathbf{AB}) = R(\mathbf{B}^T\mathbf{A}^T) \leqslant R(\mathbf{A}^T) = R(\mathbf{A}).$$

The assertion therefore follows.

(ii) Let $R(\mathbf{A}) = r$ and let n now denote the number of rows of $\mathbf{A}$. Suppose that $\lambda_1,\ldots,\lambda_n$ are numbers such that

$$\lambda_1 \mathbf{A}_{1*} + \ldots + \lambda_n \mathbf{A}_{n*} = \mathbf{0}. \tag{5.6.2}$$

Then, by Theorem 3.3.5 (i) (p. 84),

$$\lambda_1(\mathbf{AB})_{1*} + \ldots + \lambda_n(\mathbf{AB})_{n*} = \mathbf{0}. \tag{5.6.3}$$

If $r < n$, then any $r+1$ rows of $\mathbf{A}$ are connected by a linear relation in which not all the coefficients are zero; and, in view of (5.6.2) and (5.6.3), it follows that the same is true of the rows of $\mathbf{AB}$. Hence $R(\mathbf{AB}) \leqslant r = R(\mathbf{A})$, and when $r = n$ this inequality holds trivially. The proof is now completed in the same way as in (i).

(iii) A longer but more direct proof is as follows. Suppose, in the first place, that $\mathbf{A}$, $\mathbf{B}$ (and therefore $\mathbf{AB}$) are square matrices of order k, and write $\mathbf{A} = (a_{ij})$, $\mathbf{B} = (b_{ij})$. The case when $R(\mathbf{AB}) = 0$ is trivial and we shall assume that $R(\mathbf{AB}) = r$, where $1 \leqslant r \leqslant k$. Then at least one r-rowed minor of $\mathbf{AB}$ (say Δ) does not vanish, and there is no loss of generality in supposing that Δ lies in the top left-hand corner of $\mathbf{AB}$. Thus

$$\Delta = \begin{vmatrix} a_{11}b_{11}+\ldots+a_{1k}b_{k1} & . & . & . & a_{11}b_{1r}+\ldots+a_{1k}b_{kr} \\ . & . & . & . & . \\ a_{r1}b_{11}+\ldots+a_{rk}b_{k1} & . & . & . & a_{r1}b_{1r}+\ldots+a_{rk}b_{kr} \end{vmatrix},$$

and so, by the corollary to Theorem 1.2.5 (p. 11), we obtain

$$\Delta = \sum_{\nu_1,\ldots,\nu_r=1}^{k} \begin{vmatrix} a_{1\nu_1} b_{\nu_1 1} & . & . & . & a_{1\nu_r} b_{\nu_r r} \\ . & . & . & . & . \\ a_{r\nu_1} b_{\nu_1 1} & . & . & . & a_{r\nu_r} b_{\nu_r r} \end{vmatrix}$$

$$= \sum_{\nu_1,\ldots,\nu_r=1}^{k} b_{\nu_1 1} \ldots b_{\nu_r r} \begin{vmatrix} a_{1\nu_1} & . & . & . & a_{1\nu_r} \\ . & . & . & . & . \\ a_{r\nu_1} & . & . & . & a_{r\nu_r} \end{vmatrix}.$$

But $\Delta \neq 0$ and therefore, for some values of $\nu_1,\ldots,\nu_r$ chosen from $1, 2,\ldots, k$, we must have

$$\begin{vmatrix} a_{1\nu_1} & . & . & . & a_{1\nu_r} \\ . & . & . & . & . \\ a_{r\nu_1} & . & . & . & a_{r\nu_r} \end{vmatrix} \neq 0.$$

Hence $\mathbf{A}$ possesses at least one non-vanishing r-rowed minor, and so

$$R(\mathbf{AB}) \leqslant R(\mathbf{A}). \tag{5.6.4}$$

Next, let $\mathbf{A}$, $\mathbf{B}$ be of type $l \times m$, $m \times n$ respectively, write $\mathbf{C} = \mathbf{AB}$, and put $k = \max(l, m, n)+1$. To each of the three matrices $\mathbf{A}, \mathbf{B}, \mathbf{C}$ we add a suitable number of zero rows and zero columns so as to convert them all into square matrices of order k. The new matrices $\mathbf{A}'$, $\mathbf{B}'$, $\mathbf{C}'$ thus obtained can be expressed most conveniently in partitioned form:

$$\mathbf{A}' = \begin{pmatrix} \mathbf{A} & \mathbf{O}_l^{k-m} \\ \mathbf{O}_{k-l}^m & \mathbf{O}_{k-l}^{k-m} \end{pmatrix}, \quad \mathbf{B}' = \begin{pmatrix} \mathbf{B} & \mathbf{O}_m^{k-n} \\ \mathbf{O}_{k-m}^n & \mathbf{O}_{k-m}^{k-n} \end{pmatrix}, \quad \mathbf{C}' = \begin{pmatrix} \mathbf{AB} & \mathbf{O}_l^{k-n} \\ \mathbf{O}_{k-l}^n & \mathbf{O}_{k-l}^{k-n} \end{pmatrix}.$$

Since $\mathbf{AB} = \mathbf{C}$, it follows easily by Theorem 3.8.1 (p. 104), that $\mathbf{A}'\mathbf{B}' = \mathbf{C}'$. Hence, by (5.6.4), $R(\mathbf{C}') \leqslant R(\mathbf{A}')$. But clearly

$$R(\mathbf{A}') = R(\mathbf{A}), \quad R(\mathbf{B}') = R(\mathbf{B}), \quad R(\mathbf{C}') = R(\mathbf{AB});$$

and therefore (5.6.4) continues to hold in the general case.

The result just proved establishes an inequality between the ranks of two matrices and the rank of their product. In one important case this inequality may be sharpened to an equality.

THEOREM 5.6.3. *The rank of a matrix remains unchanged if the matrix is premultiplied or postmultiplied by a non-singular square matrix.*

Let $\mathbf{A}$ be an $m \times n$ matrix and let $\mathbf{X}$, $\mathbf{Y}$ be non-singular square matrices of order m, n respectively. Then, by Theorem 5.6.2, $R(\mathbf{XA}) \leqslant R(\mathbf{A})$ and also

$$R(\mathbf{A}) = R(\mathbf{X}^{-1}.\mathbf{XA}) \leqslant R(\mathbf{XA}).$$

Hence $R(\mathbf{XA}) = R(\mathbf{A})$. This implies

$$R(\mathbf{AY}) = R(\mathbf{Y}^T\mathbf{A}^T) = R(\mathbf{A}^T) = R(\mathbf{A}).$$

COROLLARY. *Rank is invariant under similarity transformations, i.e.*

$$R(\mathbf{X}^{-1}\mathbf{AX}) = R(\mathbf{A}).$$

In view of Theorem 4.2.2 (p. 116), the interpretation of Theorem 5.6.3 in terms of linear transformations is clear. It shows, in fact, that the rank of a matrix is a number which is characteristic not of that matrix only but of the entire class of matrices representing the same linear transformation. The corollary expresses the same idea for the special case of square matrices and linear transformations of a linear manifold into itself.†

Theorem 5.6.2 gives an upper bound for the rank of the product of two matrices. To obtain a lower bound we first need a preliminary result.

THEOREM 5.6.4. *The vectors of the form* $\mathbf{Bx}$, *subject to the condition* $\mathbf{ABx} = \mathbf{0}$, *constitute a vector space of dimensionality* $R(\mathbf{B})-R(\mathbf{AB})$.

We note that when $\mathbf{B} = \mathbf{I}$, this reduces to the dimensionality theorem (Theorem 5.4.2, p. 149). However, the argument below does not contain a new proof of Theorem 5.4.2 since it depends itself on that theorem.

Let $\mathfrak{B}$ be the set of all $\mathbf{Bx}$ such that $\mathbf{ABx} = \mathbf{0}$. It is at once obvious that $\mathfrak{B}$ is a vector space. Let $\mathbf{A}$, $\mathbf{B}$ be of type $m \times n$, $n \times p$ respectively and write $p-R(\mathbf{B}) = q$, $p-R(\mathbf{AB}) = r$. Then, by Theorem 5.6.2, $0 \leqslant q \leqslant r$. We shall, in the first place, assume that $0 < q < r$.

By Theorem 5.4.2, the vector space of vectors $\mathbf{x}$ satisfying $\mathbf{Bx} = \mathbf{0}$ is q-dimensional, and so has a basis $\{\mathbf{x}_1,..., \mathbf{x}_q\}$, say. Again, the vector space of vectors $\mathbf{x}$ satisfying $\mathbf{ABx} = \mathbf{0}$ is r-dimensional and, in view of the corollary to Theorem 2.3.5 (p. 54), there exist vectors $\mathbf{x}_{q+1},..., \mathbf{x}_r$ such that $\{\mathbf{x}_1,..., \mathbf{x}_q, \mathbf{x}_{q+1},..., \mathbf{x}_r\}$ is a basis of this vector space. The $r-q$ vectors $\mathbf{Bx}_{q+1},..., \mathbf{Bx}_r$ are linearly independent; for if

$$\mathbf{0} = \alpha_{q+1}.\mathbf{Bx}_{q+1}+...+\alpha_r.\mathbf{Bx}_r = \mathbf{B}(\alpha_{q+1}\mathbf{x}_{q+1}+...+\alpha_r\mathbf{x}_r),$$

then there exist scalars $\alpha_1,..., \alpha_q$ such that

$$\alpha_{q+1}\mathbf{x}_{q+1}+...+\alpha_r\mathbf{x}_r = \alpha_1\mathbf{x}_1+...+\alpha_q\mathbf{x}_q,$$

and this implies that $\alpha_{q+1} = ... = \alpha_r = 0$. Thus $\mathfrak{B}$ possesses $r-q$ linearly independent vectors $\mathbf{Bx}_{q+1},..., \mathbf{Bx}_r$. Suppose now that $\mathbf{Bx} \in \mathfrak{B}$. Then $\mathbf{ABx} = \mathbf{0}$, and so $\mathbf{x}$ can be writen in the form

$$\mathbf{x} = \beta_1\mathbf{x}_1+...+\beta_q\mathbf{x}_q+\beta_{q+1}\mathbf{x}_{q+1}+...+\beta_r\mathbf{x}_r.$$

Hence $$\mathbf{Bx} = \beta_{q+1}.\mathbf{Bx}_{q+1}+...+\beta_r.\mathbf{Bx}_r,$$

† Cf. Theorem 4.2.5 (i) (p. 119).

and therefore $\mathbf{Bx}_{q+1},\ldots,\mathbf{Bx}_r$ are generators of $\mathfrak{B}$. Thus, when $0 < q < r$, $d(\mathfrak{B}) = r-q = R(\mathbf{B})-R(\mathbf{AB})$.

The remaining cases can be disposed of without difficulty. If $0 = q = r$, then only the zero vector satisfies $\mathbf{ABx} = \mathbf{0}$. Hence $\mathfrak{B}$ consists of $\mathbf{0}$ only, and $d(\mathfrak{B}) = 0 = q-r$. If $0 < q = r$, then the q linearly independent solutions $\mathbf{x}_1,\ldots,\mathbf{x}_q$ of $\mathbf{Bx} = \mathbf{0}$ form a basis of the space of vectors such that $\mathbf{ABx} = \mathbf{0}$. Hence any such vector $\mathbf{x}$ may be written as

$$\mathbf{x} = \beta_1\mathbf{x}_1+\ldots+\beta_q\mathbf{x}_q.$$

This implies that $\mathbf{Bx} = \mathbf{0}$, and hence $\mathfrak{B}$ consists of $\mathbf{0}$ only, so that $d(\mathfrak{B}) = 0 = r-q$. Finally, let $0 = q < r$. In this case let $\{\mathbf{x}_1,\ldots,\mathbf{x}_r\}$ denote a basis of the space of vectors $\mathbf{x}$ such that $\mathbf{ABx} = \mathbf{0}$. Then, arguing as above, we easily see that $\mathbf{Bx}_1,\ldots,\mathbf{Bx}_r$ are linearly independent generators of $\mathfrak{B}$. Hence $d(\mathfrak{B}) = r = r-q$.

Theorem 5.6.5. *If the product* $\mathbf{AB}$ *exists, then*

$$R(\mathbf{AB}) \geqslant R(\mathbf{A})+R(\mathbf{B})-n,$$

where n *is the number of columns in* $\mathbf{A}$ *(and of rows in* $\mathbf{B}$*).*

Theorems 5.6.2 and 5.6.5 were found, for square matrices, by Sylvester in 1884 and are known jointly as *Sylvester's law of nullity*.

The 'nullity', say $\nu(\mathbf{A})$, of an $n \times n$ matrix $\mathbf{A}$ is defined as $n-R(\mathbf{A})$. For square matrices Theorems 5.6.2 and 5.6.5 state that

$$\max\{\nu(\mathbf{A}),\nu(\mathbf{B})\} \leqslant \nu(\mathbf{AB}) \leqslant \nu(\mathbf{A})+\nu(\mathbf{B}).$$

To prove Theorem 5.6.5 we do not require the full force of Theorem 5.6.4. For our purpose it is sufficient to know that there exist $R(\mathbf{B})-R(\mathbf{AB})$ linearly independent vectors $\mathbf{y} = \mathbf{Bx}$ satisfying $\mathbf{Ay} = \mathbf{0}$. Hence, by Theorem 5.4.2,

$$R(\mathbf{B})-R(\mathbf{AB}) \leqslant n-R(\mathbf{A}),$$

and the assertion follows.

Exercise 5.6.2. Show that the sign '$\geqslant$' in Theorem 5.6.5 cannot be replaced by '$>$'.

A result proved by Frobenius in 1911 goes a little beyond Theorem 5.6.5.

Theorem 5.6.6. *If* $\mathbf{ABC}$ *exists, then*

$$R(\mathbf{AB})+R(\mathbf{BC}) \leqslant R(\mathbf{B})+R(\mathbf{ABC}).$$

This theorem contains the law of nullity, for when $\mathbf{B} = \mathbf{I}$ it reduces to Theorem 5.6.5, while the two cases $\mathbf{A} = \mathbf{O}$ and $\mathbf{C} = \mathbf{O}$ lead at once to Theorem 5.6.2.

Denote by $\mathfrak{U}$ the vector space of all vectors $\mathbf{BCx}$ such that $\mathbf{ABCx} = \mathbf{0}$, and by $\mathfrak{B}$ the vector space of all vectors $\mathbf{Bx}$ such that $\mathbf{ABx} = \mathbf{0}$. Clearly $\mathfrak{U} \subset \mathfrak{B}$, and so $d(\mathfrak{U}) \leqslant d(\mathfrak{B})$. But, by Theorem 5.6.4,

$$d(\mathfrak{U}) = R(\mathbf{BC}) - R(\mathbf{ABC}), \qquad d(\mathfrak{B}) = R(\mathbf{B}) - R(\mathbf{AB}),$$

and the required theorem follows at once.

It should be noted that in the discussion of the last three theorems we have been employing, in effect, arguments of the 'invariant' type.

PROBLEMS ON CHAPTER V

1. Solve completely the following systems of equations.

(i) $$\begin{aligned} 2x-3y+6z &= 3,\\ 4x-y+z &= 1,\\ 3x-2y+3z &= 4. \end{aligned}$$

(ii) $$\begin{aligned} x+y-2z+3w &= 0,\\ x-2y+z-w &= 0,\\ x+7y-8z+11w &= 0,\\ x-5y+4z-5w &= 0. \end{aligned}$$

(iii) $$\begin{aligned} 5x+2y-3u-v &= 11,\\ 5x-y+5z-u-2v &= 2,\\ x-2y+4z+u-v &= -5. \end{aligned}$$

(iv) $$\begin{aligned} -x+2y+z+3w &= 6,\\ 7x+7y-13z+3w &= -24,\\ 3x+y-5z-w &= -12. \end{aligned}$$

(v) $$\begin{aligned} 11x+8y-2z+3w &= 0,\\ 2x+3y-z+2w &= 0,\\ 7x-y+z-3w &= 0,\\ 4x-11y+5z-12w &= 0. \end{aligned}$$

(vi) $$\begin{aligned} x+2y-z &= 2,\\ 2x-3y+7z &= -1,\\ -x+y+3z &= 6,\\ 5x+y-2z &= 0. \end{aligned}$$

(vii) $$\begin{aligned} 4x-2y+5z &= 0,\\ x-y+z &= 0,\\ 5x-y+7z &= 0,\\ -3x+y-4z &= 0. \end{aligned}$$

(viii) $$\begin{aligned} 3x+2y-2z-w &= 9,\\ x-y+z+2w &= 7,\\ 5x-10y+10z+17w &= 47. \end{aligned}$$

2. Solve completely the system of equations

$$\begin{aligned} (n-1)x_1 &= x_2+x_3+\dots+x_n,\\ (n-1)x_2 &= x_1+x_3+\dots+x_n,\\ &\dots\dots\dots\\ (n-1)x_n &= x_1+x_2+\dots+x_{n-1}. \end{aligned}$$

3. For what values of a, b, c is the system of equations

$$\begin{aligned} (a+3)x-2y+3z &= 4,\\ 3x+(a-3)y+9z &= b,\\ 4x-8y+(a+14)z &= c \end{aligned}$$

consistent?

4. Discuss, for all values of a, the system of equations

$$\begin{aligned} x+y+z &= 2, \\ 2x+y-2z &= 2, \\ ax+y+4z &= 2. \end{aligned}$$

5. Discuss, for all values of a, the system of equations

$$\begin{aligned} ax+(3a+4)y+2(a+1)z &= 0, \\ ax+(4a+2)y+(a+4)z &= 0, \\ 2x+(3a+4)y+3az &= 0. \end{aligned}$$

6. Illustrate Theorem 5.4.2 by discussing, for all values of ρ, the system of equations

$$\begin{aligned} (2\rho+1)x+(\rho+2)y+(3\rho+3)z &= 0, \\ (5\rho+1)x+(3\rho+3)y+(7\rho+5)z &= 0, \\ 3\rho x+3\rho y+(4\rho+2)z &= 0. \end{aligned}$$

7. Determine the conditions for the consistency of the system of equations

$$\begin{aligned} x+y-2z &= 0, \\ ax+by+cz &= 0, \\ bx+cy+az &= d, \end{aligned}$$

and obtain the complete solution in each consistent case.

8. Determine the conditions for the consistency of the system of equations

$$\begin{aligned} x+y+z &= b, \\ ax+by+bz &= a^2, \\ ax+cy+dz &= ab, \end{aligned}$$

and obtain the complete solution in each consistent case.

9. Determine the conditions for the consistency of the system of equations

$$\begin{aligned} x+y+z &= 1, \\ ax+by+cz &= d, \\ a^2x+b^2y+c^2z &= d^2, \end{aligned}$$

and obtain the complete solution in each consistent case.

10. Determine the conditions for the consistency of the system of equations

$$\begin{aligned} x+y+z &= 1, \\ ax+by+cz &= d, \\ a^3x+b^3y+c^3z &= d^3, \end{aligned}$$

and obtain the complete solution in each consistent case.

11. Let $\mathfrak{M}$ be a linear manifold, of dimensionality n, over a reference field $\mathfrak{F}$. Show that there is a biunique correspondence between the set of linear transformations of $\mathfrak{M}$ into $\mathfrak{F}$, the set of vectors of order n over $\mathfrak{F}$, and the set of linear forms in n variables with coefficients in $\mathfrak{F}$.

12. Show, by an example, that linear dependence of the columns of a matrix does not imply the linear dependence of the rows.

13. Let $\mathbf{A}$, $\mathbf{B}$ be matrices of type $m \times n$, $n \times p$ respectively. Show that the inequality $R(\mathbf{AB}) \geqslant R(\mathbf{A})+p-n$ does not always hold.

14. Let $\mathbf{A}$ be a square matrix. Show that the inequality $R(\mathbf{A}) \leqslant R(\mathbf{A}^2)+1$ does not always hold.

15. Show that, if the numbers $a_1,\ldots,a_n$ are distinct, then the n vectors $(1, a_r, a_r^2,\ldots, a_r^{n-1})$ $(r = 1,\ldots,n)$ are linearly independent.

16. The vectors $\mathbf{x}_1,\ldots,\mathbf{x}_m$ are defined by the equations

$$\mathbf{x}_i = (a_{i1},\ldots,a_{in}), \quad a_{ij} = b_i+(i+j)c_i \qquad (i = 1,\ldots,m;\, j = 1,\ldots,n).$$

Show that, if $m \geqslant 3$, then $\mathbf{x}_1,\ldots,\mathbf{x}_m$ are linearly dependent.

17. Show that, given any system of m linear homogeneous equations in $m+2$ unknowns $x_1,\ldots,x_{m+2}$, there are two indices i,j in the range $1, 2,\ldots, m+2$ such that the system possesses a solution in which x_i and x_j have any prescribed values.

18. Show that, for any given constants a_{ij} and b_i $(i = 1,\ldots,m;\, j = 1,\ldots,n)$, exactly one of the two systems of equations

$$\text{(i)} \quad \sum_{j=1}^{n} a_{ij}x_j = b_i \qquad (i = 1,\ldots,m),$$

$$\text{(ii)} \quad \sum_{i=1}^{m} a_{ij}y_i = 0 \qquad (j = 1,\ldots,n), \qquad \sum_{i=1}^{m} b_i y_i = 1$$

is consistent.

19. Show that, if $\mathbf{A}$ has n columns and $\mathbf{B}$ n rows and if $\mathbf{AB} = \mathbf{O}$, then $R(\mathbf{A})+R(\mathbf{B}) \leqslant n$.

20. Show that, if the $n\times n$ matrix $\mathbf{A}$ satisfies the equation $\mathbf{A}^2 = \mathbf{A}$, then $R(\mathbf{A})+R(\mathbf{I}-\mathbf{A}) = n$.

21. Let $\mathbf{A}, \mathbf{B}$ be rectangular matrices for which the product $\mathbf{AB}$ is defined. Show that $R(\mathbf{AB}) = R(\mathbf{B})$ if and only if $\mathbf{ABx} = \mathbf{0}$ implies $\mathbf{Bx} = \mathbf{0}$.

22. Let $\mathbf{A}$ be a square matrix, p a positive integer, and $\mathbf{x}$ a non-zero vector such that $\mathbf{A}^p\mathbf{x} \neq \mathbf{0}$, $\mathbf{A}^{p+1}\mathbf{x} = \mathbf{0}$. Show that the vectors $\mathbf{x}, \mathbf{Ax},\ldots, \mathbf{A}^p\mathbf{x}$ are linearly independent.

23. Show that, if $R(\mathbf{A}) = R(\mathbf{BA})$, then $R(\mathbf{AC}) = R(\mathbf{BAC})$.

24. Show that, for $k \geqslant 2$,

$$R(\mathbf{A}_1\ldots\mathbf{A}_k) \geqslant R(\mathbf{A}_1)+\ldots+R(\mathbf{A}_k)-n(k-1),$$

where $\mathbf{A}_1,\ldots,\mathbf{A}_k$ are square matrices of order n.

25. Show that, if k is a positive integer such that $R(\mathbf{A}^k) = R(\mathbf{A}^{k+1})$, then

$$R(\mathbf{A}^{k+1}) = R(\mathbf{A}^{k+2}) = R(\mathbf{A}^{k+3}) = \ldots.$$

26. A sequence $\{a_k\}$ of real numbers is said to be *convex* if, for all k,

$$a_{k+1} \leqslant \tfrac{1}{2}(a_k+a_{k+2}).$$

Show that the sequence $\{R(\mathbf{A}^k)\}$ is convex.

27. Let $\mathbf{A}$ be any $n\times n$ matrix. Show, using either No. 22 or No. 25, that

$$R(\mathbf{A}^n) = R(\mathbf{A}^{n+1}) = R(\mathbf{A}^{n+2}) = \ldots.$$

28. Let the matrices $\mathbf{P}_1,\ldots, \mathbf{P}_k, \mathbf{Q}_1,\ldots, \mathbf{Q}_k$ commute in pairs, and suppose that $R(\mathbf{P}_\kappa) = R(\mathbf{P}_\kappa \mathbf{Q}_\kappa)$ $(\kappa = 1,\ldots, k)$. Show that

$$R(\mathbf{P}_1\ldots\mathbf{P}_k) = R(\mathbf{P}_1\ldots\mathbf{P}_k\, \mathbf{Q}_1\ldots\mathbf{Q}_k).$$

29. Let the matrices $\mathbf{A}_1,\ldots, \mathbf{A}_k$ commute in pairs and suppose that $R(\mathbf{A}_\kappa) = R(\mathbf{A}_\kappa^2)$ $(\kappa = 1,\ldots, k)$. Show that, for any positive integers $\alpha_1,\ldots, \alpha_k$,

$$R(\mathbf{A}_1\ldots\mathbf{A}_k) = R(\mathbf{A}_1^{\alpha_1}\ldots\mathbf{A}_k^{\alpha_k}).$$

30. Let $\theta_1,\ldots,\theta_n$ be given numbers and let the linear forms $L_1,\ldots,L_n$ be defined by

$$L_k = L_k(x_1,\ldots,x_n) = x_1+\theta_k x_2+\theta_k^2 x_3+\ldots+\theta_k^{n-1}x_n \quad (k = 1,\ldots,n).$$

Show that the maximum number of linearly independent forms among $L_1,\ldots,L_n$ is equal to the number of distinct numbers among $\theta_1,\ldots,\theta_n$.

31. Let $\mathbf{u}_1,\ldots,\mathbf{u}_n$ be n linearly independent complex vectors of order n. Show that there exist n vectors $\mathbf{x}_1,\ldots,\mathbf{x}_n$ such that $(\mathbf{x}_i,\mathbf{u}_j) = \delta_{ij}\ (1 \leqslant i,j \leqslant n)$.

32. Let $\mathbf{A}$ be an $m\times n$ matrix of rank r. Show that the general solution of the system of homogeneous equations $\mathbf{Ax} = \mathbf{0}$ can be written in the form $\mathbf{x} = \mathbf{P}\boldsymbol{\lambda}$, where $\mathbf{P}$ is any fixed $n\times(n-r)$ matrix such that $\mathbf{AP} = \mathbf{O}$, $R(\mathbf{P}) = n-r$, and $\boldsymbol{\lambda}$ is a variable vector of order $n-r$.

33. Let $n > 1$ and suppose that $\mathbf{A}$ is an $n\times n$ matrix. Show that (i) $R(\mathbf{A}) = n$ if and only if $R(\mathbf{A}^*) = n$; (ii) $R(\mathbf{A}) = n-1$ if and only if $R(\mathbf{A}^*)=1$; (iii) $R(\mathbf{A}) < n-1$ if and only if $\mathbf{A}^* = \mathbf{O}$.

34. Show that, if the matrix $\mathbf{A}$ is singular, then its adjugate can be expressed in the form $\mathbf{A}^* = \mathbf{x}\mathbf{y}^T$, where $\mathbf{x}$ and $\mathbf{y}$ are column vectors.

35. Let $\mathfrak{U}$ be an r-dimensional subspace of $\mathfrak{V}_n$, where $1 \leqslant r < n$. Show that an $n\times n$ matrix $\mathbf{A}$ satisfies the conditons

$$\mathbf{Ax} = \mathbf{0} \text{ (for all } \mathbf{x} \in \mathfrak{U}), \qquad \mathbf{Ax} \neq \mathbf{0} \text{ (for all } \mathbf{x} \notin \mathfrak{U})$$

if and only if it satisfies the conditions

$$\mathbf{Ax}_1 = \ldots = \mathbf{Ax}_r = \mathbf{0}, \qquad R(\mathbf{A}) = n-r,$$

where $\mathbf{x}_1,\ldots,\mathbf{x}_r$ are any r linearly independent vectors of $\mathfrak{U}$.

36. Suppose that the system of equations $\sum_{j=1}^{n} a_{ij}x_j = b_i\ (i = 1,\ldots,m)$ possesses a unique solution. Show that this solution is given by

$$x_i = \sum_{j=1}^{m} \sigma_{ij}b_j \qquad (i = 1,\ldots,n),$$

where the σ's depend only on the a's.

37. A square matrix $\mathbf{A}$, of order n, is called a $\mathbf{G}$-matrix if the transformation $\mathbf{y} = \mathbf{Ax}$ implies the identity $\mathbf{y}^T\mathbf{Gy} = \mathbf{x}^T\mathbf{Gx}$, where $\mathbf{G}$ is a given non-singular symmetric matrix. Prove that, if $\mathbf{A}$ is a $\mathbf{G}$-matrix, then $|\mathbf{A}| = \pm 1$.

If $\mathbf{A} = (a_{ij})$ and $\mathbf{B} = (b_{ij})$ are two $\mathbf{G}$-matrices and if $|\mathbf{A}|+|\mathbf{B}| = 0$, prove that the system of equations

$$\sum_{j=1}^{n}(a_{ij}x_j+b_{ij}x_j) = 0 \qquad (i = 1,\ldots,n)$$

possesses a non-trivial solution.

38. $\mathbf{A}$ is a symmetric $n\times n$ matrix of rank $n-1$, and $\boldsymbol{\xi}$ is a non-zero vector such that $\mathbf{A}\boldsymbol{\xi} = \mathbf{0}$. Show that the relations $\mathbf{u}^T\boldsymbol{\xi} = 0$ and $\mathbf{u}^T\mathbf{A}^*\mathbf{u} = 0$ imply each other.

39. Prove that a necessary and sufficient condition for the system of equations

$$\sum_{j=1}^{n} a_{ij}x_j = b_i \qquad (i = 1,\ldots,m)$$

to have a solution is that every solution of the system of equations

$$\sum_{i=1}^{m} a_{ij} y_i = 0 \qquad (j = 1,..., n)$$

should also satisfy
$$\sum_{i=1}^{m} b_i y_i = 0.$$

Find for what values of η the equations

$$x+y+z = 1,$$
$$x+2y+4z = \eta,$$
$$x+4y+10z = \eta^2$$

have a solution; and solve them completely in each case.

40. Let $k \leqslant n$ and let $L_1,..., L_k$ be linearly independent linear forms in $x_1,..., x_n$. Show that there exists a non-singular linear transformation of $x_1,..., x_n$ into $y_1,..., y_n$ which carries $L_1,..., L_k$ into $y_1,..., y_k$ respectively.

41. $\mathbf{A}$ is a square matrix of order n and rank r; $\mathbf{x}_1,..., \mathbf{x}_k$ are $n \times 1$ matrices and $\mathbf{u}_1,..., \mathbf{u}_k$ are $1 \times n$ matrices. The matrix $\mathbf{B}$ is formed by bordering $\mathbf{A}$ with columns $\mathbf{x}_1,..., \mathbf{x}_k$ and rows $\mathbf{u}_1,..., \mathbf{u}_k$, and completing the matrix with a block of k^2 zeros. Prove that, if $k < n-r$, then $|\mathbf{B}| = 0$.

42. Show that the system of equations $\sum_{j=1}^{n} a_{ij} x_j = b_i$ $(i = 1,..., m)$ is consistent if and only if

$$\sum_{i=1}^{m} p_i \sum_{j=1}^{n} a_{ij} x_j = 0 \quad \text{(identically in } x_1,..., x_n)$$

implies $\sum_{i=1}^{m} p_i b_i = 0$.

$\mathbf{A}$ and $\mathbf{B}$ are rectangular matrices having the same number of columns. Show that a necessary and sufficient condition for the existence of a matrix $\mathbf{C}$ such that $\mathbf{B} = \mathbf{CA}$ is that $\mathbf{Ax} = \mathbf{0}$ should imply $\mathbf{Bx} = \mathbf{0}$.

43. Show that, if $\mathbf{A}^2 = \mathbf{A}$, $\mathbf{AB} = \mathbf{BA} = \mathbf{O}$, then

$$R(\mathbf{A}+\mathbf{B}) = R(\mathbf{A})+R(\mathbf{B}).$$

State and prove the analogous result for a set of m matrices.

44. Let $1 \leqslant r < n$ and suppose that $\mathbf{x}_1,..., \mathbf{x}_r$ are linearly independent vectors in $\mathfrak{V}_n$. Show that there exists a non-singular $n \times n$ matrix $\mathbf{A}$ such that the last $n-r$ components of each of the vectors $\mathbf{Ax}_1,..., \mathbf{Ax}_r$ all vanish.

VI

ELEMENTARY OPERATIONS AND THE CONCEPT OF EQUIVALENCE

THOUGH the results obtained below are interesting and afford additional insight into the structure of matrices, they will not—with very few exceptions—be needed subsequently. The present chapter (apart from § 6.5) may therefore be omitted by the reader who is anxious to proceed at once to the more advanced theory of Part II.

6.1. E-operations and E-matrices

6.1.1. DEFINITION 6.1.1. *An* ELEMENTARY OPERATION *(or, more briefly, an E-*OPERATION*) on a matrix is an operation of one of the following three types.*

(i) *The interchange of two rows (or columns).*

(ii) *The multiplication of a row (or column) by a non-zero scalar.*

(iii) *The addition of a multiple of one row (or column) to another row (or column).*

We distinguish between *row operations* and *column operations* according as the E-operations in question apply to rows or to columns.

Our aim is to study the effect of E-operations upon matrices, and it is useful to note at the outset that if $\mathbf{B}$ is obtained from $\mathbf{A}$ by an E-operation, then $\mathbf{A}$ can be obtained from $\mathbf{B}$ by an E-operation.

EXERCISE 6.1.1. Prove this statement and deduce that, if $\mathbf{D}$ is obtained from $\mathbf{C}$ by a chain of E-operations, then the converse is also true, i.e. $\mathbf{C}$ can be obtained from $\mathbf{D}$ by a chain of E-operations.†

EXERCISE 6.1.2. Show that the *identical transformation*, i.e. one which leaves the matrix unchanged, is an E-operation.

We shall adopt the following notation for E-operations. The interchange of the ith and jth rows (columns) will be denoted by $R_i \leftrightarrow R_j$ ($C_i \leftrightarrow C_j$); the multiplication of the ith row (column) by a scalar $\lambda \neq 0$ by $R_i \rightarrow \lambda R_i$ ($C_i \rightarrow \lambda C_i$); and the addition of μ times the jth row (column) to the ith row (column) by $R_i \rightarrow R_i + \mu R_j$ ($C_i \rightarrow C_i + \mu C_j$).

† The term 'chain' is taken to mean 'finite chain'. A chain may, of course, consist of a single operation only.

THEOREM 6.1.1. *The rank of a matrix is invariant (i.e. remains unchanged) under elementary operations.*

Operations of types (i) and (ii) obviously do not affect rank and we need therefore only consider those of type (iii). We may, without loss of generality, restrict our attention to row operations.

Let, then, the matrix $\mathbf{B}$ be obtained as the result of an E-operation of type (iii) on the rows of $\mathbf{A}$. Then each row of $\mathbf{B}$ is a linear combination of the rows of $\mathbf{A}$. Hence the maximum number of linearly independent rows in $\mathbf{B}$ is not greater than the maximum number of linearly independent rows in $\mathbf{A}$,† i.e. $R(\mathbf{B}) \leqslant R(\mathbf{A})$. But $\mathbf{A}$ may be obtained from $\mathbf{B}$ by means of an E-operation; hence $R(\mathbf{A}) \leqslant R(\mathbf{B})$, and so $R(\mathbf{A}) = R(\mathbf{B})$. The proof is therefore complete.‡

The notion of *invariance* which appears in Theorem 6.1.1, and which we have met earlier on a number of occasions, is a key concept in algebra. Before passing on to further results on E-operations we shall, therefore, formulate this notion in general terms.

Let $\mathfrak{S}$ be a set whose elements we denote by $x, y, z,\ldots$. Let Ω be a transformation of $\mathfrak{S}$ into itself, i.e. an operator operating on the elements of $\mathfrak{S}$ in such a way that with each element $x \in \mathfrak{S}$ is associated a unique element $y = \Omega x \in \mathfrak{S}$. Suppose, furthermore, that a function f is defined on $\mathfrak{S}$ so that with each $x \in \mathfrak{S}$ is associated a functional value $f(x)$ (which need not, of course, be either a number or an element of $\mathfrak{S}$). If, for every $x \in \mathfrak{S}$,

$$f(\Omega x) = f(x), \tag{6.1.1}$$

then we say that f is *invariant under (or with respect to) the operator* Ω.

In Theorem 6.1.1 above, $\mathfrak{S}$ is the set of matrices $x, y,\ldots$; $f(x)$ is the rank of the matrix x, and Ω is any E-operation. In the corollary to Theorem 5.6.3 (p. 160), $\mathfrak{S}$ is the set of square matrices, $f(x)$ the rank of the matrix x, and Ω any similarity transformation. Again, the proposition that the value of a determinant remains unaltered when its rows and columns are interchanged, may also be expressed in the same terms by taking $\mathfrak{S}$ as the set of square matrices, $f(x)$ as the determinant of the matrix x, and Ω as the operation of transposition.

It is easy to see why the notion of invariance is important. For let Φ be a set of operators on $\mathfrak{S}$ and suppose that (6.1.1) holds for every $x \in \mathfrak{S}$ and for every $\Omega \in \Phi$; in other words, suppose that f is invariant under every operator of the set Φ. If now the determination of $f(x_0)$, for some particular $x_0 \in \mathfrak{S}$, is difficult or tedious we may be able to find a suitable operator $\Omega_0 \in \Phi$ such that the determination of $f(\Omega_0 x_0)$ can be carried out more easily. Since, however, $f(x_0) = f(\Omega_0 x_0)$, the original difficulty will then have been

† Cf. Problem II, 5.

‡ For an alternative proof depending on consideration of minors, see Bôcher, **16**, 55.

overcome. One useful application of this procedure will be found, for example, at the end of § 6.2.2.

6.1.2. We resume now our discussion of E-operations. The next step—an important one—is to represent such operations by means of matrix multiplication. This device will enable us to use matrix technique when dealing with E-operations.

DEFINITION 6.1.2. *An* ELEMENTARY MATRIX *(or, more briefly,* E-MATRIX*) is any matrix derived from a unit matrix by a single E-operation.*

Thus, for instance,

$$\begin{pmatrix} 0 & 1 & 0 \\ 1 & 0 & 0 \\ 0 & 0 & 1 \end{pmatrix}, \qquad \begin{pmatrix} 1 & 0 & 0 \\ 0 & 5 & 0 \\ 0 & 0 & 1 \end{pmatrix}, \qquad \begin{pmatrix} 1 & 0 & 0 \\ 0 & 1 & 0 \\ 3 & 0 & 1 \end{pmatrix}$$

are the E-matrices obtained from $\mathbf{I}_3$ by means of the E-operations $C_1 \leftrightarrow C_2$, $C_2 \to 5C_2$, $R_3 \to R_3 + 3R_1$ respectively.

It should be noted that any E-matrix can be obtained equally well by a row operation or a column operation on $\mathbf{I}$. For the operations $R_i \leftrightarrow R_j$ and $C_i \leftrightarrow C_j$ have the same effect on $\mathbf{I}$; so have $R_i \to \lambda R_i$ and $C_i \to \lambda C_i$; and, finally, so have $R_i \to R_i + \mu R_j$ and $C_j \to C_j + \mu C_i$.

THEOREM 6.1.2. *An elementary operation on the rows (columns) of an $m \times n$ matrix $\mathbf{A}$ is equivalent to premultiplication (post-multiplication) of $\mathbf{A}$ by the elementary matrix derived by the same operation from $\mathbf{I}_m$ ($\mathbf{I}_n$).*

Let ϕ denote any E-operation on the rows, ψ any E-operation on the columns; and write $\phi(\mathbf{X})$, $\psi(\mathbf{X})$ for the matrices obtained when $\mathbf{X}$ is operated on by ϕ, ψ respectively. The theorem then asserts that

$$\phi(\mathbf{A}) = \phi(\mathbf{I}_m)\mathbf{A}, \qquad \psi(\mathbf{A}) = \mathbf{A}\psi(\mathbf{I}_n).$$

The proof is almost immediate. For if $\mathbf{X}$, $\mathbf{Y}$ are any matrices for which the product $\mathbf{XY}$ exists, then $(\mathbf{XY})_{i*} = \mathbf{X}_{i*}\mathbf{Y}$; and consequently any E-operation on the rows of $\mathbf{X}$ effects the same operation on the rows of $\mathbf{XY}$. Thus we have

$$\phi(\mathbf{XY}) = \phi(\mathbf{X})\mathbf{Y}. \tag{6.1.2}$$

Similarly, any E-operation on the columns of $\mathbf{Y}$ effects the same operation on the columns of $\mathbf{XY}$, so that

$$\psi(\mathbf{XY}) = \mathbf{X}\psi(\mathbf{Y}). \tag{6.1.3}$$

As special cases of (6.1.2) and (6.1.3) we obtain

$$\phi(\mathbf{A}) = \phi(\mathbf{I}_m \mathbf{A}) = \phi(\mathbf{I}_m)\mathbf{A}; \qquad \psi(\mathbf{A}) = \psi(\mathbf{A}\mathbf{I}_n) = \mathbf{A}\psi(\mathbf{I}_n),$$

and the theorem is therefore proved.

EXERCISE 6.1.3. Let $\mathbf{A}$ be an $m \times n$ matrix and let $\mathbf{E}, \mathbf{E}'$ be any E-matrices of order m, n respectively. Show that $\mathbf{EA}$ can be obtained by an E-operation on the rows of $\mathbf{A}$ and $\mathbf{AE}'$ by an E-operation on the columns of $\mathbf{A}$.

To illustrate Theorem 6.1.2, consider for instance the 3×4 matrix

$$\mathbf{A} = \begin{pmatrix} a_1 & a_2 & a_3 & a_4 \\ b_1 & b_2 & b_3 & b_4 \\ c_1 & c_2 & c_3 & c_4 \end{pmatrix}.$$

If ϕ is the operation $R_2 \leftrightarrow R_3$, then

$$\phi(\mathbf{A}) = \begin{pmatrix} a_1 & a_2 & a_3 & a_4 \\ c_1 & c_2 & c_3 & c_4 \\ b_1 & b_2 & b_3 & b_4 \end{pmatrix}, \qquad \phi(\mathbf{I}_3) = \begin{pmatrix} 1 & 0 & 0 \\ 0 & 0 & 1 \\ 0 & 1 & 0 \end{pmatrix};$$

and we have, in fact,

$$\begin{pmatrix} a_1 & a_2 & a_3 & a_4 \\ c_1 & c_2 & c_3 & c_4 \\ b_1 & b_2 & b_3 & b_4 \end{pmatrix} = \begin{pmatrix} 1 & 0 & 0 \\ 0 & 0 & 1 \\ 0 & 1 & 0 \end{pmatrix} \begin{pmatrix} a_1 & a_2 & a_3 & a_4 \\ b_1 & b_2 & b_3 & b_4 \\ c_1 & c_2 & c_3 & c_4 \end{pmatrix}.$$

Again, if ψ is the operation $C_4 \to C_4 - 2C_1$, then

$$\psi(\mathbf{A}) = \begin{pmatrix} a_1 & a_2 & a_3 & a_4 - 2a_1 \\ b_1 & b_2 & b_3 & b_4 - 2b_1 \\ c & c_2 & c_3 & c_4 - 2c_1 \end{pmatrix}, \qquad \psi(\mathbf{I}_4) = \begin{pmatrix} 1 & 0 & 0 & -2 \\ 0 & 1 & 0 & 0 \\ 0 & 0 & 1 & 0 \\ 0 & 0 & 0 & 1 \end{pmatrix};$$

and we have, in fact,

$$\begin{pmatrix} a_1 & a_2 & a_3 & a_4 - 2a_1 \\ b_1 & b_2 & b_3 & b_4 - 2b_1 \\ c_1 & c_2 & c_3 & c_4 - 2c_1 \end{pmatrix} = \begin{pmatrix} a_1 & a_2 & a_3 & a_4 \\ b_1 & b_2 & b_3 & b_4 \\ c_1 & c_2 & c_3 & c_4 \end{pmatrix} \begin{pmatrix} 1 & 0 & 0 & -2 \\ 0 & 1 & 0 & 0 \\ 0 & 0 & 1 & 0 \\ 0 & 0 & 0 & 1 \end{pmatrix}.$$

THEOREM 6.1.3. (i) *Every elementary matrix is non-singular.* (ii) *The inverse of an elementary matrix is again an elementary matrix.*

The first assertion is obvious by Theorem 6.1.1 since an E-matrix is derived by an E-operation from a non-singular matrix.

To prove the second assertion let $\mathbf{E}$ denote a given E-matrix, say of order n. Then $\mathbf{I}_n$ can be derived from $\mathbf{E}$ by a row operation, i.e.

$\mathbf{I}_n = \phi(\mathbf{E})$. Writing $\mathbf{E}' = \phi(\mathbf{I}_n)$ and using Theorem 6.1.2 we obtain

$$\mathbf{E}'\mathbf{E} = \phi(\mathbf{I}_n)\mathbf{E} = \phi(\mathbf{E}) = \mathbf{I}_n.$$

The inverse of $\mathbf{E}$ is therefore the E-matrix $\mathbf{E}'$.

6.2. Equivalent matrices

6.2.1. We begin this section by introducing a general notion which is useful in the investigation of algebraic problems.

Let Ω_1, Ω_2 be operators mapping a set $\mathfrak{S}$ of elements into itself, so that if $x \in \mathfrak{S}$, then $\Omega_1 x, \Omega_2 x \in \mathfrak{S}$.

DEFINITION 6.2.1. *The* PRODUCT $\Omega_2 \Omega_1$ *of the operators* Ω_1 *and* Ω_2 *(in that order) is defined by the relation*

$$(\Omega_2\Omega_1)x = \Omega_2(\Omega_1 x) \qquad (x \in \mathfrak{S}). \tag{6.2.1}$$

In other words, $(\Omega_2\Omega_1)x$ denotes the element obtained when x is first operated on by Ω_1, and $\Omega_1 x$ then operated on by Ω_2. In view of this definition we may, without ambiguity, write $\Omega_2\Omega_1 x$ for either side in (6.2.1). The notion of multiplication of operators can, of course, be extended immediately to any number of factors. Using informal language we may say that when we multiply operators we simply apply them successively; or that the product of operators is their resultant.

THEOREM 6.2.1.† *Multiplication of operators is associative.*

Let $\Omega_1, \Omega_2, \Omega_3$ be operators mapping a set $\mathfrak{S}$ into itself and let x be any element of $\mathfrak{S}$. Writing $\Omega_3 x = y$ and using Definition 6.2.1, we obtain

$$\begin{aligned}\{\Omega_1(\Omega_2\Omega_3)\}x &= \Omega_1(\Omega_2\Omega_3 x) = \Omega_1\{\Omega_2(\Omega_3 x)\} = \Omega_1(\Omega_2 y)\\ &= (\Omega_1\Omega_2)y = (\Omega_1\Omega_2)(\Omega_3 x) = \{(\Omega_1\Omega_2)\Omega_3\}x.\end{aligned}$$

Thus

$$\Omega_1(\Omega_2\Omega_3) = (\Omega_1\Omega_2)\Omega_3, \tag{6.2.2}$$

and the theorem is proved.

It follows that we may, without ambiguity, write $\Omega_1\Omega_2\Omega_3$ for either side of (6.2.2).

6.2.2. When we are given a set Φ of operators each of which maps $\mathfrak{S}$ into itself, we sometimes require to know under what circumstances there exists an operator $\Omega \in \Phi$ such that $\Omega x = x'$, where x, x' are given elements of $\mathfrak{S}$.

† This theorem is not used in the present chapter.

Our present discussion is concerned with the set $\mathfrak{S}$ of matrices and the set Φ of operators each of which is a product of E-operations. For this special case the question just formulated in general terms takes the following form. When is it possible to pass from a matrix $\mathbf{A}$ to a matrix $\mathbf{B}$ (of the same type) by a chain of E-operations? The main object of the present section is to answer this question.

We begin with a result which asserts the possibility of reducing every matrix to some particularly simple standard form by means of E-operations.

Theorem 6.2.2. (Reduction of matrices to normal form)

Any $m \times n$ matrix of rank r can be reduced to the form

$$\begin{pmatrix} \mathbf{I}_r & \mathbf{O}_r^{n-r} \\ \mathbf{O}_{m-r}^r & \mathbf{O}_{m-r}^{n-r} \end{pmatrix} \tag{6.2.3}$$

by a chain of elementary operations. Conversely, the matrix (6.2.3) *can be transformed into the original matrix by a chain of elementary operations.*

The matrix (6.2.3) is known as the *normal form* of the original matrix. It will, for brevity, be denoted by $\mathbf{N}_r^{(m,n)}$.

The expression (6.2.3) for the normal form has, of course, to be modified in an obvious way if $r = 0, r = m$, or $r = n$. In particular, if the original matrix is a non-singular $n \times n$ matrix, then its normal form is simply $\mathbf{I}_n$.

In view of Exercise 6.1.1, the second part of the theorem is a trivial consequence of the first; and in the proof of the first we may obviously assume that the given matrix is not a zero matrix, since otherwise it already has the required form. If the leading element† vanishes it is possible, therefore, to bring a non-zero element into leading position by means of E-operations of type (i). Let the matrix so obtained be denoted by $\mathbf{A} = (a_{ij})$, where $a_{11} \neq 0$. The successive application of the E-operation $R_i \to R_i - a_{11}^{-1} a_{i1} R_1$ $(i = 2,..., m)$ then transforms $\mathbf{A}$ into the matrix

$$\begin{pmatrix} a_{11} & a_{12} & . & . & . & a_{1n} \\ 0 & b_{22} & . & . & . & b_{2n} \\ . & . & . & . & . & . \\ 0 & b_{m2} & . & . & . & b_{mn} \end{pmatrix}$$

† The *leading element* of a matrix is the element standing in the top left-hand corner.

(where $b_{ij} = a_{ij} - a_{11}^{-1} a_{i1} a_{1j}$). We next apply the operation $C_1 \to a_{11}^{-1} C_1$ and follow this by the operations $C_j \to C_j - a_{1j} C_1$ ($j = 2,..., n$). These transformations reduce the matrix to the form

$$\begin{pmatrix} 1 & 0 & . & . & . & 0 \\ 0 & c_{22} & . & . & . & c_{2n} \\ . & . & . & . & . & . \\ 0 & c_{m2} & . & . & . & c_{mn} \end{pmatrix},$$

and this may be written as

$$\begin{pmatrix} 1 & \mathbf{O}_1^{n-1} \\ \mathbf{O}_{m-1}^1 & \mathbf{C} \end{pmatrix}, \tag{6.2.4}$$

where $\mathbf{C}$ is an $(m-1) \times (n-1)$ matrix. If $\mathbf{C} = \mathbf{O}$ our reduction is complete, but if $\mathbf{C} \neq \mathbf{O}$, then we treat it in the same was as the original matrix. This does not affect the first row and first column in (6.2.4) and leads to the matrix

$$\begin{pmatrix} \mathbf{I}_2 & \mathbf{O}_2^{n-2} \\ \mathbf{O}_{m-2}^2 & \mathbf{D} \end{pmatrix},$$

where $\mathbf{D}$ is an $(m-2) \times (n-2)$ matrix. Continuing in this way as long as the bottom right-hand matrix is not zero, we ultimately obtain the matrix

$$\begin{pmatrix} \mathbf{I}_s & \mathbf{O}_s^{n-s} \\ \mathbf{O}_{m-s}^s & \mathbf{O}_{m-s}^{n-s} \end{pmatrix}.$$

Now the rank of this matrix is s; hence, by Theorem 6.1.1, $s = r$ and the proof is complete.

We shall illustrate the procedure described above by actually carrying out the reduction of the matrix

$$\begin{pmatrix} 0 & 3 & -3 & 1 \\ 5 & 9 & -10 & 3 \\ -1 & 0 & 5 & -2 \\ 2 & 1 & -3 & 1 \end{pmatrix}. \tag{6.2.5}$$

The operation $C_1 \leftrightarrow C_4$ transforms this into

$$\begin{pmatrix} 1 & 3 & -3 & 0 \\ 3 & 9 & -10 & 5 \\ -2 & 0 & 5 & -1 \\ 1 & 1 & -3 & 2 \end{pmatrix}.$$

Next, applying in succession the operations $R_2 \to R_2 - 3R_1$, $R_3 \to R_3 + 2R_1$, $R_4 \to R_4 - R_1$, $C_2 \to C_2 - 3C_1$, $C_3 \to C_3 + 3C_1$, we obtain

$$\begin{pmatrix} 1 & 0 & 0 & 0 \\ 0 & 0 & -1 & 5 \\ 0 & 6 & -1 & -1 \\ 0 & -2 & 0 & 2 \end{pmatrix}.$$

The operation $C_2 \leftrightarrow C_3$ transforms this matrix into

$$\begin{pmatrix} 1 & 0 & 0 & 0 \\ 0 & -1 & 0 & 5 \\ 0 & -1 & 6 & -1 \\ 0 & 0 & -2 & 2 \end{pmatrix}.$$

We now apply the operations $R_3 \to R_3 - R_2$, $C_2 \to -C_2$, $C_4 \to C_4 - 5C_2$, and are led to

$$\begin{pmatrix} 1 & 0 & 0 & 0 \\ 0 & 1 & 0 & 0 \\ 0 & 0 & 6 & -6 \\ 0 & 0 & -2 & 2 \end{pmatrix}.$$

The final stage consists of the operations $R_4 \to R_4 + \frac{1}{3}R_3$, $C_3 \to \frac{1}{6}C_3$, $C_4 \to C_4 + 6C_3$. This reduces the matrix to

$$\begin{pmatrix} 1 & 0 & 0 & 0 \\ 0 & 1 & 0 & 0 \\ 0 & 0 & 1 & 0 \\ 0 & 0 & 0 & 0 \end{pmatrix},$$

which is, in fact, the normal form $\mathbf{N}_3^{(4,4)}$.

Throughout the above series of operations we have followed precisely the procedure laid down in the proof of Theorem 6.2.2. But in numerical cases it is usually possible to introduce *ad hoc* modifications which simplify the work, and the reader should have no difficulty in performing the reduction of the matrix (6.2.5) in fewer steps than were required above.

The technique exemplified in the preceding example is particularly useful in the determination of rank. As we know by Theorem 6.1.1, the rank of a matrix remains unchanged when the matrix is subjected to E-operations. When, therefore, we wish to determine the rank of a given matrix $\mathbf{A}$ we apply a chain of suitable E-operations which transform $\mathbf{A}$ into some matrix $\mathbf{A}'$ whose rank can be recognized at a glance.† It is not, as a rule, necessary to carry the reduction as far as the normal form, and often a small number of steps suffices to effect the necessary transformation.

6.2.3. We are now ready to deal with the problem raised at the beginning of this section.

DEFINITION 6.2.2. *A matrix* **A** *is* EQUIVALENT *to a matrix* **B** (*in symbols*: $\mathbf{A} \sim \mathbf{B}$) *if it is possible to pass from* **A** *to* **B** *by a chain of E-operations.*‡

The relation of equivalence as here defined has three obvious but important properties.

(i) It is *reflexive*. This means that $\mathbf{A} \sim \mathbf{A}$, a result which follows at once from Exercise 6.1.2.

(ii) It is *symmetric*, i.e. $\mathbf{A} \sim \mathbf{B}$ implies $\mathbf{B} \sim \mathbf{A}$. This holds by virtue of Exercise 6.1.1.

(iii) It is *transitive*, i.e. the relations $\mathbf{A} \sim \mathbf{B}$, $\mathbf{B} \sim \mathbf{C}$ imply $\mathbf{A} \sim \mathbf{C}$. This is an immediate consequence of Definition 6.2.2.

Since, in particular, the relation of equivalence is symmetric (in other words, since it is immaterial whether we say that **A** is equivalent to **B**, or **B** equivalent to **A**) we can speak simply of equivalent matrices. Our problem is to determine in what circumstances two matrices are equivalent.

Theorem 6.2.3. (Equivalence of matrices)

Each of the following two conditions is necessary and sufficient for two $m \times n$ *matrices* **A** *and* **B** *to be equivalent.*

(i) *There exist non-singular matrices* **X**, **Y** (*of order* m, n *respectively*) *such that* $\mathbf{XAY} = \mathbf{B}$.

(ii) $R(\mathbf{A}) = R(\mathbf{B})$.

Let the condition that **A** and **B** should be equivalent be referred to as (iii). We have then to prove that each of the conditions (i), (ii), (iii) implies the other two, and it will be sufficient to infer (ii) from (i), (iii) from (ii), and (i) from (iii).

In the first place, if (i) is satisfied, then (ii) follows by Theorem 5.6.3 (p. 160).

Again, if $R(\mathbf{A}) = R(\mathbf{B}) = r$, say, then (by Theorem 6.2.2) **A** and **B** have the same normal form $\mathbf{N}_r^{(m,n)}$. Hence we can pass from **A** to

† Cf. the remarks at the end of § 6.1.1.

‡ For a theory of equivalence based on row operations only see Birkhoff and MacLane, 7, 270–9.

B by a chain of E-operations via the matrix $\mathbf{N}_r^{(m,n)}$. Thus $\mathbf{A} \sim \mathbf{B}$, and so (ii) implies (iii).

Finally, suppose that $\mathbf{A} \sim \mathbf{B}$. Then, by Theorem 6.1.2, there exist E-matrices $\mathbf{E}_1,..., \mathbf{E}_k, \mathbf{E}^{(1)},..., \mathbf{E}^{(l)}$ such that

$$\mathbf{E}_1 \dots \mathbf{E}_k \mathbf{A} \mathbf{E}^{(1)} \dots \mathbf{E}^{(l)} = \mathbf{B}.$$

Hence $\mathbf{XAY} = \mathbf{B}$, where

$$\mathbf{X} = \mathbf{E}_1 \dots \mathbf{E}_k, \qquad \mathbf{Y} = \mathbf{E}^{(1)} \dots \mathbf{E}^{(l)},$$

and the matrices **X**, **Y** are non-singular by Theorem 6.1.3 (i). Thus (iii) implies (i), and the proof is complete.

COROLLARY. *Similar matrices are equivalent.*

It is convenient to have a special term for transformations of matrices consisting of chains of E-operations.

DEFINITION 6.2.3. *An* EQUIVALENCE TRANSFORMATION *is a product of E-operations.*

The terminology introduced by this definition is a natural one since it makes equivalent matrices transformable into each other by equivalence transformations. It may be noted that E-operations are special equivalence transformations.

In view of Theorem 6.2.3 we know that **A** and **B** are connected by an equivalence transformation if and only if there exist non-singular matrices **X**, **Y** such that $\mathbf{B} = \mathbf{XAY}$. By restricting, in one way or another, the nature of **X** and **Y** we arrive at various special types of equivalence transformations; and in the course of our subsequent discussion we shall be led to consider in detail a number of such types. With two of them—the elementary operations and the similarity transformations—we are already familiar.

Some further useful results follow immediately from Theorem 6.2.3.

THEOREM 6.2.4. *Let* **A** *be an $m \times n$ matrix of rank r. Then there exist non-singular matrices* **X**, **Y** (*of order m, n respectively*) *such that*

$$\mathbf{A} = \mathbf{X}\mathbf{N}_r^{(m,n)}\mathbf{Y}.$$

The matrices **A** and $\mathbf{N}_r^{(m,n)}$ are equivalent by virtue of Theorem 6.2.2, and the assertion therefore follows by Theorem 6.2.3 (i).

THEOREM 6.2.5. *Every non-singular square matrix can be expressed as a product of elementary matrices.*

Let $\mathbf{A}$ be a non-singular $n \times n$ matrix. By Theorem 6.2.3 $\mathbf{A}$ and $\mathbf{I}_n$ are equivalent, i.e. $\mathbf{A}$ can be derived from $\mathbf{I}_n$ by a chain of E-operations. But, by Theorem 6.1.2, each such operation is effected by premultiplication or postmultiplication by an E-matrix. Hence

$$\mathbf{A} = \mathbf{E}_1 \ldots \mathbf{E}_k \mathbf{I}_n \mathbf{E}^{(1)} \ldots \mathbf{E}^{(l)} = \mathbf{E}_1 \ldots \mathbf{E}_k \mathbf{E}^{(1)} \ldots \mathbf{E}^{(l)},$$

where the $\mathbf{E}_i$ and $\mathbf{E}^{(j)}$ are suitable E-matrices.

6.3. Applications of the preceding theory

6.3.1. We begin by explaining a practical method of computing inverses of non-singular matrices. We know by Theorem 6.2.2 that such matrices can be reduced to the unit matrix by a chain of E-operations. We need, however, a slightly stronger result.

THEOREM 6.3.1. *A non-singular matrix of order n can be reduced to* $\mathbf{I}_n$ *by means of elementary row operations only.*

If a matrix $\mathbf{A}$ is non-singular at least one element in the first column is non-zero. Hence, interchanging two rows (if necessary) we can bring a non-zero element into leading position. Subtracting now suitable multiples of the first row from the other rows we obtain a matrix in which all elements in the first column, other than the leading element, are equal to zero. The leading element is then made equal to 1 by multiplying the first row by a suitable constant. In this way we obtain a matrix of the form

$$\begin{pmatrix} 1 & b_{12} & . & . & . & b_{1n} \\ 0 & b_{22} & . & . & . & b_{2n} \\ . & . & . & . & . & . \\ 0 & b_{n2} & . & . & . & b_{nn} \end{pmatrix}.$$

Applying now the same technique to the non-singular submatrix

$$\begin{pmatrix} b_{22} & . & . & . & b_{2n} \\ . & . & . & . & . \\ b_{n2} & . & . & . & b_{nn} \end{pmatrix}$$

and continuing the process for as long as is necessary, we obtain ultimately a triangular matrix of the form

$$\begin{pmatrix} 1 & p_{12} & p_{13} & . & . & . & p_{1n} \\ 0 & 1 & p_{23} & . & . & . & p_{2n} \\ 0 & 0 & 1 & . & . & . & p_{3n} \\ . & . & . & . & . & . & . \\ 0 & 0 & 0 & . & . & . & 1 \end{pmatrix}.$$

All elements above the diagonal can now be made equal to 0 by means of row operations of type (iii), and the reduction of $\mathbf{A}$ to $\mathbf{I}$ has thus been proved possible.

EXERCISE 6.3.1. Show that, by the application of elementary row operations, any rectangular matrix can be reduced to the form $\mathbf{P} = (p_{ij})$, where $p_{ij} = 0$ whenever $i > j$.

THEOREM 6.3.2. *Suppose that a non-singular matrix* $\mathbf{A}$ *is reduced to* $\mathbf{I}$ *by a sequence, say* $\phi_1,..., \phi_k$, *of elementary row operations. Then the application of these operations, in the same order, transforms* $\mathbf{I}$ *into* $\mathbf{A}^{-1}$.

By Theorem 6.1.2 the application of $\phi_1,..., \phi_k$ respectively is equivalent to premultiplication by certain E-matrices, say $\mathbf{E}_1,..., \mathbf{E}_k$. Hence, by hypothesis,

$$\mathbf{E}_k \dots \mathbf{E}_1 \mathbf{A} = \mathbf{I},$$

and so, postmultiplying by $\mathbf{A}^{-1}$ we obtain

$$\mathbf{A}^{-1} = \mathbf{E}_k \dots \mathbf{E}_1 \mathbf{I}.$$

Consequently $\mathbf{A}^{-1}$ is obtained when the operations $\phi_1,..., \phi_k$ are applied, in that order, to $\mathbf{I}$.

To illustrate the result just proved we consider the matrix

$$\mathbf{A} = \begin{pmatrix} 1 & 1 & -3 \\ -1 & 0 & 2 \\ -3 & 5 & 0 \end{pmatrix}.$$

The reader should verify that $\mathbf{A}$ can be reduced to $\mathbf{I}_3$ by applying, in turn, the operations $R_2 \rightarrow R_2 + R_1$; $R_3 \rightarrow R_3 + 3R_1$; $R_3 \rightarrow R_3 - 8R_2$; $R_3 \rightarrow -R_3$; $R_2 \rightarrow R_2 + R_3$; $R_1 \rightarrow R_1 + 3R_3$; $R_1 \rightarrow R_1 - R_2$. If these operations are applied, in the same order, to $\mathbf{I}_3$ we obtain the following sequence of matrices:

$$\begin{pmatrix} 1 & 0 & 0 \\ 1 & 1 & 0 \\ 0 & 0 & 1 \end{pmatrix}, \quad \begin{pmatrix} 1 & 0 & 0 \\ 1 & 1 & 0 \\ 3 & 0 & 1 \end{pmatrix}, \quad \begin{pmatrix} 1 & 0 & 0 \\ 1 & 1 & 0 \\ -5 & -8 & 1 \end{pmatrix}, \quad \begin{pmatrix} 1 & 0 & 0 \\ 1 & 1 & 0 \\ 5 & 8 & -1 \end{pmatrix},$$

$$\begin{pmatrix} 1 & 0 & 0 \\ 6 & 9 & -1 \\ 5 & 8 & -1 \end{pmatrix}, \quad \begin{pmatrix} 16 & 24 & -3 \\ 6 & 9 & -1 \\ 5 & 8 & -1 \end{pmatrix}, \quad \begin{pmatrix} 10 & 15 & -2 \\ 6 & 9 & -1 \\ 5 & 8 & -1 \end{pmatrix}.$$

The last matrix in the sequence is the required matrix $\mathbf{A}^{-1}$.

A procedure for finding inverses which involves both row operations and column operations can also be devised without difficulty since, given a non-singular matrix **A**, E-matrices $\mathbf{E}_1,\ldots,\mathbf{E}_k$, $\mathbf{E}^{(1)},\ldots,\mathbf{E}^{(l)}$ can be found such that

$$\mathbf{E}_k \ldots \mathbf{E}_1 \mathbf{A}\mathbf{E}^{(1)} \ldots \mathbf{E}^{(l)} = \mathbf{I}.$$

This procedure is not, however, as convenient as that given by Theorem 6.3.2.

6.3.2. An important use of E-operations (or, more precisely, of elementary row operations) can be made in the theory of linear equations.

If in a system $\mathfrak{S}$ of linear equations we interchange the position of two equations, or multiply the coefficients of an equation by some non-zero constant, or add any multiple of one equation to another, then the resulting system is plainly equivalent to $\mathfrak{S}$. Now a system of linear equations can be conveniently represented by its augmented matrix, and the transformations just listed correspond, in fact, to E-operations on the rows of this matrix. Thus, for instance, the addition of twice the third equation to the fifth equation corresponds to the operation $R_5 \to R_5 + 2R_3$ on the augmented matrix.

It follows, then, that if we subject the augmented matrix **B** of $\mathfrak{S}$ to a chain of elementary row operations, the resulting matrix will be the augmented matrix of a system of linear equations equivalent to $\mathfrak{S}$. The obvious procedure is, therefore, to operate on **B** in such a way as to obtain a matrix which shall be as simple as possible;† and Exercise 6.3.1 shows that we can always reduce **B** to some matrix $\mathbf{P} = (p_{ij})$ in which $p_{ij} = 0$ for $i > j$. The system of which **P** is the augmented matrix is then far more manageable than the original system. It is not, of course, necessary in every case to effect the complete reduction of **B** to the quasi-triangular form **P**, for often we recognize even after a few steps whether the original system is consistent, whether it possesses non-trivial solutions, or whether certain equations are redundant.

As an example we may consider the system of equations (5.3.6) discussed in § 5.3.2. Performing the operations $R_2 \leftrightarrow R_1$, $R_2 \to R_2 - 3R_1$, $R_4 \to R_4 - 2R_1$, $R_2 \to \frac{1}{5}R_2$, $R_4 \to -\frac{1}{7}R_4$ on the

† This amounts, in practice, to having as many zero elements as possible.

augmented matrix

$$\begin{pmatrix} 3 & -7 & 14 & -8 & 24 \\ 1 & -4 & 3 & -1 & -2 \\ 0 & 1 & 1 & -1 & 6 \\ 2 & -15 & -1 & 5 & -46 \end{pmatrix},$$

we reduce it to the form

$$\begin{pmatrix} 1 & -4 & 3 & -1 & -2 \\ 0 & 1 & 1 & -1 & 6 \\ 0 & 1 & 1 & -1 & 6 \\ 0 & 1 & 1 & -1 & 6 \end{pmatrix}.$$

The original system (5.3.6) is therefore equivalent to the system consisting of the two equations

$$x-4y+3z-w = -2, \qquad y+z-w = 6.$$

We can now take y and z as the disposable unknowns and complete the solution as before.

The procedure just indicated is not, of course, at all surprising since it amounts to little more than the familiar manipulation of simultaneous linear equations as described in school textbooks. However, by pursuing the idea a little farther, we could give alternative proofs of some of the results established in § 5.3 and § 5.4. We must refer the reader elsewhere for details.†

6.3.3. As another application of the theory developed in § 6.1 and § 6.2 we shall give a new proof of Theorem 5.6.2 (p. 159).

The first step is to prove Theorem 5.6.3. Let $\mathbf{S}$ be an $m \times n$ matrix and $\mathbf{X}$ a non-singular matrix of order m. By Theorem 6.2.5, $\mathbf{X}$ can be expressed as a product of E-matrices, say

$$\mathbf{X} = \mathbf{E}_1 \ldots \mathbf{E}_k.$$

Now, by Theorem 6.1.2, multiplication by an E-matrix is equivalent to an E-operation and since, by Theorem 6.1.1, E-operations preserve rank we have

$$R(\mathbf{S}) = R(\mathbf{E}_k \mathbf{S}) = R(\mathbf{E}_{k-1} \mathbf{E}_k \mathbf{S}) = \ldots = R(\mathbf{E}_1 \ldots \mathbf{E}_{k-1} \mathbf{E}_k \mathbf{S}).$$

Thus
$$R(\mathbf{XS}) = R(\mathbf{S}). \tag{6.3.1}$$

We next come to the proof of Theorem 5.6.2, i.e. effectively of the inequality $R(\mathbf{AB}) \leqslant R(\mathbf{B})$. We assume that $\mathbf{A}$, $\mathbf{B}$ are of type

† See, for instance, Birkhoff and MacLane, **7**, 50–53, and Stoll, **11**, chap. i.

$m \times n$, $n \times p$ respectively and that $R(\mathbf{A}) = r$, and we write $\mathbf{N}_r^{(m,n)} = \mathbf{N}$.

Let $\mathbf{P}$ be an $n \times p$ matrix and represent it in the partitioned form

$$\mathbf{P} = \begin{pmatrix} \mathbf{P}_1 \\ \mathbf{P}_2 \end{pmatrix},$$

where $\mathbf{P}_1$ denotes the matrix consisting of the first r rows of $\mathbf{P}$, and $\mathbf{P}_2$ that consisting of the remaining $n-r$ rows. We then have

$$\mathbf{NP} = \begin{pmatrix} \mathbf{P}_1 \\ \mathbf{O}_{m-r}^{p} \end{pmatrix}.$$

Hence $R(\mathbf{NP}) = R(\mathbf{P}_1)$, and therefore

$$R(\mathbf{NP}) \leqslant R(\mathbf{P}). \tag{6.3.2}$$

Now, by Theorem 6.2.4, $\mathbf{A}$ may be written in the form $\mathbf{A} = \mathbf{XNY}$, where $\mathbf{X}$, $\mathbf{Y}$ are non-singular matrices of order m, n respectively. Hence, by (6.3.1) and (6.3.2),

$$R(\mathbf{AB}) = R(\mathbf{X}.\mathbf{NYB}) = R(\mathbf{N}.\mathbf{YB}) \leqslant R(\mathbf{YB}) = R(\mathbf{B}),$$

and Theorem 5.6.2 is therefore established.

6.4. Congruence transformations

It was mentioned earlier that various special classes of equivalence transformations play an important part in linear algebra. In the present section we propose to consider one such class—the class of congruence transformations. Its precise significance will not, however, become fully apparent until we come to the discussion of quadratic forms in Chapter XII.

Unless the contrary is stated, all matrices considered below are assumed to be square matrices of order n.

DEFINITION 6.4.1. *If* $\mathbf{B} = \mathbf{P}^T\mathbf{AP}$, *where* $\mathbf{P}$ *is some non-singular matrix, then* $\mathbf{B}$ *is said to be obtained from* $\mathbf{A}$ *by a* CONGRUENCE TRANSFORMATION, *and* $\mathbf{B}$ *is said to be* CONGRUENT *to* $\mathbf{A}$.

We observe at once that a congruence transformation is a special equivalence transformation. We also observe that, since

$$\mathbf{Q}^T(\mathbf{P}^T\mathbf{AP})\mathbf{Q} = (\mathbf{PQ})^T\mathbf{A}(\mathbf{PQ}),$$

the product of two congruence transformations is again a congruence transformation.

EXERCISE 6.4.1. Show that the relation of congruence is reflexive, symmetric and transitive.

Since, in view of this exercise, the statements '**A** is congruent to **B**' and '**B** is congruent to **A**' imply each other, it follows that we may speak simply of two matrices as being congruent (to each other).

The problem we propose to discuss below is suggested by Theorem 6.2.2. That result shows, in particular, that every square matrix can be reduced to diagonal form by an equivalence transformation. We now ask ourselves whether this statement continues to be valid if instead of considering equivalence transformations we restrict ourselves to the narrower class of congruence transformations. The answer to this question is contained in Theorem 6.4.1 which we prove by the use of E-operations and E-matrices.

DEFINITION 6.4.2. *The square matrix* $\mathbf{A} = (a_{ij})$ *of order* n *is* SYMMETRIC *if* $a_{ij} = a_{ji}$ $(i, j = 1,..., n)$, *i.e. if* $\mathbf{A}^T = \mathbf{A}$.†

The symmetry mentioned in the definition is, of course, symmetry with respect to the diagonal.

EXERCISE 6.4.2. If **A** is a rectangular matrix, show that $\mathbf{A}^T\mathbf{A}$ is a symmetric (square) matrix.

THEOREM 6.4.1. *A matrix is congruent to a diagonal matrix if and only if it is symmetric.*

(i) One part of the theorem is trivial. For if **A** is congruent to a diagonal matrix, then there exists a non-singular matrix **P** and a diagonal matrix **D** such that $\mathbf{A} = \mathbf{P}^T\mathbf{D}\mathbf{P}$. Since diagonal matrices are obviously symmetric, this implies that

$$\mathbf{A}^T = \mathbf{P}^T\mathbf{D}^T\mathbf{P} = \mathbf{P}^T\mathbf{D}\mathbf{P} = \mathbf{A},$$

and so **A** is symmetric.

(ii) Next, assuming that $\mathbf{A} = (a_{ij})$ is symmetric, we have to show that if can be reduced to diagonal form by a congruence transformation.

We first make some preliminary observations. Suppose that $i \neq j$. Let $\mathbf{E}_{(ij)}$ be the E-matrix obtained from **I** by the operation $R_i \leftrightarrow R_j$ (or alternatively by the operation $C_i \leftrightarrow C_j$). We then clearly have

$$\mathbf{E}^T_{(ij)} = \mathbf{E}_{(ij)}.$$

† For some interesting results on the rank of symmetric matrices see Stoll, **11**, 118.

If Ω_{ij} denotes the product (in either order) of the operations $R_i \leftrightarrow R_j$, $C_i \leftrightarrow C_j$, then, by Theorem 6.1.2,

$$\begin{aligned}\Omega_{ij}(\mathbf{A}) &= \mathbf{E}_{(ij)}\mathbf{A}\mathbf{E}_{(ij)} \\ &= \mathbf{E}_{(ij)}^T\mathbf{A}\mathbf{E}_{(ij)}.\end{aligned} \tag{6.4.1}$$

Again, let $\mathbf{G}_{(ij)}$ be the E-matrix obtained from $\mathbf{I}$ by the operation $R_i \rightarrow R_i+R_j$, or, alternatively, by the operation $C_j \rightarrow C_j+C_i$. It is then at once clear that

$$\mathbf{G}_{(ji)}^T = \mathbf{G}_{(ij)}.$$

If Ω_{ij}^* denotes the product (in either order) of the operations $R_i \rightarrow R_i+R_j$, $C_i \rightarrow C_i+C_j$, then

$$\begin{aligned}\Omega_{ij}^*(\mathbf{A}) &= \mathbf{G}_{(ij)}\mathbf{A}\mathbf{G}_{(ji)} \\ &= \mathbf{G}_{(ji)}^T\mathbf{A}\mathbf{G}_{(ji)}.\end{aligned} \tag{6.4.2}$$

Equations (6.4.1) and (6.4.2) show that both Ω_{ij} and Ω_{ij}^* are congruence transformations.

We now come to the main part of the proof. Assume, as may be done, that $\mathbf{A} = (a_{ij}) \neq \mathbf{O}$. Then at least one element of $\mathbf{A}$ is non-zero, and we shall show that the leading element is either non-zero or else can be made so by a congruence transformation. If $a_{11} = 0$, but $a_{ii} \neq 0$ for some i in the range $2 \leqslant i \leqslant n$, then the operator Ω_{1i} effects the required change. If, on the other hand, $a_{ii} = 0$ for $1 \leqslant i \leqslant n$, then there exists some $a_{ij} \neq 0$ with $i \neq j$. In that case we apply Ω_{ij}^*, and it is easily seen that the (i, i)th element of $\Omega_{ij}^*(\mathbf{A})$ is $a_{ij}+a_{ji} = 2a_{ij} \neq 0$. If $i = 1$ this means that the leading element has now been made non-zero, while if $i > 1$ the required change is brought about by the additional application of Ω_{1i}.

Thus it is always possible to place a non-zero element in the leading position by applying a suitable congruence transformation. The resulting matrix will, of course, be again symmetric. Thus there exists a non-singular matrix $\mathbf{P}$ such that

$$\mathbf{P}^T\mathbf{A}\mathbf{P} = \begin{pmatrix} b & \mathbf{x}^T \\ \mathbf{x} & \mathbf{B} \end{pmatrix},$$

where $b \neq 0$, $\mathbf{x}$ is a column vector of order $n-1$, and $\mathbf{B}$ is a square matrix of order $n-1$.

The matrix

$$\mathbf{Q} = \begin{pmatrix} 1 & -b^{-1}\mathbf{x}^T \\ \mathbf{O}_{n-1}^1 & \mathbf{I}_{n-1} \end{pmatrix}$$

is plainly non-singular, and we have

$$\mathbf{Q}^T(\mathbf{P}^T\mathbf{A}\mathbf{P})\mathbf{Q} = \begin{pmatrix} 1 & \mathbf{0} \\ -b^{-1}\mathbf{x} & \mathbf{I} \end{pmatrix}\begin{pmatrix} b & \mathbf{x}^T \\ \mathbf{x} & \mathbf{B} \end{pmatrix}\begin{pmatrix} 1 & -b^{-1}\mathbf{x}^T \\ \mathbf{0} & \mathbf{I} \end{pmatrix}$$

$$= \begin{pmatrix} b & \mathbf{x}^T \\ \mathbf{0} & \mathbf{B}-b^{-1}\mathbf{x}\mathbf{x}^T \end{pmatrix}\begin{pmatrix} 1 & -b^{-1}\mathbf{x}^T \\ \mathbf{0} & \mathbf{I} \end{pmatrix} = \begin{pmatrix} b & \mathbf{0} \\ \mathbf{0} & \mathbf{B}-b^{-1}\mathbf{x}\mathbf{x}^T \end{pmatrix}.$$

Thus there exists a non-singular matrix $\mathbf{S}$ $(= \mathbf{PQ})$ such that

$$\mathbf{S}^T\mathbf{A}\mathbf{S} = \begin{pmatrix} b & \mathbf{0} \\ \mathbf{0} & \mathbf{C} \end{pmatrix},$$

where $\mathbf{C}$ is a square matrix of order $n-1$. This conclusion has been reached on the assumption that $\mathbf{A} \neq \mathbf{O}$, but for $\mathbf{A} = \mathbf{O}$ it holds, of course, trivially. We have, furthermore,

$$(\mathbf{S}^T\mathbf{A}\mathbf{S})^T = \mathbf{S}^T\mathbf{A}^T\mathbf{S} = \mathbf{S}^T\mathbf{A}\mathbf{S};$$

thus $\mathbf{S}^T\mathbf{A}\mathbf{S}$ is symmetric and so, therefore, is $\mathbf{C}$.

The proof is now easily completed by induction. For $n = 1$ the assertion is trivial. Assume that it is true for $n-1$, where $n \geqslant 2$. Then there exists a non-singular matrix $\mathbf{U}_1$ and a diagonal matrix $\mathbf{D}_1$, both of order $n-1$, such that $\mathbf{U}_1^T\mathbf{C}\mathbf{U}_1 = \mathbf{D}_1$. Hence

$$\begin{pmatrix} 1 & \mathbf{0} \\ \mathbf{0} & \mathbf{U}_1 \end{pmatrix}^T\begin{pmatrix} b & \mathbf{0} \\ \mathbf{0} & \mathbf{C} \end{pmatrix}\begin{pmatrix} 1 & \mathbf{0} \\ \mathbf{0} & \mathbf{U}_1 \end{pmatrix} = \begin{pmatrix} b & \mathbf{0} \\ \mathbf{0} & \mathbf{D}_1 \end{pmatrix}.$$

Writing $\quad \mathbf{U} = \begin{pmatrix} 1 & \mathbf{0} \\ \mathbf{0} & \mathbf{U}_1 \end{pmatrix}, \quad \mathbf{D} = \begin{pmatrix} b & \mathbf{0} \\ \mathbf{0} & \mathbf{D}_1 \end{pmatrix},$

we have, therefore, $\quad \mathbf{U}^T(\mathbf{S}^T\mathbf{A}\mathbf{S})\mathbf{U} = \mathbf{D},$

i.e. $\quad (\mathbf{SU})^T\mathbf{A}(\mathbf{SU}) = \mathbf{D}.$

Since $\mathbf{D}$ is diagonal and $\mathbf{SU}$ obviously non-singular, the proof is complete.

EXERCISE 6.4.3. Show that if $\mathbf{A}$ is a real (complex) symmetric matrix of rank r, then there exists a real (complex) matrix $\mathbf{U}$ such that $\mathbf{U}^T\mathbf{A}\mathbf{U}$ is a diagonal matrix having precisely r non-zero diagonal elements and that the value of each of these elements is 1 or -1 (is 1).

It should be noted that the reduction of $\mathbf{A}$ to diagonal form in the proof above has been carried out entirely by means of rational operations. For this reason we speak in this case of a *rational reduction*. In contrast we shall, in Chapter X, consider reductions which cannot be effected by rational operations alone.

6.5. The general concept of equivalence

6.5.1. It was pointed out after Definition 6.2.2 (p. 176) that the relation of equivalence between matrices is reflexive, symmetric, and transitive. These three properties occur so persistently in different mathematical situations that it is useful to introduce an abstract concept of equivalence, of which equivalence between matrices is a special instance.

DEFINITION 6.5.1. *Let* $\mathfrak{S}$ *be a set of elements, and let* $\sim$ *be a relation such that if* $x, y \in \mathfrak{S}$, *where* x *and* y *are not necessarily distinct, then* $x \sim y$ *either holds or does not hold. Suppose further that the relation* $\sim$ *is*

(i) REFLEXIVE, *i.e.* $x \sim x$ *for every* $x \in \mathfrak{S}$;

(ii) SYMMETRIC, *i.e.* $x \sim y$ *implies* $y \sim x$;

(iii) TRANSITIVE, *i.e.* $x \sim y$ *and* $y \sim z$ *imply* $x \sim z$.

Then $\sim$ *is said to be an* EQUIVALENCE RELATION *in* $\mathfrak{S}$, *and any two elements* x, y *such that* $x \sim y$ *are said to be* EQUIVALENT *in the sense of this particular relation.*

The simplest and in some ways the most fundamental instance of equivalence is that of *equality*.† Equivalence, as defined above, is a natural generalization of equality, and it is interesting to note that in dealing with the relation of equality we normally make use only of the three properties of reflexiveness, symmetry, and transitiveness.

Other examples of equivalence come readily to mind. Thus, if m is a positive integer, then two integers a and b are said to be *congruent modulo* m if $a-b$ is divisible by m. This relation between them is denoted symbolically by $a \equiv b \pmod{m}$, and is called *congruence modulo* m. Since it plainly satisfies the three axioms, it is an equivalence relation in the set of integers.

We are also familiar with various equivalence relations which hold between matrices. The relation defined as equivalence of matrices in Definition 6.2.2 is an instance of an equivalence relation in our present sense. Congruence of matrices is also an equivalence relation, in view of Exercise 6.4.1 (p. 182); and so, too, is similarity, as the reader should have no difficulty in verifying.

EXERCISE 6.5.1. Show that isomorphism between linear manifolds is an equivalence relation.

† Axiom 1 of Book I of Euclid's *Elements* states that two things which are equal to a third are equal to each other. This statement expresses, in effect, the transitive property of the relation of equality.

THEOREM 6.5.1. *Let* $\mathfrak{S}$ *be a set of elements for which an equivalence relation has been defined. Then* $\mathfrak{S}$ *can be subdivided into classes of elements in such a way that* (i) *each element belongs to precisely one class,* (ii) *two elements belong to the same class if and only if they are equivalent.*

The proof is almost obvious. If x is any element of $\mathfrak{S}$, let C_x denote the class of elements equivalent to x, i.e. the class of all $y \in \mathfrak{S}$ such that $x \sim y$.

We first show that if $x \sim y$, then $C_x = C_y$.† If $z \in C_x$, then $x \sim z$ and so $z \sim x$. But since $x \sim y$ this implies $z \sim y$, and hence $y \sim z$, i.e. $z \in C_y$. Similarly $z \in C_y$ implies $z \in C_x$, and we have, therefore, $C_x = C_y$.

If, on the other hand, x and y are not equivalent, then the classes C_x, C_y are disjoint, i.e. they possess no common element. For assume that $z \in C_x$, $z \in C_y$. Then $x \sim z$, $y \sim z$, and so $x \sim z$, $z \sim y$; hence $x \sim y$, contrary to hypothesis.

We have thus obtained a subdivision of $\mathfrak{S}$ into disjoint classes such that two elements belong to the same class if and only if they are equivalent. Moreover, in view of the property of reflexiveness, each element $x \in \mathfrak{S}$ belongs to at least one class, namely to C_x; and it belongs to no other class for, as we have seen, two classes are either disjoint or equal.

DEFINITION 6.5.2. *The classes of equivalent elements into which a set* $\mathfrak{S}$ *is subdivided by means of an equivalence relation* $\sim$ *are called the* EQUIVALENCE CLASSES (*of* $\mathfrak{S}$, *with respect to* $\sim$).

If an equivalence relation is defined in $\mathfrak{S}$, each equivalence class is determined completely by any one of its elements, and equivalence between elements of $\mathfrak{S}$ can be replaced by equality between equivalence classes.

The definition of equivalence classes has been given above in terms of equivalence, but the converse procedure can equally well be adopted. If a subdivision of $\mathfrak{S}$ into disjoint classes C, C',... of elements is given, we can define an equivalence relation in $\mathfrak{S}$ with respect to which C, C',... are the equivalence classes. We simply put $x \sim y$ whenever x and y belong to the same class. The relation $\sim$ then evidently satisfies the three axioms of Definition 6.5.1, and so is an equivalence relation.

† Two classes are said to be equal if they comprise precisely the same elements.

A branch of mathematics in which the idea of equivalence is particularly prominent is that part of analysis in which the construction of the number system is carried out from an initial set of axioms. Thus, in passing from the integers to the rational numbers, we consider ordered pairs (a, b) of integers a, b, and write $(a, b) \sim (c, d)$ whenever $ad = bc$. It is then easily verified that $\sim$ is an equivalence relation. Rational numbers can now be defined as equivalence classes with respect to this relation, and it then remains to be shown that they possess all the properties which we associate intuitively with rational numbers.†

Exercise 6.5.2. What are the equivalence classes when equivalence is defined as (i) equality between numbers, (ii) congruence modulo m?

Exercise 6.5.3. Let $\{x_n\}$, $\{y_n\}$ denote convergent sequences, and write $\{x_n\} \sim \{y_n\}$ whenever $x_n - y_n \to 0$ as $n \to \infty$. Show that the relation $\sim$ is an equivalence relation and determine the equivalence classes.

6.5.2. Equivalence relations may often be associated with sets of operators.

Theorem 6.5.2. *Let $\mathfrak{S}$ be a set of elements and Φ a set of transformations of $\mathfrak{S}$ into itself with the following properties.*

(i) *Φ contains the identity operator, which transforms every element of $\mathfrak{S}$ into itself.*

(ii) *If $x, y \in \mathfrak{S}$, $\Omega \in \Phi$, and $\Omega x = y$, then there exists an operator $\Omega^* \in \Phi$ such that $\Omega^* y = x$.*

(iii) *If x, y, $z \in \mathfrak{S}$, $\Omega, \Omega' \in \Phi$, and $\Omega x = y$, $\Omega' y = z$, then there exists an operator $\Omega'' \in \Phi$ such that $\Omega'' x = z$.*

Suppose further that, for any $x, y \in \mathfrak{S}$, we write $x \sim y$ if and only if there exists some $\Omega \in \Phi$ such that $\Omega x = y$. Then $\sim$ is an equivalence relation in $\mathfrak{S}$.

The relation $\sim$ is reflexive since, in view of (i), $x \sim x$ for all $x \in \mathfrak{S}$. Suppose next that $x \sim y$, i.e. $\Omega x = y$ for some $\Omega \in \Phi$. Then, by (ii), there exists some $\Omega^* \in \Phi$ such that $\Omega^* y = x$, i.e. $y \sim x$. This proves the symmetry of $\sim$. Finally, let $x \sim y$, $y \sim z$. This means that, for some Ω, $\Omega' \in \Phi$, $\Omega x = y$ and $\Omega' y = z$. Hence, by (iii), there exists some $\Omega'' \in \Phi$ such that $\Omega'' x = z$, i.e. $x \sim z$. Thus $\sim$ is transitive, and therefore it satisfies all three axioms of Definition 6.5.1.

† For details see Landau, *Foundations of Analysis*, chap. ii.

DEFINITION 6.5.3. *The equivalence relation* $\sim$ *of the preceding theorem is called* EQUIVALENCE IN $\mathfrak{S}$ WITH RESPECT TO THE SET Φ OF OPERATORS.

In the case of an equivalence with respect to a set of operators, an equivalence class simply consists of all elements which can be transformed into a specified element by means of suitably chosen operators.

Several examples of equivalence with respect to a set of operators are implicit in our earlier discussion. Thus, let $\mathfrak{S}$ be the set of $m \times n$ matrices and Φ the set of equivalence transformations (in the sense of Definition 6.2.3, p. 177). Then equivalence between matrices can be interpreted as equivalence in $\mathfrak{S}$ with respect to Φ; and the equivalence classes consist, as we know, of $m \times n$ matrices having the same rank. Again, let $\mathfrak{S}$ be the set of $n \times n$ matrices. Similarity is then an equivalence relation in $\mathfrak{S}$, the equivalence classes being sets of similar matrices. This equivalence is, in fact, equivalence in $\mathfrak{S}$ with respect to the set of similarity transformations. An analogous statement applies to congruence transformations.

EXERCISE 6.5.4. Verify these statements.

In our subsequent discussion in Parts II and III we shall meet many further instances of equivalence relations, equivalence classes, and equivalence with respect to sets of operators. The reader's attention is directed particularly to the remarks in § 9.2.3.

6.6. Axiomatic characterization of determinants

The technique evolved in studying elementary operations enables us to throw further light on the theory of determinants. In Chapter I we were concerned with deriving a series of properties of determinants, and our object now will be to prove that three of the simplest of these properties suffice to characterize completely the nature of determinants. We shall, in fact, show that if a particular numerical function of a square matrix is known to satisfy certain simple conditions stated below, then it is equal to the determinant of that matrix.

We require, in the first place, a preliminary result on certain matrix transformations which closely resemble E-operations.

THEOREM 6.6.1. *Any non-singular square matrix can be transformed into a diagonal matrix by a chain of operations of the following two types.*

(i) *Addition of any multiple of a row to another row.*

(ii) *Interchange of two rows followed by the multiplication of one of them by* -1.

It may be noted in passing that the relation between two matrices which can be transformed into each other by chains of operations enumerated above is a relation of equivalence. Theorem 6.6.1 asserts, in fact, that every non-singular matrix is equivalent (in the sense specified) to a diagonal matrix. Since, moreover, the determinant of a matrix clearly remains unaffected by either of the two operations, it follows that the diagonal matrix in question is also non-singular.

To effect the transformation whose existence is asserted by the theorem we rely essentially upon the method established in § 6.2.2 and § 6.3.1. We denote, as usual, the order of the matrices by n. Since the given matrix is non-singular the first column contains at least one non-zero element, and if necessary we can use an operation of type (ii) to bring this non-zero element into the leading position. Next, subtracting suitable multiples of the first row from the other rows we can reduce to zero the last $n-1$ elements in the first column. Proceeding in a similar manner we obtain eventually an upper triangular matrix whose diagonal elements are all non-zero. We now subtract suitable multiples of the last row from the first $n-1$ rows and reduce to zero the first $n-1$ elements of the last column. Treating the other columns in a similar manner we gradually reduce to zero all elements above the diagonal and are then left with a diagonal matrix.

EXERCISE 6.6.1. Show that the conclusion of Theorem 6.6.1 need not be valid for a singular matrix.

We can now prove the principal result of this section. We shall consider a numerical function $f(\mathbf{A})$ of a square matrix $\mathbf{A}$ and, assuming that it satisfies certain conditions which are also satisfied by $|\mathbf{A}|$, we shall show that it is equal to $|\mathbf{A}|$.

THEOREM 6.6.2. *Let* $f(\mathbf{A})$ *be a numerical function of the square matrix* $\mathbf{A}$, *and suppose that it satisfies the following conditions.*

(i) *The value of* $f(\mathbf{A})$ *remains unchanged if any operation of the type* $R_r \to R_r + R_s$ $(r \neq s)$ *is performed on* $\mathbf{A}$.

(ii) *If any row of* $\mathbf{A}$ *is multiplied by a constant* λ, *then the value of* $f(\mathbf{A})$ *is also multiplied by* λ.

(iii) $f(\mathbf{I}) = 1$.

We then have

$$f(\mathbf{A}) = |\mathbf{A}|. \tag{6.6.1}$$

Our first step consists in showing that the value of $f(\mathbf{A})$ remains unchanged when $\mathbf{A}$ is subjected to either of the operations listed in Theorem 6.6.1.

Let $r \neq s$, $\lambda \neq 0$, and suppose that $\mathbf{A}$ is subjected in turn to the operations $R_s \to \lambda R_s$, $R_r \to R_r + R_s$, $R_s \to \lambda^{-1} R_s$, the resulting matrices being denoted by $\mathbf{A}_1$, $\mathbf{A}_2$, $\mathbf{A}_3$ respectively. Then, by conditions (i) and (ii),

$$f(\mathbf{A}_3) = \lambda^{-1} f(\mathbf{A}_2) = \lambda^{-1} f(\mathbf{A}_1) = \lambda^{-1}\lambda f(\mathbf{A}) = f(\mathbf{A}).$$

Now $\mathbf{A}_3$ is obtained from $\mathbf{A}$ by adding λ times the sth row to the rth row; and it follows, therefore, that f is invariant under operations of type (i).

Again, let $r \neq s$ and suppose that $\mathbf{A}$ is subjected, in turn, to the operations $R_r \to R_r + R_s$, $R_s \to R_s - R_r$, $R_r \to R_r + R_s$, the resulting matrices being denoted by $\mathbf{A}_1$, $\mathbf{A}_2$, $\mathbf{A}_3$ respectively. Then, by the result just proved, we have

$$f(\mathbf{A}_3) = f(\mathbf{A}_2) = f(\mathbf{A}_1) = f(\mathbf{A}).$$

Now $\mathbf{A}_3$ is obtained from $\mathbf{A}$ by interchanging the rth and sth rows and then multiplying the sth row by -1. Hence f is invariant under operations of type (ii).

Now let $\mathbf{A}$ be a given non-singular matrix. By virtue of Theorem 6.6.1, $\mathbf{A}$ can be transformed by a chain of operations of types (i) and (ii) into a matrix $\mathbf{\Lambda} = \mathbf{dg}(\lambda_1, \ldots, \lambda_n)$, where $\lambda_1 \neq 0, \ldots, \lambda_n \neq 0$. Hence $f(\mathbf{A}) = f(\mathbf{\Lambda})$. But, by conditions (ii) and (iii),

$$f(\mathbf{\Lambda}) = \lambda_1 \ldots \lambda_n f(\mathbf{I}) = \lambda_1 \ldots \lambda_n = |\mathbf{\Lambda}| = |\mathbf{A}|,$$

and therefore (6.6.1) is valid for non-singular matrices.

If, on the other hand, $\mathbf{A}$ is singular then its rows are linearly dependent and one of them, say the kth, is expressible as a linear combination of the other rows. Hence a chain of suitable operations of type (i) will reduce all elements in the kth row to zero. Denoting the matrix thus obtained by $\mathbf{A}'$ we have $f(\mathbf{A}) = f(\mathbf{A}')$. But, multiplying the kth row of $\mathbf{A}'$, which consists entirely of zeros, by $\lambda \neq 0$ and making use of condition (ii), we see at once that $f(\mathbf{A}') = 0$. Hence $f(\mathbf{A}) = 0$ and (6.6.1) remains valid for singular matrices. The proof is therefore complete.

It should be noted that even if all references to determinants were suppressed, the above argument would still show that a function $f(\mathbf{A})$ which satisfies conditions (i)–(iii) is unique. This suggests that the entire theory of determinants could be built up in a way quite distinct from that adopted in Chapter I. The basis of the new method would be the *definition* of the determinant of $\mathbf{A}$ as $f(\mathbf{A})$. We do not propose here to enter into further details but may mention in conclusion that this programme was carried out about seventy years ago by Weierstrass.†

PROBLEMS ON CHAPTER VI

1. Let $f(x)$, $g(x)$,... be functions defined for $0 \leqslant x \leqslant 1$, and write $f \sim g$ whenever $\overline{\text{bd}}_{0 \leqslant x \leqslant 1} |f(x)-g(x)| \leqslant 1$. Show that $\sim$ is not an equivalence relation.

2. Criticize the following argument. 'Conditions (ii) and (iii) of Definition 6.5.1 (p. 186) imply condition (i), for $x \sim y$ implies $y \sim x$, and these two relations imply $x \sim x$.'

3. Express the matrix
$$\begin{pmatrix} 4 & 3 & -5 \\ 1 & 1 & 0 \\ 1 & 1 & 5 \end{pmatrix}$$
as a product of E-matrices.

4. Find the inverse of
$$\begin{pmatrix} 1 & 1 & 1 & -3 \\ 0 & 1 & 0 & 0 \\ 1 & 1 & 2 & -3 \\ 2 & 2 & 4 & -5 \end{pmatrix}.$$

5. Determine the normal forms of the matrices
$$\text{(i)} \quad \begin{pmatrix} 6 & -1 & 0 & 4 \\ 3 & 5 & 1 & 1 \\ -1 & 0 & -2 & 2 \end{pmatrix}; \qquad \text{(ii)} \quad \begin{pmatrix} -5 & 10 & 5 \\ 1 & -2 & -1 \\ -2 & 4 & 2 \\ 2 & -1 & 1 \end{pmatrix}.$$

6. Solve completely each of the following systems of equations.
$$\text{(i)} \qquad \begin{aligned} 4x+6y+3z+4w &= 3, \\ 2x+4y+z+w &= -1, \\ 3x+4y+3z+4w &= 4, \\ 2x+3y+z+w &= 0. \end{aligned}$$
$$\text{(ii)} \qquad \begin{aligned} 2x-y+3z-5w &= -7, \\ x+3y+w &= 3, \\ -7x+5y+4z-w &= -3. \end{aligned}$$

† See Carathéodory, *Vorlesungen über reelle Funktionen* (2nd edition), 318–26; Artin, *Galois Theory* (2nd edition), 11–20; Schreier and Sperner, **8**, 68–83; Lichnerowicz, **25**, 26–39.

$$\text{(iii)} \qquad \begin{aligned} 3x+3y-z+4u-2v &= 14, \\ x-y+7z-u &= -2, \\ 5x+y+13z+2u-2v &= 10, \\ 2x+4y-8z+5u-2v &= 16. \end{aligned}$$

7. Let **A**, **B** be two diagonal $n \times n$ matrices whose diagonal elements are identical except possibly for order. Show that there exists a non-singular symmetric matrix **H** such that $\mathbf{HAH} = \mathbf{B}$.

8. Show that the reduction to normal form of an $m \times n$ matrix can always be effected by fewer than $(m+1)(n+1)$ E-operations.

9. Show that any rectangular matrix of rank r ($\geqslant 1$) can be expressed as the sum of r matrices of rank 1.

10. By using the technique of E-operations prove that (i) a system of n homogeneous linear equations in n unknowns possesses a non-trivial solution if and only if the matrix of coefficients is singular; (ii) a system of m homogeneous linear equations in $n > m$ unknowns always possesses a non-trivial solution.

11. Show that every non-singular 2×2 matrix can be represented as a product of matrices of the following types:

$$\begin{pmatrix} t & 0 \\ 0 & 1 \end{pmatrix}, \quad \begin{pmatrix} 1 & 0 \\ 0 & t \end{pmatrix}, \quad \begin{pmatrix} 1 & t \\ 0 & 1 \end{pmatrix}, \quad \begin{pmatrix} 1 & 0 \\ t & 1 \end{pmatrix}, \quad \begin{pmatrix} 0 & 1 \\ 1 & 0 \end{pmatrix}.$$

Obtain such a representation for the matrix $\begin{pmatrix} 1 & -3 \\ 2 & 2 \end{pmatrix}$.

12. Show that an E-operation of type (i) can be expressed as a product of E-operations of types (ii) and (iii). Hence obtain a refinement of the result stated in the preceding question.

13. Show that every square matrix can be represented as a product of triangular matrices.

14. Show that if **A** is any square matrix, then there exist non-singular matrices **P** and **Q** such that **PA** and **AQ** are triangular.

15. Let the $n \times n$ matrix $\mathbf{A} = (a_{ij})$ be of rank r and suppose that it satisfies the equation $\mathbf{A}^2 = \mathbf{A}$. Use Theorem 6.2.4 (p. 177) to prove that

$$\sum_{i=1}^{n} a_{ii} = r.$$

16. Use Theorem 6.2.4 (p. 177) to prove Theorem 5.6.5 (p. 162) for the case of square matrices.

17. By a modification of the proof of Theorem 6.4.1 (p. 183) show that, if **A** is a matrix such that $\bar{\mathbf{A}}^T = \mathbf{A}$, then there exists a non-singular matrix **P** such that $\bar{\mathbf{P}}^T\mathbf{AP}$ is diagonal.

18. A rectangular matrix **A** has an $r \times r$ submatrix **B** such that (i) **B** is non-singular, (ii) every $(r+1) \times (r+1)$ submatrix of **A** which contains **B** is singular. Prove that $R(\mathbf{A}) = r$.

19. Find the rank of

$$\begin{pmatrix} 0 & c & -b & a' \\ -c & 0 & a & b' \\ b & -a & 0 & c' \\ -a' & -b' & -c' & 0 \end{pmatrix},$$

where $aa'+bb'+cc' = 0$, and a, b, c are positive numbers.

20. By an *elementary integral operation* on a matrix we shall understand an operation of one of the following types: (i) interchange of any two rows (or columns); (ii) addition of any integral multiple of one row (or column) to another row (or column); (iii) multiplication of any row (or column) by -1.

Let $\mathbf{A}$ be a rectangular non-zero matrix whose elements are integers, and let $[\mathbf{A}]$ denote the minimum of the absolute values of the non-zero elements of $\mathbf{A}$. Show that, if $[\mathbf{A}]$ does not divide every element of $\mathbf{A}$, then $\mathbf{A}$ can be transformed, by a chain of elementary integral operations, into a matrix $\mathbf{B}$ such that $[\mathbf{B}] < [\mathbf{A}]$. Deduce that any non-zero matrix with integral elements can be transformed, by a chain of elementary integral operations, into a matrix $\mathbf{C}$ such that $[\mathbf{C}]$ divides all elements of $\mathbf{C}$.

21. Prove that every rectangular matrix $\mathbf{A}$ with integral elements and of rank $r > 0$ can be transformed, by a chain of elementary integral operations, into a matrix

$$\mathbf{D} = \begin{pmatrix} d_1 & \cdot & \cdot & \cdot & \cdot & \cdot & \cdot \\ \cdot & d_2 & \cdot & \cdot & \cdot & \cdot & \cdot \\ \cdot & \cdot & \cdot & \cdot & \cdot & \cdot & \cdot \\ \cdot & \cdot & \cdot & d_r & \cdot & \cdot & \cdot \\ \cdot & \cdot & \cdot & \cdot & \cdot & \cdot & \cdot \\ \cdot & \cdot & \cdot & \cdot & \cdot & \cdot & \cdot \end{pmatrix},$$

where dots indicate zeros and $d_1,\ldots,d_r$ are positive integers such that, for $1 \leqslant k \leqslant r-1$, d_k divides d_{k+1}. Show, furthermore, that

$$d_k = h_k/h_{k-1} \qquad (k = 1,\ldots,r),$$

where $h_0 = 1$ and, for $1 \leqslant k \leqslant r$, h_k is the highest common factor of all k-rowed minors of $\mathbf{A}$. (The numbers $d_1,\ldots,d_r$ are called the *elementary divisors* of $\mathbf{A}$).

Let two matrices with integral elements be called equivalent if one can be transformed into the other by a chain of elementary integral operations. Give necessary and sufficient conditions for two matrices to be equivalent.

22. A square matrix is said to be *unimodular* if its elements are integers and its determinant is ± 1. Show that, if $\mathbf{A}$ is any given $m \times n$ matrix with integral elements and of rank r, then there exist unimodular matrices $\mathbf{P}$, $\mathbf{Q}$ of order m, n respectively, such that $\mathbf{PAQ} = \mathbf{D}$, where $\mathbf{D}$ is the matrix defined in the preceding question.

PART II

FURTHER DEVELOPMENT OF MATRIX THEORY

VII

THE CHARACTERISTIC EQUATION

ALL results obtained in Part I have been essentially simple and may be said to have followed almost automatically from the two notions of matrix and linear dependence. The introduction of a new and powerful idea, that of the characteristic equation, permits the further development of the subject at a somewhat less elementary level, and enables us to deal with a series of problems which are beyond the range of the methods employed previously. The idea of the characteristic equation will be seen to underlie virtually all subsequent discussion of matrices and quadratic forms in the remaining parts of the book.

From now on, unless the contrary is stated, all vectors are assumed to be of order n and all matrices of type $n\times n$; moreover, both vectors and matrices are assumed to be complex.

7.1. The coefficients of the characteristic polynomial

7.1.1. DEFINITION 7.1.1. *Let* $\mathbf{A} = (a_{ij})$ *be an* $n\times n$ *matrix and* λ *a scalar variable. The* CHARACTERISTIC POLYNOMIAL *of* $\mathbf{A}$ *is the polynomial* $\chi(\lambda) = \chi(\lambda;\mathbf{A})$ *given by*†

$$\chi(\lambda) = |\lambda\mathbf{I}-\mathbf{A}| = \begin{vmatrix} \lambda-a_{11} & -a_{12} & \cdot & \cdot & \cdot & -a_{1n} \\ -a_{21} & \lambda-a_{22} & \cdot & \cdot & \cdot & -a_{2n} \\ \cdot & \cdot & \cdot & \cdot & \cdot & \cdot \\ -a_{n1} & -a_{n2} & \cdot & \cdot & \cdot & \lambda-a_{nn} \end{vmatrix}.$$

The CHARACTERISTIC EQUATION *of* $\mathbf{A}$ *is the equation* $\chi(\lambda) = 0$. *Its roots*‡ *are the* CHARACTERISTIC (*or* LATENT) ROOTS *of* $\mathbf{A}$.

The polynomial $\chi(\lambda)$ is evidently of degree n; in fact, its leading term is λ^n. Thus the characteristic equation has precisely n roots;

† We have, in fact, already made use of this polynomial in § 3.5.

‡ The existence of the roots is guaranteed by the fundamental theorem of algebra.

but these need not, of course, be all distinct. If a certain root occurs precisely k times it is said to be a *k-fold root*, or a root of *multiplicity k*. A k-fold root is called *simple* or *multiple* according as $k = 1$ or $k > 1$. We shall denote by $m_\lambda(\mathbf{A})$ the multiplicity of the number λ as characteristic root of $\mathbf{A}$, with the convention that $m_\lambda(\mathbf{A}) = 0$ if λ is not a characteristic root of $\mathbf{A}$ at all.

EXERCISE 7.1.1. Show that different matrices need not have different characteristic polynomials.

The characteristic equation was first investigated for general matrices by Cayley in 1853. It occurs in many different contexts of which one may be indicated here. Consider a collineation of projective space of $n-1$ dimensions. In this space each point is represented by a non-zero vector of order n, and two vectors which only differ by a non-zero scalar multiple represent the same point. For a given choice of the coordinate system the collineation may be specified by an $n \times n$ matrix $\mathbf{A}$ such that each point $\mathbf{x}$ is transformed into $\mathbf{Ax}$. It is often important to obtain information about the united points of the collineation, i.e. points which are transformed into themselves. Now the vector $\mathbf{x} \neq \mathbf{0}$ represents such a point if and only if there exists some $\lambda \neq 0$ such that $\mathbf{Ax} = \lambda\mathbf{x}$, i.e. $(\lambda\mathbf{I}-\mathbf{A})\mathbf{x} = \mathbf{0}$. By the theory of homogeneous linear equations we know that an $\mathbf{x} \neq \mathbf{0}$ which satisfies this equation exists if and only if $|\lambda\mathbf{I}-\mathbf{A}| = 0$. Thus the analysis of collineations leads at once to the study of the characteristic equation.

There are many other topics in both pure and applied mathematics where the characteristic equation makes its appearance.† Historically, one of its first applications occurs in the theory of secular perturbation of planetary motion. It is for this reason that the equation was referred by the earlier writers as the *secular equation*.

The determination of the characteristic roots of a matrix involves the solution of an equation of the nth degree, but in a few special cases the roots can be found by inspection. Thus, if $\mathbf{A}$ is a diagonal matrix, or even a triangular matrix, its n characteristic roots are equal to the n diagonal elements.

7.1.2. A simple criterion for the singularity of a matrix can be deduced from the following result.

† See, for example, § 12.2.3.

THEOREM 7.1.1. *If* $\lambda_1,..., \lambda_n$ *are the characteristic roots of* $\mathbf{A}$, *then*

$$|\mathbf{A}| = \lambda_1...\lambda_n.$$

We have
$$|\lambda\mathbf{I}-\mathbf{A}| = (\lambda-\lambda_1)...(\lambda-\lambda_n).$$
Hence, putting $\lambda = 0$, we obtain
$$|-\mathbf{A}| = (-1)^n\lambda_1...\lambda_n,$$
and the assertion follows.

COROLLARY. *A matrix is singular if and only if at least one of its characteristic roots is equal to zero.*

The problem of determining the precise number of zero characteristic roots of a matrix will be considered in § 10.2.1.

Theorem 7.1.1 relates the value of the constant term in $\chi(\lambda)$ and of a minor of $\mathbf{A}$. It is not difficult to obtain analogous results for all coefficients of $\chi(\lambda)$.

DEFINITION 7.1.2. *A* PRINCIPAL MINOR *of a matrix* $\mathbf{A}$ *is a minor whose diagonal is part of the diagonal of* $\mathbf{A}$.

Thus a principal minor of $\mathbf{A}$ is obtained by selecting rows and columns with the same suffixes. Special cases of the principal minors of $\mathbf{A} = (a_{ij})$ are the diagonal elements $a_{11},..., a_{nn}$ and the determinant $|\mathbf{A}|$.

THEOREM 7.1.2. *For* $0 \leqslant r < n$, *the coefficient of* λ^r *in the characteristic polynomial* $\chi(\lambda)$ *of* $\mathbf{A}$ *is equal to* $(-1)^{n-r}$ *times the sum of all* $(n-r)$*-rowed principal minors of* $\mathbf{A}$.

We note that for $r = 0$ this result reduces, effectively, to Theorem 7.1.1.

We have
$$(-1)^n\chi(\lambda) = |\mathbf{A}-\lambda\mathbf{I}|.$$
Writing the determinant on the right in the form

$$\begin{vmatrix} a_{11}-\lambda & a_{12}+0 & . & . & . & a_{1n}+0 \\ a_{21}+0 & a_{22}-\lambda & . & . & . & a_{2n}+0 \\ . & . & . & . & . & . \\ a_{n1}+0 & a_{n2}+0 & . & . & . & a_{nn}-\lambda \end{vmatrix}$$

and using the corollary to Theorem 1.2.5 (p. 11), we obtain

$$(-1)^n\chi(\lambda) = |-\lambda\mathbf{I}|+\sum_{k=1}^{n}\sum_{1\leqslant i_1<...<i_k\leqslant n} D(i_1,..., i_k),$$

where $D(i_1,..., i_k)$ denotes the n-rowed determinant whose columns with suffixes $i_1,..., i_k$ are identical with the corresponding columns in $\mathbf{A}$ and whose remaining columns (if any exist) are identical with the corresponding columns in $-\lambda\mathbf{I}$. Hence, for $1 \leqslant k \leqslant n$,

$$D(i_1,\ldots,i_k) = (-\lambda)^{n-k}\Delta(i_1,\ldots,i_k),$$

where $\Delta(i_1,\ldots,i_k)$ is the k-rowed (principal) minor of $\mathbf{A}$, obtained by deleting from $\mathbf{A}$ all rows and columns other than those with suffixes $i_1,\ldots,i_k$. Thus

$$(-1)^n\chi(\lambda) = (-1)^n\lambda^n + \sum_{k=1}^{n} \sum_{1\leqslant i_1<\ldots<i_k\leqslant n} (-\lambda)^{n-k}\Delta(i_1,\ldots,i_k),$$

and so

$$\chi(\lambda) = \lambda^n + \sum_{k=1}^{n}(-1)^k\lambda^{n-k} \sum_{1\leqslant i_1<\ldots<i_k\leqslant n} \Delta(i_1,\ldots,i_k)$$

$$= \lambda^n + \sum_{k=1}^{n}(-1)^k S_k \lambda^{n-k},$$

where S_k is the sum of all k-rowed principal minors of $\mathbf{A}$. The theorem now follows at once.

Exercise 7.1.2. Let $1 \leqslant k < n$, $1 \leqslant i_1 < \ldots < i_k \leqslant n$. Show that the coefficient of $\lambda_{i_1}\ldots\lambda_{i_k}$ in

$$\begin{vmatrix} a_{11}-\lambda_1 & a_{12} & \cdot & \cdot & \cdot & a_{1n} \\ a_{21} & a_{22}-\lambda_2 & \cdot & \cdot & \cdot & a_{2n} \\ \cdot & \cdot & \cdot & \cdot & \cdot & \cdot \\ a_{n1} & a_{n2} & \cdot & \cdot & \cdot & a_{nn}-\lambda_n \end{vmatrix}$$

is equal to $(-1)^k M_{i_1\ldots i_k}$, where $M_{i_1\ldots i_k}$ denotes the $(n-k)$-rowed (principal) minor of $\mathbf{A}$ obtained by deleting the rows and columns with suffixes $i_1,\ldots,i_k$. Hence deduce Theorem 7.1.2.

Theorem 7.1.3. *Let $1 \leqslant r \leqslant n$. Then the r-th elementary symmetric function*† *of the characteristic roots of* $\mathbf{A}$ *is equal to the sum of all r-rowed principal minors of* $\mathbf{A}$.

This result follows immediately from Theorem 7.1.2 and the well-known relations between the coefficients and the roots of an algebraic equation.

Definition 7.1.3. *The* TRACE *of a matrix* $\mathbf{A}$ *is the sum of its diagonal elements. It is denoted by* $\operatorname{tr}\mathbf{A}$.

Exercise 7.1.3. Show that the function $\operatorname{tr}\mathbf{X}$ is a linear operator in the linear manifold of $n\times n$ matrices.

† The rth elementary symmetric function of the numbers $x_1,\ldots,x_n$ (where $1 \leqslant r \leqslant n$) is defined as

$$\sum_{1\leqslant i_1<\ldots<i_r\leqslant n} x_{i_1}\ldots x_{i_r}.$$

The special case $r = 1$ of Theorem 7.1.3 is worth stating explicitly.

THEOREM 7.1.4. *The trace of a matrix is equal to the sum of its characteristic roots.*

EXERCISE 7.1.4. Write out an independent proof of Theorem 7.1.4.

7.2. Characteristic polynomials and similarity transformations

The significance of similarity transformations has already been discussed in § 4.2. We shall now note some of their properties in relation to characteristic polynomials.

THEOREM 7.2.1. *The characteristic polynomial, and therefore the characteristic roots, of a matrix are invariant under similarity transformations.*

For we have

$$|\lambda\mathbf{I}-\mathbf{S}^{-1}\mathbf{AS}| = |\mathbf{S}^{-1}(\lambda\mathbf{I}-\mathbf{A})\mathbf{S}| = |\mathbf{S}^{-1}|\,|\lambda\mathbf{I}-\mathbf{A}|\,|\mathbf{S}| = |\lambda\mathbf{I}-\mathbf{A}|.$$

Theorem 7.2.1, though proved in a single line, is important. It shows, in fact, that the characteristic roots are numbers which pertain not merely to a matrix but to the linear operator of which that matrix is a representation. We may consequently speak of characteristic roots of linear operators.

Theorem 7.2.1 implies, in particular, that similar matrices have the same determinant. Moreover, in view of Theorem 7.1.4, we can make the following inference from Theorem 7.2.1.

COROLLARY. *The trace of a matrix is invariant under similarity transformations. i.e.*

$$\operatorname{tr}(\mathbf{S}^{-1}\mathbf{AS}) = \operatorname{tr}\mathbf{A}.$$

EXERCISE 7.2.1. Show that a matrix and its transpose have the same characteristic equation.

THEOREM 7.2.2. *The diagonal matrices* $\mathbf{dg}(\alpha_1,\ldots,\alpha_n)$ *and* $\mathbf{dg}(\beta_1,\ldots,\beta_n)$ *are similar if and only if* $(\alpha_1,\ldots,\alpha_n)$ *is an arrangement of* $(\beta_1,\ldots,\beta_n)$.

Denote the two matrices by $\mathbf{A}$ and $\mathbf{B}$ respectively and suppose, in the first place, that they are similar. Their characteristic polynomials are $(\lambda-\alpha_1)\ldots(\lambda-\alpha_n)$ and $(\lambda-\beta_1)\ldots(\lambda-\beta_n)$ respectively; these polynomials are, by Theorem 7.2.1, identical and hence $(\alpha_1,\ldots,\alpha_n)$ is an arrangement of $(\beta_1,\ldots,\beta_n)$.

Next, assume that $(\alpha_1,\ldots,\alpha_n)$ is an arrangement of $(\beta_1,\ldots,\beta_n)$. Denote by $\mathbf{E}$ the matrix obtained when the ith and jth rows of the unit matrix are interchanged. Then $\mathbf{E} = \mathbf{E}^{-1}$ and it follows at once from Theorem 6.1.2 (p. 170) that $\mathbf{E}^{-1}\mathbf{A}\mathbf{E}$ is the matrix obtained from $\mathbf{A}$ when the elements α_i and α_j are interchanged.† Hence matrices $\mathbf{E},\ldots,\mathbf{G}$ may be chosen such that

$$\mathbf{G}^{-1} \ldots \mathbf{E}^{-1}\mathbf{A}\mathbf{E} \ldots \mathbf{G} = \mathbf{B},$$

i.e.

$$(\mathbf{E} \ldots \mathbf{G})^{-1}\mathbf{A}(\mathbf{E} \ldots \mathbf{G}) = \mathbf{B},$$

and $\mathbf{A}$ and $\mathbf{B}$ are therefore similar.

THEOREM 7.2.3. *If* $\mathbf{A}, \mathbf{B}$ *are arbitrary matrices, then* $\mathbf{AB}$ *and* $\mathbf{BA}$ *have the same characteristic polynomial.*

In proving this result (which is due to Sylvester) we shall employ an idea whose usefulness was already recognized in § 3.5.

If $|\mathbf{A}| \neq 0$, then $\mathbf{BA} = \mathbf{A}^{-1}(\mathbf{AB})\mathbf{A}$. In that case $\mathbf{AB}$, $\mathbf{BA}$ are similar matrices and the assertion follows by Theorem 7.2.1. In the general case we have $|\mathbf{A}-t\mathbf{I}| \neq 0$ for all sufficiently large values of t, say for $t > t_0$. In view of the previous case it therefore follows that, for $t > t_0$, the matrices

$$(\mathbf{A}-t\mathbf{I})\mathbf{B}, \qquad \mathbf{B}(\mathbf{A}-t\mathbf{I})$$

have the same characteristic polynomial. Thus

$$|\lambda\mathbf{I}-(\mathbf{A}-t\mathbf{I})\mathbf{B}| = |\lambda\mathbf{I}-\mathbf{B}(\mathbf{A}-t\mathbf{I})|$$

for all $t > t_0$ and all λ. For every fixed value of λ we have, therefore, two polynomials in t which are equal whenever $t > t_0$. Hence, by the corollary to Theorem 1.6.3 (p. 31), we have equality for $t = 0$, i.e.

$$|\lambda\mathbf{I}-\mathbf{AB}| = |\lambda\mathbf{I}-\mathbf{BA}|$$

for all values of λ.

The required result can be established more rapidly if we are prepared to make use of the notion of continuity. For, if $c_r(\mathbf{X})$ denotes the coefficient of λ^r in the characteristic polynomial of $\mathbf{X}$, then clearly

$$c_r(\mathbf{AB}) = c_r(\mathbf{BA}) \qquad (r = 0, 1,\ldots, n-1) \tag{7.2.1}$$

provided that $\mathbf{A}$ is non-singular. But $c_r(\mathbf{X})$ is a continuous function of the elements of $\mathbf{X}$, and hence (7.2.1) remains valid even when $\mathbf{A}$ is singular.‡

† The reader who is not acquainted with the theory of E-operations may regard the proof of this statement as an exercise.

‡ For yet another proof see MacDuffee, **31**, 23–24.

Corollary. *If* $\mathbf{B}_1,\ldots,\mathbf{B}_k$ *is any cyclic arrangement*† *of the ordered set* $\mathbf{A}_1,\ldots,\mathbf{A}_k$ *of matrices, then* $\mathbf{A}_1\ldots\mathbf{A}_k$ *and* $\mathbf{B}_1\ldots\mathbf{B}_k$ *have the same characteristic polynomial.*

Exercise 7.2.2. (i) Deduce from Theorem 7.2.3 that

$$\operatorname{tr}(\mathbf{AB}) = \operatorname{tr}(\mathbf{BA}), \tag{7.2.2}$$

and also give an independent proof of this result. (ii) Use (7.2.2) to deduce the corollary to Theorem 7.2.1. (iii) Extend (7.2.2) to products of any number of matrices.

7.3. Characteristic roots of rational functions of matrices

Polynomials and rational functions of a matrix $\mathbf{A}$ were defined in § 3.7. We shall now discuss the relation between the characteristic roots of such matrix functions and those of the original matrix $\mathbf{A}$. The results derived below are due to Frobenius (1878).

Theorem 7.3.1. *If f is a polynomial and* $\lambda_1,\ldots,\lambda_n$ *are the characteristic roots of* $\mathbf{A}$ *then*

$$|f(\mathbf{A})| = f(\lambda_1)\ldots f(\lambda_n).$$

Denote the characteristic polynomial of $\mathbf{A}$ by $\chi(\lambda;\,\mathbf{A})$, and write

$$f(t) = c(\alpha_1-t)\ldots(\alpha_k-t).$$

Since, as we know by Theorem 3.7.2 (p. 98), any identity between scalar polynomials implies the corresponding identity between matrix polynomials, we have

$$f(\mathbf{A}) = c(\alpha_1\mathbf{I}-\mathbf{A})\ldots(\alpha_k\mathbf{I}-\mathbf{A}).$$

Hence

$$|f(\mathbf{A})| = c^n\prod_{i=1}^{k}|\alpha_i\mathbf{I}-\mathbf{A}| = c^n\prod_{i=1}^{k}\chi(\alpha_i;\,\mathbf{A})$$

$$= c^n\prod_{i=1}^{k}\prod_{j=1}^{n}(\alpha_i-\lambda_j) = \prod_{j=1}^{n}c\prod_{i=1}^{k}(\alpha_i-\lambda_j) = \prod_{j=1}^{n}f(\lambda_j),$$

and the assertion is proved.

Exercise 7.3.1. Show that Theorem 7.1.1 is a special case of Theorem 7.3.1.

Theorem 7.3.2. *The conclusion of Theorem 7.3.1 is still valid if f is a rational function, say $f = g/h$, where g, h are polynomials such that* $|h(\mathbf{A})| \neq 0$.

† A cyclic arrangement of the ordered set of objects $x_1,\ldots,x_k$ is any arrangement of the form $x_i, x_{i+1},\ldots, x_k, x_1,\ldots, x_{i-1}$.

By Theorem 7.3.1,

$$g(\mathbf{A}) = \prod_{i=1}^{n} g(\lambda_i), \qquad h(\mathbf{A}) = \prod_{i=1}^{n} h(\lambda_i).$$

Hence

$$|f(\mathbf{A})| = |g(\mathbf{A})\{h(\mathbf{A})\}^{-1}| = \frac{|g(\mathbf{A})|}{|h(\mathbf{A})|} = \prod_{i=1}^{n} \frac{g(\lambda_i)}{h(\lambda_i)} = \prod_{i=1}^{n} f(\lambda_i).$$

THEOREM 7.3.3. *If the characteristic roots of* $\mathbf{A}$ *are* $\lambda_1,\ldots,\lambda_n$ *and if* $\psi = f/g$, *where* f, g *are polynomials such that* $|g(\mathbf{A})| \neq 0$, *then the characteristic roots of* $\psi(\mathbf{A})$ *are* $\psi(\lambda_1),\ldots,\psi(\lambda_n)$.

For any value of λ write

$$\phi(x) = \lambda - \psi(x) = \frac{\lambda g(x) - f(x)}{g(x)}.$$

Since $|g(\mathbf{A})| \neq 0$, Theorem 7.3.2 may be applied to $\phi(x)$. This leads to the relation

$$|\phi(\mathbf{A})| = \phi(\lambda_1)\ldots\phi(\lambda_n),$$

i.e.

$$|\lambda\mathbf{I} - \psi(\mathbf{A})| = (\lambda - \psi(\lambda_1))\ldots(\lambda - \psi(\lambda_n)).$$

This holds for all values of λ, and the theorem now follows since the left-hand side represents the characteristic polynomial of $\psi(\mathbf{A})$.

COROLLARY. *If the characteristic roots of* $\mathbf{A}$ *are* $\lambda_1,\ldots,\lambda_n$ *and* k *is an integer (positive if* $\mathbf{A}$ *is singular), then the characteristic roots of* $\mathbf{A}^k$ *are* $\lambda_1^k,\ldots,\lambda_n^k$.

EXERCISE 7.3.2. (i) Show that if, for some positive integer m, $\mathbf{A}^m = \mathbf{I}$, then all characteristic roots of $\mathbf{A}$ are roots of unity. (ii) Show, by considering the matrix $\begin{pmatrix} 1 & 1 \\ 0 & 1 \end{pmatrix}$ that the converse inference is false.

7.4. The minimum polynomial and the theorem of Cayley and Hamilton

7.4.1. DEFINITION 7.4.1. *A polynomial* f ANNIHILATES *the matrix* $\mathbf{A}$ *if* $f(\mathbf{A}) = \mathbf{O}$.

The existence of a non-zero polynomial annihilating a given matrix $\mathbf{A}$ is an immediate consequence of the fact that the set of all $n \times n$ matrices is a linear manifold of dimensionality n^2.† This implies that the n^2+1 matrices $\mathbf{I}, \mathbf{A}, \mathbf{A}^2,\ldots, \mathbf{A}^{n^2}$ are linearly dependent; thus there exist scalars $c_0, c_1, c_2,\ldots, c_{n^2}$, not all zero, such that

$$c_0\mathbf{I} + c_1\mathbf{A} + c_2\mathbf{A}^2 + \ldots + c_{n^2}\mathbf{A}^{n^2} = \mathbf{O}. \tag{7.4.1}$$

† See Exercise 3.3.2 (p. 80).

Hence the non-zero polynomial

$$f(\lambda) = c_0+c_1\lambda+c_2\lambda^2+...+c_{n^2}\lambda^{n^2}$$

annihilates **A**.

Instead of appealing to properties of linear manifolds we may also argue directly. The matrix equation (7.4.1) is equivalent to a system of n^2 homogeneous linear equations in the n^2+1 unknowns $c_0, c_1,..., c_{n^2}$. Such a system possesses a non-trivial solution and hence the existence of a non-zero polynomial annihilating **A** is proved once again.

Among all non-zero polynomials annihilating **A** we now consider those of least degree, and by multiplying them by suitable non-zero constants we ensure that they are *monic* (i.e. they have their leading coefficients equal to 1). Any two such polynomials, say f_1 and f_2, are in fact identical. For otherwise $\phi = f_1-f_2$ is a non-zero polynomial of lower degree than f_1 and f_2, and

$$\phi(\mathbf{A}) = f_1(\mathbf{A})-f_2(\mathbf{A}) = \mathbf{O},$$

which contradicts the definition of f_1, f_2 as non-zero annihilating polynomials of minimum degree. Thus $f_1 = f_2$, and we may introduce the following definition.

Definition 7.4.2. *The* minimum polynomial *of a matrix* **A** *is the (unique) monic polynomial of minimum degree which annihilates* **A**.

The notion of a minimum polynomial is due to Frobenius (1878). Our previous remarks show that the degree of the minimum polynomial of **A** is at most equal to n^2. This crude bound will be improved in § 7.4.2.

We already know that the characteristic polynomial is invariant under similarity transformations. The corresponding result will now be established for the minimum polynomial.

Theorem 7.4.1. *Similar matrices have the same minimum polynomial.*

If $\mathbf{B} = \mathbf{S}^{-1}\mathbf{A}\mathbf{S}$, we can at once verify, by induction with respect to k, that

$$\mathbf{B}^k = \mathbf{S}^{-1}\mathbf{A}^k\mathbf{S} \qquad (k = 0, 1, 2,...).$$

It follows that, for every polynomial f,

$$f(\mathbf{B}) = \mathbf{S}^{-1}f(\mathbf{A})\mathbf{S}.$$

Hence the same polynomials annihilate **A** and **B**, and therefore **A** and **B** have the same minimum polynomial.

THEOREM 7.4.2. *The minimum polynomial of a matrix is a divisor of every polynomial which annihilates the matrix.*

Denote the minimum polynomial of $\mathbf{A}$ by $\mu(\lambda)$, and let $f(\lambda)$ be a polynomial such that $f(\mathbf{A}) = \mathbf{O}$. By the division algorithm, there exist polynomials $q(\lambda)$, $r(\lambda)$ such that

$$f(\lambda) = q(\lambda)\mu(\lambda)+r(\lambda),$$

where $r(\lambda)$ is either identically equal to 0 or else is of lower degree than $\mu(\lambda)$.†

Hence $$f(\mathbf{A}) = q(\mathbf{A})\mu(\mathbf{A})+r(\mathbf{A}),$$

and so $r(\mathbf{A}) = 0$. Thus, by the minimum property of $\mu(\lambda)$, it follows that $r(\lambda)$ vanishes identically; therefore $f(\lambda) = q(\lambda)\mu(\lambda)$, i.e. $\mu(\lambda)$ is a divisor of $f(\lambda)$.

The existence of a minimum polynomial for every matrix enables us to reduce rational functions of matrices to polynomial form.

THEOREM 7.4.3. *If $f(\lambda)$, $g(\lambda)$ are scalar polynomials and $g(\mathbf{A})$ is non-singular, then the quotient $f(\mathbf{A})/g(\mathbf{A})$ is equal to a polynomial in $\mathbf{A}$ whose coefficients depend on f, g and $\mathbf{A}$.*

It is sufficient to show that $\{g(\mathbf{A})\}^{-1}$ is equal to a polynomial in $\mathbf{A}$. We write $\mathbf{S} = g(\mathbf{A})$ and denote by

$$\mu(\lambda) = \lambda^k+\alpha_1\lambda^{k-1}+...+\alpha_{k-1}\lambda+\alpha_k$$

the minimum polynomial of $\mathbf{S}$. Hence

$$\mathbf{S}^k+\alpha_1\mathbf{S}^{k-1}+...+\alpha_{k-1}\mathbf{S}+\alpha_k\mathbf{I} = \mathbf{O},$$

and so $$\alpha_k\mathbf{S}^{-1} = -(\mathbf{S}^{k-1}+\alpha_1\mathbf{S}^{k-2}+...+\alpha_{k-1}\mathbf{I}).$$

The assumption that $\alpha_k = 0$ would clearly contradict the definition of $\mu(\lambda)$. Hence $\alpha_k \neq 0$ and we have

$$\mathbf{S}^{-1} = \alpha_0'\mathbf{S}^{k-1}+\alpha_1'\mathbf{S}^{k-2}+...+\alpha_{k-1}'\mathbf{I}.$$

Thus $\{g(\mathbf{A})\}^{-1}$ is equal to a polynomial in $\mathbf{S}$, and so in $\mathbf{A}$. The coefficients of this last polynomial depend not only on g but also on $\alpha_0',..., \alpha_{k-1}'$, i.e. on the minimum polynomial of $\mathbf{S} = g(\mathbf{A})$; they depend, therefore, on $\mathbf{A}$.

The fact that the coefficients of the polynomial whose existence is asserted by Theorem 7.4.3 depend on $\mathbf{A}$ is important. It shows that Theorem 7.4.3 does not furnish us with an *identity* for converting rational functions of matrices into polynomials. Thus, if

† We associate no degree with the identically vanishing polynomial.

$f(\mathbf{A})/g(\mathbf{A}) = \phi(\mathbf{A})$, where f, g, ϕ are polynomials, it is not legitimate to conclude that $f(\mathbf{B})/g(\mathbf{B}) = \phi(\mathbf{B})$, for the coefficients of ϕ depend on $\mathbf{A}$ as well as on f and g. On the other hand, there exists, of course, a polynomial ψ such that $f(\mathbf{B})/g(\mathbf{B}) = \psi(\mathbf{B})$.

The possibility of expressing every rational function of a matrix as a polynomial marks a vital point of difference between matrix algebra and the algebra of numbers.

7.4.2. We had observed in § 7.4.1 that the minimum polynomial of an $n \times n$ matrix is of degree not greater than n^2. We shall now improve on this statement by showing that the degree of the minimum polynomial is, in fact, not greater than n. We shall establish this result by determining an annihilating polynomial of degree n.

It will be necessary to consider polynomials in a scalar variable λ which have *matrix* coefficients, and we shall always write each power of λ to the *right* of the corresponding matrix coefficient. Let $\mathbf{f}(\lambda)$ be such a polynomial, say

$$\mathbf{f}(\lambda) = \mathbf{C}_0 + \mathbf{C}_1\lambda + \ldots + \mathbf{C}_k\lambda^k. \tag{7.4.2}$$

Then the matrix $\mathbf{f}(\mathbf{A})$ is defined by the equation

$$\mathbf{f}(\mathbf{A}) = \mathbf{C}_0 + \mathbf{C}_1\mathbf{A} + \ldots + \mathbf{C}_k\mathbf{A}^k.$$

THEOREM 7.4.4. *Let* $\mathbf{f}(\lambda)$, $\mathbf{g}(\lambda)$ *be polynomials in* λ *with matrix coefficients; and suppose that they are related by the equation*

$$\mathbf{g}(\lambda) = \mathbf{f}(\lambda).(\lambda\mathbf{I} - \mathbf{A}),$$

where $\mathbf{A}$ *is a given matrix. Then* $\mathbf{g}(\mathbf{A}) = \mathbf{O}$.

This result, though very simple to prove, is not quite as trivial as it may look at first sight. It is not, of course, legitimate merely to 'substitute' $\lambda = \mathbf{A}$ in $\mathbf{f}(\lambda).(\lambda\mathbf{I} - \mathbf{A})$, for we must multiply out the factors $\mathbf{f}(\lambda)$ and $\lambda\mathbf{I} - \mathbf{A}$ and arrange the product as a polynomial in λ before making the substitution.

Let $\mathbf{f}(\lambda)$ be given by (7.4.2). Then

$$\begin{aligned}\mathbf{g}(\lambda) &= \mathbf{f}(\lambda).(\lambda\mathbf{I} - \mathbf{A}) \\ &= -\mathbf{C}_0\mathbf{A} + (\mathbf{C}_0 - \mathbf{C}_1\mathbf{A})\lambda + (\mathbf{C}_1 - \mathbf{C}_2\mathbf{A})\lambda^2 + \ldots + \\ &\qquad + (\mathbf{C}_{k-1} - \mathbf{C}_k\mathbf{A})\lambda^k + \mathbf{C}_k\lambda^{k+1},\end{aligned}$$

and so

$$\begin{aligned}\mathbf{g}(\mathbf{A}) = -\mathbf{C}_0\mathbf{A} + (\mathbf{C}_0 - \mathbf{C}_1\mathbf{A})\mathbf{A} + (\mathbf{C}_1 - \mathbf{C}_2\mathbf{A})\mathbf{A}^2 + \ldots + \\ + (\mathbf{C}_{k-1} - \mathbf{C}_k\mathbf{A})\mathbf{A}^k + \mathbf{C}_k\mathbf{A}^{k+1} = \mathbf{O}.\end{aligned}$$

We next come to our principal result.

Theorem 7.4.5. (Theorem of Cayley and Hamilton)

Every matrix satisfies its own characteristic equation, i.e. if $\chi(\lambda)$ *is the characteristic polynomial of* $\mathbf{A}$, *then* $\chi(\mathbf{A}) = \mathbf{O}$.

This theorem was established by Hamilton for a special class of matrices in 1853. Five years later Cayley enunciated the general result without proof.

Write

$$\chi(\lambda) = |\lambda\mathbf{I}-\mathbf{A}| = \lambda^n+c_1\lambda^{n-1}+\ldots+c_{n-1}\lambda+c_n.$$

The adjugate matrix of $\lambda\mathbf{I}-\mathbf{A}$ has the form

$$(\lambda\mathbf{I}-\mathbf{A})^* = \begin{pmatrix} p_{11}(\lambda) & . & . & . & p_{1n}(\lambda) \\ . & . & . & . & . \\ p_{n1}(\lambda) & . & . & . & p_{nn}(\lambda) \end{pmatrix},$$

where the $p_{ij}(\lambda)$ are polynomials in λ. Clearly, then, $(\lambda\mathbf{I}-\mathbf{A})^*$ may be written as a polynomial in λ having matrix coefficients. Now, by Theorem 3.5.2 (p. 88),

$$\mathbf{I}\lambda^n+c_1\mathbf{I}\lambda^{n-1}+\ldots+c_{n-1}\mathbf{I}\lambda+c_n\mathbf{I} = (\lambda\mathbf{I}-\mathbf{A})^*(\lambda\mathbf{I}-\mathbf{A}),$$

and hence, by Theorem 7.4.4,

$$\mathbf{A}^n+c_1\mathbf{A}^{n-1}+\ldots+c_{n-1}\mathbf{A}+c_n\mathbf{I} = \mathbf{O},$$

i.e. $\chi(\mathbf{A}) = \mathbf{O}$.†

EXERCISE 7.4.1. Discuss the fallacy involved in 'proving' the theorem of Cayley–Hamilton by substituting $\lambda = \mathbf{A}$ in $\chi(\lambda) = |\lambda\mathbf{I}-\mathbf{A}|$.

The manipulation of matrices is often greatly facilitated by the Cayley–Hamilton theorem, which provides an easy method for expressing any polynomial in $\mathbf{A}$ as a polynomial of degree not exceeding $n-1$. For, if $f(\lambda)$ is any polynomial, then there exist polynomials $q(\lambda)$, $r(\lambda)$ such that

$$f(\lambda) = q(\lambda)\chi(\lambda)+r(\lambda),$$

where $r(\lambda)$ is either identically zero or else is of degree less than the degree of $\chi(\lambda)$, i.e. less than n. Hence, in view of the Cayley–Hamilton theorem,

$$f(\mathbf{A}) = q(\mathbf{A})\chi(\mathbf{A})+r(\mathbf{A}) = r(\mathbf{A}).$$

† For a less sophisticated version of the same proof see Ferrar, **2**, 111–12.

As an example, let us compute $2\mathbf{A}^8-3\mathbf{A}^5+\mathbf{A}^4+\mathbf{A}^2-4\mathbf{I}$, where $\mathbf{A}$ is the matrix

$$\mathbf{A} = \begin{pmatrix} 1 & 0 & 2 \\ 0 & -1 & 1 \\ 0 & 1 & 0 \end{pmatrix}.$$

The characteristic polynomial of $\mathbf{A}$ is

$$\begin{vmatrix} \lambda-1 & 0 & -2 \\ 0 & \lambda+1 & -1 \\ 0 & -1 & \lambda \end{vmatrix} = \lambda^3-2\lambda+1.$$

Now, long division of $2\lambda^8-3\lambda^5+\lambda^4+\lambda^2-4$ by $\lambda^3-2\lambda+1$ leaves the remainder

$$r(\lambda) = 24\lambda^2-37\lambda+10.$$

Hence $$2\mathbf{A}^8-3\mathbf{A}^5+\mathbf{A}^4+\mathbf{A}^2-4\mathbf{I} = 24\mathbf{A}^2-37\mathbf{A}+10\mathbf{I}.$$

But
$$\mathbf{A}^2 = \begin{pmatrix} 1 & 2 & 2 \\ 0 & 2 & -1 \\ 0 & -1 & 1 \end{pmatrix},$$

and therefore

$$2\mathbf{A}^8-3\mathbf{A}^5+\mathbf{A}^4+\mathbf{A}^2-4\mathbf{I} = \begin{pmatrix} -3 & 48 & -26 \\ 0 & 95 & -61 \\ 0 & -61 & 34 \end{pmatrix}.$$

Again, the Cayley–Hamilton theorem may be used to express rational functions of a given matrix $\mathbf{A}$ in polynomial form. This is achieved by carrying out the process explained in the proof of Theorem 7.4.3 but using the characteristic polynomial in place of the minimum polynomial.

7.4.3. Theorems 7.4.2 and 7.4.5 lead at once to the following consequence.

THEOREM 7.4.6. *The minimum polynomial of a matrix divides its characteristic polynomial.*

This theorem furnishes us with no information as to whether the minimum polynomial and the characteristic polynomial are equal or not. In fact, equality occurs in some but not in all cases. Thus, if $\mathbf{A} = \mathbf{dg}(1, 0, -1)$, the minimum polynomial $\mu(\lambda)$ and the characteristic polynomial $\chi(\lambda)$ are both equal to $\lambda(\lambda-1)(\lambda+1)$. On the other hand, if

$$\mathbf{A} = \begin{pmatrix} 0 & 0 & 1 \\ 0 & 0 & 0 \\ 0 & 0 & 0 \end{pmatrix},$$

then $\mu(\lambda) = \lambda^2$, $\chi(\lambda) = \lambda^3$. Both these examples illustrate the following general result.

THEOREM 7.4.7. *The distinct linear factors of the minimum polynomial coincide with those of the characteristic polynomial.*

Let $\lambda_1,..., \lambda_k$ be the distinct characteristic roots of $\mathbf{A}$, and denote their multiplicities by $\alpha_1,..., \alpha_k$ respectively, so that $\alpha_1,..., \alpha_k \geqslant 1$ and $\alpha_1+...+\alpha_k = n$. The characteristic polynomial $\chi(\lambda)$ is then given by

$$\chi(\lambda) = (\lambda-\lambda_1)^{\alpha_1}...(\lambda-\lambda_k)^{\alpha_k},$$

and so, by Theorem 7.4.6, the minimum polynomial $\mu(\lambda)$ must be of the form

$$\mu(\lambda) = (\lambda-\lambda_1)^{\beta_1}...(\lambda-\lambda_k)^{\beta_k},$$

where $0 \leqslant \beta_i \leqslant \alpha_i$ $(i = 1,..., k)$. Assume now that some β_j is equal to 0. Then $\mu(\lambda_j) \neq 0$, and so, by Theorem 7.3.3 (p. 202), the matrix $\mu(\mathbf{A})$ possesses at least one non-zero characteristic root. Hence $\mu(\mathbf{A}) \neq \mathbf{O}$, contrary to hypothesis. Thus $\beta_1,..., \beta_k \geqslant 1$, and the theorem is proved.

COROLLARY. *If the characteristic roots of a matrix are distinct, then its minimum polynomial and its characteristic polynomial are equal.*†

7.5. Estimates of characteristic roots

In the present section we shall establish a number of results giving information about the position of characteristic roots in the complex plane.

7.5.1. We have already met the notion of a symmetric matrix (Definition 6.4.2, p. 183). Another special class of matrices is specified by the next definition.

DEFINITION 7.5.1. *The matrix* $\mathbf{A} = (a_{rs})$ *is* SKEW-SYMMETRIC *if* $a_{rs} = -a_{sr}$ $(r, s = 1,..., n)$, *i.e. if* $\mathbf{A}^T = -\mathbf{A}$.

EXERCISE 7.5.1. Show that all the diagonal elements of a skew-symmetric matrix are equal to zero.

The notions of symmetry and skew-symmetry are generally only of interest when the matrix in question is real. For matrices over the complex field it is desirable to modify these notions by introducing complex conjugates. If $\mathbf{A} = (a_{rs})$ we shall, as previously, use the symbol $\bar{\mathbf{A}}$ to denote the conjugate matrix $(\bar{a}_{rs})$.

† For further information on the minimum polynomial see MacDuffee, **12**, 77–85.

EXERCISE 7.5.2. Show that (i) $(\overline{\mathbf{A}^T}) = (\bar{\mathbf{A}})^T$; (ii) $\overline{\mathbf{AB}} = \bar{\mathbf{A}}\bar{\mathbf{B}}$; (iii) $(\overline{\mathbf{A}^*}) = (\bar{\mathbf{A}})^*$; (iv) $(\overline{\mathbf{A}^{-1}}) = (\bar{\mathbf{A}})^{-1}$ if $|\mathbf{A}| \neq 0$.

DEFINITION 7.5.2. *The (complex) matrix* $\mathbf{A} = (a_{rs})$ *is* HERMITIAN *if* $\bar{a}_{rs} = a_{sr}$ $(r, s = 1,..., n)$, *i.e. if* $\bar{\mathbf{A}} = \mathbf{A}^T$; *it is* SKEW-HERMITIAN *if* $\bar{a}_{rs} = -a_{sr}$ $(r, s = 1,..., n)$, *i.e. if* $\bar{\mathbf{A}} = -\mathbf{A}^T$.

Hermitian matrices are so named after Hermite, who was the first to discuss their properties.

For real matrices the terms 'hermitian' and 'symmetric' evidently coincide, as do the terms 'skew-hermitian' and 'skew-symmetric'.

EXERCISE 7.5.3. Show that a hermitian matrix all of whose elements are purely imaginary† is skew-symmetric.

EXERCISE 7.5.4. Show that if $\mathbf{A}$ is skew-hermitian, then $i\mathbf{A}$ is hermitian.

Theorem 7.5.1. *The characteristic roots of a hermitian matrix, and in particular of a real symmetric matrix, are real.*

For real symmetric matrices this result was established by Cauchy in 1829; the general theorem for hermitian matrices is due to Hermite (1855).

Let $\mathbf{A}$ be a hermitian matrix and λ any characteristic root of $\mathbf{A}$. Then $|\lambda\mathbf{I}-\mathbf{A}| = 0$ and, in view of Corollary 1 to Theorem 5.4.1 (p. 149), there exists a (possibly complex) vector $\mathbf{x} \neq \mathbf{0}$ such that $(\lambda\mathbf{I}-\mathbf{A})\mathbf{x} = \mathbf{0}$. Thus $\mathbf{Ax} = \lambda\mathbf{x}$; and, premultiplying by $\bar{\mathbf{x}}^T$, we obtain

$$\bar{\mathbf{x}}^T\mathbf{Ax} = \lambda\bar{\mathbf{x}}^T\mathbf{x}. \tag{7.5.1}$$

On the other hand, $\mathbf{Ax} = \lambda\mathbf{x}$ implies $\bar{\mathbf{x}}^T\bar{\mathbf{A}}^T = \bar{\lambda}\bar{\mathbf{x}}^T$. Since $\mathbf{A}$ is hermitian, $\bar{\mathbf{A}}^T = \mathbf{A}$; and hence

$$\bar{\mathbf{x}}^T\mathbf{Ax} = \bar{\lambda}\bar{\mathbf{x}}^T\mathbf{x}. \tag{7.5.2}$$

Comparing (7.5.1) and (7.5.2) we infer that $(\lambda-\bar{\lambda})\bar{\mathbf{x}}^T\mathbf{x} = 0$. But $\bar{\mathbf{x}}^T\mathbf{x} \neq 0$, and therefore $\lambda-\bar{\lambda} = 0$, i.e. λ is real.

EXERCISE 7.5.5. Write out the proof of the preceding theorem without making use of the notion of matrix multiplication.

COROLLARY. *The characteristic roots of a skew-hermitian matrix are purely imaginary.*

For let $\mathbf{A}$ be skew-hermitian. Then, by Exercise 7.5.4, $i\mathbf{A}$ is hermitian. If λ is any characteristic root of $\mathbf{A}$, then $|i\lambda\mathbf{I}-i\mathbf{A}| = 0$ and so, by Theorem 7.5.1, $i\lambda$ is real, i.e. λ is purely imaginary.

† A complex number is said to be purely imaginary if its real part is equal to zero. In particular, the complex number 0 is purely imaginary.

The relation just proved was first noted (for real skew-symmetric matrices) by Clebsch in 1863.

A quantitative result which contains Theorem 7.5.1 for the case of real matrices was proved by Bendixson in 1902.

THEOREM 7.5.2. *Let* $\mathbf{A} = (a_{rs})$ *be a real* $n \times n$ *matrix and write*

$$\alpha = \max_{1 \leqslant r,s \leqslant n} \tfrac{1}{2}|a_{rs} - a_{sr}|.$$

If λ *is any characteristic root of* $\mathbf{A}$, *then*

$$|\mathfrak{Im}\,\lambda| \leqslant \alpha\sqrt{\{n(n-1)/2\}}.$$

The number α may be regarded as a measure of deviation of $\mathbf{A}$ from symmetry. $\mathbf{A}$ is evidently symmetric if and only if $\alpha = 0$, and in that case the asserted inequality implies the reality of the characteristic roots.

Let $\mathbf{x} = (x_1,\ldots,x_n)^T$ be a non-zero vector such that $\lambda\mathbf{x} = \mathbf{A}\mathbf{x}$ and assume, without loss of generality, that $\bar{\mathbf{x}}^T\mathbf{x} = 1$. Then

$$\lambda = \bar{\mathbf{x}}^T\mathbf{A}\mathbf{x}$$

and so
$$\bar{\lambda} = (\overline{\bar{\mathbf{x}}^T\mathbf{A}\mathbf{x}})^T = \bar{\mathbf{x}}^T\mathbf{A}^T\mathbf{x}.$$

Hence
$$2i\,\mathfrak{Im}\,\lambda = \lambda - \bar{\lambda} = \bar{\mathbf{x}}^T(\mathbf{A} - \mathbf{A}^T)\mathbf{x}.$$

It is easily verified that the expression on the right-hand side is equal to $\sum_{r,s=1}^{n} (a_{rs} - a_{sr})\bar{x}_r x_s$. Since the left-hand side is purely imaginary, it follows that

$$2i\,\mathfrak{Im}\,\lambda = \sum_{r,s=1}^{n} (a_{rs} - a_{sr})\tfrac{1}{2}(\bar{x}_r x_s - x_r \bar{x}_s),$$

$$2|\mathfrak{Im}\,\lambda| \leqslant \alpha \sum_{\substack{r,s=1 \\ r \neq s}}^{n} |\bar{x}_r x_s - x_r \bar{x}_s|. \tag{7.5.3}$$

Now for any real non-negative numbers $p_1,\ldots,p_k$, we have

$$(p_1 + \ldots + p_k)^2 \leqslant k(p_1^2 + \ldots + p_k^2).\dagger \tag{7.5.4}$$

Hence, since the right-hand side in (7.5.3) contains $n(n-1)$ terms, we have

$$4|\mathfrak{Im}\,\lambda|^2 \leqslant \alpha^2 n(n-1) \sum_{\substack{r,s=1 \\ r \neq s}}^{n} |\bar{x}_r x_s - x_r \bar{x}_s|^2. \tag{7.5.5}$$

But
$$|\bar{x}_r x_s - x_r \bar{x}_s|^2 = 2|x_r|^2|x_s|^2 - \bar{x}_r^2 x_s^2 - x_r^2 \bar{x}_s^2,$$

† This inequality is a special case of the inequality of Cauchy and Schwarz which was mentioned in the proof of Theorem 2.5.2 (p. 63).

and so

$$\sum_{r,s=1}^{n} |\bar{x}_r x_s - x_r \bar{x}_s|^2 \leqslant 2 \sum_{r=1}^{n} |x_r|^2 \sum_{s=1}^{n} |x_s|^2 - 2 \sum_{r=1}^{n} \bar{x}_r^2 \sum_{s=1}^{n} x_s^2$$

$$= 2|\mathbf{x}|^2|\mathbf{x}|^2 - 2\bar{t}t = 2-2|t|^2,$$

where $t = \sum_{s=1}^{n} x_s^2$. Hence

$$\sum_{r,s=1}^{n} |\bar{x}_r x_s - x_r \bar{x}_s|^2 \leqslant 2,$$

and the theorem follows by (7.5.5).

Other useful estimates were obtained by Hirsch in 1902.

Theorem 7.5.3. *If* $\mathbf{A} = (a_{rs})$ *is a complex* $n \times n$ *matrix and*

$$\rho = \max_{1 \leqslant r,s \leqslant n} |a_{rs}|, \qquad \sigma = \max_{1 \leqslant r,s \leqslant n} \tfrac{1}{2}|a_{rs} + \bar{a}_{sr}|,$$

$$\tau = \max_{1 \leqslant r,s \leqslant n} \tfrac{1}{2}|a_{rs} - \bar{a}_{sr}|,$$

then every characteristic root λ *of* $\mathbf{A}$ *satisfies the inequalities*

$$|\lambda| \leqslant n\rho, \qquad |\mathfrak{Rl}\,\lambda| \leqslant n\sigma, \qquad |\mathfrak{Im}\,\lambda| \leqslant n\tau.$$

The third inequality implies, of course, Theorem 7.5.1 and the second implies its corollary. All three inequalities will be sharpened in § 10.4.2.

Let $\mathbf{C} = (c_{rs})$ be any complex matrix, and suppose that $|c_{rs}| \leqslant \kappa$ $(r, s = 1,..., n)$. If $\mathbf{x}$ is any complex unit vector, then

$$|\bar{\mathbf{x}}^T \mathbf{C} \mathbf{x}| = \left| \sum_{r,s=1}^{n} c_{rs} \bar{x}_r x_s \right| \leqslant \kappa \sum_{r,s=1}^{n} |x_r||x_s| = \kappa \left(\sum_{r=1}^{n} |x_r| \right)^2.$$

Hence, by (7.5.4), $\qquad |\bar{\mathbf{x}}^T \mathbf{C} \mathbf{x}| \leqslant n\kappa \sum_{r=1}^{n} |x_r|^2,$

and so

$$|\bar{\mathbf{x}}^T \mathbf{C} \mathbf{x}| \leqslant n\kappa. \tag{7.5.6}$$

Now, by hypothesis, there exists a complex unit vector $\mathbf{x}$ such that $\lambda \mathbf{x} = \mathbf{A}\mathbf{x}$. Hence $\lambda = \bar{\mathbf{x}}^T \mathbf{A} \mathbf{x}$ and so, by (7.5.6),

$$|\lambda| = |\bar{\mathbf{x}}^T \mathbf{A} \mathbf{x}| \leqslant n\rho.$$

Again, $\bar{\lambda} = \bar{\mathbf{x}}^T \bar{\mathbf{A}}^T \mathbf{x}$, and therefore

$$2\,\mathfrak{Rl}\,\lambda = \lambda + \bar{\lambda} = \bar{\mathbf{x}}^T(\mathbf{A} + \bar{\mathbf{A}}^T)\mathbf{x}, \qquad 2i\,\mathfrak{Im}\,\lambda = \lambda - \bar{\lambda} = \bar{\mathbf{x}}^T(\mathbf{A} - \bar{\mathbf{A}}^T)\mathbf{x}.$$

Hence the remaining assertions again follow by (7.5.6).

7.5.2. THEOREM 7.5.4. *If* $\mathbf{A} = (a_{rs})$ *is a complex* $n \times n$ *matrix and*

$$\rho_k = \sum_{\substack{s=1 \\ s \neq k}}^{n} |a_{ks}|,$$

then every characteristic root λ *of* $\mathbf{A}$ *lies in at least one of the circles specified by the inequalities*

$$|z - a_{kk}| \leqslant \rho_k \quad (k = 1,..., n).$$

Let $\mathbf{x} = (x_1,..., x_n)^T \neq \mathbf{0}$ satisfy the equation $\mathbf{Ax} = \lambda\mathbf{x}$, so that

$$(\lambda - a_{rr})x_r = \sum_{\substack{s=1 \\ s \neq r}}^{n} a_{rs} x_s \quad (r = 1,..., n).$$

Denote by x_k the greatest, in modulus, among the numbers $x_1,..., x_n$. Then $x_k \neq 0$ and we have

$$|\lambda - a_{kk}||x_k| = \left| \sum_{\substack{s=1 \\ s \neq k}}^{n} a_{ks} x_s \right| \leqslant \sum_{\substack{s=1 \\ s \neq k}}^{n} |a_{ks}||x_k| = \rho_k |x_k|.$$

Hence $|\lambda - a_{kk}| \leqslant \rho_k$, as asserted. The conclusion also follows readily from Theorem 1.6.4 (p. 32).

The theorem just proved shows, in particular, that if $\mathbf{A} = (a_{rs})$ is *dominated* by its diagonal elements, i.e. if

$$|a_{kk}| > \sum_{\substack{s=1 \\ s \neq k}}^{n} |a_{ks}| \quad (k = 1,..., n),$$

then each characteristic root λ of $\mathbf{A}$ satisfies at least one of the equations

$$|\lambda - a_{kk}| < |a_{kk}| \quad (k = 1,..., n).$$

This implies, of course, that no characteristic root of $\mathbf{A}$ is equal to zero, i.e. that $\mathbf{A}$ is non-singular—a result with which we are already familiar in view of Theorem 1.6.4.

EXERCISE 7.5.6. Show that if $\mathbf{A} = (a_{rs})$ is dominated by its diagonal elements, and

$$\delta = \min_{1 \leqslant k \leqslant n} \left\{ |a_{kk}| - \sum_{\substack{s=1 \\ s \neq k}}^{n} |a_{ks}| \right\},$$

then

$$|\det \mathbf{A}| \geqslant \delta^n. \quad (7.5.7)$$

The following lower estimate for $|\det \mathbf{A}|$, more precise than (7.5.7), was given by Ostrowski (1937).

THEOREM 7.5.5. *Let* $\mathbf{A} = (a_{rs})$ *be a complex* $n \times n$ *matrix which is dominated by its diagonal elements. If*

$$d_r = |a_{rr}| - \sum_{\substack{s=1 \\ s \neq r}}^{n} |a_{rs}| \quad (r = 1,..., n),$$

then

$$|\det \mathbf{A}| \geqslant d_1 ... d_n.†$$

Consider the matrix $\mathbf{B} = (b_{rs})$ specified by the equations

$$b_{rs} = a_{rs}/d_r \quad (r, s = 1,..., n).$$

Then

$$\det \mathbf{A} = d_1 ... d_n \det \mathbf{B}. \tag{7.5.8}$$

If λ is any characteristic root of $\mathbf{B}$ we have, by Theorem 7.5.4,

$$|\lambda - b_{kk}| \leqslant \sum_{\substack{s=1 \\ s \neq k}}^{n} |b_{ks}|$$

for some value of k. Hence

$$|\lambda| \geqslant |b_{kk}| - \sum_{\substack{s=1 \\ s \neq k}}^{n} |b_{ks}| = 1,$$

and so, by Theorem 7.1.1, $|\det \mathbf{B}| \geqslant 1$. The assertion now follows by (7.5.8).

We conclude our series of estimates by obtaining an upper bound for the modulus of a determinant.

THEOREM 7.5.6. *For any complex* $n \times n$ *matrix* $\mathbf{A}$ *we have*

$$|\det \mathbf{A}| \leqslant \min(R^n, C^n),$$

where

$$R = \max_{1 \leqslant k \leqslant n} R_k, \qquad C = \max_{1 \leqslant k \leqslant n} C_k,$$

R_k (C_k) *being the sum of the absolute values of the elements in the* k*-th row* (k*-th column*) *of* $\mathbf{A}$.

Writing $\mathbf{A} = (a_{rs})$, we have

$$|\det \mathbf{A}| \leqslant \sum_{s_1,...,s_n=1}^{n} |a_{1s_1} ... a_{ns_n}|$$

$$= \prod_{r=1}^{n} \sum_{s=1}^{n} |a_{rs}| = \prod_{r=1}^{n} R_r \leqslant R^n.$$

Hence

$$|\det \mathbf{A}| = |\det \mathbf{A}^T| \leqslant C^n,$$

and the theorem follows.

† A still more precise inequality was stated, without proof, in § 1.6.4.

7.6. Characteristic vectors

The determination of the united points of a collineation leads, as we have seen in § 7.1.1, to a system of equations having the form $\mathbf{Ax} = \lambda\mathbf{x}$; and this, in turn, entails the investigation of the characteristic equation $|\lambda\mathbf{I}-\mathbf{A}| = 0$. It is, however, the vector $\mathbf{x}$ rather than the scalar λ that is needed in the geometrical problem; and the situation is similar in many other applications. Furthermore, the study of vectors of this type is indispensable in the further development of matrix theory.

DEFINITION 7.6.1. *Let λ be a characteristic root of* $\mathbf{A}$. *Then any non-zero vector* $\mathbf{x}$ *which satisfies the relation*

$$(\lambda\mathbf{I}-\mathbf{A})\mathbf{x} = \mathbf{0}$$

(*i.e.* $\mathbf{Ax} = \lambda\mathbf{x}$) *is called a* CHARACTERISTIC VECTOR *of* $\mathbf{A}$, *and it is said to be associated with the characteristic root* λ.

Since $|\lambda\mathbf{I}-\mathbf{A}| = 0$, we know that characteristic vectors associated with λ do, in fact, exist; and it is clear that the set of all these vectors, augmented by the zero vector, is a vector space. The problem of determining the dimensionality of this space (a problem which is equivalent to the determination of the rank of $\lambda\mathbf{I}-\mathbf{A}$) is important but cannot be treated at all fully at this stage. We shall, however, obtain one interesting result, and then return to the topic in § 10.2.

Theorem 7.6.1. (Rank-multiplicity theorem)

For every $n\times n$ matrix $\mathbf{A}$ *and every number ω we have*

$$R(\omega\mathbf{I}-\mathbf{A}) \geqslant n-m_{\omega}(\mathbf{A}).\dagger \tag{7.6.1}$$

When ω is not a characteristic root of $\mathbf{A}$, the theorem is true trivially. When $1 \leqslant k \leqslant n$ and ω is a k-fold characteristic root of $\mathbf{A}$, then the theorem asserts that there exist at most k linearly independent characteristic vectors associated with ω. The equivalence of the two ways of formulating the theorem is an immediate consequence of Theorem 5.4.2 (p. 149).

The assertion is obviously true for $k = n$ and we may, therefore, assume that $1 \leqslant k < n$. By Theorem 7.3.3 (p. 202), 0 is a k-fold characteristic root of $\omega\mathbf{I}-\mathbf{A}$, and hence

$$|\lambda\mathbf{I}-(\omega\mathbf{I}-\mathbf{A})| = \lambda^{n}+c_{1}\lambda^{n-1}+\ldots+c_{n-k}\lambda^{k},$$

† The reader is reminded that $m_{\omega}(\mathbf{A})$ stands for the multiplicity of ω as characteristic root of $\mathbf{A}$.

where $c_{n-k} \neq 0$. But, by Theorem 7.1.2 (p. 197), $(-1)^{n-k}c_{n-k}$ is equal to the sum of certain $(n-k)$-rowed minors of $\omega\mathbf{I}-\mathbf{A}$. Hence at least one of these minors does not vanish; thus $R(\omega\mathbf{I}-\mathbf{A}) \geqslant n-k$, and (7.6.1) is proved.

It is possible to give an alternative proof in which we make use of Theorem 2.3.5 instead of Theorem 7.1.2. The assertion is trivial when $R(\omega\mathbf{I}-\mathbf{A}) = 0$ or n, and we may, therefore, assume that $0 < s < n$, where $s = n - R(\omega\mathbf{I}-\mathbf{A})$. By Theorem 5.4.2, s linearly independent vectors $\mathbf{x}_1,\ldots,\mathbf{x}_s$ can be found such that

$$(\omega\mathbf{I}-\mathbf{A})\mathbf{x}_i = \mathbf{0} \qquad (i = 1,\ldots,s).$$

By the corollary to Theorem 2.3.5 (p. 54), there exist vectors $\mathbf{x}_{s+1},\ldots,\mathbf{x}_n$ such that $\mathbf{x}_1,\ldots,\mathbf{x}_s,\mathbf{x}_{s+1},\ldots,\mathbf{x}_n$ are linearly independent. Let $\mathbf{X}$ denote the (non-singular) matrix defined by the relations $\mathbf{X}_{*i} = \mathbf{x}_i$ $(i = 1,\ldots,n)$. We then have

$$\{\mathbf{X}^{-1}(\omega\mathbf{I}-\mathbf{A})\mathbf{X}\}_{*i} = \{\mathbf{X}^{-1}(\omega\mathbf{I}-\mathbf{A})\}\mathbf{X}_{*i} = \mathbf{X}^{-1}(\omega\mathbf{I}-\mathbf{A})\mathbf{x}_i = \mathbf{0} \quad (i = 1,\ldots,s),$$

and so the first s columns of $\mathbf{X}^{-1}(\omega\mathbf{I}-\mathbf{A})\mathbf{X}$ consist entirely of zeros. By Laplace's expansion theorem (Theorem 1.4.4, p. 21) it now follows that the characteristic polynomial of $\mathbf{X}^{-1}(\omega\mathbf{I}-\mathbf{A})\mathbf{X}$, and so of $\omega\mathbf{I}-\mathbf{A}$, has the form $\lambda^s\psi(\lambda)$, where ψ is a certain polynomial. Thus 0 is at least an s-fold characteristic root of $\omega\mathbf{I}-\mathbf{A}$, i.e. ω is at least an s-fold characteristic root of $\mathbf{A}$. Hence $k \geqslant s$, and this implies (7.6.1).

It should be noted that strict inequality in (7.6.1) may actually occur. Thus consider the case

$$\mathbf{A} = \begin{pmatrix} 1 & 0 \\ 2 & 1 \end{pmatrix}, \qquad \omega = 1, \qquad k = 2.$$

Here

$$\omega\mathbf{I}-\mathbf{A} = \begin{pmatrix} 0 & 0 \\ -2 & 0 \end{pmatrix},$$

and $1 = R(\omega\mathbf{I}-\mathbf{A}) > n-k = 0$. On the other hand, equality in (7.6.1) may also occur, as can be seen by considering the case $\mathbf{A} = \mathbf{I}_2$, $\omega = 1$, $k = 2$.

As an immediate consequence of Theorem 7.6.1 we have the following corollary.

Corollary. *If ω is a simple characteristic root of the $n \times n$ matrix $\mathbf{A}$, then*

$$R(\omega\mathbf{I}-\mathbf{A}) = n-1,$$

i.e. the vector space consisting of the characteristic vectors of $\mathbf{A}$ associated with ω, together with the zero vector, has dimensionality 1.

For, by Theorem 7.6.1, $R(\omega\mathbf{I}-\mathbf{A}) \geqslant n-1$, and since ω is a characteristic root of $\mathbf{A}$ we cannot have $R(\omega\mathbf{I}-\mathbf{A}) = n$.

EXERCISE 7.6.1. Show that an $n \times n$ matrix $\mathbf{A}$ possesses at least $n-R(\mathbf{A})$ zero characteristic roots.

From Theorem 7.1.2 we can also deduce the following useful consequence.

THEOREM 7.6.2. *If* $\mathbf{A}$ *is a non-zero matrix such that*

$$R(\mathbf{A}) = n-m_0(\mathbf{A}),$$

then $\mathbf{A}$ *possesses a critical principal minor.*

If 0 is not a characteristic root of $\mathbf{A}$, the assertion is obvious; and if 0 is an n-fold characteristic root of $\mathbf{A}$, then $\mathbf{A} = \mathbf{O}$, contrary to hypothesis. We assume, therefore, that 0 is a k-fold characteristic root of $\mathbf{A}$, where $0 < k < n$. The characteristic polynomial of $\mathbf{A}$ then has the form $\lambda^n+c_1\lambda^{n-1}+...+c_{n-k}\lambda^k$, where $c_{n-k} \neq 0$. But $(-1)^{n-k}c_{n-k}$ is equal to the sum of $(n-k)$-rowed principal minors. Hence at least one of these minors does not vanish; and since $n-k = R(\mathbf{A})$, the theorem follows.

PROBLEMS ON CHAPTER VII

1. If $\mathbf{A}$ is any rectangular matrix, show that $\mathbf{A}^T\mathbf{A}$ is symmetric and $\bar{\mathbf{A}}^T\mathbf{A}$ hermitian.

2. Is either symmetry or skew-symmetry preserved by similarity transformations?

3. Establish the following results. (i) The sum and the difference of two hermitian matrices are again hermitian matrices. (ii) The product of two hermitian matrices is hermitian if and only if the matrices commute. (iii) If $\mathbf{A}$ and $\mathbf{B}$ are hermitian, then so is $\mathbf{AB}+\mathbf{BA}$.

4. Show that the determinant of every hermitian matrix is real.

5. Find the characteristic roots of the matrices

$$\text{(i)}\quad \begin{pmatrix} \cos\theta & -\sin\theta \\ \sin\theta & \cos\theta \end{pmatrix}; \qquad \text{(ii)}\quad \begin{pmatrix} \cosh\theta & \sinh\theta \\ \sinh\theta & \cosh\theta \end{pmatrix};$$

$$\text{(iii)}\quad \begin{pmatrix} 1/\sqrt{3} & (1+i)/\sqrt{3} \\ (1-i)/\sqrt{3} & -1/\sqrt{3} \end{pmatrix}.$$

6. Find the characteristic roots of the matrix

$$\begin{pmatrix} 1 & 1 & 1 \\ 1 & \omega & \omega^2 \\ 1 & \omega^2 & \omega \end{pmatrix},$$

where $\omega = e^{2\pi i/3}$.

7. Let $\omega_1,..., \omega_n$ be the characteristic roots of the matrix $\mathbf{A} = (a_{rs})$. Show that

$$\sum_{r=1}^{n} \omega_r^2 = \sum_{r,s=1}^{n} a_{rs}\, a_{sr},$$

and also express $\sum_{r=1}^{n} \omega_r^3$ in terms of the elements of $\mathbf{A}$.

8. Find the cube of $$\begin{pmatrix} 1 & -18 & 5 \\ 3 & 2 & 1 \\ 5 & 22 & -3 \end{pmatrix}.$$

9. Evaluate the matrix $\mathbf{A}^6 - 25\mathbf{A}^2 + 112\mathbf{A}$, where
$$\mathbf{A} = \begin{pmatrix} 0 & 0 & 2 \\ 2 & 1 & 0 \\ -1 & -1 & 3 \end{pmatrix}.$$

10. Evaluate the matrix $\mathbf{A}^4 - 4\mathbf{A}^3 - \mathbf{A}^2 + 2\mathbf{A} - 5\mathbf{I}$, where $\mathbf{A} = \begin{pmatrix} 2 & -1 \\ 1 & 3 \end{pmatrix}$.

11. Let $\mathbf{A}$ be an $n \times n$ matrix, λ a characteristic root of $\mathbf{A}$, and $\mathbf{x}$ a characteristic vector associated with λ. Show that, for every positive integer k, $\mathbf{A}^k\mathbf{x} = \lambda^k\mathbf{x}$. Deduce that, if f is any rational function for which $f(\mathbf{A})$ is defined, then $f(\lambda)$ is a characteristic root of $f(\mathbf{A})$.

12. Find the characteristic equation of the matrix
$$\mathbf{A} = \begin{pmatrix} b & c & a \\ c & a & b \\ a & b & c \end{pmatrix},$$
and prove that the matrices
$$\mathbf{B} = \begin{pmatrix} c & a & b \\ a & b & c \\ b & c & a \end{pmatrix}, \qquad \mathbf{C} = \begin{pmatrix} a & b & c \\ b & c & a \\ c & a & b \end{pmatrix}$$
have the same characteristic equations as $\mathbf{A}$. Show also that, if $\mathbf{BC} = \mathbf{CB}$, then at least two roots of this equation are equal to zero.

13. If $\mathbf{A} = \begin{pmatrix} 1 & -1 \\ 2 & 5 \end{pmatrix}$, express $(2\mathbf{A}^4 - 12\mathbf{A}^3 + 19\mathbf{A}^2 - 29\mathbf{A} + 37\mathbf{I})^{-1}$ as a linear polynomial in $\mathbf{A}$.

14. If $\mathbf{A} = \begin{pmatrix} 2 & -5 \\ -1 & 3 \end{pmatrix}$, express $(\mathbf{A}^4 + 5\mathbf{A}^3 - 48\mathbf{A}^2 - \mathbf{I})^{-1}$ as a linear polynomial in $\mathbf{A}$.

15. The characteristic roots of the 3×3 matrix $\mathbf{A}$ are $1, -1, 2$. Express $\mathbf{A}^{2n}$ as a quadratic polynomial in $\mathbf{A}$.

16. Let
$$\mathbf{A} = \begin{pmatrix} 1 & 0 & 0 \\ 1 & 0 & 1 \\ 0 & 1 & 0 \end{pmatrix}.$$
Show that, for every integer $n \geqslant 3$, $\mathbf{A}^n = \mathbf{A}^{n-2} + \mathbf{A}^2 - \mathbf{I}$. Hence find $\mathbf{A}^{100}$.

17. If $\operatorname{tr}(\mathbf{AX}) = 0$ for all matrices $\mathbf{X}$, show that $\mathbf{A} = \mathbf{O}$.

18. Find the characteristic roots of the $n \times n$ matrix $\mathbf{A}$ all of whose elements are equal to 1. Also find the minimum polynomial of $\mathbf{A}$.

19. Express the sum of the diagonal elements of $\mathbf{A}^T\mathbf{A}$ in terms of the elements of $\mathbf{A}$.

Show that, if $\mathbf{A}$, $\mathbf{B}$ are real and symmetric and $\mathbf{C}$ is real and skew-symmetric, then $\mathbf{A}^2 + \mathbf{B}^2 = \mathbf{C}^2$ implies $\mathbf{A} = \mathbf{B} = \mathbf{C} = \mathbf{O}$. Does this conclusion still hold if $\mathbf{A}$ is not necessarily symmetric?

20. Give an easy proof of the theorem of Cayley and Hamilton for the case of matrices which are similar to diagonal matrices.

21. Show that the following statements relating to an $n \times n$ matrix are equivalent. (i) $\mathbf{A}^n = \mathbf{O}$; (ii) $\mathbf{A}^k = \mathbf{O}$ for some positive integer k; (iii) all characteristic roots of $\mathbf{A}$ are equal to 0.

22. Show that there exist no matrices $\mathbf{A}$, $\mathbf{B}$ such that $\mathbf{AB}-\mathbf{BA} = \mathbf{I}$.

23. Show that the constant term in the minimum polynomial of $\mathbf{A}$ vanishes if and only if $\mathbf{A}$ is singular.

24. Show that the zeros of the minimum polynomial of a diagonal matrix are distinct. Deduce that, if $\mathbf{A}$ is similar to a diagonal matrix, then the zeros of the minimum polynomial of $\mathbf{A}$ are distinct.

25. Prove the identity

$$M(\bar{\mathbf{A}}^T\mathbf{B}-\mathbf{B}\bar{\mathbf{A}}^T)-M(\mathbf{AB}-\mathbf{BA}) = \operatorname{tr}\{(\bar{\mathbf{A}}^T\mathbf{A}-\mathbf{A}\bar{\mathbf{A}}^T)(\bar{\mathbf{B}}^T\mathbf{B}-\mathbf{B}\bar{\mathbf{B}}^T)\},$$

where $M(\mathbf{X})$ is defined by the equation $M(\mathbf{X}) = \operatorname{tr}(\bar{\mathbf{X}}^T\mathbf{X})$.

26. Suppose that all characteristic roots of $\mathbf{I}-\mathbf{A}$ are less than 1 in modulus. Prove that $0 < |\det \mathbf{A}| < 2^n$; and show that this result is best possible.

27. Let $\mathbf{A} = (a_{rs})$ be a real $n \times n$ matrix and suppose that

$$a_{rr} > \sum_{\substack{s=1 \\ s\neq r}}^{n} |a_{rs}| \quad (r = 1,\ldots,n).$$

Show that all characteristic roots of $\mathbf{A}$ have positive real parts.

28. A real matrix $\mathbf{A}$ has all its elements equal to zero except those of the form a_{ii}, $a_{i,i+1}$, $a_{i+1,i}$. If, for each value of i, $a_{i,i+1}$ and $a_{i+1,i}$ have the same sign, show that there exists a real diagonal matrix $\mathbf{D}$ such that $\mathbf{D}^{-1}\mathbf{AD}$ is symmetric. Deduce that all characteristic roots of $\mathbf{A}$ are real.

29. Suppose that 0 is a simple characteristic root of $\mathbf{A}$. Show that $\operatorname{tr}\mathbf{A}^* \neq 0$. Show also that, if $|\mathbf{A}| = 0$ and $\operatorname{tr}\mathbf{A}^* \neq 0$, then 0 is a simple characteristic root of $\mathbf{A}$.

30. Let the distinct characteristic roots of the $n \times n$ matrix $\mathbf{A}$ be $\lambda_1,\ldots,\lambda_k$. Show that

$$n(k-1) \leqslant \sum_{i=1}^{k} R(\lambda_i \mathbf{I}-\mathbf{A}) \leqslant k(n-1).$$

31. The characteristic roots of an $(n+1)\times(n+1)$ matrix $\mathbf{A}$ are 0 and the nth roots of unity. Prove that

$$\frac{2\mathbf{I}}{2\mathbf{I}-\mathbf{A}} = \mathbf{I}+\frac{1}{2^n-1}(2^{n-1}\mathbf{A}+2^{n-2}\mathbf{A}^2+\ldots+2\mathbf{A}^{n-1}+\mathbf{A}^n).$$

32. Find the characteristic polynomials of the matrices

$$\text{(i)} \quad \begin{pmatrix} a_1 & -1 & 0 & 0 & . & . & . & 0 & 0 \\ a_2 & 0 & -1 & 0 & . & . & . & 0 & 0 \\ a_3 & 0 & 0 & -1 & . & . & . & 0 & 0 \\ . & . & . & . & . & . & . & . & . \\ a_{n-1} & 0 & 0 & 0 & . & . & . & 0 & -1 \\ a_n & 0 & 0 & 0 & . & . & . & 0 & 0 \end{pmatrix};$$

(ii)
$$\begin{pmatrix} 0 & 0 & 0 & . & . & . & 0 & 0 & 0 & -a_n \\ 1 & 0 & 0 & . & . & . & 0 & 0 & 0 & -a_{n-1} \\ 0 & 1 & 0 & . & . & . & 0 & 0 & 0 & -a_{n-2} \\ . & . & . & . & . & . & . & . & . & . \\ 0 & 0 & 0 & . & . & . & 0 & 0 & 1 & -a_1 \end{pmatrix};$$

(iii)
$$\begin{pmatrix} a_1 & 1 & 0 & 0 & . & . & . & 0 & 0 \\ a_2 & 0 & 1 & 0 & . & . & . & 0 & 0 \\ a_3 & 0 & 0 & 1 & . & . & . & 0 & 0 \\ . & . & . & . & . & . & . & . & . \\ a_{n-1} & 0 & 0 & 0 & . & . & . & 0 & 1 \\ a_n & 0 & 0 & 0 & . & . & . & 0 & 0 \end{pmatrix}.$$

33. Show that if $\mathbf{A}^2 = \mathbf{I}$, then every characteristic root of $\mathbf{A}$ is $+1$ or -1.

Let the 3×3 matrix $\mathbf{A}$ satisfy the equation $\mathbf{A}^2 = \mathbf{I}_3$ and suppose that all characteristic roots of $\mathbf{A}$ are equal to 1. Show that $\mathbf{A} = \mathbf{I}_3$.

34. Show that any real 2×2 matrix $\mathbf{A}$ which satisfies the equation $\mathbf{A}^2 = -\mathbf{I}$ is similar to the matrix $\mathbf{dg}(i, -i)$.

35. Show that, if $\mathbf{A}$ is a non-singular $n\times n$ matrix, then the coefficient of λ in the characteristic polynomial $\chi(\lambda)$ of $\mathbf{A}$ is equal to $(-1)^{n-1}|\mathbf{A}|\operatorname{tr}(\mathbf{A}^{-1})$.

36. Establish the identity
$$|\lambda\mathbf{I}-\mathbf{A}| = \lambda^3-\lambda^2\operatorname{tr}\mathbf{A}+\lambda\operatorname{tr}\mathbf{A}^*-|\mathbf{A}|$$
for every 3×3 matrix $\mathbf{A}$.

37. Show that, for any 3×3 matrices $\mathbf{A}$ and $\mathbf{B}$,
$$|\lambda\mathbf{A}-\mathbf{B}| = \lambda^3|\mathbf{A}|-\lambda^2\operatorname{tr}(\mathbf{A}^*\mathbf{B})+\lambda\operatorname{tr}(\mathbf{B}^*\mathbf{A})-|\mathbf{B}|.$$
Deduce that $$|\mathbf{A}+\mathbf{B}| = |\mathbf{A}|+\operatorname{tr}(\mathbf{A}^*\mathbf{B})+\operatorname{tr}(\mathbf{B}^*\mathbf{A})+|\mathbf{B}|.$$

38. Let $\lambda^n+c_1\lambda^{n-1}+...+c_{n-1}\lambda+c_n$ denote the characteristic polynomial of $\mathbf{A}$. Deduce, from the Cayley–Hamilton theorem, that
$$\mathbf{A}^* = (-1)^{n-1}\{\mathbf{A}^{n-1}+c_1\mathbf{A}^{n-2}+...+c_{n-1}\mathbf{I}\}.$$
Also, derive the Cayley–Hamilton theorem from this identity.

39. Let $\mathbf{A}$ be a matrix which satisfies the relation $\mathbf{A}^2 = \mathbf{I}$. Let $\mathfrak{U}$ be the space of vectors $\mathbf{x}$ such that $\mathbf{Ax} = \mathbf{x}$ and $\mathfrak{U}'$ the space of vectors $\mathbf{x}$ such that $\mathbf{Ax} = -\mathbf{x}$. Show that $\mathfrak{U}$ and $\mathfrak{U}'$ are complements.

40. Let $\omega_1,...,\omega_k$ be a set of distinct characteristic roots of the $n\times n$ matrix $\mathbf{A}$ and let, for $1 \leqslant i \leqslant k$, $\mathfrak{U}_i$ denote the space of vectors $\mathbf{x}$ such that $\mathbf{Ax} = \omega_i\mathbf{x}$. Show that, if $\mathbf{x}_1+...+\mathbf{x}_k = \mathbf{0}$, where $\mathbf{x}_1 \in \mathfrak{U}_1,...,\mathbf{x}_k \in \mathfrak{U}_k$, then $\mathbf{x}_1 = ... = \mathbf{x}_k = \mathbf{0}$.

41. Let
$$\mathbf{A} = \begin{pmatrix} \mathbf{A}_1 & \mathbf{O} \\ \mathbf{O} & \mathbf{A}_2 \end{pmatrix},$$
and let χ, χ_1, χ_2 be the characteristic polynomials of $\mathbf{A}$, $\mathbf{A}_1$, $\mathbf{A}_2$ respectively and μ, μ_1, μ_2 the minimum polynomials of $\mathbf{A}$, $\mathbf{A}_1$, $\mathbf{A}_2$ respectively. Show that $\chi = \chi_1\chi_2$ and that μ is the least common multiple of μ_1 and μ_2. Find the characteristic polynomial and the minimum polynomial of the matrix
$$\begin{pmatrix} 5 & -1 & 0 & 0 & 0 \\ 6 & 0 & -1 & 0 & 0 \\ 0 & 0 & 0 & 0 & 0 \\ 0 & 0 & 0 & 3 & 1 \\ 0 & 0 & 0 & -1 & 1 \end{pmatrix}.$$

42. Let

$$\mathbf{A} = \begin{pmatrix} 1 & 1 & 1 & . & . & . & 1 \\ 1 & \omega & \omega^2 & . & . & . & \omega^{n-1} \\ 1 & \omega^2 & \omega^4 & . & . & . & \omega^{2(n-1)} \\ . & . & . & . & . & . & . \\ 1 & \omega^{n-1} & \omega^{2(n-1)} & . & . & . & \omega^{(n-1)(n-1)} \end{pmatrix},$$

where $\omega = e^{2\pi i/n}$. By considering $\bar{\mathbf{A}}^T\mathbf{A}$, determine the value of $|\det \mathbf{A}|$.

43. Let $f(x)$ be a monic polynomial and let $\theta_1,\ldots,\theta_n$ denote the roots of the equation $f(x) = 0$. Show that the discriminant Δ of $f(x)$ is given by

$$\Delta = (-1)^{\frac{1}{2}n(n-1)} \prod_{r=1}^{n} f'(\theta_r).$$

Deduce that $(\det \mathbf{A})^2 = (-1)^{\frac{1}{2}(n-1)(n-2)} n^n$, where $\mathbf{A}$ denotes the matrix defined in the preceding question.

44. Show that the matrix $\mathbf{A}$ defined in No. 42 satisfies the relation

$$n^{-1}\mathbf{A}^2 = \begin{pmatrix} 1 & 0 & 0 & . & . & . & 0 & 0 & 0 \\ 0 & 0 & 0 & . & . & . & 0 & 0 & 1 \\ 0 & 0 & 0 & . & . & . & 0 & 1 & 0 \\ . & . & . & . & . & . & . & . & . \\ 0 & 0 & 1 & . & . & . & 0 & 0 & 0 \\ 0 & 1 & 0 & . & . & . & 0 & 0 & 0 \end{pmatrix}.$$

Hence obtain the expression for $(\det \mathbf{A})^2$ given in the preceding question. Also deduce that $\mathbf{A}^4$ is a scalar matrix and show that all characteristic roots of $\mathbf{A}$ are to be found among the numbers $\pm\sqrt{n}$, $\pm i\sqrt{n}$.

45. Suppose that the elements of $\mathbf{A}^m$ are bounded as $m \to \infty$. Show that the modulus of every characteristic root of $\mathbf{A}$ is less than or equal to 1. Show also that the converse inference is false.

46. Let $\mathbf{G}$ be a 2×2 matrix; let (x_1, x_2), (y_1, y_2) be any two row vectors; and write

$$(\xi_1, \xi_2) = (x_1, x_2)\mathbf{G}, \qquad (\eta_1, \eta_2) = (y_1, y_2)\mathbf{G}.$$

Show that the three components of the row vector on the right of the equation

$$(x_1 y_1,\ x_1 y_2 + x_2 y_1,\ x_2 y_2)\mathbf{\Gamma} = (\xi_1 \eta_1,\ \xi_1 \eta_2 + \xi_2 \eta_1,\ \xi_2 \eta_2)$$

are all linear combinations of the three components of the row vector on the left, so that the equation thereby defines the 3×3 matrix $\mathbf{\Gamma}$. Obtain the relation

$$\operatorname{tr}\mathbf{\Gamma} = (\operatorname{tr}\mathbf{G})^2 - |\mathbf{G}|.$$

If the characteristic roots of $\mathbf{G}$ are λ and μ, find the characteristic roots of $\mathbf{\Gamma}$; and deduce that $|\mathbf{\Gamma}| = |\mathbf{G}|^3$.

47. Show that the $n\times n$ matrix $\mathbf{A} = (a_{rs})$ is non-singular if and only if there exists a matrix $\mathbf{B} = (b_{rs})$ such that

$$h_r = \left|\sum_{k=1}^{n} a_{rk} b_{kr}\right| - \sum_{\substack{s=1 \\ s\neq r}}^{n} \left|\sum_{k=1}^{n} a_{rk} b_{ks}\right| > 0 \quad (r = 1,\ldots, n).$$

Show, furthermore, that for any such matrix $\mathbf{B}$,

$$|\det \mathbf{A}|\,|\det \mathbf{B}| \geqslant h_1 \ldots h_n.$$

Obtain theorems arising from the special cases $\mathbf{B} = \mathbf{I}$ and $\mathbf{B} = \bar{\mathbf{A}}^T$.

48. The characteristic roots of the matrix $\mathbf{A}$ are $\omega_1, \omega_2,..., \omega_n$, and $|\omega_1| > |\omega_k|$ $(k = 2,..., n)$. Show that

$$\lim_{m\to\infty} \{\mathrm{tr}(\mathbf{A}^m)\}^{1/m} = \omega_1.$$

49. Let $\mathbf{A}$ be a given $n \times n$ matrix. Show that, for every value of λ,

$$(\mathbf{A}-\lambda\mathbf{I})^* = \psi_0\mathbf{I}+\psi_1\mathbf{A}+...+\psi_{n-1}\mathbf{A}^{n-1},$$

where $\psi_0, \psi_1,..., \psi_{n-1}$ are polynomials in λ. Find these polynomials when

$$\mathbf{A} = \begin{pmatrix} 0 & 1 & 0 \\ -1 & 0 & 1 \\ 0 & -1 & 0 \end{pmatrix}.$$

50. $\mathbf{A} = (a_{rs})$ is a complex $n \times n$ matrix; $p_1,..., p_n$ are positive numbers; and

$$K = \max_{1\leqslant r\leqslant n}\left(\frac{|a_{r1}|p_1+...+|a_{rn}|p_n}{p_r}\right).$$

Show that every characteristic root ω of $\mathbf{A}$ satisfies the inequality $|\omega| \leqslant K$.

51. Let $\mathbf{A}_1$, $\mathbf{B}_1$ be 3×3 matrices and suppose that, for all values of λ and μ,

$$(\lambda\mathbf{A}_1+\mu\mathbf{B}_1)^* = \lambda^2\mathbf{A}_2+\lambda\mu\mathbf{C}_2+\mu^2\mathbf{B}_2,$$

$$|\lambda\mathbf{A}_1+\mu\mathbf{B}_1| = \lambda^3\Delta+\lambda^2\mu\Theta+\lambda\mu^2\Theta'+\mu^3\Delta'.$$

Show that $\quad \mathbf{C}_2\mathbf{A}_1+\mathbf{A}_2\mathbf{B}_1 = \Theta\mathbf{I}, \qquad \mathbf{C}_2\mathbf{B}_1+\mathbf{B}_2\mathbf{A}_1 = \Theta'\mathbf{I},$

and deduce that $|\mathbf{C}_2| = \Theta\Theta'-\Delta\Delta'$.

Also show that, if $(\lambda\mathbf{A}_2+\mu\mathbf{B}_2)^* = \lambda^2\mathbf{A}_3+\lambda\mu\mathbf{C}_3+\mu^2\mathbf{B}_3$, then

$$\mathbf{A}_2\mathbf{C}_3+\Delta\mathbf{B}_2\mathbf{A}_1 = \Delta\Theta'\mathbf{I}, \qquad \mathbf{B}_2\mathbf{C}_3+\Delta'\mathbf{A}_2\mathbf{B}_1 = \Delta'\Theta\mathbf{I}.$$

52. Show that, for any matrix $\mathbf{A}$, any distinct numbers $\omega_1,..., \omega_k$, and any polynomial f of degree $\leqslant k-1$,

$$f(\mathbf{A}) = \sum_{i=1}^{k} f(\omega_i) \prod_{\substack{1\leqslant\kappa\leqslant k \\ \kappa\neq i}} \left(\frac{\mathbf{A}-\omega_\kappa\mathbf{I}}{\omega_i-\omega_\kappa}\right).$$

Show also that this identity continues to hold for a polynomial f of any degree, provided that $\omega_1,..., \omega_k$ are taken as the distinct characteristic roots of $\mathbf{A}$ and provided further that the minimum polynomial of $\mathbf{A}$ is a product of distinct linear factors. (This result is known as *Sylvester's interpolation formula*.)

53. Show that $R(\mathbf{A}) = R(\mathbf{A}^2)$ if and only if $R(\mathbf{A}) = n-m_0(\mathbf{A})$.

54. Show that $R(\mathbf{A}^m) = R(\mathbf{A}^{m+1}) = R(\mathbf{A}^{m+2}) = ...$, where $m = m_0(\mathbf{A})$.

55. The elements of an $n \times n$ matrix $\mathbf{M}$ are the integers $1, 2,..., n^2$ arranged in such a way that the sum of the elements in each row is the same. Show that $\frac{1}{2}n(n^2+1)$ is a characteristic root of $\mathbf{M}$.

56. Show that any characteristic root λ of an $n \times n$ matrix $\mathbf{A}$ satisfies the inequality

$$|\lambda|^2 \leqslant \sum_{k=1}^{n} |\mathbf{A}_{k*}| \cdot |\mathbf{A}_{*k}|.$$

VIII

ORTHOGONAL AND UNITARY MATRICES

THE first two sections of this chapter are devoted to the investigation of two special classes of matrices, namely orthogonal matrices and unitary matrices. The remaining two sections deal with the use of orthogonal matrices in the algebraic manipulation of rotations in two and three dimensions.

8.1. Orthogonal matrices

8.1.1. DEFINITION 8.1.1. *A matrix* $\mathbf{A}$ *is* ORTHOGONAL *if it is real and*

$$\mathbf{A}^T\mathbf{A} = \mathbf{I}. \tag{8.1.1}$$

Equation (8.1.1) implies that $\mathbf{A}$ is non-singular. It may therefore be written as $\mathbf{A}^{-1} = \mathbf{A}^T$, and hence also as

$$\mathbf{A}\mathbf{A}^T = \mathbf{I}. \tag{8.1.2}$$

By equating corresponding elements on both sides in (8.1.1) and (8.1.2), we can rewrite these relations in the more explicit form

$$\sum_{k=1}^{n} a_{kr}\, a_{ks} = \delta_{rs} \quad (r, s = 1,..., n), \tag{8.1.3}$$

$$\sum_{k=1}^{n} a_{rk}\, a_{sk} = \delta_{rs} \quad (r, s = 1,..., n), \tag{8.1.4}$$

where $\mathbf{A} = (a_{rs})$.

THEOREM 8.1.1. *If* $\mathbf{A}$ *is orthogonal, then* $|\mathbf{A}| = \pm 1$.

For, by definition, $\mathbf{A}^T\mathbf{A} = \mathbf{I}$. Hence $|\mathbf{A}^T||\mathbf{A}| = 1$, i.e. $|\mathbf{A}|^2 = 1$.

Examples of orthogonal matrices with determinants 1 and -1 respectively are easily given. Two such matrices are $\mathbf{I}_2$ and $\mathbf{dg}(1, -1)$.

EXERCISE 8.1.1. Show that if $\mathbf{A} = (a_{rs})$ is an orthogonal matrix, then the (unique) solution of the system of equations

$$\begin{aligned} a_{11}x_1 + ... + a_{1n}x_n &= b_1, \\ &\cdots \\ a_{n1}x_1 + ... + a_{nn}x_n &= b_n \end{aligned}$$

is

$$\begin{aligned} x_1 &= a_{11}b_1 + ... + a_{n1}b_n, \\ &\cdots \\ x_n &= a_{1n}b_1 + ... + a_{nn}b_n. \end{aligned}$$

THEOREM 8.1.2. *A real matrix is orthogonal if and only if its columns (or rows) form an orthonormal set of vectors.*†

A real matrix $\mathbf{A} = (a_{rs})$ is orthogonal if and only if the relations (8.1.3) hold, and that is precisely the condition for the columns of $\mathbf{A}$ to form an orthonormal set. Similarly, (8.1.4) is precisely the condition for the rows of $\mathbf{A}$ to form an orthonormal set.

By Theorem 8.1.2 it is, for instance, obvious that the matrix

$$\begin{pmatrix} 1/\sqrt{2} & -1/\sqrt{2} \\ 1/\sqrt{2} & 1/\sqrt{2} \end{pmatrix}$$

is orthogonal.

Two immediate inferences from Theorem 8.1.2 are as follows.

COROLLARY 1. *If the columns of a real matrix form an orthonormal set, then so do the rows; and conversely.*

COROLLARY 2. *If the order of the columns (or rows) of an orthogonal matrix is changed, then the resulting matrix is again orthogonal.*

EXERCISE 8.1.2. Deduce Theorem 8.1.2 from Definition 8.1.1 and Exercise 3.3.3 (p. 80).

EXERCISE 8.1.3. Show that if any rows (or columns) of an orthogonal matrix are multiplied by -1, then the resulting matrix is again orthogonal.

EXERCISE 8.1.4. Let

$$\mathbf{T} = \begin{pmatrix} \lambda & \lambda' & \lambda'' \\ \mu & \mu' & \mu'' \\ \nu & \nu' & \nu'' \end{pmatrix}$$

be an orthogonal matrix with positive determinant. Using the equation $\mathbf{T}^T = \mathbf{T}^{-1}$, show that

$$\lambda = \mu'\nu'' - \mu''\nu', \qquad \mu = \nu'\lambda'' - \nu''\lambda', \qquad \nu = \lambda'\mu'' - \lambda''\mu'.$$

THEOREM 8.1.3. *Let $\mathbf{x}_1$ be a real unit vector. Then there exists an orthogonal matrix $\mathbf{A}$ having $\mathbf{x}_1$ as its first column. Furthermore, the sign of $|\mathbf{A}|$ can be chosen at will.*

In view of the theorem on orthonormal bases (Theorem 2.5.5, p. 66) there exist vectors $\mathbf{x}_2,\ldots,\mathbf{x}_n$ such that $\mathbf{x}_1,\mathbf{x}_2,\ldots,\mathbf{x}_n$ is an orthonormal set. The matrix $\mathbf{A}$ having $\mathbf{x}_1,\mathbf{x}_2,\ldots,\mathbf{x}_n$ as its columns is then orthogonal by Theorem 8.1.2. If necessary, we can adjust the sign of $|\mathbf{A}|$ by replacing $\mathbf{x}_2$ by $-\mathbf{x}_2$.

† In view of this result, the term 'orthonormal matrix' might be regarded as more appropriate than 'orthogonal matrix', but custom has firmly established the latter usage.

THEOREM 8.1.4. *If* $\mathbf{A}$ *and* $\mathbf{B}$ *are orthogonal matrices, then so are* $\mathbf{A}^T$, $\mathbf{A}^{-1}$, *and* $\mathbf{AB}$.

Since $\mathbf{A}$ is orthogonal, $\mathbf{AA}^T = \mathbf{I}$. Hence

$$(\mathbf{A}^T)^T\mathbf{A}^T = \mathbf{I},$$

and so $\mathbf{A}^T$ is orthogonal. Moreover, $\mathbf{A}^{-1} = \mathbf{A}^T$ and so $\mathbf{A}^{-1}$ is orthogonal. Again

$$(\mathbf{AB})^T\mathbf{AB} = \mathbf{B}^T\mathbf{A}^T\mathbf{AB} = \mathbf{B}^T\mathbf{B} = \mathbf{I},$$

and therefore $\mathbf{AB}$ is orthogonal.

An important property of orthogonal matrices can be formulated in terms of linear substitutions.

THEOREM 8.1.5. *A real matrix* $\mathbf{A}$ *is orthogonal if and only if the substitution* $\mathbf{x} = \mathbf{Ay}$ *transforms the polynomial* $x_1^2+...+x_n^2$ *into* $y_1^2+...+y_n^2$, *where* $\mathbf{x} = (x_1,..., x_n)^T$, $\mathbf{y} = (y_1,..., y_n)^T$.

The substitution $\mathbf{x} = \mathbf{Ay}$ transforms $x_1^2+...+x_n^2 = \mathbf{x}^T\mathbf{x}$ into $(\mathbf{Ay})^T\mathbf{Ay} = \mathbf{y}^T\mathbf{A}^T\mathbf{Ay}$. Now if $\mathbf{A}$ is orthogonal, this polynomial is clearly equal to $\mathbf{y}^T\mathbf{y} = y_1^2+...+y_n^2$. On the other hand, if the polynomials $\mathbf{y}^T\mathbf{A}^T\mathbf{Ay}$ and $\mathbf{y}^T\mathbf{y}$ are equal, then $\mathbf{A}^T\mathbf{A} = \mathbf{I}$, i.e. $\mathbf{A}$ is orthogonal. The theorem is therefore proved.

Consider, for example, the substitution $\mathbf{x} = \mathbf{Ay}$, where $\mathbf{A}$ is the 2×2 matrix given on p. 223. Then

$$x_1 = (y_1-y_2)/\sqrt{2}, \qquad x_2 = (y_1+y_2)/\sqrt{2},$$

and it can be verified at once that this substitution transforms $x_1^2+x_2^2$ into $y_1^2+y_2^2$.

THEOREM 8.1.6. *A real matrix* $\mathbf{A}$ *is orthogonal if and only if it preserves length, i.e. if and only if*

$$|\mathbf{Ax}| = |\mathbf{x}| \tag{8.1.5}$$

for every real vector $\mathbf{x}$.

Equation (8.1.5) is equivalent to

$$(\mathbf{Ax}, \mathbf{Ax}) = (\mathbf{x}, \mathbf{x}), \tag{8.1.6}$$

i.e.

$$\mathbf{x}^T(\mathbf{A}^T\mathbf{A}-\mathbf{I})\mathbf{x} = 0. \tag{8.1.7}$$

If $\mathbf{A}$ is orthogonal, this relation clearly holds for all $\mathbf{x}$. Suppose, on the other hand, that (8.1.7), and so (8.1.5), holds for all $\mathbf{x}$. Writing $\mathbf{B} = (b_{rs}) = \mathbf{A}^T\mathbf{A}-\mathbf{I}$ and putting $\mathbf{x} = \mathbf{e}_k$ in (8.1.7) we obtain, by Exercise 3.3.9 (p. 85),

$$b_{kk} = 0 \quad (k = 1,..., n). \tag{8.1.8}$$

Next, let $k \neq l$ and put $\mathbf{x} = \mathbf{e}_k + \mathbf{e}_l$ in (8.1.7). Again using Exercise 3.3.9 we obtain

$$b_{kk} + b_{ll} + b_{kl} + b_{lk} = 0 \quad (k, l = 1,..., n;\ k \neq l).$$

Hence, in view of (8.1.8) and the obvious symmetry of $\mathbf{B}$,

$$b_{kl} = 0 \quad (k, l = 1,..., n;\ k \neq l). \tag{8.1.9}$$

By (8.1.8) and (8.1.9) it follows that $\mathbf{B} = \mathbf{O}$, i.e. $\mathbf{A}$ is orthogonal This completes the proof.

COROLLARY 1. *A real matrix* $\mathbf{A}$ *is orthogonal if and only if it preserves separation i.e. if and only if*

$$|\mathbf{Ax} - \mathbf{Ay}| = |\mathbf{x} - \mathbf{y}| \tag{8.1.10}$$

for all real vectors $\mathbf{x}$, $\mathbf{y}$.

If $\mathbf{A}$ is orthogonal, then, by (8.1.5),

$$|\mathbf{Ax} - \mathbf{Ay}| = |\mathbf{A}(\mathbf{x} - \mathbf{y})| = |\mathbf{x} - \mathbf{y}|.$$

On the other hand, if (8.1.10) holds for all real $\mathbf{x}$, $\mathbf{y}$, then (8.1.5) holds *a fortiori*, and $\mathbf{A}$ is orthogonal.

EXERCISE 8.1.5. Give a geometrical interpretation of Corollary 1 for the case $n = 3$.

COROLLARY 2. *A real matrix* $\mathbf{A}$ *is orthogonal if and only if it preserves inner products, i.e. if and only if*

$$(\mathbf{Ax}, \mathbf{Ay}) = (\mathbf{x}, \mathbf{y}) \tag{8.1.11}$$

for all real vectors $\mathbf{x}$, $\mathbf{y}$.

If $\mathbf{A}$ is orthogonal, then

$$(\mathbf{Ax}, \mathbf{Ay}) = (\mathbf{Ay})^T\mathbf{Ax} = \mathbf{y}^T\mathbf{A}^T\mathbf{Ax} = \mathbf{y}^T\mathbf{x} = (\mathbf{x}, \mathbf{y}).$$

On the other hand, if (8.1.11) holds for all real $\mathbf{x}$, $\mathbf{y}$, then (8.1.6), and so (8.1.5), holds *a fortiori*. Hence $\mathbf{A}$ is orthogonal.

8.1.2. The next series of results is concerned with the values of characteristic roots of orthogonal matrices.

THEOREM 8.1.7. *If* λ *is a characteristic root of the orthogonal matrix* $\mathbf{A}$, *then so is* λ^{-1}.

It should be noted that, since $|\mathbf{A}| \neq 0$, none of the characteristic roots of $\mathbf{A}$ is equal to zero.

By Theorem 7.3.3 (p. 202), we know that λ^{-1} is a characteristic root of $\mathbf{A}^{-1}$, and so of $\mathbf{A}^T$. Hence

$$0 = |\lambda^{-1}\mathbf{I}-\mathbf{A}^T| = |(\lambda^{-1}\mathbf{I}-\mathbf{A})^T| = |\lambda^{-1}\mathbf{I}-\mathbf{A}|,$$

and therefore λ^{-1} is a characteristic root of $\mathbf{A}$.

The next theorem we prove is due to Brioschi (1854).

THEOREM 8.1.8. *Every characteristic root of an orthogonal matrix has unit modulus.*

Let λ be a characteristic root of the orthogonal matrix $\mathbf{A}$, and let $\mathbf{x}$ be a non-zero vector such that

$$\lambda\mathbf{x} = \mathbf{A}\mathbf{x}. \tag{8.1.12}$$

Then $\bar{\lambda}\bar{\mathbf{x}} = \mathbf{A}\bar{\mathbf{x}}$, and so

$$\bar{\lambda}\bar{\mathbf{x}}^T = \bar{\mathbf{x}}^T\mathbf{A}^T. \tag{8.1.13}$$

By (8.1.12) and (8.1.13) we obtain

$$\bar{\lambda}\lambda\bar{\mathbf{x}}^T\mathbf{x} = \bar{\mathbf{x}}^T\mathbf{A}^T\mathbf{A}\mathbf{x} = \bar{\mathbf{x}}^T\mathbf{x}.$$

Since $\bar{\mathbf{x}}^T\mathbf{x} \neq 0$, this implies that $|\lambda|^2 = \bar{\lambda}\lambda = 1$, i.e. $|\lambda| = 1$.

EXERCISE 8.1.6. Give a proof of the above theorem without making use of matrix notation.

As an immediate consequence of Theorem 8.1.8 we have the following result.

COROLLARY. *The non-real characteristic roots of an orthogonal matrix occur in conjugate pairs of the type* $e^{i\alpha}$, $e^{-i\alpha}$, *where* $0 < \alpha < \pi$.

We may also note some further relations concerning characteristic roots.

THEOREM 8.1.9. *Let* $\mathbf{A}$ *be an orthogonal* $n \times n$ *matrix.* (i) *If* $|\mathbf{A}| = 1$ *and* n *is odd or if* $|\mathbf{A}| = -1$ *and* n *is even, then* 1 *is a characteristic root of* $\mathbf{A}$. (ii) *If* $|\mathbf{A}| = -1$, *then* -1 *is a characteristic root of* $\mathbf{A}$.

We have $$\mathbf{A}^T(\mathbf{I}-\mathbf{A}) = \mathbf{A}^T-\mathbf{I} = -(\mathbf{I}-\mathbf{A})^T,$$

and therefore $$|\mathbf{A}|\,.\,|\mathbf{I}-\mathbf{A}| = (-1)^n|\mathbf{I}-\mathbf{A}|,$$

$$|\mathbf{I}-\mathbf{A}|\{|\mathbf{A}|-(-1)^n\} = 0.$$

Hence (i) follows. Again,

$$\mathbf{A}^T(\mathbf{I}+\mathbf{A}) = \mathbf{A}^T+\mathbf{I} = (\mathbf{I}+\mathbf{A})^T,$$

$$|\mathbf{A}|.|\mathbf{I}+\mathbf{A}| = |\mathbf{I}+\mathbf{A}|,$$

and (ii) follows.

8.1.3. In § 8.1.1 we obtained a number of conditions necessary and sufficient to ensure that a matrix should be orthogonal. These conditions do not lead to a convenient method for constructing orthogonal matrices, but the next theorem (discovered by Cayley in 1846) provides us with such a method.

THEOREM 8.1.10. *If* $\mathbf{S}$ *is a real skew-symmetric matrix, then* $\mathbf{I}+\mathbf{S}$ *is non-singular, and the matrix*

$$\mathbf{A} = (\mathbf{I}-\mathbf{S})(\mathbf{I}+\mathbf{S})^{-1}$$

is orthogonal.

By the corollary to Theorem 7.5.1 (p. 209), all characteristic roots of $\mathbf{S}$ are purely imaginary, and therefore $|\mathbf{I}+\mathbf{S}| \neq 0$.

Since $\mathbf{S}^T = -\mathbf{S}$ we have

$$(\mathbf{I}-\mathbf{S})^T = \mathbf{I}+\mathbf{S}, \qquad (\mathbf{I}+\mathbf{S})^T = \mathbf{I}-\mathbf{S},$$

and therefore

$$\begin{aligned}\mathbf{A}^T &= \{(\mathbf{I}-\mathbf{S})(\mathbf{I}+\mathbf{S})^{-1}\}^T = \{(\mathbf{I}+\mathbf{S})^{-1}\}^T(\mathbf{I}-\mathbf{S})^T \\ &= \{(\mathbf{I}+\mathbf{S})^T\}^{-1}(\mathbf{I}+\mathbf{S}) = (\mathbf{I}-\mathbf{S})^{-1}(\mathbf{I}+\mathbf{S}).\end{aligned}$$

Hence

$$\begin{aligned}\mathbf{A}^T\mathbf{A} &= (\mathbf{I}-\mathbf{S})^{-1}(\mathbf{I}+\mathbf{S})(\mathbf{I}-\mathbf{S})(\mathbf{I}+\mathbf{S})^{-1} \\ &= (\mathbf{I}-\mathbf{S})^{-1}(\mathbf{I}-\mathbf{S})(\mathbf{I}+\mathbf{S})(\mathbf{I}+\mathbf{S})^{-1} = \mathbf{I},\end{aligned}$$

and the theorem is proved.†

EXERCISE 8.1.7. Obtain an orthogonal matrix of order 3 involving 3 independent parameters.

8.1.4. We know, by Corollary 1 to Theorem 8.1.6 that every transformation $\mathbf{x}' = \mathbf{A}\mathbf{x}$, where $\mathbf{A}$ is an orthogonal matrix, preserves separation. We shall show next that the converse statement is also, in a sense, true.

† For further results bearing on the relation between orthogonal and skew-symmetric matrices, see Ferrar, **2**, 164–7. For a generalization of Theorem 8.1.10, see Turnbull, **26**, 158–9.

THEOREM 8.1.11. *Let f be a transformation of the real total vector space $\mathfrak{B}_n$ into itself. If*

$$f(\mathbf{0}) = \mathbf{0}, \tag{8.1.14}$$

and, for all $\mathbf{x}, \mathbf{y} \in \mathfrak{B}_n$,

$$|f(\mathbf{x})-f(\mathbf{y})| = |\mathbf{x}-\mathbf{y}|, \tag{8.1.15}$$

then $f(\mathbf{x}) = \mathbf{A}\mathbf{x}$, *where* $\mathbf{A}$ *is an orthogonal matrix.*

To prove the theorem we begin by showing that f preserves inner products. For, by (8.1.14) and (8.1.15), we have

$$|f(\mathbf{x})| = |f(\mathbf{x})-f(\mathbf{0})| = |\mathbf{x}|$$

for all $\mathbf{x} \in \mathfrak{B}_n$, and so

$$(f(\mathbf{x}), f(\mathbf{x})) = (\mathbf{x}, \mathbf{x}). \tag{8.1.16}$$

Again, by (8.1.15) we have, for all $\mathbf{x}$, $\mathbf{y} \in \mathfrak{B}_n$,

$$(f(\mathbf{x})-f(\mathbf{y}), f(\mathbf{x})-f(\mathbf{y})) = (\mathbf{x}-\mathbf{y}, \mathbf{x}-\mathbf{y}).$$

Therefore, in view of (8.1.16),

$$(f(\mathbf{x}), f(\mathbf{y})) = (\mathbf{x}, \mathbf{y}) \tag{8.1.17}$$

for all $\mathbf{x}$, $\mathbf{y} \in \mathfrak{B}_n$. In particular

$$(f(\mathbf{e}_i), f(\mathbf{e}_j)) = (\mathbf{e}_i, \mathbf{e}_j) = \delta_{ij} \quad (i, j = 1,..., n),$$

and the vectors $f(\mathbf{e}_1),..., f(\mathbf{e}_n)$ thus form an orthonormal set. Hence the matrix $\mathbf{A}$, defined by the relations $\mathbf{A}_{*i} = f(\mathbf{e}_i)$ $(i = 1,..., n)$, is orthogonal by Theorem 8.1.2. Furthermore,

$$f(\mathbf{e}_i) = \mathbf{A}\mathbf{e}_i \qquad (i = 1,..., n). \tag{8.1.18}$$

Now, if the vectors $\mathbf{u}_1,..., \mathbf{u}_n$ constitute an orthonormal set, then, by Exercise 2.5.4 (p. 66), any vector $\mathbf{v}$ can be expressed in the form

$$\mathbf{v} = \sum_{i=1}^{n} (\mathbf{u}_i, \mathbf{v})\mathbf{u}_i.$$

Hence, by (8.1.17) and (8.1.18),

$$f(\mathbf{x}) = \sum_{i=1}^{n} (f(\mathbf{e}_i), f(\mathbf{x}))f(\mathbf{e}_i) = \sum_{i=1}^{n} (\mathbf{e}_i, \mathbf{x})\mathbf{A}\mathbf{e}_i$$

$$= \mathbf{A} \sum_{i=1}^{n} (\mathbf{e}_i, \mathbf{x})\mathbf{e}_i = \mathbf{A}\mathbf{x},$$

and the assertion is therefore established.

Theorem 8.1.11 states that a transformation which preserves separation and leaves the zero vector invariant is an orthogonal transformation. By dispensing with the second requirement we are led to the following modified statement.

COROLLARY. *Let f be a transformation of the real total vector space $\mathfrak{V}_n$ into itself. If, for all $\mathbf{x}, \mathbf{y} \in \mathfrak{V}_n$, $|f(\mathbf{x})-f(\mathbf{y})| = |\mathbf{x}-\mathbf{y}|$, then $f(\mathbf{x}) = \mathbf{Ax}+\mathbf{c}$, where $\mathbf{A}$ is an orthogonal matrix and $\mathbf{c}$ a fixed vector.*

Putting $g(\mathbf{x}) = f(\mathbf{x})-f(\mathbf{0})$, we see that $g(\mathbf{0}) = \mathbf{0}$ and $|g(\mathbf{x})-g(\mathbf{y})| = |\mathbf{x}-\mathbf{y}|$ for all real $\mathbf{x}$, $\mathbf{y}$. Hence, by Theorem 8.1.11, $g(\mathbf{x}) = \mathbf{Ax}$, where $\mathbf{A}$ is an orthogonal matrix, i.e.

$$f(\mathbf{x}) = \mathbf{Ax}+f(\mathbf{0}).$$

The result just proved has a simple geometrical interpretation for the case $n = 3$. Taking $\mathfrak{V}_3$ to represent ordinary three-dimensional space we can infer from the corollary, the remark following Definition 2.5.2 (p. 63), and Theorems 8.4.7 and 8.4.13 below, that a transformation of space which preserves distance is either the product of a rotation and a translation or else the product of a rotation, a reflection, and a translation.†

8.2. Unitary matrices

In considering complex matrices, it is desirable to generalize the notion of orthogonality.

DEFINITION 8.2.1. *The complex matrix $\mathbf{U}$ is* UNITARY *if*

$$\overline{\mathbf{U}}^T\mathbf{U} = \mathbf{I}. \tag{8.2.1}$$

Thus a real unitary matrix is simply an orthogonal matrix. Unitary matrices stand in roughly the same relation to orthogonal matrices as hermitian matrices to real symmetric ones.

The defining relation (8.2.1) may, of course, be restated in many alternative ways, such as $\overline{\mathbf{U}}^T = \mathbf{U}^{-1}$, $\overline{\mathbf{U}} = (\mathbf{U}^T)^{-1}$, and $\mathbf{U}\overline{\mathbf{U}}^T = \mathbf{I}$. A simple example of a unitary matrix is given by

$$\begin{pmatrix} (1+i)/2 & (-1+i)/2 \\ (1+i)/2 & (1-i)/2 \end{pmatrix}.$$

EXERCISE 8.2.1. Show that if $\mathbf{A}$ is hermitian and $\mathbf{U}$ unitary, then $\mathbf{U}^{-1}\mathbf{AU}$ is hermitian.

The theory of unitary matrices closely resembles that of orthogonal matrices. We shall, therefore, deal with it rather summarily and in many cases leave the details to the reader.

† We recall that by the product of a number of transformations we mean their resultant.

THEOREM 8.2.1. *If* $\mathbf{U}$ *is unitary, then* $|\det \mathbf{U}| = 1$.

THEOREM 8.2.2. *If* $\mathbf{U}, \mathbf{V}$ *are unitary, then so are* $\overline{\mathbf{U}}, \mathbf{U}^T, \mathbf{U}^{-1}$ *and* $\mathbf{UV}$.

These results follow immediately from (8.2.1). The next theorem involves the notion of inner product of complex vectors, introduced in Definition 2.5.1 (p. 62).

THEOREM 8.2.3. *A (complex) matrix is unitary if and only if its columns (or rows) form an orthonormal set of vectors.*

The defining relation (8.2.1) is equivalent to

$$(\overline{\mathbf{U}}^T\mathbf{U})_{rs} = \delta_{rs} \qquad (r, s = 1,..., n),$$

and this means that

$$\delta_{rs} = (\overline{\mathbf{U}}^T)_{r*}\mathbf{U}_{*s} = (\overline{\mathbf{U}}_{*r})^T\mathbf{U}_{*s} = (\mathbf{U}_{*r}, \mathbf{U}_{*s}) \quad (r, s = 1,..., n).$$

In other words, the columns of $\mathbf{U}$ form an orthonormal set. Again, (8.2.1) is equivalent to $\overline{\mathbf{U}}\mathbf{U}^T = \mathbf{I}$, i.e.

$$\delta_{rs} = (\overline{\mathbf{U}}\mathbf{U}^T)_{rs} = \overline{\mathbf{U}}_{r*}(\mathbf{U}^T)_{*s} = \overline{\mathbf{U}}_{r*}(\mathbf{U}_{s*})^T \qquad (r, s = 1,..., n).$$

Since $\mathbf{U}_{r*}, \mathbf{U}_{s*}$ are row vectors, this means

$$(\mathbf{U}_{r*}, \mathbf{U}_{s*}) = \delta_{rs} \qquad (r, s = 1,..., n),$$

i.e. the rows of $\mathbf{U}$ form an orthonormal set.

EXERCISE 8.2.2. If $\mathbf{X}$ is a unitary (orthogonal) matrix of order $n-1$, show that

$$\begin{pmatrix} 1 & \mathbf{O}_1^{n-1} \\ \mathbf{O}_{n-1}^1 & \mathbf{X} \end{pmatrix}$$

is a unitary (orthogonal) matrix of order n.

EXERCISE 8.2.3. Show that Corollaries 1 and 2 to Theorem 8.1.2. remain valid if the term 'orthogonal' is replaced by 'unitary'.

THEOREM 8.2.4. *Let* $\mathbf{x}_1$ *be a (complex) unit vector. Then there exists a unitary matrix having* $\mathbf{x}_1$ *as its first column.*

This theorem is proved in the same way as Theorem 8.1.3.

THEOREM 8.2.5. *A (complex) matrix* $\mathbf{U}$ *is unitary if and only if the substitution* $\mathbf{x} = \mathbf{Uy}$ *transforms the expression* $\bar{x}_1 x_1 + ... + \bar{x}_n x_n$ *into* $\bar{y}_1 y_1 + ... + \bar{y}_n y_n$, *where* $\mathbf{x} = (x_1,..., x_n)^T$, $\mathbf{y} = (y_1,..., y_n)^T$.

The proof is immediate.

THEOREM 8.2.6. *A (complex) matrix* $\mathbf{U}$ *is unitary if and only if*

$$|\mathbf{Ux}| = |\mathbf{x}| \tag{8.2.2}$$

for every complex vector $\mathbf{x}$.

Equation (8.2.2) is equivalent to

$$\bar{\mathbf{x}}^T(\bar{\mathbf{U}}^T\mathbf{U}-\mathbf{I})\mathbf{x} = 0. \tag{8.2.3}$$

If $\mathbf{U}$ is unitary, then (8.2.3) is clearly satisfied for all $\mathbf{x}$. Suppose, on the other hand, that (8.2.3) is satisfied for all $\mathbf{x}$. Writing $\mathbf{V} = (v_{rs}) = \bar{\mathbf{U}}^T\mathbf{U}-\mathbf{I}$ and putting $\mathbf{x} = \mathbf{e}_k$ in (8.2.3), we obtain

$$v_{kk} = 0 \qquad (k = 1,\dots,n). \tag{8.2.4}$$

Next, let $k \neq l$ and put $\mathbf{x} = \mathbf{e}_k+\mathbf{e}_l$ in (8.2.3). Making use of (8.2.4) we then obtain

$$v_{kl}+v_{lk} = 0 \qquad (k,l = 1,\dots,n;\ k \neq l).$$

Similarly, putting $\mathbf{x} = \mathbf{e}_k+i\mathbf{e}_l$, we obtain

$$v_{kl}-v_{lk} = 0 \qquad (k,l = 1,\dots,n;\ k \neq l).$$

Hence $\mathbf{V} = \mathbf{O}$, and $\mathbf{U}$ is unitary.

Corollary 1. *A (complex) matrix* $\mathbf{U}$ *is unitary if and only if*

$$|\mathbf{U}\mathbf{x}-\mathbf{U}\mathbf{y}| = |\mathbf{x}-\mathbf{y}|$$

for all complex vectors $\mathbf{x}$, $\mathbf{y}$.

Corollary 2. *A (complex) matrix* $\mathbf{U}$ *is unitary if and only if*

$$(\mathbf{U}\mathbf{x}, \mathbf{U}\mathbf{y}) = (\mathbf{x}, \mathbf{y})$$

for all complex vectors $\mathbf{x}$, $\mathbf{y}$.

Theorem 8.2.7. *If* λ *is a characteristic root of the unitary matrix* $\mathbf{U}$, *then so is* $1/\bar{\lambda}$.

For $1/\lambda$ is a characteristic root of $\mathbf{U}^{-1} = \bar{\mathbf{U}}^T$, and therefore

$$0 = |(1/\lambda)\mathbf{I}-\bar{\mathbf{U}}^T| = |(1/\bar{\lambda})\mathbf{I}-\mathbf{U}^T| = |(1/\bar{\lambda})\mathbf{I}-\mathbf{U}|.$$

The following generalization of Theorem 8.1.8 is due to Frobenius (1883).

Theorem 8.2.8. *Every characteristic root of a unitary matrix has unit modulus.*

Let λ be a characteristic root of the unitary matrix $\mathbf{U}$, and let the vector $\mathbf{x} \neq \mathbf{0}$ satisfy the equation $\lambda\mathbf{x} = \mathbf{U}\mathbf{x}$. Then $\bar{\lambda}\bar{\mathbf{x}}^T = \bar{\mathbf{x}}^T\bar{\mathbf{U}}^T$, and so

$$\bar{\lambda}\lambda\bar{\mathbf{x}}^T\mathbf{x} = \bar{\mathbf{x}}^T\bar{\mathbf{U}}^T\mathbf{U}\mathbf{x} = \bar{\mathbf{x}}^T\mathbf{x}.$$

Hence $|\lambda|^2 = 1$, and the assertion follows. Alternatively, it may be established by using Theorem 8.2.6 and noting that

$$|\mathbf{x}| = |\mathbf{U}\mathbf{x}| = |\lambda\mathbf{x}| = |\lambda||\mathbf{x}|.$$

The reader will have observed the similarity between the proofs of Theorems 8.2.8 and 7.5.1 (p. 209). These theorems are, in fact, included in the following more general result.

THEOREM 8.2.9. *If the matrix* $\mathbf{A}$ *satisfies the equation*

$$\sum_{\nu=1}^{N} c_\nu (\bar{\mathbf{A}}^T)^{r_\nu} \mathbf{A}^{s_\nu} = \mathbf{O},$$

and if λ *is any characteristic root of* $\mathbf{A}$, *then*

$$\sum_{\nu=1}^{N} c_\nu \bar{\lambda}^{r_\nu} \lambda^{s_\nu} = 0.$$

Let $\mathbf{x}$ be a characteristic vector of $\mathbf{A}$, associated with λ, so that $\mathbf{Ax} = \lambda\mathbf{x}$. Then, for $k \geqslant 1$,

$$\mathbf{A}^k\mathbf{x} = \mathbf{A}^{k-1}\mathbf{Ax} = \lambda\mathbf{A}^{k-1}\mathbf{x},$$

and therefore $$\mathbf{A}^k\mathbf{x} = \lambda^k\mathbf{x} \quad (k \geqslant 0).$$

This implies that $$\bar{\mathbf{x}}^T(\bar{\mathbf{A}}^T)^k = \bar{\lambda}^k\bar{\mathbf{x}}^T \quad (k \geqslant 0).$$

Hence $$\bar{\mathbf{x}}^T(\bar{\mathbf{A}}^T)^r\mathbf{A}^s\mathbf{x} = \bar{\lambda}^r\lambda^s\bar{\mathbf{x}}^T\mathbf{x},$$

and so

$$0 = \bar{\mathbf{x}}^T\left\{\sum_{\nu=1}^{N} c_\nu(\bar{\mathbf{A}}^T)^{r_\nu}\mathbf{A}^{s_\nu}\right\}\mathbf{x} = \sum_{\nu=1}^{N} c_\nu \bar{\mathbf{x}}^T(\bar{\mathbf{A}}^T)^{r_\nu}\mathbf{A}^{s_\nu}\mathbf{x}$$

$$= \left(\sum_{\nu=1}^{N} c_\nu \bar{\lambda}^{r_\nu}\lambda^{s_\nu}\right)\bar{\mathbf{x}}^T\mathbf{x}.$$

The assertion now follows since $\bar{\mathbf{x}}^T\mathbf{x} \neq 0$.

We may restate Theorem 8.2.9 in a convenient manner by saying that if $f(x,y)$ is a polynomial in (the non-commuting variables) x and y, and $f(\bar{\mathbf{A}}^T, \mathbf{A}) = \mathbf{O}$, then $f(\bar{\lambda}, \lambda) = 0$. If we take $f(x,y) = xy-1$, we obtain the result on the characteristic roots of unitary matrices (Theorem 8.2.8), while the choice $f(x,y) = x-y$ leads to the result on hermitian matrices (Theorem 7.5.1), and the choice $f(x,y) = x+y$ to that on skew-hermitian matrices (corollary to Theorem 7.5.1).

Theorem 8.1.10 has an obvious analogue for unitary matrices.

THEOREM 8.2.10. *If* $\mathbf{S}$ *is a skew-hermitian matrix, then* $\mathbf{I}+\mathbf{S}$ *is non-singular, and the matrix* $(\mathbf{I}-\mathbf{S})(\mathbf{I}+\mathbf{S})^{-1}$ *is unitary.*

EXERCISE 8.2.4. Write out a proof of this theorem.

8.3. Rotations in the plane

The theory of orthogonal matrices owes its interest, in the first place, to the part it plays in the study of rotations in the plane and in space. Its relevance to this problem is not surprising, for orthogonal substitutions preserve separation and may therefore be expected to occur in the analysis of rigid motion.

In this section and the next we shall discuss in some detail the relation between orthogonal matrices and rotations.† We shall find it necessary to distinguish between orthogonal matrices with positive and negative determinants.

DEFINITION 8.3.1. *The orthogonal matrix* $\mathbf{A}$ *is* PROPER *or* IMPROPER *according as* $|\mathbf{A}| = 1$ *or* $|\mathbf{A}| = -1$.

Throughout the present section all matrices will be assumed to be of type 2×2. We shall choose some point O in the plane and take it as the origin of a system of rectangular coordinates, which will then be kept fixed throughout the section. A vector $\mathbf{x} = (x,y)^T$ will be said to represent the point $P(x,y)$ with respect to the given coordinate system. When we speak of a transformation $\mathbf{x}' = f(\mathbf{x})$ we shall understand a transformation which changes the point represented by $\mathbf{x}$ into that represented by $f(\mathbf{x})$. The term 'rotation' will be taken to mean a rotation of the plane about the origin. As usual, a rotation will be reckoned as positive if its sense is counterclockwise.

THEOREM 8.3.1. *Any rotation of the plane can be represented by a proper orthogonal matrix.*

Stated more explicitly this means that there exists a proper orthogonal matrix $\mathbf{A}$ such that, if P is any point and P' the point into which P is carried by the given rotation, then $\mathbf{x}' = \mathbf{A}\mathbf{x}$, where $\mathbf{x}$, $\mathbf{x}'$ represent P, P' respectively.

Let the plane be rotated through an angle α in a counterclockwise sense, while the coordinate axes remain fixed. If the point (x,y) is carried into position (x',y'), then, as is well known,

$$x' = x\cos\alpha - y\sin\alpha,$$

$$y' = x\sin\alpha + y\cos\alpha.$$

† A much fuller account of two-dimensional and three-dimensional rigid motion will be found in Schreier and Sperner, 8, 153–78. See also Schwerdtfeger, 9, 237–44.

In matrix form these relations may be written as

$$\begin{pmatrix} x' \\ y' \end{pmatrix} = \begin{pmatrix} \cos\alpha & -\sin\alpha \\ \sin\alpha & \cos\alpha \end{pmatrix} \begin{pmatrix} x \\ y \end{pmatrix},$$

i.e.

$$\mathbf{x}' = \mathbf{A}\mathbf{x},$$

where $\mathbf{x} = (x, y)^T$, $\mathbf{x}' = (x', y')^T$, and $\mathbf{A}$ is a proper orthogonal matrix.

To establish the converse of the theorem just proved we need a simple result on proper orthogonal matrices.

THEOREM 8.3.2. *Let* $\mathbf{A}$ *be a proper orthogonal matrix. Then there exists a unique angle* α *such that* $0 \leqslant \alpha < 2\pi$ *and*

$$\mathbf{A} = \begin{pmatrix} \cos\alpha & -\sin\alpha \\ \sin\alpha & \cos\alpha \end{pmatrix}. \tag{8.3.1}$$

We write

$$\mathbf{A} = \begin{pmatrix} p & q \\ r & s \end{pmatrix}.$$

Since $p^2+q^2 = 1$, $r^2+s^2 = 1$, there exist angles α, β such that

$$p = \cos\alpha, \quad q = -\sin\alpha, \quad 0 \leqslant \alpha < 2\pi;$$

$$r = \sin\beta, \quad s = \cos\beta, \quad 0 \leqslant \beta < 2\pi.$$

But $1 = |\mathbf{A}| = ps-qr = \cos(\alpha-\beta)$, and so $\alpha-\beta$ is an integral multiple of 2π. Hence $\beta = \alpha$, and $\mathbf{A}$ is of the form (8.3.1) with $0 \leqslant \alpha < 2\pi$. Moreover, this representation of $\mathbf{A}$ is obviously unique.

EXERCISE 8.3.1. Show that, if $\mathbf{A}$ is a proper orthogonal matrix, then $\mathbf{x}' = \mathbf{A}\mathbf{x}$ represents a rotation.

THEOREM 8.3.3. *For every proper orthogonal matrix* $\mathbf{A}$ *there exists a unique angle* α *such that* $0 \leqslant \alpha \leqslant \pi$ *and the characteristic roots of* $\mathbf{A}$ *are* $e^{i\alpha}$, $e^{-i\alpha}$. *The equation* $\mathbf{x}' = \mathbf{A}\mathbf{x}$ *then represents a rotation either through* α *or through* $2\pi-\alpha$.

By Theorem 8.1.8 we know that the characteristic roots of $\mathbf{A}$ have unit modulus, and it is obvious that they are either both real or else are conjugate complex numbers. Furthermore, their product is equal to $|\mathbf{A}| = 1$, and so they cannot be 1 and -1. They may therefore be taken as $e^{i\alpha}$, $e^{-i\alpha}$, where $0 \leqslant \alpha < 2\pi$. It is, however possible to restrict α to the range $0 \leqslant \alpha \leqslant \pi$, for, since

$$e^{i(2\pi-\alpha)} = e^{-i\alpha} \quad \text{and} \quad e^{-i(2\pi-\alpha)} = e^{i\alpha},$$

we can replace α by $2\pi-\alpha$ whenever $\pi < \alpha < 2\pi$. The characteristic roots of $\mathbf{A}$ are thus $e^{i\alpha}$, $e^{-i\alpha}$, where $0 \leqslant \alpha \leqslant \pi$, and the value of α satisfying these requirements is plainly unique. The first part of the theorem is therefore proved.

By Theorem 8.3.2, $\mathbf{A}$ may be written in the form

$$\mathbf{A} = \begin{pmatrix} \cos\beta & -\sin\beta \\ \sin\beta & \cos\beta \end{pmatrix},$$

where $0 \leqslant \beta < 2\pi$. The characteristic roots of $\mathbf{A}$ are therefore $e^{i\beta}$, $e^{-i\beta}$, and hence $e^{i\beta} = e^{i\alpha}$ or $e^{i\beta} = e^{-i\alpha}$, i.e. $\beta = \alpha$ or $\beta = 2\pi-\alpha$. Hence the equation $\mathbf{x}' = \mathbf{A}\mathbf{x}$, which can be written as

$$x' = x\cos\beta - y\sin\beta,$$

$$y' = x\sin\beta + y\cos\beta,$$

represents a rotation either through α or through $2\pi-\alpha$.

Finally, we give a geometrical interpretation of improper orthogonal matrices.

THEOREM 8.3.4. *Let* $\mathbf{A}$ *be an improper orthogonal matrix. Then the equation* $\mathbf{x}' = \mathbf{A}\mathbf{x}$ *represents the product of a rotation and a reflection in a line through the origin.*

Write
$$\mathbf{B} = \mathbf{A}\begin{pmatrix} 1 & 0 \\ 0 & -1 \end{pmatrix},$$

so that
$$\mathbf{A} = \mathbf{B}\begin{pmatrix} 1 & 0 \\ 0 & -1 \end{pmatrix}.$$

Then $\mathbf{B}$ is a proper orthogonal matrix. The transformation $\mathbf{x}' = \mathbf{A}\mathbf{x}$ is the product (in that order) of the transformations

$$\mathbf{x}' = \begin{pmatrix} 1 & 0 \\ 0 & -1 \end{pmatrix}\mathbf{x}, \qquad \mathbf{x}' = \mathbf{B}\mathbf{x}.$$

The second of these is, by Theorem 8.3.3, a rotation. The first may be written as $x' = x$, $y' = -y$, and is therefore a reflection in the x-axis. Thus the transformation $\mathbf{x}' = \mathbf{A}\mathbf{x}$ is the product (in that order) of a reflection and a rotation. By considering the matrix

$$\mathbf{C} = \begin{pmatrix} 1 & 0 \\ 0 & -1 \end{pmatrix}\mathbf{A}$$

in place of $\mathbf{B}$, we see that $\mathbf{x}' = \mathbf{A}\mathbf{x}$ is equally the product of a rotation and a reflection.

It is interesting to observe that (in view of Theorem 8.3.3) the characteristic roots of a proper orthogonal matrix virtually determine its geometrical significance. This is no longer the case for improper orthogonal matrices, since the characteristic roots of any such matrix are 1 and -1.

EXERCISE 8.3.2. Prove the last statement.

8.4. Rotations in space

In this section all matrices are assumed to be of type 3×3. We shall find it convenient to write

$$\mathbf{R}_\alpha = \begin{pmatrix} 1 & 0 & 0 \\ 0 & \cos\alpha & -\sin\alpha \\ 0 & \sin\alpha & \cos\alpha \end{pmatrix}.$$

8.4.1. We begin with a number of purely algebraic results concerning properties of proper orthogonal matrices.

THEOREM 8.4.1. *For every proper orthogonal matrix* $\mathbf{A}$ *there exists a unique angle* α *such that* $0 \leqslant \alpha \leqslant \pi$ *and the characteristic roots of* $\mathbf{A}$ *are* $1, e^{i\alpha}, e^{-i\alpha}$.

By Theorem 8.1.9 (i) (p. 226) at least one characteristic root of $\mathbf{A}$ is equal to 1. The remaining two roots are, of course, of unit modulus and are either both real or else are conjugate complex numbers. Since $|\mathbf{A}| = 1$, the characteristic roots of $\mathbf{A}$ cannot be $1, 1, -1$. Hence they are $1, e^{i\alpha}, e^{-i\alpha}$, where it may, of course, be assumed that $0 \leqslant \alpha < 2\pi$. The angle α can now be made unique by being restricted to the range $0 \leqslant \alpha \leqslant \pi$.†

DEFINITION 8.4.1. *The unique angle* α *of the preceding theorem will be called the* ANGLE *of the* (*proper orthogonal*) *matrix* $\mathbf{A}$.

THEOREM 8.4.2. *If* $\mathbf{A}$ *is a proper orthogonal matrix with angle* α, $\boldsymbol{\xi}$ *is a unit characteristic vector of* $\mathbf{A}$ *associated with the characteristic root* 1, *and* $\mathbf{T}$ *is any proper orthogonal matrix with* $\boldsymbol{\xi}$ *as its first column,*‡ *then*

$$\mathbf{A} = \mathbf{T}\mathbf{R}_\beta \mathbf{T}^{-1},$$

where $\beta = \alpha$ or $2\pi - \alpha$.

We have, by hypothesis,§

$$\mathbf{T}\mathbf{e}_1 = \boldsymbol{\xi}.$$

† Cf. the proof of Theorem 8.3.3.

‡ Such a matrix exists by Theorem 8.1.3 (p. 223).

§ We write, of course, $\mathbf{e}_1 = (1, 0, 0)^T$, $\mathbf{e}_2 = (0, 1, 0)^T$, $\mathbf{e}_3 = (0, 0, 1)^T$.

Putting $\mathbf{B} = (b_{rs}) = \mathbf{T}^{-1}\mathbf{A}\mathbf{T}$, we obtain

$$\mathbf{B}\mathbf{e}_1 = \mathbf{T}^{-1}\mathbf{A}\mathbf{T}\mathbf{e}_1 = \mathbf{T}^{-1}\mathbf{A}\boldsymbol{\xi} = \mathbf{T}^{-1}\boldsymbol{\xi} = \mathbf{e}_1$$

and, since $\mathbf{B}\mathbf{e}_1$ is simply the first column of $\mathbf{B}$, this means that

$$\mathbf{B} = \begin{pmatrix} 1 & b_{12} & b_{13} \\ 0 & b_{22} & b_{23} \\ 0 & b_{32} & b_{33} \end{pmatrix}.$$

Now $\mathbf{B}$ is, by definition, a (proper) orthogonal matrix. Hence

$$1+b_{12}^2+b_{13}^2 = 1,$$

and so $b_{12} = b_{13} = 0$. We therefore have

$$\mathbf{B} = \begin{pmatrix} 1 & 0 & 0 \\ 0 & b_{22} & b_{23} \\ 0 & b_{32} & b_{33} \end{pmatrix}.$$

The submatrix
$$\begin{pmatrix} b_{22} & b_{23} \\ b_{32} & b_{33} \end{pmatrix}$$
must again be proper orthogonal and so, by Theorem 8.3.2, there exists an angle β such that $0 \leqslant \beta < 2\pi$ and $\mathbf{B} = \mathbf{R}_\beta$, i.e. $\mathbf{A} = \mathbf{T}\mathbf{R}_\beta\mathbf{T}^{-1}$. Now the characteristic roots of $\mathbf{R}_\beta$ are 1, $e^{i\beta}$, $e^{-i\beta}$, while those of $\mathbf{A}$ are 1, $e^{i\alpha}$, $e^{-i\alpha}$. But, by Theorem 7.2.1 (p. 199), $\mathbf{R}_\beta$ and $\mathbf{A}$ have the same characteristic roots. Hence $\beta = \alpha$ or $2\pi-\alpha$, and the theorem is proved.

DEFINITION 8.4.2. *If* $\mathbf{A}$ *is a proper orthogonal matrix, any unit vector* $\boldsymbol{\xi}$ *satisfying the equation* $\mathbf{A}\boldsymbol{\xi} = \boldsymbol{\xi}$ *will be called a* PRINCIPAL VECTOR *of* $\mathbf{A}$.

THEOREM 8.4.3. *If* $\mathbf{A}$ *is a proper orthogonal matrix other than* $\mathbf{I}$, *then its principal vector is uniquely determined to within a scalar multiple* ± 1.

In view of the corollary to the rank-multiplicity theorem (Theorem 7.6.1, p. 214), it suffices to show that 1 is a *simple* characteristic root of $\mathbf{A}$. Denote the angle of $\mathbf{A}$ by α. If $\alpha = 0$, then the angle β of Theorem 8.4.2 is 0 or 2π, and then $\mathbf{A} = \mathbf{I}$ contrary to hypothesis. Hence $0 < \alpha \leqslant \pi$, and as the characteristic roots of $\mathbf{A}$ are 1, $e^{i\alpha}$, $e^{-i\alpha}$, it follows that 1 is, in fact, a simple root.

8.4.2. We next consider the geometrical aspect of the theory of orthogonal matrices. We shall choose a system S of rectangular coordinate axes, to be kept fixed throughout the discussion, and we shall refer all measurements to this system unless the contrary is stated. If $\tilde{S}$ is another system of rectangular coordinates with the same origin O as S, we can associate with it the unique orthogonal matrix whose columns are the vectors representing (with respect to S) the points on the positive axes of $\tilde{S}$ and at a unit distance from O. This matrix will be called the matrix of $\tilde{S}$. In this way a biunique correspondence is established between orthogonal matrices and systems of rectangular coordinates with O as origin. The matrix of the fundamental system S has, of course, $\mathbf{e}_1$, $\mathbf{e}_2$, $\mathbf{e}_3$ as its columns and is therefore the unit matrix.

THEOREM 8.4.4. *If* $\mathbf{P}$ *is the matrix of the coordinate system* $\tilde{S}$, *and if a point* X *is represented by the vectors* $\mathbf{x}$, $\tilde{\mathbf{x}}$ *with respect to* S, $\tilde{S}$ *respectively, then*

$$\tilde{\mathbf{x}} = \mathbf{P}^T\mathbf{x}.$$

Write $\mathbf{P} = (p_{rs})$. Then the unit vectors $\tilde{\mathbf{e}}_1$, $\tilde{\mathbf{e}}_2$, $\tilde{\mathbf{e}}_3$ along the positive axes of $\tilde{S}$ are given by

$$\left.\begin{aligned} \tilde{\mathbf{e}}_1 &= p_{11}\,\mathbf{e}_1+p_{21}\,\mathbf{e}_2+p_{31}\,\mathbf{e}_3 \\ \tilde{\mathbf{e}}_2 &= p_{12}\,\mathbf{e}_1+p_{22}\,\mathbf{e}_2+p_{32}\,\mathbf{e}_3 \\ \tilde{\mathbf{e}}_3 &= p_{13}\,\mathbf{e}_1+p_{23}\,\mathbf{e}_2+p_{33}\,\mathbf{e}_3 \end{aligned}\right\}. \tag{8.4.1}$$

Since $\mathbf{P}$ is orthogonal, it follows easily that†

$$\left.\begin{aligned} \mathbf{e}_1 &= p_{11}\,\tilde{\mathbf{e}}_1+p_{12}\,\tilde{\mathbf{e}}_2+p_{13}\,\tilde{\mathbf{e}}_3 \\ \mathbf{e}_2 &= p_{21}\,\tilde{\mathbf{e}}_1+p_{22}\,\tilde{\mathbf{e}}_2+p_{23}\,\tilde{\mathbf{e}}_3 \\ \mathbf{e}_3 &= p_{31}\,\tilde{\mathbf{e}}_1+p_{32}\,\tilde{\mathbf{e}}_2+p_{33}\,\tilde{\mathbf{e}}_3 \end{aligned}\right\}.$$

Writing $\mathbf{x} = (x, y, z)^T$ and $\tilde{\mathbf{x}} = (\tilde{x}, \tilde{y}, \tilde{z})^T$, we therefore have

$$\begin{aligned} \mathbf{x} &= x\mathbf{e}_1+y\mathbf{e}_2+z\mathbf{e}_3 \\ &= x(p_{11}\,\tilde{\mathbf{e}}_1+p_{12}\,\tilde{\mathbf{e}}_2+p_{13}\,\tilde{\mathbf{e}}_3)+y(p_{21}\,\tilde{\mathbf{e}}_1+p_{22}\,\tilde{\mathbf{e}}_2+p_{23}\,\tilde{\mathbf{e}}_3)+ \\ &\qquad +z(p_{31}\,\tilde{\mathbf{e}}_1+p_{32}\,\tilde{\mathbf{e}}_2+p_{33}\,\tilde{\mathbf{e}}_3) \\ &= (p_{11}\,x+p_{21}\,y+p_{31}\,z)\tilde{\mathbf{e}}_1+(p_{12}\,x+p_{22}\,y+p_{32}\,z)\tilde{\mathbf{e}}_2+ \\ &\qquad +(p_{13}\,x+p_{23}\,y+p_{33}\,z)\tilde{\mathbf{e}}_3. \end{aligned}$$

Thus

$$\left.\begin{aligned} \tilde{x} &= p_{11}\,x+p_{21}\,y+p_{31}\,z \\ \tilde{y} &= p_{12}\,x+p_{22}\,y+p_{32}\,z \\ \tilde{z} &= p_{13}\,x+p_{23}\,y+p_{33}\,z \end{aligned}\right\},$$

† Cf. Exercise 8.1.1 (p. 222).

i.e.
$$\tilde{\mathbf{x}} = \begin{pmatrix} \tilde{x} \\ \tilde{y} \\ \tilde{z} \end{pmatrix} = \begin{pmatrix} p_{11} & p_{21} & p_{31} \\ p_{12} & p_{22} & p_{32} \\ p_{13} & p_{23} & p_{33} \end{pmatrix} \begin{pmatrix} x \\ y \\ z \end{pmatrix} = \mathbf{P}^T\mathbf{x},$$
as asserted.

It may be noted in passing that the relations between x, y, z and $\tilde{x}$, $\tilde{y}$, $\tilde{z}$ (i.e. between the coordinates of the same point with respect to the two systems S, $\tilde{S}$) may be exhibited in a convenient form by means of the following table:

	x	y	z
$\tilde{x}$	p_{11}	p_{21}	p_{31}
$\tilde{y}$	p_{12}	p_{22}	p_{32}
$\tilde{z}$	p_{13}	p_{23}	p_{33}

The rows of this array give the direction cosines of the axes of $\tilde{S}$ with respect to the axes of S, while the columns give the direction cosines of the axes of S with respect to those of $\tilde{S}$. For, by (8.4.1),

$$(\tilde{\mathbf{e}}_r, \mathbf{e}_s) = p_{sr} \qquad (r, s = 1, 2, 3),$$

and $(\tilde{\mathbf{e}}_r, \mathbf{e}_s)$ is clearly the cosine of the angle between the rth axis of $\tilde{S}$ and the sth axis of S.

In discussing spatial relations we cannot avoid referring, in one form or another, to right-handedness and left-handedness of coordinate systems. This intuitive notion (like the notion of a counterclockwise sense) is unambiguous but does not admit of a precise verbal definition. We are able, however, to express mathematically the *distinction* between right-handed and left-handed systems.

DEFINITION 8.4.3. *The coordinate system $\tilde{S}$ will be called proper or improper according as its matrix is proper or improper orthogonal.*

The property of being proper or improper is not an intrinsic feature of $\tilde{S}$ but one that characterizes the relation of $\tilde{S}$ to the arbitrarily chosen fundamental system S. In fact, if S is right-handed, then $\tilde{S}$ is right-handed or left-handed according as it is proper or improper, while if S is left-handed, then $\tilde{S}$ is right-handed or left-handed according as it is improper or proper. We can justify these assertions by a continuity argument. Thus, if S and $\tilde{S}$ are both right-handed, then $\tilde{S}$ can be obtained from S by a continuous rigid motion of the coordinate axes, in which the determinant of the matrix associated with the coordinate system cannot change its value discontinuously. Since its value is 1 for S, it is also 1 for $\tilde{S}$; and $\tilde{S}$ is therefore proper.

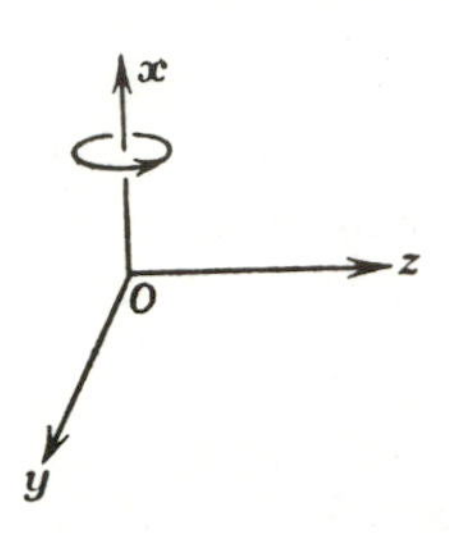

8.4.3. We now turn to the mathematical analysis of rotation. By 'rotation' we shall understand a rotation of space about a directed line through the origin. This line will be called the *axis of rotation*. A rotation through an angle α about the positive x-axis clearly carries any point $\mathbf{x} = (x, y, z)^T$ into the point $\mathbf{x}' = (x', y', z')^T$, where

$$\left.\begin{aligned} x' &= x \\ y' &= y\cos\alpha - z\sin\alpha \\ z' &= y\sin\alpha + z\cos\alpha \end{aligned}\right\}, \tag{8.4.2}$$

or, in matrix notation,

$$\mathbf{x}' = \mathbf{R}_\alpha \mathbf{x}. \tag{8.4.3}$$

In these formulae all components are assumed to be measured with respect to a right-handed system of coordinates.

Consider now a rotation, through α, about any directed line l. This rotation can be represented by the equation (8.4.3) with respect to a right-handed system of coordinates having l as the positive x-axis. Accordingly, a convenient mathematical definition of rotation can be given as follows.

DEFINITION 8.4.4. *A rotation through an angle α about a directed line l (through the origin) is the transformation represented by* (8.4.3) *with respect to a proper coordinate system $\tilde{S}$ having l as its positive x-axis.*

For the sake of brevity we shall contract the phrase 'rotation through an angle α' into 'an α-rotation'.

If S is right-handed, then $\tilde{S}$ is also right-handed and the definition just given agrees with our intuitive notion of rotation. In interpreting geometrically the theorems deduced below, we shall accordingly take a right-handed system of axes as the fundamental system S.

THEOREM 8.4.5. *Let l be a directed line through the origin and let l' be the same line with its sense reversed. Then an α-rotation about l is equivalent to a $(2\pi - \alpha)$-rotation about l'.*

An α-rotation about l is represented by the equations (8.4.2) with respect to a proper coordinate system in which the positive x-axis lies along l. We now take a new system of coordinates by reversing the directions of the x-axis and the y-axis. This new system is again proper, and with respect to it the rotation in question assumes the form

$$x' = x,$$
$$y' = y\cos\alpha + z\sin\alpha,$$
$$z' = -y\sin\alpha + z\cos\alpha,$$

i.e.
$$\mathbf{x}' = \mathbf{R}_{-\alpha}\mathbf{x} = \mathbf{R}_{2\pi-\alpha}\mathbf{x}.$$

This transformation is, by definition, a $(2\pi - \alpha)$-rotation about l'.

The next step is to obtain the matrix representation of a prescribed rotation.

Theorem 8.4.6. *Let $P = (\lambda, \mu, \nu)$ be a point at unit distance from the origin. Then the equation representing the α-rotation of space about the line $\overrightarrow{OP}$ is given by*

$$\mathbf{x}' = \mathbf{T}\mathbf{R}_\alpha \mathbf{T}^{-1}\mathbf{x},$$

where $\mathbf{T}$ is any proper orthogonal matrix with $(\lambda, \mu, \nu)^T$ as its first column.

Let $\mathbf{T}$ be a proper orthogonal matrix with $(\lambda, \mu, \nu)^T$ as its first column, and let $\tilde{S}$ be the coordinate system whose matrix is $\mathbf{T}$. Then $\tilde{S}$ is proper and $\overrightarrow{OP}$ is its positive x-axis. If X is any point and if the vectors representing it with respect to S, $\tilde{S}$ are $\mathbf{x}$, $\tilde{\mathbf{x}}$ respectively, then, by Theorem 8.4.4,

$$\tilde{\mathbf{x}} = \mathbf{T}^{-1}\mathbf{x}.$$

The α-rotation about $\overrightarrow{OP}$ carries X into a point X', say, and the vectors $\tilde{\mathbf{x}}$, $\tilde{\mathbf{x}}'$ representing these points with respect to $\tilde{S}$ are, in view of (8.4.3), connected by the equation

$$\tilde{\mathbf{x}}' = \mathbf{R}_\alpha \tilde{\mathbf{x}}.$$

Again, if $\mathbf{x}'$ is the vector representing X' with respect to S, then

$$\tilde{\mathbf{x}}' = \mathbf{T}^{-1}\mathbf{x}'.$$

Combining these results, we obtain

$$\mathbf{x}' = \mathbf{T}\tilde{\mathbf{x}}' = \mathbf{T}\mathbf{R}_\alpha \tilde{\mathbf{x}} = \mathbf{T}\mathbf{R}_\alpha \mathbf{T}^{-1}\mathbf{x},$$

as asserted.

It is sometimes useful to break up the formula of Theorem 8.4.6 into expressions for the separate components of $\mathbf{x}'$. Writing

$$\mathbf{T} = \begin{pmatrix} \lambda & \lambda' & \lambda'' \\ \mu & \mu' & \mu'' \\ \nu & \nu' & \nu'' \end{pmatrix}$$

and using the orthogonality relations for $\mathbf{T}$ as well as the equations of Exercise 8.1.4 (p. 223) we are led to the identity

$$\mathbf{T}\mathbf{R}_\alpha \mathbf{T}^{-1} = \begin{pmatrix} c+\lambda^2(1-c) & \lambda\mu(1-c)-\nu s & \nu\lambda(1-c)+\mu s \\ \lambda\mu(1-c)+\nu s & c+\mu^2(1-c) & \mu\nu(1-c)-\lambda s \\ \nu\lambda(1-c)-\mu s & \mu\nu(1-c)+\lambda s & c+\nu^2(1-c) \end{pmatrix},$$

where $c = \cos\alpha$, $s = \sin\alpha$. Writing $\mathbf{x}' = (x', y', z')^T$, $\mathbf{x} = (x, y, z)^T$ in Theorem 8.4.6 we now obtain

$$x' = x\cos\alpha + (\mu z - \nu y)\sin\alpha + \lambda(1-\cos\alpha)(\lambda x + \mu y + \nu z),$$
$$y' = y\cos\alpha + (\nu x - \lambda z)\sin\alpha + \mu(1-\cos\alpha)(\lambda x + \mu y + \nu z),$$
$$z' = z\cos\alpha + (\lambda y - \mu x)\sin\alpha + \nu(1-\cos\alpha)(\lambda x + \mu y + \nu z).$$

These formulae are known as *Euler's equations of transformation.*†

EXERCISE 8.4.1. Obtain the equations (8.4.2) as a special case of Euler's equations of transformation.

EXERCISE 8.4.2. Show that in Euler's equations of transformation it is permissible to interchange the accented and the unaccented coordinates provided that, at the same time, α is replaced by $-\alpha$.

Theorem 8.4.6 shows how a matrix representation for a given rotation can be determined. The next result deals with the converse problem of interpreting geometrically a given orthogonal transformation.

Theorem 8.4.7. *Let* $\mathbf{A}$ *be a proper orthogonal matrix other than* $\mathbf{I}$, *and let* α *be its angle. Then the equation* $\mathbf{x}' = \mathbf{A}\mathbf{x}$ *represents an* α*-rotation about a suitably directed line specified by a principal vector of* $\mathbf{A}$.

Let $\boldsymbol{\xi}$ be a principal vector of $\mathbf{A}$ and write $\boldsymbol{\xi} = (\lambda, \mu, \nu)^T$. Then $\boldsymbol{\xi}$ specifies the directed line $l = \overrightarrow{OP}$, where $P = (\lambda, \mu, \nu)$. We denote by l' the same line with its sense reversed. Since, by Theorem 8.4.3, the principal vector of $\mathbf{A}$ ($\neq \mathbf{I}$) is determined to within a scalar multiple ± 1, it follows that the line specified by the principal vector of $\mathbf{A}$ is unique except for its sense.

† For a simple geometrical derivation see Sommerville, *Analytical Geometry of Three Dimensions*, 38–39.

Let $\mathbf{T}$ be any proper orthogonal matrix having $\boldsymbol{\xi}$ as its first column. Then, by Theorem 8.4.2,

$$\mathbf{A} = \mathbf{T}\mathbf{R}_\beta \mathbf{T}^{-1},$$

where $\beta = \alpha$ or $2\pi - \alpha$; and the equation $\mathbf{x}' = \mathbf{A}\mathbf{x}$ can therefore be written in the form

$$\mathbf{T}^{-1}\mathbf{x}' = \mathbf{R}_\beta . \mathbf{T}^{-1}\mathbf{x}. \tag{8.4.4}$$

If $\tilde{S}$ is the (proper) coordinate system associated with the matrix $\mathbf{T}$, then (by Theorem 8.4.4) the transformation (8.4.4) assumes, with respect to $\tilde{S}$, the form

$$\mathbf{x}' = \mathbf{R}_\beta \mathbf{x};$$

and this is a β-rotation about the positive x-axis of $\tilde{S}$, i.e. about l. Thus, by virtue of Theorem 8.4.5, the transformation in question is an α-rotation about l or about l'.

Theorem 8.4.8. *The product of two rotations is again a rotation.*

Suppose that the first (second) rotation carries the point represented by $\mathbf{x}$ into that represented by $\mathbf{x}'$ ($\mathbf{x}''$). Then it follows by Theorem 8.4.6 that there exist proper orthogonal matrices $\mathbf{A}$, $\mathbf{B}$ such that

$$\mathbf{x}' = \mathbf{A}\mathbf{x}, \qquad \mathbf{x}'' = \mathbf{B}\mathbf{x}.$$

Hence the product of the two rotations is a transformation of space which carries the point represented by $\mathbf{x}$ into that represented by $\mathbf{B}\mathbf{A}\mathbf{x}$. Now $\mathbf{B}\mathbf{A}$ is again a proper orthogonal matrix and the transformation in question is, by Theorem 8.4.7, once more a rotation.

Theorem 8.4.8 will presently be superseded by the more general Theorem 8.4.12.

8.4.4. We have so far used orthogonal matrices to represent rotations. A different mode of representation in terms of (real) skew-symmetric matrices is to be considered next.

Let R be a rotation through an angle α ($0 \leqslant \alpha < 2\pi$) about the line $\overrightarrow{OP}$, where $P = (\lambda, \mu, \nu)$ is a point at unit distance from the origin. If $\alpha \neq \pi$, we associate with R the skew-symmetric matrix

$$\mathbf{S} = \begin{pmatrix} 0 & \nu \tan \tfrac{1}{2}\alpha & -\mu \tan \tfrac{1}{2}\alpha \\ -\nu \tan \tfrac{1}{2}\alpha & 0 & \lambda \tan \tfrac{1}{2}\alpha \\ \mu \tan \tfrac{1}{2}\alpha & -\lambda \tan \tfrac{1}{2}\alpha & 0 \end{pmatrix}. \tag{8.4.5}$$

Conversely, any skew-symmetric matrix $\mathbf{S}$ can be written in the form (8.4.5), where $\lambda^2+\mu^2+\nu^2 = 1$ and $0 \leqslant \alpha < 2\pi$. This expression for $\mathbf{S}$ is not unique, since $\alpha, \lambda, \mu, \nu$ can be replaced by $2\pi - \alpha$, $-\lambda$,

$-\mu$, $-\nu$ respectively, but these two sets of parameters correspond to the same rotation.† We have thus set up a biunique correspondence between the set of all rotations (other that π-rotations) and the set of all skew-symmetric matrices. The matrix $\mathbf{S}$ associated with the rotation R will be called the *skew-symmetric matrix of* R. Similarly, the (proper) orthogonal matrix $\mathbf{A}$ which represents R in the sense that R is specified by the equation $\mathbf{x}' = \mathbf{A}\mathbf{x}$, will be called the *orthogonal matrix of* R. The advantage of associating with R a skew-symmetric rather than an orthogonal matrix is that from the former the geometrical character of R can be inferred almost immediately.

EXERCISE 8.4.3. Show that the skew-symmetric matrix

$$\begin{pmatrix} 0 & -\surd 2 & -1/\surd 2 \\ \surd 2 & 0 & 1/\surd 2 \\ 1/\surd 2 & -1/\surd 2 & 0 \end{pmatrix}$$

represents the rotation through 120° about the directed line joining the origin to the point $(1, 1, -2)$.

THEOREM 8.4.9. *Let R be a rotation other than a π-rotation, and let $\mathbf{A}$ and $\mathbf{S}$ be its orthogonal matrix and its skew-symmetric matrix respectively. Then*

$$\mathbf{A} = (\mathbf{I}-\mathbf{S})(\mathbf{I}+\mathbf{S})^{-1}, \qquad \mathbf{S} = (\mathbf{I}-\mathbf{A})(\mathbf{I}+\mathbf{A})^{-1}.$$

It is, of course, sufficient to establish either of these equations. We may note at once that, since $\mathbf{S}$ is a real skew-symmetric matrix, $\mathbf{I}+\mathbf{S}$ is non-singular and, by Theorem 8.1.10 (p. 227) $(\mathbf{I}-\mathbf{S})(\mathbf{I}+\mathbf{S})^{-1}$ is orthogonal.

Let the given rotation R be specified by the numbers α, λ, μ, ν defined above. Then, by Theorem 8.4.6, $\mathbf{A} = \mathbf{T}\mathbf{R}_\alpha\mathbf{T}^{-1}$, where $\mathbf{T}$ is any proper orthogonal matrix with $(\lambda, \mu, \nu)^T$ as its first column, say

$$\mathbf{T} = \begin{pmatrix} \lambda & \lambda' & \lambda'' \\ \mu & \mu' & \mu'' \\ \nu & \nu' & \nu'' \end{pmatrix}.$$

Now $\alpha \neq \pi$ and therefore -1 is not a characteristic root of $\mathbf{A}$. Hence $\mathbf{I}+\mathbf{A}$ is non-singular, and

$$\begin{aligned}(\mathbf{I}-\mathbf{A})(\mathbf{I}+\mathbf{A})^{-1} &= (\mathbf{I}-\mathbf{T}\mathbf{R}_\alpha\mathbf{T}^{-1})(\mathbf{I}+\mathbf{T}\mathbf{R}_\alpha\mathbf{T}^{-1})^{-1} \\ &= \mathbf{T}(\mathbf{I}-\mathbf{R}_\alpha)\mathbf{T}^{-1}\{\mathbf{T}(\mathbf{I}+\mathbf{R}_\alpha)\mathbf{T}^{-1}\}^{-1} \\ &= \mathbf{T}(\mathbf{I}-\mathbf{R}_\alpha)(\mathbf{I}+\mathbf{R}_\alpha)^{-1}\mathbf{T}^{-1}.\end{aligned}$$

† Cf. Theorem 8.4.5.

Now it is easy to verify that

$$(\mathbf{I}-\mathbf{R}_\alpha)(\mathbf{I}+\mathbf{R}_\alpha)^{-1} = \tan\tfrac{1}{2}\alpha\begin{pmatrix}0 & 0 & 0\\ 0 & 0 & 1\\ 0 & -1 & 0\end{pmatrix}.$$

Hence, using Exercise 8.1.4 (p. 223), we obtain

$$(\mathbf{I}-\mathbf{A})(\mathbf{I}+\mathbf{A})^{-1} = \tan\tfrac{1}{2}\alpha\begin{pmatrix}\lambda & \lambda' & \lambda''\\ \mu & \mu' & \mu''\\ \nu & \nu' & \nu''\end{pmatrix}\begin{pmatrix}0 & 0 & 0\\ 0 & 0 & 1\\ 0 & -1 & 0\end{pmatrix}\begin{pmatrix}\lambda & \mu & \nu\\ \lambda' & \mu' & \nu'\\ \lambda'' & \mu'' & \nu''\end{pmatrix}$$

$$= \tan\tfrac{1}{2}\alpha\begin{pmatrix}0 & \nu & -\mu\\ -\nu & 0 & \lambda\\ \mu & -\lambda & 0\end{pmatrix} = \mathbf{S}.$$

The theorem is therefore proved.

Theorem 8.4.9 is concerned with the representation of a single rotation, but the skew-symmetric matrix of the product of two rotations is also easy to evaluate.

THEOREM 8.4.10. *Let R be the product of the two rotations R_1, R_2, carried out in that order, and suppose that none of the three rotations is a π-rotation. If the skew-symmetric matrices of R, R_1, R_2 are $\mathbf{S}$, $\mathbf{S}_1$, $\mathbf{S}_2$ respectively, then*

$$\mathbf{S} = (\mathbf{I}+\mathbf{S}_1)(\mathbf{I}+\mathbf{S}_2\mathbf{S}_1)^{-1}(\mathbf{S}_1+\mathbf{S}_2)(\mathbf{I}+\mathbf{S}_1)^{-1}.$$

We know, by Theorem 8.4.9, that the orthogonal matrices of R_1, R_2 are

$$(\mathbf{I}-\mathbf{S}_1)(\mathbf{I}+\mathbf{S}_1)^{-1}, \qquad (\mathbf{I}-\mathbf{S}_2)(\mathbf{I}+\mathbf{S}_2)^{-1} = (\mathbf{I}+\mathbf{S}_2)^{-1}(\mathbf{I}-\mathbf{S}_2).$$

It follows that the orthogonal matrix $\mathbf{A}$ of R is given by

$$\mathbf{A} = (\mathbf{I}+\mathbf{S}_2)^{-1}(\mathbf{I}-\mathbf{S}_2)(\mathbf{I}-\mathbf{S}_1)(\mathbf{I}+\mathbf{S}_1)^{-1}.$$

Hence

$$\mathbf{I}+\mathbf{A} = (\mathbf{I}+\mathbf{S}_2)^{-1}\{(\mathbf{I}+\mathbf{S}_2)(\mathbf{I}+\mathbf{S}_1)+(\mathbf{I}-\mathbf{S}_2)(\mathbf{I}-\mathbf{S}_1)\}(\mathbf{I}+\mathbf{S}_1)^{-1}$$
$$= 2(\mathbf{I}+\mathbf{S}_2)^{-1}(\mathbf{I}+\mathbf{S}_2\mathbf{S}_1)(\mathbf{I}+\mathbf{S}_1)^{-1},$$

and similarly

$$\mathbf{I}-\mathbf{A} = 2(\mathbf{I}+\mathbf{S}_2)^{-1}(\mathbf{S}_1+\mathbf{S}_2)(\mathbf{I}+\mathbf{S}_1)^{-1}.$$

But, by Theorem 8.4.9, $\mathbf{S} = (\mathbf{I}+\mathbf{A})^{-1}(\mathbf{I}-\mathbf{A})$, and so

$$\mathbf{S} = \tfrac{1}{2}(\mathbf{I}+\mathbf{S}_1)(\mathbf{I}+\mathbf{S}_2\mathbf{S}_1)^{-1}(\mathbf{I}+\mathbf{S}_2).2(\mathbf{I}+\mathbf{S}_2)^{-1}(\mathbf{S}_1+\mathbf{S}_2)(\mathbf{I}+\mathbf{S}_1)^{-1}$$
$$= (\mathbf{I}+\mathbf{S}_1)(\mathbf{I}+\mathbf{S}_2\mathbf{S}_1)^{-1}(\mathbf{S}_1+\mathbf{S}_2)(\mathbf{I}+\mathbf{S}_1)^{-1}.$$

8.4.5. The last topic to be considered in the present discussion is that of rigid motion. We restrict ourselves to motions of space for which one point remains fixed, and we naturally choose this point as the origin of coordinate systems.

DEFINITION 8.4.5. *Let $\tilde{S}$ be a system of coordinates with the same origin O as the fundamental system S. A transformation of space which carries every point X into a point X', whose coordinates with respect to $\tilde{S}$ are the same as those of X with respect to S, is called a* (RIGID) MOTION. *The motion will be said to be proper or improper according as $\tilde{S}$ is proper or improper.*

To visualize a proper motion we can imagine the axes of S forming a stiff framework and being moved, without deformation, about the origin until they come into coincidence with the corresponding axes of $\tilde{S}$. We then obtain a proper motion if the entire space is made to move together with the coordinate axes as one rigid body.

THEOREM 8.4.11. (i) *If* $\mathbf{P}$ *is the matrix of the coordinate system* $\tilde{S}$, *then the motion specified by Definition* 8.4.5 *can be represented, with respect to the system* S, *by the equation*

$$\mathbf{x}' = \mathbf{P}\mathbf{x}. \tag{8.4.6}$$

(ii) *Conversely, any transformation of type* (8.4.6), *where* $\mathbf{P}$ *is orthogonal, is a proper or improper motion according as* $\mathbf{P}$ *is proper or improper.*

Consider a point X, represented by the vector $\mathbf{x}$ with respect to S. If the motion in question transforms X into X', then (by definition) X' is represented by $\mathbf{x}$ with respect to $\tilde{S}$ and so (in view of Theorem 8.4.4, p. 238) by $\mathbf{P}\mathbf{x}$ with respect to S. Thus $\mathbf{x}' = \mathbf{P}\mathbf{x}$, where $\mathbf{x}'$ is the vector representing X' with respect to S.

Again, given the transformation (8.4.6), let us denote by $\tilde{S}$ the system of coordinates associated with the matrix $\mathbf{P}$. Then, by (i), (8.4.6) represents the motion specified in Definition 8.4.5. This motion is proper or improper according as $\tilde{S}$ is proper or improper, i.e. according as $|\mathbf{P}| = 1$ or $|\mathbf{P}| = -1$.

We are now able to prove an important result due substantially to Euler (1776).

Theorem 8.4.12. (Euler's theorem on rigid motion)
Every proper motion is a rotation, and conversely.

By virtue of Theorem 8.4.11 (i), every proper motion can be represented by an equation $\mathbf{x}' = \mathbf{A}\mathbf{x}$, where $\mathbf{A}$ is a proper orthogonal matrix. Hence, by Theorem 8.4.7, every proper motion is a rotation. Conversely, we know by Theorem 8.4.6 that every rotation can be represented by an equation $\mathbf{x}' = \mathbf{A}\mathbf{x}$, where $\mathbf{A}$ is a proper orthogonal matrix. Hence, by Theorem 8.4.11 (ii), every rotation is a proper motion.

In geometrical terms Theorem 8.4.12 means that if a system of coordinate axes is moved in any way about the origin, subject only to the requirement that it is treated as a rigid body, then its motion is equivalent to a single rotation about a suitable axis. This result is of considerable importance in the dynamics of rigid bodies.†

The case of transformations involving improper orthogonal matrices is now easily disposed of.

THEOREM 8.4.13. *If* $\mathbf{A}$ *is an improper orthogonal matrix, then the equation* $\mathbf{x}' = \mathbf{A}\mathbf{x}$ *represents the product of a rotation and a reflection in the origin. Conversely, the product of a rotation and a reflection in the origin can be represented by an equation* $\mathbf{x}' = \mathbf{A}\mathbf{x}$, *where* $\mathbf{A}$ *is an improper orthogonal matrix.*

The transformation $\mathbf{x}' = \mathbf{A}\mathbf{x}$ is the product, in either order, of the transformations

$$\mathbf{x}' = (-\mathbf{A})\mathbf{x}, \qquad \mathbf{x}' = -\mathbf{x}. \tag{8.4.7}$$

If $\mathbf{A}$ is an improper orthogonal matrix, then the first of these is a rotation, while the second is evidently a reflection in the origin.

Again, consider a rotation and a reflection in the origin. These transformations can be represented by equations (8.4.7), where $\mathbf{A}$ is an improper orthogonal matrix. Hence their product, in either order, is the transformation $\mathbf{x}' = \mathbf{A}\mathbf{x}$.

EXERCISE 8.4.4. Arguing as in the proof of Theorem 8.3.4 (p. 235), show that the equation $\mathbf{x}' = \mathbf{A}\mathbf{x}$, where $\mathbf{A}$ is an improper orthogonal matrix, represents the product of a rotation and a reflection in a plane through the origin.

EXERCISE 8.4.5. Show that every improper motion is the product of a rotation and a reflection in the origin.

† For a geometrical proof see Lamb, *Higher Mechanics*, 2–3.

PROBLEMS ON CHAPTER VIII

1. Verify that
$$\begin{pmatrix} \dfrac{1+i}{2\sqrt{2}} & \dfrac{\sqrt{3}(1-i)}{2\sqrt{2}} \\ \dfrac{\sqrt{2}+i}{2} & \dfrac{-1+i\sqrt{2}}{2\sqrt{3}} \end{pmatrix}$$
is a unitary matrix, and evaluate its determinant.

2. Let $\alpha_1,..., \alpha_n$ be complex numbers of modulus 1. Show that, if the rows of a unitary matrix are multiplied by $\alpha_1,..., \alpha_n$ respectively, then the resulting matrix is again unitary.

3. The rows of a matrix $\mathbf{A}$ form an orthogonal set. Show by an example that the columns of $\mathbf{A}$ need not form an orthogonal set.

4. Show that, if an orthogonal matrix is triangular, then it is diagonal; and that all its diagonal elements are equal to ± 1.

5. Let $\mathbf{U}$ be a complex matrix, and write $\mathbf{U} = \mathbf{P}+i\mathbf{Q}$, where $\mathbf{P}, \mathbf{Q}$ are real. Show that $\mathbf{U}$ is unitary if and only if $\mathbf{P}^T\mathbf{Q}$ is symmetric and $\mathbf{P}^T\mathbf{P}+\mathbf{Q}^T\mathbf{Q} = \mathbf{I}$.

6. Show that any two of the following three statements relating to a matrix $\mathbf{A}$ imply the third: (i) $\mathbf{A}$ is hermitian; (ii) $\mathbf{A}$ is unitary; (iii) $\mathbf{A}^2 = \mathbf{I}$.

7. Show that Theorem 8.1.11 (p. 228) is no longer valid if the words 'real' and 'orthogonal' are replaced by 'complex' and 'unitary' respectively.

8. Let α be an angle such that $\cos\frac{1}{2}\alpha \neq 0$, and let
$$\mathbf{A} = \begin{pmatrix} 1 & 0 & 0 \\ 0 & \cos\alpha & -\sin\alpha \\ 0 & \sin\alpha & \cos\alpha \end{pmatrix}.$$
Verify that $\mathbf{I}+\mathbf{A}$ is non-singular, and show that
$$(\mathbf{I}-\mathbf{A})(\mathbf{I}+\mathbf{A})^{-1} = \tan\tfrac{1}{2}\alpha \begin{pmatrix} 0 & 0 & 0 \\ 0 & 0 & 1 \\ 0 & -1 & 0 \end{pmatrix}.$$

9. Let $\mathbf{A}$ be a proper orthogonal matrix of order 3. Show that there exists a number t such that $-1 \leqslant t \leqslant 3$ and
$$\mathbf{A}^3 - t\mathbf{A}^2 + t\mathbf{A} - \mathbf{I} = \mathbf{O}.$$

10. Let $a_{rs} = n^{-\frac{1}{2}}e^{2\pi i(r-1)(s-1)/n}$ $(r, s = 1,..., n)$. Show that the matrix (a_{rs}) is unitary.

11. A matrix $\mathbf{A}$ satisfies the equation $\bar{\mathbf{A}}^T\mathbf{A} = -\mathbf{A}$. Show that the value of each characteristic root of $\mathbf{A}$ is 0 or -1.

12. $\mathbf{U}$ is a unitary matrix such that $|\mathbf{U}-\mathbf{I}| \neq 0$, and $\mathbf{H}$ is defined by the equation $i\mathbf{H} = (\mathbf{U}+\mathbf{I})(\mathbf{U}-\mathbf{I})^{-1}$. Prove that $\mathbf{H}$ is a hermitian matrix. If the characteristic roots of $\mathbf{U}$ are $e^{i\theta_1},..., e^{i\theta_n}$, find the characteristic roots of $\mathbf{H}$.

13. If $\mathbf{S}$ is real symmetric and $\mathbf{T}$ real skew-symmetric, show that
$$|\mathbf{I}-\mathbf{T}-i\mathbf{S}| \neq 0$$
and that the matrix
$$\mathbf{U} = (\mathbf{I}+\mathbf{T}+i\mathbf{S})(\mathbf{I}-\mathbf{T}-i\mathbf{S})^{-1}$$

is unitary. Find the characteristic roots of $\mathbf{U}$, when

$$\mathbf{S} = \begin{pmatrix} 1 & 1 \\ 1 & 1 \end{pmatrix}, \qquad \mathbf{T} = \begin{pmatrix} 0 & -1 \\ 1 & 0 \end{pmatrix}.$$

14. Let $\mathbf{A}$ be a real symmetric and $\mathbf{S}$ a real skew-symmetric matrix, and suppose that $\mathbf{AS} = \mathbf{SA}$, $|\mathbf{A}-\mathbf{S}| \neq 0$. Prove that $(\mathbf{A}+\mathbf{S})(\mathbf{A}-\mathbf{S})^{-1}$ is orthogonal.

Construct an orthogonal matrix by taking $\mathbf{A} = \mathbf{I}_3$ and

$$\mathbf{S} = \begin{pmatrix} 0 & 1 & 0 \\ -1 & 0 & 2 \\ 0 & -2 & 0 \end{pmatrix}.$$

15. Let $\mathbf{A}$ and $\mathbf{B}$ be commuting orthogonal matrices such that $\mathbf{A}+\mathbf{I}$ and $\mathbf{B}+\mathbf{I}$ are non-singular. Prove that

$$(\mathbf{AB}-\mathbf{I})(\mathbf{AB}+\mathbf{A}+\mathbf{B}+\mathbf{I})^{-1}$$

is a skew-symmetric matrix.

16. Let $\mathbf{A}$ and $\mathbf{B}$ be orthogonal matrices and suppose that $|\mathbf{A}| = -|\mathbf{B}|$. Show that $\mathbf{A}+\mathbf{B}$ is singular.

17. Let $\begin{pmatrix} \mathbf{A} & \mathbf{x} \\ \mathbf{y}^T & a \end{pmatrix}$ be an orthogonal matrix and suppose that $\mathbf{x} \neq 0$. Show that $-1 < a < 1$, and prove that $\mathbf{A}+b\mathbf{x}\mathbf{y}^T$ is orthogonal if and only if $b = (1-a)^{-1}$ or $b = -(1+a)^{-1}$.

18. Writing $N(\mathbf{A}) = \{\operatorname{tr}(\bar{\mathbf{A}}^T\mathbf{A})\}^{\frac{1}{2}}$, show that, for every matrix $\mathbf{A}$ and every unitary matrix $\mathbf{U}$,

$$N(\mathbf{UA}) = N(\mathbf{AU}) = N(\mathbf{A}), \qquad N(\mathbf{A}-\mathbf{U}) = N(\mathbf{I}-\mathbf{U}^{-1}\mathbf{A}).$$

19. Let $\mathbf{U}$ be a unitary matrix, and $\mathbf{A} = \mathbf{dg}(a_1,\ldots,a_n)$. Show that any characteristic root ω of $\mathbf{UA}$ satisfies the inequalities

$$m \leqslant |\omega| \leqslant M,$$

where $M = \max\limits_{1\leqslant r\leqslant n} |a_r|$ and $m = \min\limits_{1\leqslant r\leqslant n} |a_r|$.

20. Let $\mathbf{S}$, $\mathbf{T}$ be proper orthogonal 3×3 matrices with the same first column. Show that, for any angle α, $\mathbf{SR}_\alpha\mathbf{S}^{-1} = \mathbf{TR}_\alpha\mathbf{T}^{-1}$; and interpret this result geometrically.

21. A rotation through an angle α is said to be *infinitesimal* if α^2 is negligible. Use Euler's equations of transformation to show that any two infinitesimal rotations commute with each other.

22. Let $P = (\lambda,\mu,\nu)$ be a point at unit distance from the origin; let $\mathbf{A}$ be the orthogonal matrix of the α-rotation about $\overrightarrow{OP}$; and put

$$\mathbf{T} = \begin{pmatrix} 0 & \nu & -\mu \\ -\nu & 0 & \lambda \\ \mu & -\lambda & 0 \end{pmatrix}.$$

Show that

$$\mathbf{A} = \mathbf{I}-\sin\alpha.\mathbf{T}+(1-\cos\alpha)\mathbf{T}^2, \qquad \tfrac{1}{2}(\mathbf{A}-\mathbf{A}^T) = -\sin\alpha.\mathbf{T}.$$

23. If P is the point $(1, 1, 1)$, find the orthogonal matrix which represents the rotation of space through $45°$ about the line $\overrightarrow{OP}$.

24. The orthogonal matrix of a rotation is given by

$$\begin{pmatrix} 1/\sqrt{2} & 1/\sqrt{2} & 0 \\ 1/2 & -1/2 & 1/\sqrt{2} \\ 1/2 & -1/2 & -1/\sqrt{2} \end{pmatrix}.$$

Find the angle and the axis of rotation.

25. Let **A** be the proper orthogonal 3×3 matrix corresponding to the θ-rotation of space about a straight line through the origin. Show that

$$\cos \tfrac{1}{2}\theta = \tfrac{1}{2}\sqrt{(1+\operatorname{tr}\mathbf{A})}.$$

26. Let a fixed coordinate system be given. The entire space is made to carry out, in turn, an α-rotation about the x-axis, a β-rotation about the y-axis, and a γ-rotation about the z-axis. Show that the angle θ of the resultant rotation is given by

$$2\cos\theta = \cos\beta\cos\gamma+\cos\gamma\cos\alpha+\cos\alpha\cos\beta+\sin\alpha\sin\beta\sin\gamma-1.$$

27. The entire space is made to carry out an α-rotation about the x-axis and then a β-rotation about the y-axis of a fixed coordinate system. Show that, if $0 < \alpha, \beta < \pi$, then the resultant transformation of space is a rotation through the angle $2\cos^{-1}(\cos\frac{1}{2}\alpha\cos\frac{1}{2}\beta)$ about a line whose direction ratios are $\tan\frac{1}{2}\alpha$, $\tan\frac{1}{2}\beta$, $-\tan\frac{1}{2}\alpha\tan\frac{1}{2}\beta$.

28. Show that, if **S** is the skew-symmetric matrix of a θ-rotation and $0 < \theta < \pi$, then $2\tan^2\frac{1}{2}\theta = -\operatorname{tr}(\mathbf{S}^2)$.

29. Let $\mathbf{S}_1$, $\mathbf{S}_2$ be the skew-symmetric matrices of two rotations carried out successively, and suppose that $\mathbf{I}+\mathbf{S}_2\mathbf{S}_1$ is non-singular. Show that $\sigma = \tan^2\frac{1}{2}\theta$, where θ is the angle of the resultant rotation and σ is the sum of 2-rowed principal minors of $(\mathbf{S}_1+\mathbf{S}_2)(\mathbf{I}+\mathbf{S}_2\mathbf{S}_1)^{-1}$.

30. Let the directed lines l_1, l_2 through the origin make an angle ϕ with each other, and suppose that an α_1-rotation of space about l_1 is followed by an α_2-rotation about l_2. Show that the angle θ of the resultant rotation is given by

$$\cos\tfrac{1}{2}\theta = \cos\tfrac{1}{2}\alpha_1\cos\tfrac{1}{2}\alpha_2-\cos\phi\sin\tfrac{1}{2}\alpha_1\sin\tfrac{1}{2}\alpha_2.$$

31. Let $\sigma(\mathbf{A})$ denote the sum of the squares of the elements of **A**. Show that, for every orthogonal matrix **P**, $\sigma(\mathbf{P}^T\mathbf{A}\mathbf{P}) = \sigma(\mathbf{A})$.

32. Let $\sigma(\mathbf{A})$ be defined as in the preceding question. If the matrix **S** is such that, for *all* **A**,

$$\sigma(\mathbf{S}^{-1}\mathbf{A}\mathbf{S}) = \sigma(\mathbf{A}),$$

show that **S** is a scalar multiple of an orthogonal matrix.

33. Show that every unitary 2×2 matrix can be expressed in the form

$$\mathbf{dg}(e^{i\alpha}, e^{i\beta})\begin{pmatrix} \cos\theta & -\sin\theta \\ \sin\theta & \cos\theta \end{pmatrix}\mathbf{dg}(e^{i\gamma}, e^{-i\gamma}),$$

where α, β, γ, θ are real numbers.

34. Use Theorem 2.5.6 (Schmidt's orthogonalization process) to show that every non-singular matrix **A** can be expressed in the form $\mathbf{A} = \mathbf{U}_1\mathbf{\Delta}_1$, where $\mathbf{U}_1$ is unitary and $\mathbf{\Delta}_1$ triangular, and also in the form $\mathbf{A} = \mathbf{\Delta}_2\mathbf{U}_2$, where $\mathbf{U}_2$ is unitary and $\mathbf{\Delta}_2$ triangular.

35. Show that, if

$$\mathbf{X}^T\mathbf{A}\mathbf{X} = \mathbf{A}, \qquad |\mathbf{I}+\mathbf{X}| \neq 0, \qquad \mathbf{Y} = (\mathbf{I}-\mathbf{X})(\mathbf{I}+\mathbf{X})^{-1},$$

then $\mathbf{A}\mathbf{Y}+\mathbf{Y}^T\mathbf{A} = \mathbf{O}$.

Find the most general real matrix $\mathbf{X}$ satisfying the relations $\mathbf{X}^T\mathbf{A}\mathbf{X} = \mathbf{A}$, $|\mathbf{I}+\mathbf{X}| \neq 0$, in each of the following cases: (i) $\mathbf{A} = \mathbf{dg}(a, b)$ $(a \neq 0, b \neq 0)$; (ii) $\mathbf{A} = \mathbf{I}_2$; (iii) $\mathbf{A} = \mathbf{dg}(1, -1)$; (iv) $\mathbf{A} = \mathbf{I}_3$; (v) $\mathbf{A} = \mathbf{dg}(1, 1, -1)$.

36. Show, by using Laplace's expansion theorem (Theorem 1.4.4) and Jacobi's theorem (Theorem 1.5.3), that the sum of the squares of all r-rowed minors formed from r given rows of an orthogonal matrix is equal to 1.

37. Show that all characteristic roots of any square submatrix of a unitary matrix are, in modulus, less than or equal to 1. Also show that any minor of a unitary matrix is, in modulus, less than or equal to 1.

38. Let $\mathbf{z}_1,..., \mathbf{z}_n$ and $\mathbf{w}_1,..., \mathbf{w}_n$ be two orthonormal sets of vectors of order n. Prove that there exists a unitary matrix $\mathbf{U}$ such that $\mathbf{U}\mathbf{z}_k = \mathbf{w}_k$ $(1 \leqslant k \leqslant n)$.

39. Let $\mathbf{A}$ be an $n \times n$ matrix. The *range* of $\mathbf{A}$ is defined as the set of all vectors $\mathbf{A}\mathbf{x}$, where $\mathbf{x} \in \mathfrak{V}_n$; the *null-space* of $\mathbf{A}$ is defined as the set of all vectors $\mathbf{x} \in \mathfrak{V}_n$ such that $\mathbf{A}\mathbf{x} = \mathbf{0}$.

If $\mathbf{U}$ is a unitary matrix, show that the range and null-space of $\mathbf{I}-\mathbf{U}$ are orthogonal complements.

40. Let $\mathbf{x}_1,..., \mathbf{x}_n$ be an orthonormal set of vectors, and suppose that the matrix $\mathbf{P}$ satisfies the relations

$$\mathbf{P}\mathbf{x}_1 = \mathbf{x}_n, \quad \mathbf{P}\mathbf{x}_2 = \mathbf{x}_1, \quad \mathbf{P}\mathbf{x}_3 = \mathbf{x}_2, \quad ..., \quad \mathbf{P}\mathbf{x}_n = \mathbf{x}_{n-1}.$$

By considering the matrix $\bar{\mathbf{U}}^T\mathbf{P}\mathbf{U}$, where $\mathbf{U}$ is the matrix having $\mathbf{x}_1,..., \mathbf{x}_n$ as its columns, show that $\mathbf{P}$ is unitary.

IX

GROUPS

In the course of our previous discussion we have repeatedly met classes of matrices—such as the class of non-singular diagonal matrices or that of orthogonal matrices—which have the property that if two matrices belong to the class in question, then so do their inverses and their product. Since sets of objects having a similar structure occur frequently both in algebra and in other branches of mathematics, it is important to isolate and to study the common properties of all such sets. We are, in this way, led to introduce the concept of a group; and the object of the present chapter is to explain and illustrate this concept. We do not intend here to prove any general theorems on groups, but merely to exhibit the basic notions so as to make the language of the theory of groups available in the subsequent discussion of matrices and transformations.†

9.1. The axioms of group theory

9.1.1. Let $\mathfrak{S}$ be a set of elements, and suppose that some rule is given whereby with each *ordered* pair of elements a, b in $\mathfrak{S}$ we associate a new object, which we denote by ab and which may or may not be an element of $\mathfrak{S}$. Such a rule is known as a *rule of composition*, and since we use the product notation for denoting the object constructed from a and b, we also call it *multiplication*. It is, of course, important to bear in mind that ab is, in general, distinct from ba.

Definition 9.1.1. *Let G be a set of elements $a, b, c,...,$ and let R be a rule of composition (called multiplication), defined for all ordered pairs of elements of G. Then G is a* GROUP *(with respect to R) if the following four axioms—the* GROUP AXIOMS*—are satisfied.*

(i) *If a, b are any two (distinct or equal) elements of G, then ab is also an element of G.*

(ii) *For all $a, b, c \in G$,*

$$(ab)c = a(bc).$$

† For a systematic exposition of the theory of groups see Ledermann, *Introduction to the Theory of Finite Groups*, or, if a more advanced treatment is required, van der Waerden, *Modern Algebra*, vol. i, chaps. ii and vi.

(iii) *G contains an element e, called a* UNIT ELEMENT, *such that, for all* $a \in G$,

$$ae = ea = a.$$

(iv) *To every element* $a \in G$ *there corresponds an element* a^{-1}, *called an* INVERSE ELEMENT *of* a, *such that*

$$aa^{-1} = a^{-1}a = e.$$

Axiom (i) states simply that G is closed with respect to multiplication, and axiom (ii) that multiplication is associative. Axiom (iii) asserts the existence of an element having certain special properties, and axiom (iv) the existence of a special element associated with each element of the group.

A group, then, is a set G of elements with a structure imposed on it by the rule of composition. This rule is often referred to as the *group operation*. One of the simplest instances of a group is provided by the set of all real numbers other than zero; this set is a group with ordinary multiplication as its group operation.

EXERCISE 9.1.1. Verify this statement.

It is easy to see that the elements whose existence is postulated by axioms (iii) and (iv) are, in fact, unique.

THEOREM 9.1.1. *Each group contains precisely one unit element, and each element of a group possesses precisely one inverse element.*

The validity of this theorem will enable us to speak of *the* unit element and of *the* inverse of any given element.

Let both e and e' be unit elements of a group G, so that, for all $a \in G$,

$$ae = ea = a, \qquad ae' = e'a = a.$$

This implies, in particular, that

$$e = e'e = e',$$

and the uniqueness of the unit element is therefore established.

Next, let both x and y be inverses of an element a. Then

$$ax = e, \qquad ya = e.$$

Premultiplying the first equation by y, we obtain

$$y(ax) = ye.$$

Hence, by axioms (ii) and (iii),

$$(ya)x = y,$$

and so

$$x = ex = (ya)x = y.$$

The inverse of a is therefore unique, and the symbol a^{-1} is thus unambiguously defined.

EXERCISE 9.1.2. Show that the unit element is its own inverse.

EXERCISE 9.1.3. Let a and b be any elements of a group G. Show that the equation $ax = b$ possesses the unique solution $x = a^{-1}b$, and obtain the corresponding result for the equation $xa = b$.

DEFINITION 9.1.2 *A group G is* ABELIAN *if any two elements commute with each other, i.e. if, for all $a, b \in G$, $ab = ba$.*

DEFINITION 9.1.3. *A group is* FINITE *or* INFINITE *according as it possesses a finite or infinite number of elements. In the former case the number of elements is called the* ORDER *of the group.*

DEFINITION 9.1.4. *Let G be a group and H a subset of G. If H is itself a group with respect to the same rule of composition as G, then H is said to be a* SUBGROUP *of G.*

Trivial examples of subgroups of G are G itself and the group consisting solely of the unit element of G.

EXERCISE 9.1.4. Let G be a group, and let Z be the set of all those elements of G which commute with every element of G. Show that Z—which is known as the *centre* of G—is an abelian subgroup of G.

9.1.2. Mathematics affords many instances of concrete systems which possess the group structure, and in order to give some idea of the importance of the theory of groups we shall now mention a few examples.

The most obvious instances of group structure are provided by groups whose elements are numbers. Thus the rational numbers, the real numbers, and the complex numbers, with 0 excluded in each case, form infinite abelian groups with respect to ordinary multiplication. In each case the unit element is 1, and the inverse of a is $1/a = a^{-1}$. It is precisely this fact which gives rise to the notation and terminology in axioms (iii) and (iv) of Definition 9.1.1. It may be noted in passing that the first of the three groups mentioned above is a subgroup of the second, and the second is a subgroup of the third. Again, the integers, the rational numbers, the real numbers, and the complex numbers all form infinite abelian groups with respect to addition as group operation.† The unit element is now 0, and the inverse of a is $-a$. Each group is again a subgroup of all those mentioned later in the list. All groups

† Thus the term 'multiplication' in the sense of the group operation is here interpreted as ordinary addition.

enumerated so far have been infinite. An instance of a finite group of numbers is furnished by the mth roots of unity $e^{2\pi ik/m}$ $(k = 0, 1,..., m-1)$. These numbers form an abelian group of order m with respect to ordinary multiplication. The unit element here is $1 = e^{2\pi i0/m}$, and the inverse of $e^{2\pi ik/m}$ is $e^{-2\pi ik/m}$. In particular, the numbers $1, -1$ form a multiplicative group, as do the numbers $1, i, -1, -i$.

Again, let m be a positive integer and denote by C_x the residue class consisting of all integers congruent to $x \pmod m$. Addition and multiplication of residue classes can be specified by the formulae

$$C_x+C_y = C_{x+y}, \qquad C_x C_y = C_{xy},$$

which can easily be shown to lead to unambiguous definitions. The m residue classes $(\text{mod}\, m)$ then form an abelian group with respect to addition. In this group C_0 is the unit element, and C_{-x} is the inverse of C_x. However, we obtain a much more interesting group by taking multiplication of residue classes as the rule of composition. A residue class $(\text{mod}\, m)$ is said to be prime to m if some number (and therefore every number) in it is prime to m. It is easy to show that the residue classes $(\text{mod}\, m)$ prime to m form an abelian group with respect to multiplication as defined above. The unit element is now C_1, and the inverse of C_x is the residue class $C_{x'}$, where x' is any number such that $x'x \equiv 1 \pmod m$. This group is of great importance in the theory of numbers.

Many groups have functions as their elements. In this case the most natural rule of composition for the functions $f(x)$, $g(x)$ is the formation of the composite function $g(f(x))$; which, in accordance with Definition 6.2.1 (p. 172) we call the product of $f(x)$ and $g(x)$ and denote by the symbol gf. In view of Theorem 6.2.1, we have $f(gh) = (fg)h$, so that with our definition of products the associative law is satisfied. Consider, for instance, the set G of all functions $f(x)$ continuous and strictly increasing in the interval $0 \leqslant x \leqslant 1$, and such that $f(0) = 0$, $f(1) = 1$. Then $f, g \in G$ implies $fg \in G$. Again, the function $e(x) = x$ belongs to G and satisfies the equation $fe = ef = f$ for all $f \in G$. Finally, if f^{-1} denotes the (unique) functional inverse of $f \in G$, then clearly $f^{-1} \in G$ and $ff^{-1} = f^{-1}f = e$. Hence G is a group and it is evidently infinite and non-abelian.

Some of the most important groups have as their elements geometrical transformations of various types. In accordance with Definition 6.2.1 we shall continue to refer to the resultant of two transformations as their product. More precisely, if t_1 and t_2 are transformations, we shall denote by $t_2 t_1$ the transformation obtained by first applying t_1 and then t_2.

The translations of a plane (or of space) clearly constitute a group, but a more interesting group is formed by rotations. If α is any real number, let us denote by α' the unique number such that $0 \leqslant \alpha' < 2\pi$ and $\alpha' - \alpha$ is an integral multiple of 2π. Furthermore, if $0 \leqslant \alpha < 2\pi$, let us denote by $R(\alpha)$ the rotation of the plane, in a counterclockwise sense, through the angle α about a fixed origin. Then, for $0 \leqslant \alpha, \beta < 2\pi$, we have

$$R(\beta)R(\alpha) = R((\alpha+\beta)').$$

The rotations of the plane thus form an infinite abelian group; its unit element is $R(0)$ and the inverse of $R(\alpha)$ is $R((-\alpha)')$. Further, it is easy to see that the set of all *euclidean collineations* (i.e. the set of all transformations composed of translations, rotations, and reflections) is an infinite non-abelian group.

EXERCISE 9.1.5. Show that the rotations of space about a fixed point form an infinite non-abelian group.

9.1.3. By a *permutation of degree n* is meant the operation of changing the order of n given distinct objects, say the numbers $1, 2, \dots, n$. Such a permutation is, in other words, the operation of replacing one arrangement $(\lambda_1, \dots, \lambda_n)$ of $(1, \dots, n)$ by a second arrangement $(\mu_1, \dots, \mu_n)$. We represent this permutation by the symbol

$$\begin{pmatrix} \lambda_1, \dots, \lambda_n \\ \mu_1, \dots, \mu_n \end{pmatrix}.$$

If in this symbol the λ's and μ's are rearranged in the same way, then the new symbol will clearly represent the same permutation; that is, if $(k_1, \dots, k_n) = \mathscr{A}\,(1, \dots, n)$, then

$$\begin{pmatrix} \lambda_{k_1}, \dots, \lambda_{k_n} \\ \mu_{k_1}, \dots, \mu_{k_n} \end{pmatrix} = \begin{pmatrix} \lambda_1, \dots, \lambda_n \\ \mu_1, \dots, \mu_n \end{pmatrix}.$$

For example

$$\begin{pmatrix} 3 & 2 & 4 & 1 \\ 1 & 3 & 2 & 4 \end{pmatrix} = \begin{pmatrix} 2 & 4 & 3 & 1 \\ 3 & 2 & 1 & 4 \end{pmatrix} = \begin{pmatrix} 1 & 2 & 3 & 4 \\ 4 & 3 & 1 & 2 \end{pmatrix}.$$

It is clearly always possible to take $1, \dots, n$ (or any other prescribed arrangement of $1, \dots, n$) as the upper row of the symbol representing any given permutation of degree n.

The *identical permutation* e is the permutation which leaves unaltered every arrangement, i.e.

$$e = \begin{pmatrix} 1, 2,..., n \\ 1, 2,..., n \end{pmatrix}.$$

The product qp of the two permutations p and q is defined as the permutation resulting from first carrying out p and then q. Thus if

$$p = \begin{pmatrix} 1,..., n \\ \lambda_1,..., \lambda_n \end{pmatrix}, \qquad q = \begin{pmatrix} \lambda_1,..., \lambda_n \\ \mu_1,..., \mu_n \end{pmatrix},$$

then

$$qp = \begin{pmatrix} 1,..., n \\ \mu_1,..., \mu_n \end{pmatrix}.$$

The $n!$ permutations of degree n form a group with respect to multiplication, the *symmetric group* S_n. For the identity permutation acts as the unit element; the inverse of

$$p = \begin{pmatrix} \lambda_1,..., \lambda_n \\ \mu_1,..., \mu_n \end{pmatrix}$$

is the permutation

$$p^{-1} = \begin{pmatrix} \mu_1,..., \mu_n \\ \lambda_1,..., \lambda_n \end{pmatrix};$$

and it is easy to verify that the associative law is satisfied. For $n > 2$ the group S_n is non-abelian.

EXERCISE 9.1.6. Prove the last statement by considering permutations of the form

$$\begin{pmatrix} 1, 2, 3, 4, 5,..., n \\ \lambda, \mu, \nu, 4, 5,..., n \end{pmatrix},$$

where $(\lambda, \mu, \nu) = \mathscr{A}(1, 2, 3)$.

In addition to the symmetric group, which comprises all permutations (of degree n), there are other groups of permutations, all of them naturally subgroups of the symmetric group. Thus, for instance, the four permutations of degree 4,

$$e = \begin{pmatrix} 1 & 2 & 3 & 4 \\ 1 & 2 & 3 & 4 \end{pmatrix}, \quad a = \begin{pmatrix} 1 & 2 & 3 & 4 \\ 2 & 3 & 4 & 1 \end{pmatrix}, \quad b = \begin{pmatrix} 1 & 2 & 3 & 4 \\ 3 & 4 & 1 & 2 \end{pmatrix}, \quad c = \begin{pmatrix} 1 & 2 & 3 & 4 \\ 4 & 1 & 2 & 3 \end{pmatrix}, \tag{9.1.1}$$

form an abelian subgroup of S_4, in which all elements are powers of a single element,† since

$$b = a^2, \qquad c = a^3, \qquad e = a^4.$$

† Powers of elements of a group are unambiguously defined in view of the associativeness of multiplication.

Another abelian subgroup of S_4 is formed by the permutations

$$e = \begin{pmatrix} 1 & 2 & 3 & 4 \\ 1 & 2 & 3 & 4 \end{pmatrix}, \quad a = \begin{pmatrix} 1 & 2 & 3 & 4 \\ 2 & 1 & 4 & 3 \end{pmatrix}, \quad b = \begin{pmatrix} 1 & 2 & 3 & 4 \\ 3 & 4 & 1 & 2 \end{pmatrix}, \quad c = \begin{pmatrix} 1 & 2 & 3 & 4 \\ 4 & 3 & 2 & 1 \end{pmatrix}, \tag{9.1.2}$$

and here we have $a^2 = b^2 = c^2 = e$, $bc = a$, $ca = b$, $ab = c$.

We can now give an interesting interpretation, in terms of permutations, of the ϵ-symbol of § 1.1. Let $(\lambda_1,\dots,\lambda_n)$ and $(\mu_1,\dots,\mu_n)$ be arrangements of $(1,\dots,n)$. Then, in view of Theorem 1.1.2 (p. 3),

$$\epsilon\begin{pmatrix} \lambda_1,\dots,\lambda_n \\ \mu_1,\dots,\mu_n \end{pmatrix} \tag{9.1.3}$$

is a function of the permutation

$$p = \begin{pmatrix} \lambda_1,\dots,\lambda_n \\ \mu_1,\dots,\mu_n \end{pmatrix}, \tag{9.1.4}$$

and we therefore simply write $\epsilon(p)$ for the symbol (9.1.3). The function $\epsilon(p)$ assumes the values $+1$ and -1; and it satisfies, moreover, the equation

$$\epsilon(qp) = \epsilon(q)\epsilon(p) \tag{9.1.5}$$

for all permutations p, q. For let p be given by (9.1.4) and write

$$q = \begin{pmatrix} \mu_1,\dots,\mu_n \\ \nu_1,\dots,\nu_n \end{pmatrix}.$$

Then

$$\begin{aligned} \epsilon(q)\epsilon(p) &= \operatorname{sgn} \prod_{1 \leqslant r < s \leqslant n} (\mu_s - \mu_r) . \operatorname{sgn} \prod_{1 \leqslant r < s \leqslant n} (\nu_s - \nu_r) \times \\ &\qquad \times \operatorname{sgn} \prod_{1 \leqslant r < s \leqslant n} (\lambda_s - \lambda_r) . \operatorname{sgn} \prod_{1 \leqslant r < s \leqslant n} (\mu_s - \mu_r) \\ &= \operatorname{sgn} \prod_{1 \leqslant r < s \leqslant n} (\lambda_s - \lambda_r) . \operatorname{sgn} \prod_{1 \leqslant r < s \leqslant n} (\nu_s - \nu_r) \\ &= \epsilon\begin{pmatrix} \lambda_1,\dots,\lambda_n \\ \nu_1,\dots,\nu_n \end{pmatrix} = \epsilon(qp). \end{aligned}$$

A particularly simple type of permutation is a *transposition*, i.e. the interchange of two numbers of an arrangement. Any transposition t can clearly be written in the form

$$t = \begin{pmatrix} 1,\dots,r,\dots,s,\dots,n \\ 1,\dots,s,\dots,r,\dots,n \end{pmatrix},$$

where $r < s$; and it therefore follows by Theorem 1.1.3 (p. 4) that

$$\epsilon(t) = -1. \tag{9.1.6}$$

It is easy to see that every permutation can be expressed (though not in a unique way) as a product of transpositions. Let, then, a permutation p be written in the form $p = t_1 \dots t_s$, where $t_1,\dots, t_s$ are transpositions. By (9.1.5) and (9.1.6) we have

$$\epsilon(p) = \epsilon(t_1) \dots \epsilon(t_s) = (-1)^s. \tag{9.1.7}$$

This shows that, whenever a given permutation is expressed as a product of transpositions, the number of factors in the product is either always even or always odd.† It is natural, therefore, to call a permutation *even* or *odd* according as it is the product of an even or an odd number of transpositions. Equation (9.1.7) then shows that $\epsilon(p) = +1$ or -1 according as p is even or odd.

We now call the *arrangement* $(\lambda_1,\dots, \lambda_n)$ of $(1,\dots, n)$ even or odd according as the permutation

$$\begin{pmatrix} \lambda_1,\dots, \lambda_n \\ 1,\dots, n \end{pmatrix}$$

is even or odd. It follows that in the expression

$$D = \sum_{(\lambda_1,\dots,\lambda_n) = \mathscr{A}(1,\dots,n)} \epsilon(\lambda_1,\dots, \lambda_n)\, a_{1\lambda_1} \dots a_{n\lambda_n}$$

for the determinant $D = |a_{ij}|_n$, the symbol $\epsilon(\lambda_1,\dots, \lambda_n)$ has the value $+1$ or -1 according as $(\lambda_1,\dots, \lambda_n)$ is an even or odd arrangement.

9.1.4. We recall that structural identity of two linear manifolds was referred to as an isomorphism between them.‡ A similar terminology is used in the theory of groups.

DEFINITION 9.1.5. *Two groups G, G' are* ISOMORPHIC *(in symbols: $G \simeq G'$) if a biunique correspondence $a \leftrightarrow a'$ ($a \in G$, $a' \in G'$) can be set up between them in such a way that, whenever $a \leftrightarrow a'$, $b \leftrightarrow b'$, we have $ab \leftrightarrow a'b'$. The correspondence itself is then called an* ISOMORPHISM *between G and G'.*

In other words, $G \simeq G'$ if a biunique correspondence can be set up between G and G' such that, for all $a, b \in G$, $(ab)' = a'b'$, where x' denotes the (unique) element of G' which corresponds to $x \in G$. As in the case of linear manifolds we can also express the same idea in terms of the functional notation. An isomorphism between G and G' is a function $\phi(a)$ defined uniquely for all $a \in G$ and having the following properties: (i) $\phi(a)$ is an element of G' for all $a \in G$;

† Cf. Exercise 1.1.3 (p. 3). ‡ Definition 2.4.1 (p. 58).

(ii) given any element $a' \in G'$, there exists precisely one element $a \in G$ such that $\phi(a) = a'$; (iii) for all $a, b \in G$ we have $\phi(ab) = \phi(a)\phi(b)$. The function ϕ with the above properties is also referred to as an 'isomorphic mapping of G onto G''.

Definition 9.1.5, or either of the two equivalent definitions, express in precise language the requirement that the groups G and G' should possess the same structure. Two isomorphic groups will, of course, have in common all properties which are purely structural in character and are independent of the nature of the elements. For example, if one of them possesses exactly 3 subgroups of order 4, then so does the other.

Exercise 9.1.7. Show that isomorphism between groups is an equivalence relation.

Exercise 9.1.8. Let G, G' be isomorphic groups with unit elements e, e' respectively. (i) Show that $e \leftrightarrow e'$. (ii) Show that, if $a \leftrightarrow a'$ and $a^2 = e$, then $a'^2 = e'$.

A useful device for exhibiting the structure of a finite group is the *multiplication table*, which is a square array displaying the product of every pair of elements. Consider, for example, the symmetric group S_3. Its elements are

$$e = \begin{pmatrix} 1 & 2 & 3 \\ 1 & 2 & 3 \end{pmatrix}, \quad p = \begin{pmatrix} 1 & 2 & 3 \\ 1 & 3 & 2 \end{pmatrix}, \quad q = \begin{pmatrix} 1 & 2 & 3 \\ 3 & 2 & 1 \end{pmatrix}, \quad r = \begin{pmatrix} 1 & 2 & 3 \\ 2 & 1 & 3 \end{pmatrix},$$

$$s = \begin{pmatrix} 1 & 2 & 3 \\ 2 & 3 & 1 \end{pmatrix}, \quad t = \begin{pmatrix} 1 & 2 & 3 \\ 3 & 1 & 2 \end{pmatrix},$$

and its multiplication table is as follows:

	e	p	q	r	s	t
e	e	p	q	r	s	t
p	p	e	s	t	q	r
q	q	t	e	s	r	p
r	r	s	t	e	p	q
s	s	r	p	q	t	e
t	t	q	r	p	e	s

Here the product xy is the element standing in the row corresponding to x and the column corresponding to y, e.g. $qt = p$, $tq = r$.

It is clear that two finite groups are isomorphic if and only if they have the same multiplication table, provided, of course, that the elements in each group are suitably labelled. Thus the set of the six functions

$$e(x) = x, \quad p(x) = \frac{1}{x}, \quad q(x) = 1-x, \quad r(x) = \frac{x}{x-1},$$

$$s(x) = \frac{1}{1-x}, \quad t(x) = \frac{x-1}{x}$$

is a group, if multiplication of functions is defined as on p. 255. It is, in fact, easy to verify that this group has the same multiplication table as S_3, and so is isomorphic to S_3.

Again, the group of permutations given by (9.1.2) has the following multiplication table:

	e	a	b	c
e	e	a	b	c
a	a	e	c	b
b	b	c	e	a
c	c	b	a	e

Any group whose structure is specified by this table is known as a *Klein four-group*. Further instances of groups of this type are provided by the group of functions

$$e(x) = x, \quad a(x) = -x, \quad b(x) = \frac{1}{x}, \quad c(x) = -\frac{1}{x},$$

and by the multiplicative group of residue classes (mod 8) prime to 8. On the other hand, the group of permutations (9.1.1) is not a Klein four-group.

EXERCISE 9.1.9. Verify these statements.

9.2. Matrix groups and operator groups

9.2.1. Among the groups that have been studied most extensively are groups of matrices.† Obvious rules of composition for such groups are matrix addition and matrix multiplication. Thus the set of all $m \times n$ matrices is a group with respect to addition, with $\mathbf{O}_m^n$ as the unit element and $-\mathbf{A}$ as the inverse of $\mathbf{A}$. Again, a vector space (or, more generally, a linear manifold) is a group with respect to addition. However, in almost all important examples of matrix groups, multiplication is the group operation.

DEFINITION 9.2.1. *A* MATRIX GROUP *is a set of matrices which form a group with respect to matrix multiplication.*

† See, for instance, van der Waerden, 37.

It is thus to be understood that, unless the contrary is stated explicitly, *the rule of composition is matrix multiplication.*

It is plain, in the first place, that a matrix group must consist of square matrices of the same order. Secondly, either all matrices of a group are non-singular or else all are singular. For let $\mathbf{A}$ be an element of a matrix group Γ and suppose that $|\mathbf{A}| \neq 0$. If $\mathbf{X}$ is another element of Γ, and $\mathbf{X}'$ is the inverse element of $\mathbf{X}$, then

$$0 \neq |\mathbf{A}| = |\mathbf{XX}'\mathbf{A}| = |\mathbf{X}|.|\mathbf{X}'\mathbf{A}|,$$

and so $|\mathbf{X}| \neq 0$. Hence, if one element is non-singular, then so is every element; and the assertion follows.

We must therefore distinguish between groups of non-singular and those of singular matrices. All important matrix groups are of the former type, and we shall for the present confine ourselves to these groups. In § 9.4 we shall turn to groups of singular matrices and shall show that, in a sense to be explained, their structure does not differ essentially from the structure of groups of non-singular matrices.

EXERCISE 9.2.1. Discuss the fallacy in the following argument: 'Groups of singular matrices cannot exist since such matrices do not possess inverses.'

EXERCISE 9.2.2. Show that the real matrices of the type $\begin{pmatrix} x & 0 \\ 0 & 0 \end{pmatrix}$, where $x \neq 0$, form a group which is isomorphic to the multiplicative group of real non-zero numbers.

The discussion a few lines above enables us not only to recognize that a matrix group cannot contain both singular and non-singular matrices, but also to draw the following conclusion.

THEOREM 9.2.1. *In a group of non-singular matrices the unit element is the unit matrix, and the group inverse coincides with the matrix inverse.*

The second part of the theorem means simply that the inverse element $\mathbf{X}'$ of $\mathbf{X}$ is actually equal to the inverse matrix $\mathbf{X}^{-1}$—a conclusion which must not be taken for granted solely on the ground that the term 'inverse' is used in both matrix theory and group theory.

If we wish to show that any particular set of matrices constitutes a group, there is no need to verify every one of the axioms of Definition 9.1.1. We have, in fact, the following useful criterion.

THEOREM 9.2.2. *A set G of non-singular matrices (of the same order) is a matrix group if and only if, whenever* $\mathbf{A}$ *and* $\mathbf{B}$ *belong to G, so also do* $\mathbf{AB}$ *and* $\mathbf{A}^{-1}$.

In other words, G is a matrix group if and only if it is closed with respect to matrix multiplication and matrix inversion.

The proof is immediate. The stated condition is obviously necessary. It is also sufficient, for axiom (i) is implied by this condition, axiom (ii) holds automatically since matrix multiplication is associative, and axioms (iii) and (iv) follow since, if $\mathbf{A} \in G$, then $\mathbf{I} = \mathbf{AA}^{-1} \in G$.

Of the matrix groups mentioned below we shall recognize several as already familiar to us. Possibly the most important of these is the *full linear group* $GL(n)$, which consists of all non-singular $n \times n$ matrices (over a specified reference field). In view of the discussion in § 4.2, it is clear that the full linear group can be used for studying linear transformation of vector spaces. When we are dealing with projective space, however, a slightly modified group is required. This is the *projective group* $PGL(n-1)$ which is defined as the group of all non-singular $n \times n$ matrices, with the convention that two matrices which only differ by a scalar multiple are to be regarded as identical.†

The following sets of $n \times n$ matrices evidently form groups: (i) unimodular matrices, i.e. matrices whose determinants are equal to ± 1; (ii) unitary matrices; (iii) orthogonal matrices; (iv) proper orthogonal matrices. These groups are known respectively as the *unimodular group*, the *unitary group*, the *orthogonal group*, and the *rotation group*.‡ It is clear that the third of these groups is a subgroup of the first and also of the second, and the fourth is a subgroup of the third.

EXERCISE 9.2.3. Show that the non-singular upper triangular matrices form a matrix group, but that the non-singular triangular matrices do not.

An important matrix group is provided by matrices of the type

$$\begin{pmatrix} a & b \\ -b & a \end{pmatrix}, \tag{9.2.1}$$

† To make this definition precise we need the notion of a *factor group* (see Ledermann, op. cit., 102). Indeed, $PLG(n-1)$ is the factor group $GL(n)/S(n)$, where $S(n)$ is the group of non-singular scalar matrices of order n.

‡ These and several other important groups have been the subject of very detailed study. For an account of the resulting theory see Murnaghan, **40**, Dieudonné, **41**, and Weyl, **42**.

where a and b are real. We have

$$\begin{pmatrix} a & b \\ -b & a \end{pmatrix} + \begin{pmatrix} a' & b' \\ -b' & a' \end{pmatrix} = \begin{pmatrix} a+a' & b+b' \\ -(b+b') & a+a' \end{pmatrix},$$

and from this it is easily seen that the matrices (9.2.1) form a group with respect to addition. The group is isomorphic to the additive group of complex numbers, with the correspondence given by the scheme

$$\begin{pmatrix} a & b \\ -b & a \end{pmatrix} \leftrightarrow a+ib. \tag{9.2.2}$$

Again, we have

$$\begin{pmatrix} a & b \\ -b & a \end{pmatrix}\begin{pmatrix} a' & b' \\ -b' & a' \end{pmatrix} = \begin{pmatrix} aa'-bb' & ab'+a'b \\ -(ab'+a'b) & aa'-bb' \end{pmatrix}.$$

Moreover, if a and b are not both zero, then the matrix (9.2.1) is non-singular, and

$$\begin{pmatrix} a & b \\ -b & a \end{pmatrix}^{-1} = \begin{pmatrix} a/(a^2+b^2) & -b/(a^2+b^2) \\ b/(a^2+b^2) & a/(a^2+b^2) \end{pmatrix}.$$

It follows by Theorem 9.2.2 that the matrices (9.2.1), with a, b not both zero, form a group with respect to matrix multiplication. It is easy to verify that this group is isomorphic to the multiplicative group of non-zero complex numbers, and that the correspondence is once again given by the scheme (9.2.2).

The analogy, exhibited by the two isomorphisms, between the algebra of complex numbers and that of matrices of type (9.2.1), makes it possible to discuss the extension of the system of real numbers to the system of complex numbers by means of such matrices.† Furthermore, the second isomorphism implies that with any set of complex numbers forming a group with respect to multiplication there is associated an isomorphic group of real 2×2 matrices. Thus, for instance, the numbers $1, i, -1, -i$ form a multiplicative group; and an isomorphic matrix group consists of the matrices

$$\begin{pmatrix} 1 & 0 \\ 0 & 1 \end{pmatrix}, \quad \begin{pmatrix} 0 & 1 \\ -1 & 0 \end{pmatrix}, \quad \begin{pmatrix} -1 & 0 \\ 0 & -1 \end{pmatrix}, \quad \begin{pmatrix} 0 & -1 \\ 1 & 0 \end{pmatrix}.$$

The two groups are also isomorphic to the multiplicative group of residue classes (mod 5) prime to 5, and also to the group of permutations listed in (9.1.1) (p. 257).

† See, for example, Copson, *An Introduction to the Theory of Functions of a Complex Variable*, 2–5.

Next, consider the matrices

$$\mathbf{Q}(z,w) = \begin{pmatrix} z & w \\ -\bar{w} & \bar{z} \end{pmatrix},$$

where z and w are complex numbers. These matrices are easily seen to form a group with respect to addition. Moreover

$$\mathbf{Q}(z,w)\mathbf{Q}(z',w') = \mathbf{Q}(zz'-w\bar{w}', zw'+\bar{z}'w),$$

and, when z, w are not both zero,

$$\{\mathbf{Q}(z,w)\}^{-1} = \mathbf{Q}(\bar{z}/(z\bar{z}+w\bar{w}),\ -w/(z\bar{z}+w\bar{w})).$$

From these formulae it follows that the matrices $\mathbf{Q}(z,w)$, with z, w not both zero, form a group with respect to multiplication. The significance of these matrices lies in the fact that they constitute a system isomorphic (as regards both addition and multiplication) to an important system known as the *ring of real quaternions*.†

9.2.2. The majority of groups mentioned earlier in this chapter may be regarded as groups of operators. Thus $n \times n$ matrices are operators in the vector space $\mathfrak{V}_n$, functions are operators in the system of numbers, geometrical transformations are operators in geometrical space, and permutations are operators in the set of arrangements. We shall now introduce a few further groups of operators which arise in the theory of matrices.

We begin by considering similarity transformations; these may be regarded as operators in the set of matrices. If $\mathbf{X}$ is a non-singular matrix, we denote by $t_{\mathbf{X}}$ the similarity transformation having $\mathbf{X}$ as the transforming matrix, so that, for every square matrix $\mathbf{A}$,

$$t_{\mathbf{X}}(\mathbf{A}) = \mathbf{X}^{-1}\mathbf{A}\mathbf{X}.$$

We observe that

$$(t_{\mathbf{Y}}t_{\mathbf{X}})(\mathbf{A}) = t_{\mathbf{Y}}\{t_{\mathbf{X}}(\mathbf{A})\} = t_{\mathbf{Y}}(\mathbf{X}^{-1}\mathbf{A}\mathbf{X}) = \mathbf{Y}^{-1}(\mathbf{X}^{-1}\mathbf{A}\mathbf{X})\mathbf{Y}$$
$$= (\mathbf{X}\mathbf{Y})^{-1}\mathbf{A}(\mathbf{X}\mathbf{Y}) = t_{\mathbf{X}\mathbf{Y}}(\mathbf{A}),$$

and so

$$t_{\mathbf{Y}}t_{\mathbf{X}} = t_{\mathbf{X}\mathbf{Y}}. \tag{9.2.3}$$

In particular, therefore,

$$t_{\mathbf{X}}t_{\mathbf{X}^{-1}} = t_{\mathbf{X}^{-1}}t_{\mathbf{X}} = t_{\mathbf{I}}; \tag{9.2.4}$$

and $t_{\mathbf{I}}$ is, of course, the identical transformation which carries every matrix into itself. It is now clear that similarity transformations form a group. For axiom (i) is satisfied by virtue of (9.2.3), which also implies axiom (ii); axiom (iii) is satisfied since $t_{\mathbf{I}}$ has

† For details see McCoy, *Rings and Ideals*, 9 and 21.

the properties of a unit element; and axiom (iv) is satisfied since, by (9.2.4), $t_{\mathbf{X}^{-1}}$ is seen to be the inverse of $t_{\mathbf{X}}$.

DEFINITION 9.2.2. *The group of similarity transformations*† *is known as the* SIMILARITY GROUP.

EXERCISE 9.2.4. Show that $t_{\alpha\mathbf{X}} = t_{\mathbf{X}}$, where α is any non-zero scalar.

EXERCISE 9.2.5. Discuss the fallacy in the following argument. 'The elements of the similarity group can be put into biunique correspondence, $t_{\mathbf{X}} \leftrightarrow \mathbf{X}^{-1}$, with those of the full linear group. Hence, by (9.2.3), the relations $t_{\mathbf{X}} \leftrightarrow \mathbf{X}^{-1}$, $t_{\mathbf{Y}} \leftrightarrow \mathbf{Y}^{-1}$ imply $t_{\mathbf{X}} t_{\mathbf{Y}} \leftrightarrow \mathbf{X}^{-1}\mathbf{Y}^{-1}$; and the two groups are therefore isomorphic.'

Certain subgroups of the similarity group play an important part in the discussion of the later chapters.

DEFINITION 9.2.3. *If the matrices* $\mathbf{A}$ *and* $\mathbf{B}$ *are connected by the relation* $\mathbf{B} = \mathbf{U}^{-1}\mathbf{A}\mathbf{U}$, *where* $\mathbf{U}$ *is unitary (orthogonal), then* $\mathbf{B}$ *is said to be obtained from* $\mathbf{A}$ *by a* UNITARY (ORTHOGONAL) SIMILARITY TRANSFORMATION; *and* $\mathbf{A}$ *and* $\mathbf{B}$ *are said to be* UNITARILY (ORTHOGONALLY) SIMILAR *to each other.*‡

Using Theorems 8.1.4 (p. 224) and 8.2.2 (p. 230) the reader can readily convince himself that the unitary similarity transformations and the orthogonal similarity transformations form groups.

DEFINITION 9.2.4. *The group of unitary (orthogonal) similarity transformations is called the* UNITARY (ORTHOGONAL) SIMILARITY GROUP.

An analogous situation arises in the case of congruence transformations. We recall that a matrix $\mathbf{A}$ is said to be subjected to a congruence transformation if it is transformed into $\mathbf{P}^T\mathbf{A}\mathbf{P}$, where $\mathbf{P}$ is some non-singular matrix. It is easy to verify that the congruence transformations form a group. If $\mathbf{P}$ is restricted to orthogonal matrices, we call the transformation an *orthogonal congruence transformation*; such transformations again form a group.

DEFINITION 9.2.5. *The group of congruence (orthogonal congruence) transformations is called the* CONGRUENCE (ORTHOGONAL CONGRUENCE) GROUP.

The reader will have noticed that the terms 'orthogonal congruence transformation' and 'orthogonal congruence group' are,

† We speak here, of course, of transformations of matrices of a fixed order.

‡ The relation between $\mathbf{A}$ and $\mathbf{B}$ is, of course, symmetric and is, indeed, an equivalence relation.

in fact, synonymous with the terms 'orthogonal similarity transformation' and 'orthogonal similarity group' respectively. Nevertheless, it is useful to retain both pairs of terms.

Another type of transformation that arises in the discussion of Part III is that specified by the scheme $\overline{\mathbf{P}}^T\mathbf{A}\mathbf{P}$, where $|\mathbf{P}| \neq 0$. Such a transformation is called *conjunctive*, and the set of conjunctive transformations is a group. If $\mathbf{P}$ is unitary, then the transformation is simply a unitary similarity transformation.

9.2.3. Groups of operators are important in the study of equivalence relations. Thus it is clear that if Φ is a group of operators in a set $\mathfrak{S}$ of objects, then it satisfies conditions (i)–(iii) of Theorem 6.5.2 (p. 188). Hence, in view of Theorem 6.5.2 and Definition 6.5.3 we can speak of *equivalence in* $\mathfrak{S}$ *with respect to the group* Φ *of operators*. This notion enables us, in particular, to gain a unified view of several branches of geometry. Thus, for example, euclidean geometry is seen to be the study of properties of figures from the point of view of equivalence with respect to the group of euclidean operations—translations, rotations, and reflections. More generally, any group of operators in a space defines a corresponding geometry.†

9.3. Representation of groups by matrices

9.3.1. In investigating the structural properties of a group G we may find that the problems we wish to solve are intractable if G is given in an unmanageable form, say in the form of a multiplication table. In such circumstances it can be advantageous to replace G by some isomorphic group G', chosen in such a way that it can be handled easily and efficiently. It is often particularly useful to take as G' a group of matrices, for then the highly developed matrix technique at once becomes available in our investigation.

DEFINITION 9.3.1. *An isomorphic mapping of a group G onto a matrix group G' is called a* FAITHFUL REPRESENTATION *of G.*

The branch of algebra which deals with the representation of groups by matrix groups is known as the *theory of representations*. It was initiated by Frobenius at about the turn of the century and has since been developed more systematically and more intensively

† For further elaboration of this remark see Semple and Kneebone, *Algebraic Projective Geometry*, chap. i.

than any other part of group theory. Here we can do little more than give some examples of matrix representations.†

It has already been noted that the matrices

$$\begin{pmatrix} a & b \\ -b & a \end{pmatrix}, \tag{9.3.1}$$

where a, b are real and not both zero, provide a representation of the multiplicative group of all non-zero complex numbers. This implies, in particular, that any multiplicative group of complex numbers possesses a representation by real matrices; an example of such a representation was given on p. 264. We can use the same idea to obtain a representation, in terms of *real* matrices, for the group Q of matrices

$$\begin{pmatrix} z & w \\ -\bar{w} & \bar{z} \end{pmatrix}, \tag{9.3.2}$$

where z, w are complex and are not both zero. If we replace each element in (9.3.2) by the appropriate matrix of type (9.3.1), we recognize easily that the 4×4 matrices

$$\begin{pmatrix} a & b & c & d \\ -b & a & -d & c \\ -c & d & a & -b \\ -d & -c & b & a \end{pmatrix},$$

where a, b, c, d are real and not all zero, represent the group Q.

Again, the group of plane rotations about the origin is represented by the rotation group of 2×2 matrices. More precisely, in this representation the matrix

$$\begin{pmatrix} \cos\alpha & -\sin\alpha \\ \sin\alpha & \cos\alpha \end{pmatrix}$$

corresponds to the rotation $R(\alpha)$ through an angle α. The group of space rotations about the origin is represented by the rotation group of 3×3 matrices.

The intimate connexion between matrices and linear transformations of linear manifolds has already been noted in Chapter IV. We can now show that certain sets of linear transformations form groups which admit of matrix representations.

THEOREM 9.3.1. *The set of all automorphisms of a linear manifold of dimensionality n is a group isomorphic to the full linear group $GL(n)$.*

† For a systematic discussion see e.g. Boerner, **36**; Speiser, **38**, chaps. xi–xv; and Murnaghan, **40**.

Let $\mathfrak{M}$ be the linear manifold and let L_1, L_2 be two linear transformations of $\mathfrak{M}$ into itself. If $\mathfrak{B}$ is any basis of $\mathfrak{M}$, write

$$\mathbf{A}_1 = \mathscr{R}(L_1;\, \mathfrak{B}), \qquad \mathbf{A}_2 = \mathscr{R}(L_2;\, \mathfrak{B}).\dagger$$

For any $X \in \mathfrak{M}$, put

$$X' = L_1(X), \qquad X'' = (L_2 L_1)(X) = L_2(X'),$$

$$\mathbf{x} = \mathscr{R}(X;\, \mathfrak{B}), \qquad \mathbf{x}' = \mathscr{R}(X';\, \mathfrak{B}), \qquad \mathbf{x}'' = \mathscr{R}(X'';\, \mathfrak{B}).$$

Then
$$\mathbf{x}' = \mathbf{A}_1\mathbf{x}, \qquad \mathbf{x}'' = \mathbf{A}_2\mathbf{x}',$$
and so
$$\mathbf{x}'' = \mathbf{A}_2\mathbf{A}_1\mathbf{x}.$$
This means that $L_2 L_1$ is a linear transformation of $\mathfrak{M}$ into itself, and
$$\mathbf{A}_2\mathbf{A}_1 = \mathscr{R}(L_2 L_1;\, \mathfrak{B}). \tag{9.3.3}$$
Now, if L_1, L_2 are automorphisms of $\mathfrak{M}$, then, by Theorem 4.3.3 (p. 125), $\mathbf{A}_1$ and $\mathbf{A}_2$ are non-singular. Hence $\mathbf{A}_2\mathbf{A}_1$ is non-singular and so, again by Theorem 4.3.3, $L_2 L_1$ is an automorphism of $\mathfrak{M}$.

To show that the set of all automorphisms of $\mathfrak{M}$ is a group, we have to verify the four group axioms. Axiom (i) is satisfied, as we have just observed; and axiom (ii) holds by virtue of the associative law for multiplication of operators.‡ The identical automorphism, say L_0, which transforms every element into itself acts as the unit element, and so axiom (iii) is satisfied. Finally, let L be any automorphism of $\mathfrak{M}$, and write
$$\mathbf{A} = \mathscr{R}(L;\, \mathfrak{B}). \tag{9.3.4}$$
If the automorphism L^* is defined by the equation
$$\mathbf{A}^{-1} = \mathscr{R}(L^*;\, \mathfrak{B}),$$
then, by (9.3.3),
$$\mathbf{I} = \mathscr{R}(LL^*;\, \mathfrak{B}) = \mathscr{R}(L^*L;\, \mathfrak{B}).$$
In other words, $LL^* = L^*L = L_0$. Thus L^* is the inverse of L, and axiom (iv) is satisfied.

The automorphisms of $\mathfrak{M}$ thus form a group H. Consider now the biunique correspondence, given by (9.3.4), between the elements of H and those of $GL(n)$. In view of (9.3.3) it follows at once that $H \simeq GL(n)$.

Exercise 9.3.1. Show that any group of automorphisms of a linear manifold of dimensionality n is isomorphic to a group of non-singular $n \times n$ matrices.

† This notation is explained in Definition 4.2.4, p. 118.

‡ Alternatively, axiom (ii) follows from (9.3.3) and the associative law for matrix multiplication.

EXERCISE 9.3.2. Let Γ be a set of $n \times n$ matrices and Γ' the set of linear transformations of $\mathfrak{B}_n$ into itself represented by the matrices of Γ with respect to some fixed basis in $\mathfrak{B}_n$. Show that if Γ is a group, then so is Γ', and conversely; and that $\Gamma \simeq \Gamma'$.

9.3.2. We have been concerned, so far, with matrix representations of infinite groups. The possibility of representing finite groups by matrices is, however, equally interesting and we shall show that every finite group can, in fact, be so represented.

Theorem 9.3.2. *Every finite group is isomorphic to some matrix group.*†

Let the elements of a group G of order n be denoted by $a_1,..., a_n$. With each element a_r let us associate the $n \times n$ matrix $\mathbf{M}(a_r)$ defined by the equations

$$\{\mathbf{M}(a_r)\}_{ij} = \begin{cases} 1 \ (a_i^{-1}a_j = a_r) \\ 0 \ (a_i^{-1}a_j \neq a_r) \end{cases} \quad (i,j = 1,..., n). \qquad (9.3.5)$$

In particular, if e denotes the unit element of G (so that e is one of the a_r), then

$$\{\mathbf{M}(e)\}_{ij} = \begin{Bmatrix} 1 \ (a_i = a_j) \\ 0 \ (a_i \neq a_j) \end{Bmatrix} = \delta_{ij},$$

so that

$$\mathbf{M}(e) = \mathbf{I}.$$

Since evidently $\mathbf{M}(a_r) \neq \mathbf{M}(a_s)$ for $r \neq s$, it follows that the definition (9.3.5) sets up a biunique correspondence between G and the set G' of matrices $\mathbf{M}(a_r)$.

We have, for $r, s, i, j = 1,..., n$,

$$\{\mathbf{M}(a_r)\mathbf{M}(a_s)\}_{ij} = \sum_{k=1}^{n} \{\mathbf{M}(a_r)\}_{ik}\{\mathbf{M}(a_s)\}_{kj} = \sum_{k} 1,$$

where the conditions under the sign of summation on the right-hand side are $a_i^{-1}a_k = a_r$, $a_k^{-1}a_j = a_s$, $1 \leqslant k \leqslant n$, i.e.

$$a_k = a_i a_r = a_j a_s^{-1}, \qquad 1 \leqslant k \leqslant n. \qquad (9.3.6)$$

If $a_i a_r \neq a_j a_s^{-1}$, then there exists no value of k satisfying (9.3.6), and the sum has the value 0. If, on the other hand, $a_i a_r = a_j a_s^{-1}$, then precisely one value of k satisfies (9.3.6), and the sum has the value 1. Thus

$$\{\mathbf{M}(a_r)\mathbf{M}(a_s)\}_{ij} = \begin{Bmatrix} 1 \ (a_i^{-1}a_j = a_r a_s) \\ 0 \ (a_i^{-1}a_j \neq a_r a_s) \end{Bmatrix} = \{\mathbf{M}(a_r a_s)\}_{ij},$$

and so

$$\mathbf{M}(a_r)\mathbf{M}(a_s) = \mathbf{M}(a_r a_s) \quad (r, s = 1,..., n). \qquad (9.3.7)$$

† This result should be compared with Cayley's theorem that every finite group is isomorphic to some group of permutations. See Ledermann, op. cit., 78.

In particular, this implies

$$\mathbf{M}(a_r)\mathbf{M}(a_r^{-1}) = \mathbf{M}(e) = \mathbf{I},$$

i.e.

$$\{\mathbf{M}(a_r)\}^{-1} = \mathbf{M}(a_r^{-1}) \quad (r = 1,..., n). \tag{9.3.8}$$

By (9.3.7) and (9.3.8), we see that the set G' is closed with respect to multiplication and inversion of matrices. Hence, by Theorem 9.2.2 (p. 263), G' is a group; and, by (9.3.7), $G \simeq G'$.

It may be noted that each matrix $\mathbf{M}(a_r)$ has precisely one 1 in each row and precisely one 1 in each column, all other elements being equal to 0. It is thus plainly orthogonal and we see that every group of order n is isomorphic to a subgroup of the group of orthogonal $n \times n$ matrices.

The proof of Theorem 9.3.2 not merely establishes the existence of a matrix representation for every finite group but describes an actual procedure for constructing such a representation. In particular, it provides a method for representing groups of permutations by means of matrices. However, the resulting representations are generally unwieldy, the symmetric group S_n requiring, for instance, matrices of order $n!$. It is therefore useful to note that a slight modification in the procedure leads to a more convenient representation of groups of permutations.

THEOREM 9.3.3. *Any group of permutations of degree n can be represented by $n \times n$ matrices.*

Let G be a group of permutations of degree n, and let $p \in G$ be the permutation which changes the arrangement $(1,..., n)$ into $(a_1,..., a_n)$. We can regard p as a function of a single variable such that

$$p(1) = a_1, \quad ..., \quad p(n) = a_n.$$

With the permutation p we associate the $n \times n$ matrix $\mathbf{N}(p)$ defined by the equations

$$\{\mathbf{N}(p)\}_{ij} = \begin{cases} 1 \ (i = p(j)) \\ 0 \ (i \neq p(j)) \end{cases} \quad (i,j = 1,..., n). \tag{9.3.9}$$

Thus the first column of $\mathbf{N}(p)$ contains a 1 in the a_1th place, the second column contains a 1 in the a_2th place, and so on, while all the remaining elements are equal to zero. Matrices of this type are known as *permutation matrices*.

The definition (9.3.9) sets up a biunique correspondence between G and the set G' of matrices $\mathbf{N}(p)$. The reader should have no difficulty in verifying that, for any two permutations $p, q \in G$,

$$\mathbf{N}(p)\mathbf{N}(q) = \mathbf{N}(pq).$$

This implies, in particular, that $\{\mathbf{N}(p)\}^{-1} = \mathbf{N}(p^{-1})$. From these two relations we recognize that G' is a group and that it is isomorphic to G.

As an example consider the symmetric group S_3 which consists of the six permutations

$$\begin{pmatrix}1&2&3\\1&2&3\end{pmatrix}, \quad \begin{pmatrix}1&2&3\\1&3&2\end{pmatrix}, \quad \begin{pmatrix}1&2&3\\2&1&3\end{pmatrix}, \quad \begin{pmatrix}1&2&3\\2&3&1\end{pmatrix}, \quad \begin{pmatrix}1&2&3\\3&1&2\end{pmatrix}, \quad \begin{pmatrix}1&2&3\\3&2&1\end{pmatrix}.$$

The proof just given shows that S_3 is isomorphic to the group of the six 3×3 matrices

$$\begin{pmatrix}1&0&0\\0&1&0\\0&0&1\end{pmatrix}, \quad \begin{pmatrix}1&0&0\\0&0&1\\0&1&0\end{pmatrix}, \quad \begin{pmatrix}0&1&0\\1&0&0\\0&0&1\end{pmatrix}, \quad \begin{pmatrix}0&0&1\\1&0&0\\0&1&0\end{pmatrix}, \quad \begin{pmatrix}0&1&0\\0&0&1\\1&0&0\end{pmatrix}, \quad \begin{pmatrix}0&0&1\\0&1&0\\1&0&0\end{pmatrix}.$$

9.4. Groups of singular matrices

The existence of groups of singular matrices has been indicated in § 9.2.1. As an example of such a group, a little less obvious than that given in Exercise 9.2.2 (p. 262), we may mention the set of matrices of the form

$$\mathbf{A}(x) = \begin{pmatrix}x & x\\x & x\end{pmatrix},$$

where $x \neq 0$. This set is a matrix group, with $\mathbf{A}(\frac{1}{2})$ as its unit element and $\mathbf{A}\left(\frac{1}{4x}\right)$ as the inverse of $\mathbf{A}(x)$.

We propose now to investigate more systematically groups of singular matrices with a view to elucidating the relation between the structure of such groups and that of groups of non-singular matrices. In the course of our discussion we shall also discover certain criteria for deciding whether a given matrix is an element, or even the unit element, of a matrix group. It is, at any rate, clear that not every matrix is an element of a suitable matrix group. For if the matrix

$$\mathbf{A} = \begin{pmatrix}0 & 1\\0 & 0\end{pmatrix}$$

were an element of a group Γ, then so would be $\mathbf{A}^2$. Now $\mathbf{A}^2 = \mathbf{O}$, and a group which contains the zero matrix can contain no other element.

Exercise 9.4.1. Prove the last statement.

DEFINITION 9.4.1. (i) *A* GROUP MATRIX *is a matrix which is an element of at least one matrix group.* (ii) *A* GROUP UNIT MATRIX *is a matrix which is the unit element of at least one matrix group.*

We begin with an almost obvious result.

THEOREM 9.4.1. (i) *If* $\mathbf{A}$ *is a group matrix, then so is every matrix similar to* $\mathbf{A}$. (ii) *If* $\mathbf{A}$ *is a group unit matrix, then so is every matrix similar to* $\mathbf{A}$.

Let $\mathbf{A}$ be an element of a matrix group Γ and let $\mathbf{A}' = \mathbf{S}^{-1}\mathbf{A}\mathbf{S}$. We have to show that $\mathbf{A}'$ is also an element of some matrix group. Consider the set Γ' of matrices

$$\mathbf{S}^{-1}\mathbf{X}\mathbf{S} \quad (\mathbf{X} \in \Gamma).$$

There is clearly a biunique correspondence $\mathbf{X} \leftrightarrow \mathbf{S}^{-1}\mathbf{X}\mathbf{S}$ between the elements of Γ and those of Γ'. Now

$$\mathbf{S}^{-1}\mathbf{X}\mathbf{S}.\mathbf{S}^{-1}\mathbf{Y}\mathbf{S} = \mathbf{S}^{-1}\mathbf{X}\mathbf{Y}\mathbf{S},$$

and hence, if $\mathbf{X}', \mathbf{Y}' \in \Gamma'$, then $\mathbf{X}'\mathbf{Y}' \in \Gamma'$. Again, let $\mathbf{E}$ be the unit matrix of Γ and let $\mathbf{X}^*$ be the inverse element of $\mathbf{X} \in \Gamma$. Then

$$\mathbf{E}\mathbf{X} = \mathbf{X}\mathbf{E} = \mathbf{X}, \qquad \mathbf{X}^*\mathbf{X} = \mathbf{X}\mathbf{X}^* = \mathbf{E},$$

and so

$$\mathbf{S}^{-1}\mathbf{E}\mathbf{S}.\mathbf{S}^{-1}\mathbf{X}\mathbf{S} = \mathbf{S}^{-1}\mathbf{X}\mathbf{S}.\mathbf{S}^{-1}\mathbf{E}\mathbf{S} = \mathbf{S}^{-1}\mathbf{X}\mathbf{S},$$
$$\mathbf{S}^{-1}\mathbf{X}^*\mathbf{S}.\mathbf{S}^{-1}\mathbf{X}\mathbf{S} = \mathbf{S}^{-1}\mathbf{X}\mathbf{S}.\mathbf{S}^{-1}\mathbf{X}^*\mathbf{S} = \mathbf{S}^{-1}\mathbf{E}\mathbf{S}.$$

Thus the set Γ' satisfies all the necessary axioms and so is a group. Moreover, $\mathbf{A}' = \mathbf{S}^{-1}\mathbf{A}\mathbf{S}$ is an element of this group.

The second part of the theorem follows in exactly the same way, for if $\mathbf{A}$ is the unit element of Γ, then $\mathbf{A}'$ is seen to be the unit element of Γ'.

EXERCISE 9.4.2. Show that if Γ is a group of matrices, $\mathbf{S}$ a non-singular matrix, and Γ' the group of matrices of the form $\mathbf{S}^{-1}\mathbf{X}\mathbf{S}$ $(\mathbf{X} \in \Gamma)$, then $\Gamma' \simeq \Gamma$.

THEOREM 9.4.2. *The following four statements are equivalent.*

(i) $\mathbf{A}$ *is a group matrix.*

(ii) $R(\mathbf{A}) = R(\mathbf{A}^2)$.

(iii) $\mathbf{A}$ *is similar to a matrix of the form*

$$\begin{pmatrix} \mathbf{A}_1 & \mathbf{O} \\ \mathbf{O} & \mathbf{O} \end{pmatrix}, \tag{9.4.1}$$

where $\mathbf{A}_1$ *is a non-singular square matrix.*

(iv) $R(\mathbf{A}) = n - m_0(\mathbf{A})$.†

† The reader is reminded that $m_0(\mathbf{A})$ denotes the number of zero characteristic roots of $\mathbf{A}$.

Write $R(\mathbf{A}) = r$. If $\mathbf{A}$ is similar to the matrix (9.4.1) and $\mathbf{A}_1$ is non-singular, then $\mathbf{A}_1$ is of type $r \times r$. The form (9.4.1) must, of course, be interpreted in an obvious sense when $r = 0$ or $r = n$; but in these cases the theorem is evidently valid, and we shall therefore assume that $0 < r < n$.

Let (i) be given, i.e. let $\mathbf{A}$ be an element of a matrix group Γ. Denoting by $\mathbf{B}$ the group inverse of $\mathbf{A}$ and by $\mathbf{E}$ the unit element of Γ, we have

$$\mathbf{E} = \mathbf{AB}, \quad \mathbf{A} = \mathbf{AE}.$$

Therefore, by Theorem 5.6.2 (p. 159),

$$R(\mathbf{E}) = R(\mathbf{AB}) \leqslant R(\mathbf{A}) = R(\mathbf{AE}) \leqslant R(\mathbf{E}),$$

and so $R(\mathbf{A}) = R(\mathbf{E})$. All matrices in Γ have therefore equal rank, and so $R(\mathbf{A}) = R(\mathbf{A}^2)$. Thus (i) implies (ii).

Next, let (iii) be given. The matrix (9.4.1) is evidently a group matrix, since it belongs to the group of all matrices of the form

$$\begin{pmatrix} \mathbf{X} & \mathbf{O} \\ \mathbf{O} & \mathbf{O} \end{pmatrix} \quad (|\mathbf{X}| \neq 0), \tag{9.4.2}$$

where $\mathbf{X}$ is of order r. Hence, by Theorem 9.4.1 (i), $\mathbf{A}$, too, is a group matrix; and so (iii) implies (i).

Again, if $\mathbf{A}$ satisfies (iii), then, by Theorem 7.2.1 (p. 199), $m_0(\mathbf{A}) = m_0(\mathbf{C})$, where $\mathbf{C}$ denotes the matrix (9.4.1). Hence

$$m_0(\mathbf{A}) = n - r = n - R(\mathbf{A}),$$

i.e. (iii) implies (iv).

If (iv) is given, then by Theorem 7.6.1 (p. 214) and the corollary to Theorem 7.3.3 (p. 202), we have

$$R(\mathbf{A}^2) \geqslant n - m_0(\mathbf{A}^2) = n - m_0(\mathbf{A}) = R(\mathbf{A}),$$

so that $R(\mathbf{A}) \leqslant R(\mathbf{A}^2)$. But, by Theorem 5.6.2, $R(\mathbf{A}) \geqslant R(\mathbf{A}^2)$. Hence $R(\mathbf{A}) = R(\mathbf{A}^2)$, and (iv) implies (ii).

To complete the proof of the theorem it is sufficient to show that (ii) implies (iii). We denote by $\mathfrak{U}$ the vector space of all vectors of the form $\mathbf{Ax}$ ($\mathbf{x} \in \mathfrak{V}_n$), and by $\mathfrak{U}'$ the vector space of all vectors $\mathbf{x} \in \mathfrak{V}_n$ which satisfy $\mathbf{Ax} = \mathbf{0}$. Then $d(\mathfrak{U}) = r$, $d(\mathfrak{U}') = n - r$; and moreover, since $R(\mathbf{A}) = R(\mathbf{A}^2)$, it follows by Theorem 5.6.4 (p. 161) (with $\mathbf{B} = \mathbf{A}$) that the only vector common to $\mathfrak{U}$ and $\mathfrak{U}'$ is $\mathbf{0}$. Hence, by Theorem 2.3.7 (p. 55), $\mathfrak{U}$ and $\mathfrak{U}'$ are complements. If, then, $\{\mathbf{x}_1,\ldots,\mathbf{x}_r\}$ is a basis of $\mathfrak{U}$ and $\{\mathbf{x}_{r+1},\ldots,\mathbf{x}_n\}$ a basis of $\mathfrak{U}'$, it follows by Theorem 2.3.8 that $\mathfrak{X} = \{\mathbf{x}_1,\ldots,\mathbf{x}_r, \mathbf{x}_{r+1},\ldots,\mathbf{x}_n\}$ is a basis of $\mathfrak{V}_n$.

Let the linear transformation L of $\mathfrak{B}_n$ into itself be defined by the equation $\mathbf{A} = \mathscr{R}(L; \mathfrak{E})$, where $\mathfrak{E}$ denotes the basis $\{\mathbf{e}_1,..., \mathbf{e}_n\}$ of $\mathfrak{B}_n$. Let the matrix $\mathbf{B}$ be defined by the equation $\mathbf{B} = \mathscr{R}(L; \mathfrak{X})$.† Now, by Theorem 4.2.6 (p. 121), $L(\mathbf{x}) = \mathbf{Ax}$, and hence each of $L(\mathbf{x}_1),..., L(\mathbf{x}_r)$ is a linear combination of $\mathbf{x}_1,..., \mathbf{x}_r$, while

$$L(\mathbf{x}_{r+1}) = ... = L(\mathbf{x}_n) = \mathbf{0}.$$

Moreover, by equation (4.2.13) (p. 121),

$$L(\mathbf{x}_j) = \sum_{i=1}^{n} b_{ij}\mathbf{x}_i \qquad (j = 1,..., n),$$

where $\mathbf{B} = (b_{ij})$. Hence $b_{ij} = 0$ for $i = r+1,..., n$, $j = 1,..., r$, and for $i = 1,..., n$; $j = r+1,..., n$. Thus $\mathbf{B}$ has the form (9.4.1), and the theorem is proved since $\mathbf{A}$ and $\mathbf{B}$ are similar.

THEOREM 9.4.3. *The following three statements are equivalent.*

(i) $\mathbf{A}$ *is a group unit matrix.*

(ii) $\mathbf{A}^2 = \mathbf{A}$.

(iii) $\mathbf{A}$ *is similar to a matrix of the form*

$$\begin{pmatrix} \mathbf{I} & \mathbf{O} \\ \mathbf{O} & \mathbf{O} \end{pmatrix}. \tag{9.4.3}$$

Here again the form (9.4.3) must be interpreted suitably for $r = 0$ and $r = n$, where $r = R(\mathbf{A})$. In these two cases the theorem is true trivially and we shall therefore assume that $0 < r < n$.

If (i) is given, then (ii) follows at once. Again, if $\mathbf{A}$ satisfies (ii), then, by Theorem 9.4.2, it is a group matrix and so is similar to a matrix of the form (9.4.1), where $\mathbf{A}_1$ is non-singular and of order r. In view of (ii) we have $\mathbf{A}_1^2 = \mathbf{A}_1$, and therefore $\mathbf{A}_1 = \mathbf{I}_r$. Thus (ii) implies (iii).

Finally, let (iii) be given. Now the matrix (9.4.3) is obviously the unit element in the group of matrices of type (9.4.2). Hence, by Theorem 9.4.1 (ii), $\mathbf{A}$ is a group unit matrix. Thus (iii) implies (i), and the proof is complete.

Theorem 9.4.4. *For $r > 0$, every group of matrices of rank r*‡ *is isomorphic to a group of non-singular $r \times r$ matrices.*

† In point of fact we know, by Theorem 4.2.8 (p. 121), that $\mathbf{B} = \mathbf{X}^{-1}\mathbf{AX}$, where $\mathbf{X}_{*i} = \mathbf{x}_i$ $(i = 1..., n)$.

‡ The fact that all matrices of a matrix group have equal rank was established in the proof of Theorem 9.4.2.

This theorem shows that all results on the structure of matrix groups can be obtained from the study of non-singular matrices alone.

Let $\mathbf{E}$ be the unit element of a matrix group Γ. Then, by Theorem 9.4.3, there exists a (non-singular) matrix $\mathbf{S}$ such that

$$\mathbf{S}^{-1}\mathbf{E}\mathbf{S} = \begin{pmatrix} \mathbf{I}_r & \mathbf{O} \\ \mathbf{O} & \mathbf{O} \end{pmatrix}.$$

For any $\mathbf{A} \in \Gamma$ we write

$$\mathbf{S}^{-1}\mathbf{A}\mathbf{S} = \begin{pmatrix} \mathbf{A}_1 & \mathbf{A}_2 \\ \mathbf{A}_3 & \mathbf{A}_4 \end{pmatrix},$$

where $\mathbf{A}_1$ is of type $r \times r$. Then

$$\mathbf{S}^{-1}\mathbf{E}\mathbf{A}\mathbf{S} = \mathbf{S}^{-1}\mathbf{E}\mathbf{S}.\mathbf{S}^{-1}\mathbf{A}\mathbf{S} = \begin{pmatrix} \mathbf{A}_1 & \mathbf{A}_2 \\ \mathbf{O} & \mathbf{O} \end{pmatrix},$$

$$\mathbf{S}^{-1}\mathbf{A}\mathbf{E}\mathbf{S} = \mathbf{S}^{-1}\mathbf{A}\mathbf{S}.\mathbf{S}^{-1}\mathbf{E}\mathbf{S} = \begin{pmatrix} \mathbf{A}_1 & \mathbf{O} \\ \mathbf{A}_3 & \mathbf{O} \end{pmatrix}.$$

But $$\mathbf{S}^{-1}\mathbf{A}\mathbf{S} = \mathbf{S}^{-1}\mathbf{A}\mathbf{E}\mathbf{S} = \mathbf{S}^{-1}\mathbf{E}\mathbf{A}\mathbf{S},$$

and so $$\mathbf{A}_2 = \mathbf{O}, \quad \mathbf{A}_3 = \mathbf{O}, \quad \mathbf{A}_4 = \mathbf{O}.$$

Thus $$\mathbf{S}^{-1}\mathbf{A}\mathbf{S} = \begin{pmatrix} \mathbf{A}_1 & \mathbf{O} \\ \mathbf{O} & \mathbf{O} \end{pmatrix},$$

and here the $r \times r$ matrix $\mathbf{A}_1$ must be non-singular since $R(\mathbf{A}) = R(\mathbf{A}_1)$. Now $\mathbf{S}$ is independent of $\mathbf{A}$ and only depends on Γ. Hence, by Exercise 9.4.2, Γ is isomorphic to a group of matrices having the form (9.4.1) and this, in turn, is obviously isomorphic to a group of non-singular $r \times r$ matrices. Since isomorphism is an equivalence relation the theorem is proved.

EXERCISE 9.4.3. Deduce Theorem 9.4.4 without making use of Theorem 9.4.3.

9.5. Invariant spaces and groups of linear transformations

In developing the theory of linear transformations we are bound to inquire into the relations that exist between vector spaces and their image spaces under such transformations. We propose in the present section to study this problem by invariant methods; that is, by making use only of the intrinsic properties of linear transformations and not of their matrix representations. The most important new concept we shall introduce is that of an invariant space. In the course of the discussion we shall also touch on the problems considered in § 9.4, but now regard them from the point of view of invariant properties.

When we speak below about a transformation (or linear transformation) it is to be understood that we mean a linear transformation of $\mathfrak{B}_n$ into itself. All vectors spaces we shall consider are subspaces of $\mathfrak{B}_n$. If $\mathbf{x} \in \mathfrak{B}_n$ and L is a transformation we shall, for the sake of simplicity, write $L\mathbf{x}$ in place of $L(\mathbf{x})$. If $\mathfrak{U}$ is a vector space we shall denote by $L\mathfrak{U}$ the *image space* of $\mathfrak{U}$ under L, i.e. the vector space of all vectors of the form $L\mathbf{x}$, where $\mathbf{x} \in \mathfrak{U}$. A transformation will be called singular or non-singular according as its representing matrices are singular or non-singular.

Although the theory presented below is stated in terms of vector spaces, it might equally well have been stated in terms of linear manifolds. Indeed, since components of vectors are never mentioned, the difference between statements involving vector spaces and those involving linear manifolds is not one of substance but merely of language and notation.

9.5.1. DEFINITION 9.5.1. *The rank $R(L)$ of the transformation L is defined by the equation $R(L) = d(L\mathfrak{B}_n)$.*

Let $\mathbf{A} = \mathscr{R}(L; \mathfrak{E})$. Then $L\mathbf{x} = \mathbf{A}\mathbf{x}$, and hence $L\mathfrak{B}_n$ consists of all vectors of the form $\mathbf{A}\mathbf{x}$, where $\mathbf{x} \in \mathfrak{B}_n$. Hence $d(L\mathfrak{B}_n) = R(\mathbf{A})$, and thus the rank of a transformation is equal to the rank of each of its representing matrices.

DEFINITION 9.5.2. *If L is a transformation and $\mathfrak{U}$ a vector space, and if $L\mathfrak{U} = \mathfrak{U}$, then $\mathfrak{U}$ is said to be an* INVARIANT SPACE *of L.*†

Thus a vector space $\mathfrak{U}$ is said to be an invariant space of L if its image space, under L, coincides with the original space. Every transformation possesses at least one invariant space, namely, that which consists of the zero vector only.

DEFINITION 9.5.3. *If $\mathfrak{U}$ is an invariant space of a transformation L, and if every invariant space of L is a subspace of $\mathfrak{U}$, then $\mathfrak{U}$ is called the* MAXIMAL INVARIANT SPACE *of L.*

The uniqueness of this space (if it exists) is inherent in the definition. If the maximal invariant space of L exists, it will be denoted by $\mathfrak{S}(L)$. We shall see presently that it exists, in fact, for every L.

† In the literature the term 'invariant space' is normally used in a slightly different sense. Cf. MacDuffee, **12**, 116, or Julia, **34**, 69.

THEOREM 9.5.1. *Let L be a linear transformation and suppose that, for some non-negative integer k,*†

$$L^k\mathfrak{B}_n = L^{k+1}\mathfrak{B}_n.$$

Then $L^k\mathfrak{B}_n$ is the maximal invariant space of L.

Write $\mathfrak{U} = L^k\mathfrak{B}_n$. In view of our hypothesis we have $L\mathfrak{U} = \mathfrak{U}$, and so $\mathfrak{U}$ is an invariant space of L. Moreover, if $\mathfrak{B}$ is an invariant space of L, i.e. if $L\mathfrak{B} = \mathfrak{B}$, then $L^k\mathfrak{B} = \mathfrak{B}$. Hence, since $\mathfrak{B} \subset \mathfrak{B}_n$,

$$\mathfrak{B} = L^k\mathfrak{B} \subset L^k\mathfrak{B}_n = \mathfrak{U}.$$

Thus $\mathfrak{B} \subset \mathfrak{U}$, and therefore $\mathfrak{S}(L) = \mathfrak{U}$.

THEOREM 9.5.2. *Every linear transformation L possesses a maximal invariant space. Moreover, there exists a non-negative integer $k_0 = k_0(L)$ such that, for all $k \geqslant k_0$,*

$$\mathfrak{S}(L) = L^k\mathfrak{B}_n.$$

We have
$$\mathfrak{B}_n \supset L\mathfrak{B}_n \supset L^2\mathfrak{B}_n \supset \ldots.$$

Suppose that, for all $k \geqslant 0$, $L^k\mathfrak{B}_n \neq L^{k+1}\mathfrak{B}_n$. Then, in view of Theorem 2.3.4 (p. 53),

$$n = d(\mathfrak{B}_n) > d(L\mathfrak{B}_n) > d(L^2\mathfrak{B}_n) > \ldots.$$

Since each of the terms involved is a non-negative integer, this infinite chain of strict inequalities cannot be satisfied, and we must therefore have, for some k_0,

$$L^{k_0}\mathfrak{B}_n = L^{k_0+1}\mathfrak{B}_n. \tag{9.5.1}$$

Hence, by Theorem 9.5.1, L possesses a maximal invariant space, which is given by $\mathfrak{S}(L) = L^{k_0}\mathfrak{B}_n$. Moreover, in view of (9.5.1), we clearly have $L^{k_0}\mathfrak{B}_n = L^k\mathfrak{B}_n$ for all $k \geqslant k_0$. The proof of the theorem is therefore complete.

EXERCISE 9.5.1. Show that $\mathfrak{S}(L) = L^n\mathfrak{B}_n$.

EXERCISE 9.5.2. Show that, if $L^h\mathfrak{B}_n$ is an invariant space of L, then it is the maximal invariant space of L.

EXERCISE 9.5.3. Show that a linear transformation has $\mathfrak{B}_n$ as its maximal invariant space if and only if it is non-singular.

The argument above shows that if we consider first the image space of $\mathfrak{B}_n$ under L, then the image space of the image space, and so on, we obtain a series of progressively contracting spaces, each contained in the preceding one. After a certain number of steps this process terminates and the renewed application of L does not

† L^0 is defined as the identical transformation.

result in any further contraction. The space reached at this stage is the maximal invariant space of L.

We may also note that, if k_0 denotes the least non-negative integer satisfying (9.5.1), then†

$$n > d(L\mathfrak{V}_n) > ... > d(L^{k_0}\mathfrak{V}_n) = d(L^{k_0+1}\mathfrak{V}_n) = d(L^{k_0+2}\mathfrak{V}_n) = ... \tag{9.5.2}$$

Now, let $\mathbf{A}$ be any matrix and let L be the transformation represented by $\mathbf{A}$ with respect to any given basis in $\mathfrak{V}_n$. Then $\mathbf{A}^k$ represents L^k, and in view of (9.5.2) we have

$$n > R(\mathbf{A}) > ... > R(\mathbf{A}^{k_0}) = R(\mathbf{A}^{k_0+1}) = R(\mathbf{A}^{k_0+2}) = ...,$$

where the case $k_0 = 0$ (characterizing non-singular matrices) must again be interpreted in the obvious way. It follows by Theorem 9.4.2 that $\mathbf{A}$ is a group matrix if and only if $k_0 = 0$ or 1. Moreover, if $\mathbf{A}$ is any matrix and k is a sufficiently large integer (for example, $k \geqslant n$), then $\mathbf{A}^k$ is a group matrix, since $R(\mathbf{A}^k) = R(\mathbf{A}^{2k})$.

THEOREM 9.5.3. *Every linear transformation effects an automorphism of each of its invariant spaces.*

Let $\mathfrak{U}$ be an invariant space of the transformation L, so that $L\mathfrak{U} = \mathfrak{U}$. To prove the theorem it is sufficient to show that the correspondence $\mathbf{x} \rightarrow L\mathbf{x}$ of $\mathfrak{U}$ with itself is biunique. Now, if $\mathbf{y} \in \mathfrak{U}$, then (since $L\mathfrak{U} = \mathfrak{U}$) there exists a vector $\mathbf{x} \in \mathfrak{U}$ such that $L\mathbf{x} = \mathbf{y}$. It remains, therefore, to show that if $\mathbf{x}, \mathbf{x}' \in \mathfrak{U}$, $\mathbf{x} \neq \mathbf{x}'$, then $L\mathbf{x} \neq L\mathbf{x}'$. This means, in fact, that if $\mathbf{z} \in \mathfrak{U}$ and $L\mathbf{z} = \mathbf{0}$, then $\mathbf{z} = \mathbf{0}$. To prove this, write $d(\mathfrak{U}) = r$ and let $\{\mathbf{z}_1,..., \mathbf{z}_r\}$ be a basis of $\mathfrak{U}$. The vectors $L\mathbf{z}_1,..., L\mathbf{z}_r$ then obviously span $L\mathfrak{U} = \mathfrak{U}$. Hence these vectors constitute a basis of $\mathfrak{U}$ and so are linearly independent. Now, if

$$\mathbf{z} = \alpha_1 \mathbf{z}_1 + ... + \alpha_r \mathbf{z}_r,$$

then
$$\mathbf{0} = L\mathbf{z} = \alpha_1 . L\mathbf{z}_1 + ... + \alpha_r . L\mathbf{z}_r.$$

Hence $\alpha_1 = ... = \alpha_r = 0$ and so $\mathbf{z} = \mathbf{0}$. The required conclusion is therefore established.

EXERCISE 9.5.4. Show that if L effects an automorphism of $\mathfrak{U}$, then $\mathfrak{U}$ is an invariant space of L.

COROLLARY. *A linear transformation effects an automorphism of its maximal invariant space.*

† For $k_0 = 0$ these relations must be interpreted as

$$n = d(L\mathfrak{V}_n) = d(L^2\mathfrak{V}_n) = ...$$

DEFINITION 9.5.4. *The transformation L* ANNIHILATES *the vector space $\mathfrak{U}$ if $L\mathbf{x} = \mathbf{0}$ for all $\mathbf{x} \in \mathfrak{U}$.*

THEOREM 9.5.4. *If $\mathfrak{U} = L^k\mathfrak{B}_n$ is the maximal invariant space of the linear transformation L, then L effects an automorphism of $\mathfrak{U}$, and L^k annihilates one and only one complement of $\mathfrak{U}$.*

The first part of the assertion holds by the corollary to Theorem 9.5.3. To prove the second, write $d(\mathfrak{U}) = r$, and let $\{\mathbf{x}_1,\ldots,\mathbf{x}_r\}$ be a basis of $\mathfrak{U}$. Take $\mathbf{x}_{r+1},\ldots,\mathbf{x}_n$ such that the vectors

$$\mathbf{x}_1,\ldots,\mathbf{x}_r, \quad \mathbf{x}_{r+1},\ldots,\mathbf{x}_n \tag{9.5.3}$$

constitute a basis of $\mathfrak{B}_n$, and put

$$\mathbf{y}_i = L^k\mathbf{x}_i \quad (i = r+1,\ldots,n).$$

Then
$$\mathbf{y}_i \in \mathfrak{U} \qquad (i = r+1,\ldots,n).$$

Now $L^k\mathfrak{B}_n$ is, by hypothesis, an invariant space of L. Hence it follows that $L^{2k}\mathfrak{B}_n = L^k\mathfrak{B}_n$, i.e. $L^k\mathfrak{U} = \mathfrak{U}$. Thus $\mathfrak{U}$ is an invariant space of L^k and so, by Theorem 9.5.3, L^k effects an automorphism of $\mathfrak{U}$. Hence there exist vectors $\mathbf{z}_{r+1},\ldots,\mathbf{z}_n$ such that

$$L^k\mathbf{z}_i = \mathbf{y}_i, \quad \mathbf{z}_i \in \mathfrak{U} \quad (i = r+1,\ldots,n).$$

Put
$$\mathbf{x}'_i = \mathbf{x}_i - \mathbf{z}_i \quad (i = r+1,\ldots,n)$$

and consider the vectors

$$\mathbf{x}_1,\ldots,\mathbf{x}_r, \quad \mathbf{x}'_{r+1},\ldots,\mathbf{x}'_n. \tag{9.5.4}$$

Since $\mathbf{z}_{r+1},\ldots,\mathbf{z}_n \in \mathfrak{U}$, it follows that every vector in (9.5.3) is a linear combination of vectors in (9.5.4). Hence the vectors in (9.5.4) constitute a basis of $\mathfrak{B}_n$. Therefore, in view of Theorem 2.3.8 (p. 56), the vector space $\mathfrak{U}'$ spanned by $\mathbf{x}'_{r+1},\ldots,\mathbf{x}'_n$ is a complement of $\mathfrak{U}$. Moreover, for $i = r+1,\ldots,n$,

$$L^k\mathbf{x}'_i = L^k\mathbf{x}_i - L^k\mathbf{z}_i = \mathbf{y}_i - \mathbf{y}_i = \mathbf{0}.$$

Hence L^k annihilates $\mathfrak{U}'$. To show that L^k annihilates no other complement of $\mathfrak{U}$, let $\mathbf{x} \in \mathfrak{B}_n$ and write

$$\mathbf{x} = \alpha_1\mathbf{x}_1 + \ldots + \alpha_r\mathbf{x}_r + \alpha_{r+1}\mathbf{x}'_{r+1} + \ldots + \alpha_n\mathbf{x}'_n.$$

Then
$$L^k\mathbf{x} = \alpha_1 L^k\mathbf{x}_1 + \ldots + \alpha_r L^k\mathbf{x}_r.$$

But L^k is an automorphism of $\mathfrak{U}$, and so $L^k\mathbf{x}_1,\ldots, L^k\mathbf{x}_r$ are linearly independent. Hence $L^k\mathbf{x} = \mathbf{0}$ implies $\alpha_1 = \ldots = \alpha_r = 0$, i.e. $\mathbf{x} \in \mathfrak{U}'$. This completes the proof.

It may be noted at this stage that Theorem 9.4.2 follows easily from the results of the present section. We recall that the critical step in the proof of Theorem 9.4.2 consists in showing that, if $R(\mathbf{A}) = R(\mathbf{A}^2)$, then $\mathbf{A}$ is similar to a matrix of type (9.4.1). This inference can be made as follows. Let L be defined by the equation $\mathbf{A} = \mathscr{R}(L; \mathfrak{E})$. We then have $R(L) = R(L^2)$ and therefore $L\mathfrak{V}_n = L^2\mathfrak{V}_n$, so that (by Theorem 9.5.1), $\mathfrak{U} = L\mathfrak{V}_n$ is the maximal invariant space of L. Hence, by Theorem 9.5.4, L effects an automorphism of $\mathfrak{U}$ and annihilates a certain complement $\mathfrak{U}'$ of $\mathfrak{U}$. Let $\{\mathbf{x}_1,\ldots,\mathbf{x}_r\}$ be a basis of $\mathfrak{U}$ and $\{\mathbf{x}'_{r+1},\ldots,\mathbf{x}'_n\}$ a basis of $\mathfrak{U}'$. Then $\mathfrak{X} = \{\mathbf{x}_1,\ldots,\mathbf{x}_r, \mathbf{x}'_{r+1},\ldots,\mathbf{x}'_n\}$ is a basis of $\mathfrak{V}_n$; and the matrix representing L with respect to $\mathfrak{X}$ is of type (9.4.1) and is also similar to $\mathbf{A}$.

The next result may be regarded as a converse of Theorem 9.5.4.

Theorem 9.5.5. *Let $\mathfrak{U}$ be a vector space, $\mathfrak{U}'$ a complement of $\mathfrak{U}$, L a linear transformation, and k a non-negative integer. If L effects an automorphism of $\mathfrak{U}$ and L^k annihilates $\mathfrak{U}'$, then $\mathfrak{U}$ is the maximal invariant space of L, and $\mathfrak{U} = L^k\mathfrak{V}_n$.*

Let $\mathbf{x} \in \mathfrak{V}_n$ and write $\mathbf{x} = \mathbf{y}+\mathbf{y}'$, where $\mathbf{y} \in \mathfrak{U}$, $\mathbf{y}' \in \mathfrak{U}'$. Then

$$L^k\mathbf{x} = L^k\mathbf{y}+L^k\mathbf{y}' = L^k\mathbf{y}.$$

Thus $L^k\mathfrak{V}_n \subset L^k\mathfrak{U}$; and since trivially $L^k\mathfrak{U} \subset L^k\mathfrak{V}_n$, we have

$$L^k\mathfrak{V}_n = L^k\mathfrak{U}.$$

But, since L effects an automorphism of $\mathfrak{U}$, we know (by Exercise 9.5.4) that $\mathfrak{U}$ is an invariant space of L, i.e. $L\mathfrak{U} = \mathfrak{U}$. Hence $L^k\mathfrak{U} = \mathfrak{U}$ and therefore

$$\mathfrak{U} = L^k\mathfrak{V}_n.$$

In exactly the same way we obtain $\mathfrak{U} = L^{k+1}\mathfrak{V}_n$ and therefore, by Theorem 9.5.1, $\mathfrak{S}(L) = \mathfrak{U} = L^k\mathfrak{V}_n$.

We know by Theorem 9.5.2 that each transformation possesses a unique maximal invariant space. On the other hand, if a vector space $\mathfrak{U}$ is given, there exist infinitely many transformations having $\mathfrak{U}$ as their maximal invariant space. The procedure for constructing all such transformations is as follows. We take an arbitrary complement $\mathfrak{U}'$ of $\mathfrak{U}$ and an arbitrary non-negative integer k, and consider all transformations L having the property that L effects an automorphism of $\mathfrak{U}$ while L^k annihilates $\mathfrak{U}'$. Then, by Theorem 9.5.5, for each such transformation L we have $\mathfrak{S}(L) = \mathfrak{U}$. The set of all possible transformations of the type described (with $\mathfrak{U}'$ and k varying) is, by Theorem 9.5.4, identical with the set of all transformations having $\mathfrak{U}$ as maximal invariant space.

9.5.2. So far we have been concerned with relations between transformations and their invariant spaces. A good deal of further information can be obtained by considering not individual transformations but groups of transformations.

THEOREM 9.5.6. *All elements of a group of linear transformations have the same maximal invariant space.*

Let Γ be a group of transformations, and let $L, M \in \Gamma$. We know, by Theorem 9.5.2, that there exists a positive integer k such that

$$\mathfrak{S}(M) = M^k\mathfrak{V}_n.$$

Hence

$$\mathfrak{S}(M) \subset M\mathfrak{V}_n.$$

Now there exists a transformation $N \in \Gamma$ such that $L = M^kN$. Hence

$$\mathfrak{S}(L) \subset L\mathfrak{V}_n = M^kN\mathfrak{V}_n \subset M^k\mathfrak{V}_n = \mathfrak{S}(M). \tag{9.5.5}$$

Thus $\mathfrak{S}(L) \subset \mathfrak{S}(M)$, and by symmetry $\mathfrak{S}(M) \subset \mathfrak{S}(L)$. Therefore

$$\mathfrak{S}(L) = \mathfrak{S}(M) \qquad (L, M \in \Gamma), \tag{9.5.6}$$

and the theorem is established.

In view of the result just proved, we may speak of the *maximal invariant space of a group* Γ *of transformations*, and we shall denote this space by $\mathfrak{S}(\Gamma)$.

THEOREM 9.5.7. *If* Γ *is a group of linear transformations and* $L \in \Gamma$, *then*

$$\mathfrak{S}(\Gamma) = L\mathfrak{V}_n.$$

This result is an immediate consequence of (9.5.5) and (9.5.6).

THEOREM 9.5.8. *The unit element of a group* Γ *of linear transformations effects the identical automorphism of* $\mathfrak{S}(\Gamma)$.

Denote by E the unit element of Γ and let $\mathbf{x} \in \mathfrak{S}(\Gamma)$. Since, by Theorem 9.5.3, E effects an automorphism of $\mathfrak{S}(\Gamma)$, there exists a vector $\mathbf{y} \in \mathfrak{S}(\Gamma)$ such that $E\mathbf{y} = \mathbf{x}$. Now $E^2 = E$, and so

$$\mathbf{x} = E\mathbf{y} = E^2\mathbf{y} = E(E\mathbf{y}) = E\mathbf{x}.$$

The theorem is therefore proved.

THEOREM 9.5.9. *Let* $\mathfrak{U}$ *be a vector space and* $\mathfrak{U}'$ *a complement of* $\mathfrak{U}$. *The set* Γ *of all linear transformations which effect automorphisms of* $\mathfrak{U}$ *and annihilate* $\mathfrak{U}'$ *is a group with* $\mathfrak{S}(\Gamma) = \mathfrak{U}$.

We know, by Theorem 9.5.5, that all transformations of the given type have $\mathfrak{U}$ as their maximal invariant space. Hence we need only to show that Γ is a group.

We first verify that Γ is closed under multiplication of transformations. Let $L, M \in \Gamma$. Then LM obviously annihilates $\mathfrak{U}'$. Moreover, since L and M effect automorphisms of $\mathfrak{U}$, it follows that $\mathfrak{U}$ is an invariant space of both these transformations. Then $LM\mathfrak{U} = L\mathfrak{U} = \mathfrak{U}$, and so $\mathfrak{U}$ is an invariant space of LM. Hence, by Theorem 9.5.3, LM effects an automorphism of $\mathfrak{U}$, and so $LM \in \Gamma$.

Next, let E be the transformation in Γ which effects the identical automorphism of $\mathfrak{U}$.† If $\mathbf{x} \in \mathfrak{V}_n$, we write $\mathbf{x} = \mathbf{y}+\mathbf{y}'$, where $\mathbf{y} \in \mathfrak{U}$, $\mathbf{y}' \in \mathfrak{U}'$. Then

$$E\mathbf{x} = E\mathbf{y}+E\mathbf{y}' = \mathbf{y};$$

and if $L \in \Gamma$ we have, similarly, $L\mathbf{x} = L\mathbf{y}$. But, by Theorem 9.5.5, $\mathfrak{U} = L\mathfrak{V}_n$, and so $L\mathbf{x} \in \mathfrak{U}$ $(\mathbf{x} \in \mathfrak{V}_n)$. Therefore

$$EL\mathbf{x} = L\mathbf{x} = L\mathbf{y} = LE\mathbf{x}.$$

Thus

$$EL = LE = L \qquad (L \in \Gamma). \tag{9.5.7}$$

Again, denote by L^* the automorphism which L effects in $\mathfrak{U}$. The set of all automorphisms L^*, where $L \in \Gamma$, is, in fact, the set of *all* automorphisms of $\mathfrak{U}$; and by Theorem 9.3.1 (p. 268) this set is a group. The unit element of this group is evidently E^*, and with each element L^* there is associated its inverse element $(L^*)^{-1}$. We define the transformation $L^{-1} \in \Gamma$ by the requirements

$$L^{-1}\mathbf{x} = \begin{cases} (L^*)^{-1}\mathbf{x} & (\mathbf{x} \in \mathfrak{U}), \\ \mathbf{0} & (\mathbf{x} \in \mathfrak{U}'). \end{cases}$$

If $\mathbf{x} \in \mathfrak{U}$, then $L\mathbf{x} \in \mathfrak{U}$ and we have

$$L^{-1}L\mathbf{x} = (L^*)^{-1}L^*\mathbf{x} = \mathbf{x} = E\mathbf{x},$$
$$LL^{-1}\mathbf{x} = L^*(L^*)^{-1}\mathbf{x} = \mathbf{x} = E\mathbf{x}.$$

Also, if $\mathbf{x} \in \mathfrak{U}'$, then $L\mathbf{x} = \mathbf{0} = L^{-1}\mathbf{x}$, and so

$$L^{-1}L\mathbf{x} = \mathbf{0} = E\mathbf{x}, \qquad LL^{-1}\mathbf{x} = \mathbf{0} = E\mathbf{x}.$$

Thus

$$L^{-1}L = LL^{-1} = E \quad (L \in \Gamma). \tag{9.5.8}$$

Equations (9.5.7) and (9.5.8) complete the argument showing that Γ is a group.

† It should be borne in mind that a linear transformation is determined completely if it is defined on two complements.

We shall denote the group of the theorem just proved by $\Gamma(\mathfrak{U}; \mathfrak{U}')$.

Next, consider any group Γ of linear transformations. If $L \in \Gamma$, then, by Theorem 9.5.7, $\mathfrak{S}(\Gamma) = L\mathfrak{V}_n$. Hence, by Theorem 9.5.4, L annihilates a certain definite complement of $\mathfrak{S}(\Gamma)$. We therefore wish to inquire whether different elements of Γ annihilate different complements of $\mathfrak{S}(\Gamma)$. It is easy to see that this is not the case.

THEOREM 9.5.10. *All elements of a group Γ of linear transformations annihilate the same complement of $\mathfrak{S}(\Gamma)$.*

Let E, the unit element of Γ, annihilate a complement $\mathfrak{U}'$ of $\mathfrak{S}(\Gamma)$. If $L \in \Gamma$ and $\mathbf{x} \in \mathfrak{U}'$, then

$$L\mathbf{x} = LE\mathbf{x} = L\mathbf{0} = \mathbf{0}.$$

Thus L annihilates $\mathfrak{U}'$ and, in view of Theorem 9.5.4, it annihilates no other complement of $\mathfrak{S}(\Gamma)$.

COROLLARY. *Every group of linear transformations having the vector space $\mathfrak{U}$ as its maximal invariant space is a subgroup of one of the groups $\Gamma(\mathfrak{U}; \mathfrak{U}')$.*

We are now in a position to describe the construction of all groups having a preassigned maximal invariant space.

THEOREM 9.5.11. *Let $\mathfrak{U}$ be a vector space. The set of all groups $\Gamma(\mathfrak{U}; \mathfrak{U}')$—where $\mathfrak{U}'$ is a variable complement of $\mathfrak{U}$—together with all their subgroups is identical with the set $S(\mathfrak{U})$ of all groups of linear transformations having $\mathfrak{U}$ as their maximal invariant space.*

If $\mathfrak{U}'$ is any complement of $\mathfrak{U}$, then we know by Theorem 9.5.9 that $\Gamma(\mathfrak{U}; \mathfrak{U}')$ and all its subgroups are elements of $S(\mathfrak{U})$. Conversely, if $G \in S(\mathfrak{U})$, then, by the corollary just proved, G is a subgroup of one of the groups $\Gamma(\mathfrak{U}; \mathfrak{U}')$.

THEOREM 9.5.12. *Any group Γ of linear transformations is isomorphic to a group of automorphisms of $\mathfrak{S}(\Gamma)$.*

By the corollary to Theorem 9.5.3 every $L \in \Gamma$ effects an automorphism, say L^*, of $\mathfrak{S}(\Gamma)$. The set Γ^* of all automorphisms L^*, where $L \in \Gamma$, is a group. For if E is the unit element of Γ, then E^* is the unit element of Γ^*; and the element $(L^*)^{-1}$, defined by the equation

$$(L^*)^{-1}\mathbf{x} = L^{-1}\mathbf{x} \quad (\mathbf{x} \in \mathfrak{S}(\Gamma)),$$

is the inverse of L^*. Since, then, Γ^* is a group and since obviously

$$(LM)^* = L^*M^* \quad (L, M \in \Gamma),$$

it only remains to show that the correspondence $L \leftrightarrow L^*$ between Γ and Γ^* is biunique. This will be done by showing that, if $L^* = M^*$, then $L = M$. Now $L^* = M^*$ means that

$$L^*\mathbf{x} = M^*\mathbf{x} \quad (\mathbf{x} \in \mathfrak{S}(\Gamma)),$$

i.e.

$$L\mathbf{x} = M\mathbf{x} \quad (\mathbf{x} \in \mathfrak{S}(\Gamma)).$$

Hence, for $\mathbf{x} \in \mathfrak{S}(\Gamma)$,

$$L^{-1}M\mathbf{x} = L^{-1}L\mathbf{x} = E\mathbf{x},$$

and so, by Theorem 9.5.8,

$$L^{-1}M\mathbf{x} = \mathbf{x} \quad (\mathbf{x} \in \mathfrak{S}(\Gamma)). \tag{9.5.9}$$

Now, if $\mathbf{x} \in \mathfrak{V}_n$, then $E\mathbf{x} \in E\mathfrak{V}_n$, and so, by Theorem 9.5.7, $E\mathbf{x} \in \mathfrak{S}(\Gamma)$. Hence, by (9.5.9), $L^{-1}ME\mathbf{x} = E\mathbf{x}$, and so $L\mathbf{x} = M\mathbf{x}$ $(\mathbf{x} \in \mathfrak{V}_n)$. Thus $L = M$, and the proof is complete.

If in the preceding argument we put $\Gamma = \Gamma(\mathfrak{U};\, \mathfrak{U}')$, then Γ^* becomes the group of all automorphisms of $\mathfrak{S}(\Gamma)$. In view of Theorem 9.3.1 (p. 268) we therefore have the following result.

Corollary 1. *If $\mathfrak{U}$ is a vector space of dimensionality r, and $\mathfrak{U}'$ is any complement of $\mathfrak{U}$, then $\Gamma(\mathfrak{U};\, \mathfrak{U}')$ is isomorphic to $GL(r)$.*

As an immediate inference we obtain:

Corollary 2. *If $\mathfrak{U}$ is a vector space and $\mathfrak{U}'$, $\mathfrak{U}''$ are any two complements of $\mathfrak{U}$, then $\Gamma(\mathfrak{U};\, \mathfrak{U}') \simeq \Gamma(\mathfrak{U};\, \mathfrak{U}'')$.*

From Theorem 9.5.12 we may easily deduce Theorem 9.4.4. Let Γ be a group of matrices of rank r, and denote by Γ' the set of transformations represented by these matrices with respect to any fixed basis in $\mathfrak{V}_n$. Then, by Exercise 9.3.2 (p. 270), Γ' is a group isomorphic to Γ. Moreover, if $L \in \Gamma'$ and L is represented by $\mathbf{A} \in \Gamma$, then, by Theorem 9.5.7, $\mathfrak{S}(\Gamma') = L\mathfrak{V}_n$, and therefore

$$d\{\mathfrak{S}(\Gamma')\} = d(L\mathfrak{V}_n) = R(\mathbf{A}) = r.$$

Hence, in view of Theorem 9.5.12, Γ' is isomorphic to a group Γ^* of automorphisms of the r-dimensional vector space $\mathfrak{S}(\Gamma')$. But, by Exercise 9.3.1, Γ^* is isomorphic to a group Γ'' of non-singular $r \times r$ matrices. Thus $\Gamma \simeq \Gamma''$, and Theorem 9.4.4 is established once again.

It is easy to see that not every transformation is an element in a group of transformations. By making use of the results of § 9.4 it would, of course, be easy to obtain criteria for deciding whether a given transformation is an element (or possibly the unit element) of some group of transformations. However, we prefer to deduce these criteria without appealing to the theory of matrices.

THEOREM 9.5.13. *The following statements are equivalent to each other.*

(i) *L is an element in a group of linear transformations.*

(ii) $R(L) = R(L^2)$.

(iii) $\mathfrak{S}(L) = L\mathfrak{V}_n$.

If (i) is given, let L be an element of a group Γ. Then, by Theorem 9.5.7, $\mathfrak{S}(L) = L\mathfrak{V}_n$ and so $L\mathfrak{V}_n = L^2\mathfrak{V}_n$. Hence $R(L) = R(L^2)$, and so (i) implies (ii).

Next, let (ii) be given, i.e $d(L\mathfrak{V}_n) = d(L^2\mathfrak{V}_n)$. Then $L\mathfrak{V}_n = L^2\mathfrak{V}_n$ and therefore, by Theorem 9.5.1, $\mathfrak{S}(L) = L\mathfrak{V}_n$. Thus (ii) implies (iii).

Finally, let (iii) be given and write $\mathfrak{U} = L\mathfrak{V}_n$. Then, by Theorem 9.5.4, L effects an automorphism of $\mathfrak{U}$ and annihilates a certain complement $\mathfrak{U}'$ of $\mathfrak{U}$. Therefore L is an element of the group $\Gamma(\mathfrak{U}; \mathfrak{U}')$. The condition (iii) therefore implies (i), and the theorem is proved.

EXERCISE 9.5.5. Deduce from the results just proved the equivalence of the conditions (i) and (ii) in Theorem 9.4.2 (p. 273).

THEOREM 9.5.14. *The linear transformation L is the unit element of a group of linear transformations if and only if $L^2 = L$.*

If L is the unit element of a group, then obviously $L^2 = L$. If, on the other hand, $L^2 = L$, then, by Theorem 9.5.13, L is an element in a group Γ of linear transformations. Hence it possesses an inverse element L^{-1}, and so $L^{-1}L^2 = L^{-1}L$, i.e. $L = L^{-1}L$. Hence L is the unit element of Γ.

EXERCISE 9.5.6. Deduce from the result just proved the equivalence of the conditions (i) and (ii) of Theorem 9.4.3 (p. 275).

PROBLEMS ON CHAPTER IX

1. Show that the group of translations of a plane is isomorphic to the additive group of complex numbers.

2. Show that the additive group of residue classes (mod m) is isomorphic to the multiplicative group of mth roots of unity.

3. H is a subset of a group G. Show that H is a subgroup of G if and only if both the following conditions are satisfied: (i) if $a, b \in H$, then $ab \in H$; (ii) if $a \in H$, then $a^{-1} \in H$. Verify that conditions (i) and (ii) can be replaced by the single equivalent condition: if $a, b \in H$, then $ab^{-1} \in H$.

Show also that in the case of *finite* groups condition (ii) can be omitted.

4. H is a subset of a group G and, for any elements a, b of G, we write $a \sim b$ whenever $a^{-1}b \in H$. Show that $\sim$ is an equivalence relation if and only if H is a subgroup of G. When $\sim$ is an equivalence relation, show that one equivalence class and one only is a subgroup of G.

5. An *automorphism* of a group G is an isomorphic mapping of G into itself. Show that the set of all automorphisms of G is a group Γ. Show further that, when G is a Klein four-group, Γ is isomorphic to the symmetric group S_3.

6. Let G be a group and x a fixed element of G. The transformation which maps $a \in G$ into xax^{-1} is called an *inner automorphism* of G. Show that an inner automorphism is an automorphism, and that the set all inner automorphisms of G is a group.

7. Let G be a group. If $a, b \in G$, write $a \sim b$ whenever there exists some $x \in G$ such that $b = xax^{-1}$. Show that $\sim$ is an equivalence relation, and interpret it in the case when G is the full linear group of degree n.

8. Show that among the $n!$ permutations of degree n precisely $\frac{1}{2}n!$ are even and that these permutations form a group—the *alternating group* of degree n. Show also that the odd permutations do not form a group.

9. Show that each of the two sets of matrices

$$\text{(i)}\quad \begin{pmatrix}1 & 0\\ 0 & 1\end{pmatrix},\quad \begin{pmatrix}-1 & 0\\ 0 & -1\end{pmatrix},\quad \begin{pmatrix}1 & 0\\ 0 & -1\end{pmatrix},\quad \begin{pmatrix}-1 & 0\\ 0 & 1\end{pmatrix},$$

$$\text{(ii)}\quad \begin{pmatrix}1 & 0\\ 0 & 1\end{pmatrix},\quad \begin{pmatrix}-1 & 0\\ 0 & -1\end{pmatrix},\quad \begin{pmatrix}0 & 1\\ -1 & 0\end{pmatrix},\quad \begin{pmatrix}0 & -1\\ 1 & 0\end{pmatrix},$$

is a matrix group, and that the two groups are not isomorphic with each other.

10. Show that the matrices $\begin{pmatrix}a & b\\ 0 & 1\end{pmatrix}$ $(a \neq 0)$ constitute a group.

11. Show that all non-singular matrices which commute with a given matrix constitute a matrix group.

12. Show that the set of matrices of one or the other of the types

$$\begin{pmatrix}\alpha & 0\\ 0 & \beta\end{pmatrix},\quad \begin{pmatrix}0 & \alpha\\ \beta & 0\end{pmatrix}\qquad (\alpha \neq 0, \beta \neq 0)$$

is a matrix group.

13. Let $\mathbf{A}$ be a given non-singular matrix. Show that the set G of all matrices $\mathbf{P}$ such that $\mathbf{P}^T\mathbf{A}\mathbf{P} = \mathbf{A}$ is a matrix group. Interpret this result for the case $\mathbf{A} = \mathbf{I}$.

14. Let Γ be a matrix group, $\mathbf{S}$ a fixed non-singular matrix, and Γ^* the group of matrices $\mathbf{S}^{-1}\mathbf{A}\mathbf{S}$ $(\mathbf{A} \in \Gamma)$. Show that, if $\mathbf{\Omega}$ is a matrix such that, for all $\mathbf{A} \in \Gamma$, $\mathbf{A}^T\mathbf{\Omega}\mathbf{A} = \mathbf{\Omega}$, then, for a suitable matrix $\mathbf{\Omega}'$ and all $\mathbf{B} \in \Gamma^*$, $\mathbf{B}^T\mathbf{\Omega}'\mathbf{B} = \mathbf{\Omega}'$.

15. Let Γ_1 and Γ_2 be groups of non-singular $n \times n$ matrices and suppose that every matrix in Γ_1 commutes with every matrix in Γ_2. Show that the set of all matrices of the form $\mathbf{A}_1\mathbf{A}_2$ $(\mathbf{A}_1 \in \Gamma_1, \mathbf{A}_2 \in \Gamma_2)$ is again a group. Show also that this conclusion is not valid if the condition of commutativity is dispensed with.

16. Show that if a non-singular matrix is an element of a *finite* matrix group, then its characteristic roots are roots of unity. Show also that the condition of finiteness cannot be dispensed with.

17. Show that the characteristic roots of every permutation matrix are roots of unity.

18. Obtain a representation, in terms of real matrices, for the multiplicative group of mth roots of unity.

19. Obtain a matrix representation for the group of permutations

$$\begin{pmatrix}1234\\1234\end{pmatrix}, \quad \begin{pmatrix}1234\\2143\end{pmatrix}, \quad \begin{pmatrix}1234\\3412\end{pmatrix}, \quad \begin{pmatrix}1234\\4321\end{pmatrix}.$$

Also show that this group is isomorphic to the multiplicative group of residue classes (mod 8) prime to 8.

20. Show that the multiplicative group G of the numbers $1, i, -1, -i$ is isomorphic to the multiplicative group of residue classes (mod 5) prime to 5, and also to the group of permutations

$$\begin{pmatrix}1234\\1234\end{pmatrix}, \quad \begin{pmatrix}1234\\2341\end{pmatrix}, \quad \begin{pmatrix}1234\\3412\end{pmatrix}, \quad \begin{pmatrix}1234\\4123\end{pmatrix}.$$

Obtain a matrix representation of G in terms of 2×2 matrices and another matrix representation in terms of 4×4 matrices.

21. Let $\mathbf{M}_1,\ldots,\mathbf{M}_6$ be the six matrices given on p. 272, which represent the symmetric group S_3. By considering the matrices $\mathbf{T}^{-1}\mathbf{M}_i\mathbf{T}$ $(i = 1,\ldots,6)$, where

$$\mathbf{T} = \begin{pmatrix}1 & -1 & 1\\ 1 & 2 & 1\\ 1 & -1 & -2\end{pmatrix},$$

obtain a representation of S_3 in terms of 2×2 matrices.

22. Let Γ be a finite matrix group. Show that there exists a hermitian matrix $\mathbf{H}$ such that, for all $\mathbf{A} \in \Gamma$, $\bar{\mathbf{A}}^T\mathbf{H}\mathbf{A} = \mathbf{H}$.

23. Show that the set of all upper triangular $n\times n$ matrices, all of whose diagonal elements are equal to 1, is a matrix group.

24. Let G be the set of all $n\times n$ matrices each of which has precisely one non-zero element in each row and each column. Show that G is a matrix group.

25. Show that the set B of bilinear transformations $w = (az+b)/(cz+d)$, where a, b, c, d are complex numbers such that $ad \neq bc$, is a group. Discuss the relation between B and $GL(2)$.

26. Let $\mathbf{A}$ be an $m\times n$ matrix, and denote by $t(\mathbf{P}, \mathbf{Q})$ the operation which transforms $\mathbf{A}$ into $\mathbf{PAQ}$, where $\mathbf{P}$, $\mathbf{Q}$ are non-singular matrices of order m, n respectively. Show that the set of all operations $t(\mathbf{P}, \mathbf{Q})$ is a group. Deduce that the set of equivalence transformations (in the sense of Definition 6.2.3, p. 177) is a group.

27. Show that, if $\mathbf{A} \neq \mathbf{O}$ is an element of a matrix group, then no positive power of $\mathbf{A}$ can be equal to $\mathbf{O}$.

28. Show that

$$\begin{pmatrix} 1 & 0 & 0 & 0 & 0 \\ 0 & 1 & 0 & 0 & 0 \\ 0 & 0 & 0 & 1 & 0 \\ 0 & 0 & 0 & 0 & 1 \\ 0 & 0 & 0 & 0 & 0 \end{pmatrix}$$

is not a group matrix.

29. Let $\mathbf{A}$ and $\mathbf{B}$ be commuting matrices such that $R(\mathbf{A}^2) = R(\mathbf{A})$, $R(\mathbf{B}^2) = R(\mathbf{B})$. Show that $R(\mathbf{ABAB}) = R(\mathbf{AB})$.

30. L is a linear transformation of $\mathfrak{B}_n$ such that $L\mathfrak{B}_n = L^2\mathfrak{B}_n$, and $\mathfrak{N}$ is the subspace of $\mathfrak{B}_n$ annihilated by L. Show that $L\mathfrak{B}_n$ and $\mathfrak{N}$ are complements.

31. L is a linear transformation of $\mathfrak{B}_n$; $\mathfrak{U}$ is an r-dimensional invariant space of L; and $\mathfrak{U}'$ is a complement of $\mathfrak{U}$. Show that, if $\mathbf{x}_1,\ldots,\mathbf{x}_r$ are linearly independent vectors in $\mathfrak{U}$ and $\mathbf{x}_{r+1},\ldots,\mathbf{x}_n$ are linearly independent vectors in $\mathfrak{U}'$, then the matrix representing L with respect to the basis

$$\{\mathbf{x}_1,\ldots,\mathbf{x}_r,\ \mathbf{x}_{r+1},\ldots,\mathbf{x}_n\}$$

is of the form

$$\begin{pmatrix} \mathbf{A}_1 & \mathbf{A}_2 \\ \mathbf{O} & \mathbf{A}_3 \end{pmatrix},$$

where $\mathbf{A}_1$ is a non-singular $r \times r$ matrix.

32. Show that any matrix $\mathbf{A}$ is similar to a matrix of the form

$$\begin{pmatrix} \mathbf{A}_1 & \mathbf{O} \\ \mathbf{O} & \mathbf{A}_2 \end{pmatrix},$$

where $\mathbf{A}_1$ is non-singular and, for some value of k, $\mathbf{A}_2^k = \mathbf{O}$. Show further that $\mathbf{A}$ is a group matrix if and only if $\mathbf{A}_2 = \mathbf{O}$.

33. Show that, if the matrices $\mathbf{A}_1,\ldots,\mathbf{A}_k$ constitute a matrix group, then

$$kR(\mathbf{A}_1+\ldots+\mathbf{A}_k) = \operatorname{tr}(\mathbf{A}_1+\ldots+\mathbf{A}_k).$$

34. Prove that $\mathbf{A}$ is a group matrix if and only if there exists a matrix $\mathbf{E}$ such that

$$\mathbf{E}^2 = \mathbf{E}, \quad \mathbf{AE} = \mathbf{EA} = \mathbf{A}, \quad R(\mathbf{A}) = R(\mathbf{E}).$$

X

CANONICAL FORMS

10.1. The idea of a canonical form

10.1.1. We have already had occasion to note the importance of similarity transformations.† We shall now investigate the manner in which given matrices can be changed by similarity transformations into matrices of particularly simple types. Thus, given a matrix $\mathbf{A}$, we shall consider the transformed matrix $\mathbf{S}^{-1}\mathbf{A}\mathbf{S}$ and shall try to choose $\mathbf{S}$ in such a way that $\mathbf{S}^{-1}\mathbf{A}\mathbf{S}$ is as simple as possible. The significance of this procedure in terms of linear mappings is clear in view of Theorem 4.2.5 (p. 119). When a matrix $\mathbf{A}$ is given we interpret it as the matrix representing the linear mapping L (of $\mathfrak{B}_n$ into itself) with respect to the basis $\mathfrak{E} = \{\mathbf{e}_1,..., \mathbf{e}_n\}$, and we attempt to find a second basis $\mathfrak{B}$ with respect to which L is represented by a matrix having a simple form. The discussion below provides no more than a slight sketch of a large and important subject, since a more systematic treatment would fall outside the scope of the present work.‡

DEFINITION 10.1.1. *Let* $\mathbf{A}$ *be a given matrix. If there exists a (non-singular) matrix* $\mathbf{S}$ *such that* $\mathbf{S}^{-1}\mathbf{A}\mathbf{S} = \mathbf{\Lambda}$ *is a diagonal (triangular) matrix, then* $\mathbf{\Lambda}$ *is called a* DIAGONAL (TRIANGULAR) CANONICAL FORM OF $\mathbf{A}$ UNDER THE SIMILARITY GROUP.

Thus $\mathbf{A}$ possesses a diagonal (triangular) canonical form under the similarity group if and only if it is similar to a diagonal (triangular) matrix.

It is often desirable to operate not with the entire similarity group but with certain of its subgroups.

DEFINITION 10.1.2. *Let* $\mathbf{A}$ *be a given matrix. If there exists a unitary matrix* $\mathbf{S}$ *such that* $\mathbf{S}^{-1}\mathbf{A}\mathbf{S} = \mathbf{\Lambda}$ *is a diagonal (triangular) matrix, then* $\mathbf{\Lambda}$ *is called a* DIAGONAL (TRIANGULAR) CANONICAL FORM OF $\mathbf{A}$ UNDER THE UNITARY SIMILARITY GROUP.

† See § 4.2.

‡ For further information see Schreier and Sperner, **8**, chap. v; MacDuffee, **12**, chaps. vi–viii; Halmos, **20**, § 58; Ferrar, **21**, chap. iv; Turnbull and Aitken, **27**, chap. vi; Jacobson, **28**, chap. iii; Hamburger and Grimshaw, **30**, 122–32; MacDuffee, **31**, chaps, iv–vi; MacDuffee, **32**, 240–2; Albert, **33**, chap. iv; Wedderburn, **39**, chap. iii.

An analogous terminology applies if the term 'orthogonal' is substituted for 'unitary'.

The invariant interpretation of orthogonal and unitary canonical forms is, of course, clear in view of Theorem 4.2.8 (p. 121). If L is the linear transformation represented by $\mathbf{A}$ with respect to the orthonormal basis $\mathfrak{E}$, then we seek a new orthonormal basis $\mathfrak{B}$ of $\mathfrak{B}_n$ with respect to which we wish L to be represented by a diagonal or a triangular matrix.

Fundamentally, the study of canonical forms is the study of intrinsic properties of matrices, and the purpose of obtaining a canonical form $\mathbf{\Lambda}$ of $\mathbf{A}$ is to have a matrix at our disposal which shall be easier to manipulate than $\mathbf{A}$ but possess the same invariant characteristics. We have, in fact,

$$\mathbf{A} = \mathbf{S}\mathbf{\Lambda}\mathbf{S}^{-1}, \tag{10.1.1}$$

where $\mathbf{\Lambda}$ and $\mathbf{S}$ are subject to certain prescribed restrictions. The representation of $\mathbf{A}$ in the form (10.1.1) is very convenient for many purposes. Thus, for instance, it may enable us to determine the rank of $\mathbf{A}$, since $R(\mathbf{A}) = R(\mathbf{\Lambda})$, and $\mathbf{\Lambda}$ can generally be dealt with more easily than $\mathbf{A}$.† A similar use can be made of (10.1.1) for the evaluation of the trace. Again, if f is a polynomial, then

$$f(\mathbf{A}) = \mathbf{S}f(\mathbf{\Lambda})\mathbf{S}^{-1}.$$

This identity is indispensable for dealing with many problems which might otherwise prove insoluble.‡

In investigating canonical forms we deal with representations of matrices in certain standard forms which exhibit structure and facilitate calculation. This procedure may be compared with the representation of integers as products of powers of primes or the representation of conics in terms of their standard equations.

In the present chapter we are concerned with canonical forms under similarity transformations but it will be recalled that an analogous problem for congruence transformations was considered in § 6.4, where it was found that if $\mathbf{A}$ is symmetric, then a non-singular matrix $\mathbf{P}$ can be found such that $\mathbf{P}^T\mathbf{A}\mathbf{P}$ is diagonal. The study of 'canonical forms under congruence transformations' will be resumed in Chapter XII.

† See, for example, the proof of Theorem 10.2.3.

‡ See, for example, the proof of Theorem 11.1.1.

10.1.2. Two simple results may be noted at once.

THEOREM 10.1.1. *If* $\mathbf{\Lambda}$ *is any canonical form of* $\mathbf{A}$, *then the diagonal elements of* $\mathbf{\Lambda}$ *are the characteristic roots of* $\mathbf{A}$.†

By Theorem 7.2.1 (p. 199) $\mathbf{A}$ and $\mathbf{\Lambda}$ have the same characteristic roots, and since $\mathbf{\Lambda}$ is diagonal or triangular its diagonal elements are precisely its characteristic roots.

EXERCISE 10.1.1. Let $\mathbf{A}$ be similar to a diagonal matrix. Show that all characteristic roots of $\mathbf{A}$ are equal if and only if $\mathbf{A}$ is a scalar matrix.

Theorem 10.1.1 naturally leads us to inquire in what *order* the characteristic roots of $\mathbf{A}$ appear on the diagonal of a canonical form. The next theorem shows that for diagonal canonical forms this order may be preassigned arbitrarily.

THEOREM 10.1.2. *Suppose that* $\mathbf{A}$ *possesses a diagonal canonical form* $\mathbf{\Lambda}$ *under a group* G *of transformations. If* $\mathbf{\Lambda}'$ *is a diagonal matrix whose diagonal elements are the characteristic roots of* $\mathbf{A}$ *in any preassigned order, then* $\mathbf{\Lambda}'$ *is also a diagonal canonical form of* $\mathbf{A}$ *under* G.

We have, by hypothesis,

$$\mathbf{A} = \mathbf{S\Lambda S}^{-1},$$

where $\mathbf{S}$ is of a certain prescribed type (non-singular, unitary, or orthogonal). In view of Theorem 10.1.1, $\mathbf{\Lambda}$ and $\mathbf{\Lambda}'$ differ at most in the arrangement of their diagonal elements. Hence, by Theorem 7.2.2 (p. 199), there exists a matrix $\mathbf{T}$ such that

$$\mathbf{\Lambda} = \mathbf{T\Lambda' T}^{-1},$$

and it is, in fact, clear from the proof of Theorem 7.2.2 that $\mathbf{T}$ is orthogonal. Thus

$$\mathbf{A} = (\mathbf{ST})\mathbf{\Lambda}'(\mathbf{ST})^{-1},$$

and $\mathbf{\Lambda}'$ is therefore a canonical form of $\mathbf{A}$ under G.

Theorem 10.1.2 remains true (with obvious verbal adjustments) for triangular canonical forms. The argument just used is no longer effective in this case, but the required conclusion will be seen to be an obvious consequence of the proof of Theorem 10.4.1.

10.2. Diagonal canonical forms under the similarity group

10.2.1. We begin with a theorem which not only asserts the existence of diagonal canonical forms for certain types of matrices

† It is, of course, implied in this statement that each characteristic root of $\mathbf{A}$ occurs among the diagonal elements of $\mathbf{\Lambda}$ with its correct multiplicity.

but provides an explicit method for arriving at such canonical forms.

THEOREM 10.2.1. *If* $\mathbf{x}_1,\ldots, \mathbf{x}_n$ *are linearly independent characteristic vectors of an* $n \times n$ *matrix* $\mathbf{A}$, *and* $\mathbf{S}$ *is the (non-singular) matrix having* $\mathbf{x}_1,\ldots, \mathbf{x}_n$ *as its columns, then* $\mathbf{S}^{-1}\mathbf{AS}$ *is a diagonal matrix.*

Denote by ω_j the value of the characteristic root of $\mathbf{A}$ with which the vector $\mathbf{x}_j$ is associated.† Thus

$$\mathbf{A}\mathbf{x}_j = \omega_j \mathbf{x}_j \qquad (j = 1,\ldots, n); \tag{10.2.1}$$

and using Theorem 3.3.5 (iii) (p. 84), we obtain

$$\begin{aligned}(\mathbf{S}^{-1}\mathbf{AS})_{ij} &= (\mathbf{S}^{-1})_{i*}\mathbf{AS}_{*j} = (\mathbf{S}^{-1})_{i*}\mathbf{A}\mathbf{x}_j = (\mathbf{S}^{-1})_{i*}\omega_j \mathbf{x}_j \\ &= \omega_j(\mathbf{S}^{-1})_{i*}\mathbf{S}_{*j} = \omega_j(\mathbf{S}^{-1}\mathbf{S})_{ij} = \omega_j \mathbf{I}_{ij} = \omega_j \delta_{ij}.\end{aligned}$$

Hence

$$\mathbf{S}^{-1}\mathbf{AS} = \mathbf{dg}(\omega_1,\ldots, \omega_n), \tag{10.2.2}$$

and the theorem is proved. We may arrive at the same conclusion without any calculation by interpreting $\mathbf{A}$ as the matrix representing a linear transformation L with respect to the basis $\mathfrak{E}$. Since $\mathbf{x}_1,\ldots,\mathbf{x}_n$ are linearly independent, $\mathfrak{X} = \{\mathbf{x}_1,\ldots,\mathbf{x}_n\}$ is a basis of $\mathfrak{V}_n$ and, by Theorem 4.2.8 (p. 121), the matrix $\mathbf{S}^{-1}\mathbf{AS}$ represents L with respect to $\mathfrak{X}$. We have, by Theorem 4.2.6 and (10.2.1),

$$L\mathbf{x}_j = \omega_j \mathbf{x}_j \qquad (j = 1,\ldots, n).$$

Moreover, if $\mathbf{S}^{-1}\mathbf{AS} = (b_{ij})$, then, by (4.2.13),

$$L\mathbf{x}_j = \sum_{i=1}^{n} b_{ij}\mathbf{x}_i \qquad (j = 1,\ldots, n).$$

Hence $b_{ij} = \omega_j \delta_{ij}$ $(i,j = 1,\ldots, n)$, and this is equivalent to (10.2.2).

Consider, for instance, the matrix

$$\mathbf{A} = \begin{pmatrix} 0 & 1 & 0 \\ 0 & 0 & 1 \\ 1 & 0 & 0 \end{pmatrix} \tag{10.2.3}$$

whose characteristic roots are $1, \rho, \rho^2$, where $\rho = e^{2\pi i/3}$. If $(x, y, z)^T$ is a characteristic vector associated with the characteristic root λ, then

$$\lambda(x, y, z)^T = \mathbf{A}(x, y, z)^T = (y, z, x)^T.$$

† We are not entitled to *assume* that the numbers $\omega_1,\ldots, \omega_n$ give us all the characteristic roots with correct multiplicities. This is, however, the case as is seen by (10.2.2.)

Hence $(1,1,1)^T$, $(1,\rho,\rho^2)^T$, $(1,\rho^2,\rho)^T$ are three characteristic vectors associated with $1,\rho,\rho^2$ respectively. It is easy to verify that these vectors are linearly independent. Hence the matrix

$$\mathbf{S} = \begin{pmatrix} 1 & 1 & 1 \\ 1 & \rho & \rho^2 \\ 1 & \rho^2 & \rho \end{pmatrix}$$

is non-singular, and $\mathbf{S}^{-1}\mathbf{AS} = \mathbf{dg}(1,\rho,\rho^2)$.

EXERCISE 10.2.1. Write down matrices **T** and **U** such that

$$\mathbf{T}^{-1}\mathbf{AT} = \mathbf{dg}(1,\rho^2,\rho), \qquad \mathbf{U}^{-1}\mathbf{AU} = \mathbf{dg}(\rho^2,\rho,1),$$

where **A** is the matrix defined by (10.2.3).

Theorem 10.2.2. *An $n \times n$ matrix is similar to a diagonal matrix if and only if it possesses n linearly independent characteristic vectors.*

An alternative way to express this result is to say that an $n \times n$ matrix is similar to a diagonal matrix if and only if its characteristic vectors span the total vector space $\mathfrak{V}_n$.

In view of Theorem 10.2.1 we need only prove that if **A** is similar to a diagonal matrix, then it possesses n linearly independent characteristic vectors. Suppose, then, that

$$\mathbf{S}^{-1}\mathbf{AS} = \mathbf{\Lambda} = \mathbf{dg}(\lambda_1,\ldots,\lambda_n),$$

where $\lambda_1,\ldots,\lambda_n$ are, of course, the characteristic roots of **A**. Then, for $i = 1,\ldots,n$, $(\mathbf{AS})_{*i} = (\mathbf{S\Lambda})_{*i}$, and so

$$\mathbf{AS}_{*i} = \mathbf{S\Lambda}_{*i} = \lambda_i\,\mathbf{Se}_i = \lambda_i\,\mathbf{S}_{*i}.$$

The columns of **S** are therefore characteristic vectors of **A**; and since $|\mathbf{S}| \neq 0$, they are linearly independent.

EXERCISE 10.2.2. Prove Theorem 10.2.2. by an invariant argument.

DEFINITION 10.2.1. *A characteristic root of a matrix is* REGULAR *if its multiplicity is equal to the maximum number of linearly independent characteristic vectors associated with it.*

Thus, a k-fold characteristic root λ of **A** is said to be regular if

$$R(\lambda\mathbf{I}-\mathbf{A}) = n-k.$$

It will be recalled that, by the rank-multiplicity theorem (p. 214), we have in every case

$$R(\lambda\mathbf{I}-\mathbf{A}) \geqslant n-k.$$

Theorem 10.2.3. *A matrix is similar to a diagonal matrix if and only if all its characteristic roots are regular.*

This theorem states, in fact, that **A** is similar to a diagonal matrix if and only if $R(\lambda\mathbf{I}-\mathbf{A}) = n-m_\lambda(\mathbf{A})$ for every value of λ.

Suppose, in the first place, that $\mathbf{A}$ is similar to a diagonal matrix, say

$$\mathbf{S}^{-1}\mathbf{AS} = \mathbf{\Lambda} = \mathbf{dg}(\lambda_1,\ldots,\lambda_n).$$

If λ is a k-fold characteristic root of $\mathbf{A}$, then precisely k among the numbers $\lambda_1,\ldots,\lambda_n$ are equal to λ; and we have

$$R(\lambda\mathbf{I}-\mathbf{A}) = R\{\mathbf{S}(\lambda\mathbf{I}-\mathbf{\Lambda})\mathbf{S}^{-1}\} = R(\lambda\mathbf{I}-\mathbf{\Lambda})$$
$$= R\{\mathbf{dg}(\lambda-\lambda_1,\ldots,\lambda-\lambda_n)\} = n-k.$$

Thus every characteristic root of $\mathbf{A}$ is regular.

Suppose, next, that all characteristic roots of $\mathbf{A}$ are regular. Denote the *distinct* characteristic roots by $\omega_1,\ldots,\omega_m$ and let $r_1,\ldots,r_m$ be their multiplicities, so that $r_1+\ldots+r_m = n$. For every i in the range $1 \leqslant i \leqslant m$ there exist, by hypothesis, r_i linearly independent vectors $\mathbf{x}^{(ij)}$ $(j = 1,\ldots,r_i)$ such that

$$\mathbf{Ax}^{(ij)} = \omega_i \mathbf{x}^{(ij)} \quad (i = 1,\ldots,m;\, j = 1,\ldots,r_i). \tag{10.2.4}$$

We now assert that the n characteristic vectors $\mathbf{x}^{(ij)}$ $(i = 1,\ldots,m;$ $j = 1,\ldots,r_i)$ are, in fact, linearly independent. To prove this, let t_{ij} be scalars such that

$$\sum_{i,j} t_{ij}\mathbf{x}^{(ij)} = \mathbf{0};$$

here i ranges over the values $1,\ldots,m$ and, for each i, j ranges over the values $1,\ldots,r_i$. Let κ be any number such that $1 \leqslant \kappa \leqslant m$. Then, using (10.2.4), we obtain

$$\mathbf{0} = \prod_{\substack{k=1\\k\neq\kappa}}^{m} (\omega_k \mathbf{I}-\mathbf{A}) \sum_{i,j} t_{ij}\mathbf{x}^{(ij)} = \sum_{i,j} t_{ij} \prod_{\substack{k=1\\k\neq\kappa}}^{m} (\omega_k \mathbf{I}-\mathbf{A}).\mathbf{x}^{(ij)}$$
$$= \sum_{i,j} t_{ij} \prod_{\substack{k=1\\k\neq\kappa}}^{m} (\omega_k-\omega_i).\mathbf{x}^{(ij)}.$$

Now the product in the expression on the right-hand side certainly vanishes if $\kappa \neq i$. Hence

$$\mathbf{0} = \sum_{j=1}^{r_\kappa} t_{\kappa j} \prod_{\substack{k=1\\k\neq\kappa}}^{m} (\omega_k-\omega_\kappa).\mathbf{x}^{(\kappa j)}.$$

But the vectors $\mathbf{x}^{(\kappa j)}$ $(j = 1,\ldots,r_\kappa)$ are linearly independent, and therefore

$$t_{\kappa j} \prod_{\substack{k=1\\k\neq\kappa}}^{m} (\omega_k-\omega_\kappa) = 0 \qquad (j = 1,\ldots,r_\kappa).$$

Since the ω's are distinct, it follows that $t_{\kappa j} = 0$ $(j = 1,\ldots,r_\kappa)$. This holds for every κ in the range $1 \leqslant \kappa \leqslant m$; and therefore the n characteristic vectors $\mathbf{x}^{(ij)}$ are linearly independent. Hence, by Theorem 10.2.2, $\mathbf{A}$ is similar to a diagonal matrix.

COROLLARY. *Suppose that* $\mathbf{A}$ *possesses a diagonal canonical form under the similarity group. Then the number of its zero characteristic roots is equal to* $n-R(\mathbf{A})$.

For, by Theorem 10.2.3, $R(\mathbf{A}) = R(-\mathbf{A}) = n-m_0(\mathbf{A})$, and the assertion follows at once.

EXERCISE 10.2.3. Show that $\mathbf{A}$ is similar to a diagonal matrix if and only if $\lambda\mathbf{I}-\mathbf{A}$ is a group matrix for every value of λ.

The criteria of Theorems 10.2.2 and 10.2.3, though valuable theoretically, are of very little use in practice; and it is, therefore, important to determine some easily recognizable classes of matrices which possess diagonal canonical forms. The next theorem deals with one such class; others will be found in § 10.3.

THEOREM 10.2.4. *If the n characteristic roots of the* $n\times n$ *matrix* $\mathbf{A}$ *are distinct, then* $\mathbf{A}$ *is similar to a diagonal matrix.*

The proof is immediate for, since every characteristic root of $\mathbf{A}$ is simple, every one is regular by the corollary to Theorem 7.6.1 (p. 215). Hence the assertion follows by Theorem 10.2.3. We shall, however, also give an easy alternative proof.

Let $\lambda_1,\ldots,\lambda_n$ be the n (distinct) characteristic roots of $\mathbf{A}$, and let $\mathbf{x}_1,\ldots,\mathbf{x}_n$ be any characteristic vectors of $\mathbf{A}$ associated with $\lambda_1,\ldots,\lambda_n$ respectively. Thus

$$\mathbf{A}\mathbf{x}_i = \lambda_i\mathbf{x}_i, \quad \mathbf{x}_i \neq \mathbf{0} \quad (i = 1,\ldots,n).$$

It is sufficient to show that $\mathbf{x}_1,\ldots,\mathbf{x}_n$ are linearly independent, for then the assertion will follow by Theorem 10.2.1. Assume, on the contrary, that $\mathbf{x}_1,\ldots,\mathbf{x}_n$ are linearly dependent. Then there exists an integer k in the range $2 \leqslant k \leqslant n$ such that $\mathbf{x}_1,\ldots,\mathbf{x}_{k-1}$ are linearly independent while $\mathbf{x}_1,\ldots,\mathbf{x}_{k-1},\mathbf{x}_k$ are linearly dependent. Hence there exist scalars $\alpha_1,\ldots,\alpha_k$, not all zero, such that

$$\alpha_1\mathbf{x}_1+\ldots+\alpha_{k-1}\mathbf{x}_{k-1}+\alpha_k\mathbf{x}_k = \mathbf{0}. \qquad (10.2.5)$$

It follows that

$$\alpha_1\mathbf{A}\mathbf{x}_1+\ldots+\alpha_{k-1}\mathbf{A}\mathbf{x}_{k-1}+\alpha_k\mathbf{A}\mathbf{x}_k = \mathbf{0},$$

i.e.

$$\alpha_1\lambda_1\mathbf{x}_1+\ldots+\alpha_{k-1}\lambda_{k-1}\mathbf{x}_{k-1}+\alpha_k\lambda_k\mathbf{x}_k = \mathbf{0}.$$

Hence, in view of (10.2.5),

$$\alpha_1(\lambda_1-\lambda_k)\mathbf{x}_1+\ldots+\alpha_{k-1}(\lambda_{k-1}-\lambda_k)\mathbf{x}_{k-1} = \mathbf{0}.$$

But $\lambda_1-\lambda_k \neq 0,\ldots,\lambda_{k-1}-\lambda_k \neq 0$; and $\alpha_1,\ldots,\alpha_{k-1}$ are not all zero (for, by (10.2.5), this would imply $\alpha_k \mathbf{x}_k = \mathbf{0}$ and so $\mathbf{x}_k = \mathbf{0}$). Hence $\mathbf{x}_1,..., \mathbf{x}_{k-1}$ are linearly dependent, and this gives us the required contradiction.†

An alternative proof of Theorem 10.2.4 runs as follows. Let $\alpha_1,..., \alpha_n$ be scalars such that $\alpha_1 \mathbf{x}_1 + ... + \alpha_n \mathbf{x}_n = \mathbf{0}$. Premultiplying by $\mathbf{A}^k$ (where $0 \leqslant k < n$) and using repeatedly the relation $\mathbf{A}\mathbf{x}_i = \lambda_i \mathbf{x}_i$, we infer that

$$\sum_{i=1}^{n} \lambda_i^k (\alpha_i \mathbf{x}_i) = \mathbf{0} \quad (k = 0, 1,..., n-1)$$

Now the λ's are distinct and so, by the identity (1.4.5) (p. 17), the determinant of the numbers λ_i^k $(0 \leqslant k < n; 1 \leqslant i \leqslant n)$ is different from zero. Hence, by Exercise 5.1.2 (p. 135), $\alpha_i \mathbf{x}_i = \mathbf{0}$ $(1 \leqslant i \leqslant n)$ and therefore $\alpha_1 = ... = \alpha_n = 0$, so that $\mathbf{x}_1,..., \mathbf{x}_n$ are linearly independent.

10.2.2. A further criterion for the existence of diagonal canonical forms establishes a link between canonical forms and minimum polynomials.

THEOREM 10.2.5. *A matrix is similar to a diagonal matrix if and only if the linear factors of its minimum polynomial are distinct.*

We shall denote the matrix in question by $\mathbf{A}$. Let $\omega_1,..., \omega_k$ be its distinct characteristic roots and $r_1,..., r_k$ their respective multiplicities.

Suppose that $\mathbf{A}$ is similar to a diagonal matrix so that, for some $\mathbf{S}$, $\mathbf{S}^{-1}\mathbf{A}\mathbf{S} = \mathbf{dg}(\alpha_1,..., \alpha_n)$, where, of course, each α is equal to some ω. Denoting by $\mu(\lambda)$ the polynomial

$$\mu(\lambda) = (\lambda - \omega_1)...(\lambda - \omega_k)$$

we have
$$\mu(\mathbf{A}) = \mathbf{S}.\mathbf{dg}\{\mu(\alpha_1),..., \mu(\alpha_n)\}\mathbf{S}^{-1} = \mathbf{O}.$$

Hence, in view of Theorem 7.4.7 (p. 208), $\mu(\lambda)$ is the minimum polynomial of $\mathbf{A}$, and its linear factors are evidently distinct.

Next, suppose that $\mu(\lambda)$, as defined above, is the minimum polynomial of $\mathbf{A}$. Then

$$(\omega_1 \mathbf{I} - \mathbf{A})...(\omega_k \mathbf{I} - \mathbf{A}) = \mu(\mathbf{A}) = \mathbf{O},$$

and by repeated application of Theorem 5.6.5 (p. 162), we infer that

$$\sum_{i=1}^{k} R(\omega_i \mathbf{I} - \mathbf{A}) \leqslant (k-1)n. \tag{10.2.6}$$

† Cf. the argument on p. 409.

Now we know that

$$R(\omega_i \mathbf{I}-\mathbf{A}) \geqslant n-r_i \quad (i = 1,..., k).$$

Assume that $\mathbf{A}$ is not similar to a diagonal matrix. Then, by Theorem 10.2.3, we have, for some κ,

$$R(\omega_\kappa \mathbf{I}-\mathbf{A}) > n-r_\kappa.$$

Hence $$\sum_{i=1}^{k} R(\omega_i \mathbf{I}-\mathbf{A}) > nk-(r_1+...+r_k) = (k-1)n,$$

and this is contrary to (10.2.6). The proof of the theorem is therefore complete.

A matrix $\mathbf{A}$ may be said to be *periodic* if, for some $k \geqslant 1$, $\mathbf{A}^k = \mathbf{I}$. We already know by Exercise 7.3.2 (p. 202) that if a matrix is periodic, then its characteristic roots are roots of unity, but that the converse of this is false. Necessary and sufficient conditions for periodicity of matrices can now be deduced.

THEOREM 10.2.6. *A matrix is periodic if and only if it is similar to a diagonal matrix and has all its characteristic roots equal to roots of unity.*

Suppose that $\mathbf{A}^k = \mathbf{I}$, where $k \geqslant 1$. Then, writing $f(\lambda) = \lambda^k-1$, we have $f(\mathbf{A}) = \mathbf{O}$. Hence, by Theorem 7.4.2 (p. 204), the minimum polynomial $\mu(\lambda)$ of $\mathbf{A}$ divides $f(\lambda)$. Now the linear factors of $f(\lambda)$ are distinct, and so are therefore the linear factors of $\mu(\lambda)$. Hence, by Theorem 10.2.5, $\mathbf{A}$ is similar to a diagonal matrix. Moreover, every characteristic root of $\mathbf{A}$ is, of course, a kth root of unity.

Next, suppose that $\mathbf{S}^{-1}\mathbf{A}\mathbf{S} = \mathbf{\Lambda}$, where $\mathbf{\Lambda}$ is a diagonal matrix, and that each characteristic root of $\mathbf{A}$ is a root of unity. Then there exists an integer $k \geqslant 1$ such that each characteristic root of $\mathbf{A}$ is a kth root of unity. We therefore have

$$\mathbf{S}^{-1}\mathbf{A}^k\mathbf{S} = (\mathbf{S}^{-1}\mathbf{A}\mathbf{S})^k = \mathbf{\Lambda}^k = \mathbf{I};$$

hence $\mathbf{A}^k = \mathbf{I}$, and the theorem is proved.

10.2.3. One of the many applications of diagonal canonical forms is in the study of matrix equations. Thus, for instance, we might ask for solutions of the equation

$$\mathbf{X}^2 = \mathbf{A},$$

where $\mathbf{A}$ is a given matrix. Any such solution may, by analogy with the corresponding problem in scalar analysis, be referred to as a square root of $\mathbf{A}$. However, while every complex number possesses at least one square root, there exist matrices which have none. For example, it is easy to verify that the equation

$$\mathbf{X}^2 = \begin{pmatrix} 0 & 1 \\ 0 & 0 \end{pmatrix}$$

has no solution.

A more general problem is concerned with the equation

$$f(\mathbf{X}) = \mathbf{A}, \tag{10.2.7}$$

where f is any given polynomial. We are naturally interested in conditions which will ensure the solubility of this equation.

THEOREM 10.2.7. *If* $\mathbf{A}$ *is similar to a diagonal matrix and* f *is a non-constant polynomial, then the equation* (10.2.7) *is soluble for* $\mathbf{X}$.

Let $\mathbf{S}$ be a matrix such that

$$\mathbf{A} = \mathbf{S}.\mathbf{dg}(\lambda_1,..., \lambda_n).\mathbf{S}^{-1}, \tag{10.2.8}$$

and let $\mu_1,..., \mu_n$ be any (complex) numbers such that

$$f(\mu_1) = \lambda_1, \quad ..., \quad f(\mu_n) = \lambda_n.$$

Then the matrix $\mathbf{X} = \mathbf{S}.\mathbf{dg}(\mu_1,..., \mu_n).\mathbf{S}^{-1}$ clearly satisfies (10.2.7).

Each μ_r in the above proof can, in general, be chosen in k distinct ways, where k is the degree of f. Hence our construction yields, in general, k^n solutions of (10.2.7). This reasoning shows, in particular, that a non-singular matrix with distinct characteristic roots possesses at least 2^n distinct square roots. More precise information is given by our next result.

THEOREM 10.2.8. *An* $n \times n$ *matrix with distinct characteristic roots possesses precisely* 2^n *or* 2^{n-1} *distinct square roots according as it is non-singular or singular.*

The given matrix $\mathbf{A}$ is similar to a diagonal matrix by Theorem 10.2.4 and may, therefore, be written in the form (10.2.8). Then every matrix

$$\mathbf{X} = \mathbf{S}.\mathbf{dg}(\surd\lambda_1,..., \surd\lambda_n).\mathbf{S}^{-1} \tag{10.2.9}$$

satisfies the equation $\mathbf{X}^2 = \mathbf{A}$. If $\mathbf{A}$ is non-singular, all λ's are different from 0 and there are 2^n such matrices; if $\mathbf{A}$ is singular, then precisely one λ is equal to 0 and the number of matrices is 2^{n-1}. It remains to show that there are no other solutions of $\mathbf{X}^2 = \mathbf{A}$. Now, if $\mathbf{X}$ is any solution, we have

$$(\mathbf{S}^{-1}\mathbf{X}\mathbf{S})^2 = \mathbf{S}^{-1}\mathbf{X}^2\mathbf{S} = \mathbf{dg}(\lambda_1,..., \lambda_n) = \mathbf{\Lambda},$$

say. Hence

$$(\mathbf{S}^{-1}\mathbf{X}\mathbf{S})^3 = (\mathbf{S}^{-1}\mathbf{X}\mathbf{S})\mathbf{\Lambda} = \mathbf{\Lambda}(\mathbf{S}^{-1}\mathbf{X}\mathbf{S}).$$

Since $\mathbf{S}^{-1}\mathbf{X}\mathbf{S}$ commutes with a diagonal matrix $\mathbf{\Lambda}$ whose diagonal elements are distinct, if follows by Exercise 3.3.4 (p. 81), that $\mathbf{S}^{-1}\mathbf{X}\mathbf{S} = \mathbf{M}$, where $\mathbf{M}$ is diagonal. But

$$\mathbf{M}^2 = \mathbf{S}^{-1}\mathbf{X}^2\mathbf{S} = \mathbf{\Lambda},$$

and therefore

$$\mathbf{M} = \mathbf{dg}(\sqrt{\lambda_1},..., \sqrt{\lambda_n}),$$

when suitable values of the square roots are taken. Thus $\mathbf{X}$ is of the form (10.2.9), and the proof is complete.

When the characteristic roots of $\mathbf{A}$ are not all distinct, various possibilities may occur. We have already seen that there exist matrices which possess no square roots. There also exist matrices which possess infinitely many. Thus every matrix of the form

$$\mathbf{X} = \begin{pmatrix} x & 1 \\ 1-x^2 & -x \end{pmatrix}$$

satisfies the equation $\mathbf{X}^2 = \mathbf{I}_2$, and so $\mathbf{I}_2$ has infinitely many square roots.

10.3. Diagonal canonical forms under the orthogonal similarity group and the unitary similarity group

10.3.1. We begin this section by proving some rather special results which will presently be superseded.

THEOREM 10.3.1. *If an* $n \times n$ *matrix* $\mathbf{A}$ *possesses a set of complex (real) orthonormal characteristic vectors* $\mathbf{x}_1,..., \mathbf{x}_n$, *and* $\mathbf{S}$ *is the unitary (orthogonal) matrix having* $\mathbf{x}_1,..., \mathbf{x}_n$ *as its columns, then* $\mathbf{S}^{-1}\mathbf{A}\mathbf{S}$ *is a diagonal matrix.*

The vectors $\mathbf{x}_1,..., \mathbf{x}_n$ are orthonormal and therefore linearly independent. The assertion now follows at once by Theorem 10.2.1.

EXERCISE 10.3.1. Discuss the fallacy in the following argument. 'Let $\mathbf{x}_1,..., \mathbf{x}_n$ be linearly independent characteristic vectors of $\mathbf{A}$. The space spanned by the characteristic vectors is then the total vector space $\mathfrak{B}_n$, and we can select an orthonormal basis $\{\mathbf{x}'_1,..., \mathbf{x}'_n\}$ of $\mathfrak{B}_n$ consisting of characteristic vectors of $\mathbf{A}$. Hence, by Theorem 10.3.1, it follows that every $n \times n$ matrix which possesses n linearly independent characteristic vectors is unitarily similar to a diagonal matrix.'

THEOREM 10.3.2. *If* $\mathbf{A}$ *is a hermitian matrix, then any two characteristic vectors associated with two distinct characteristic roots of* $\mathbf{A}$ *are orthogonal to each other.*

Let λ, μ be two distinct characteristic roots of $\mathbf{A}$, and let $\mathbf{x}, \mathbf{y}$ be vectors such that $\mathbf{A}\mathbf{x} = \lambda\mathbf{x}$, $\mathbf{A}\mathbf{y} = \mu\mathbf{y}$. Since, by Theorem 7.5.1 (p. 209), λ and μ are real, we have

$$\lambda\bar{\mathbf{x}}^T\mathbf{y} = (\overline{\lambda\mathbf{x}})^T\mathbf{y} = (\overline{\mathbf{A}\mathbf{x}})^T\mathbf{y} = \bar{\mathbf{x}}^T\bar{\mathbf{A}}^T\mathbf{y} = \bar{\mathbf{x}}^T\mathbf{A}\mathbf{y} = \mu\bar{\mathbf{x}}^T\mathbf{y}.$$

Now $\lambda \neq \mu$, and therefore $(\mathbf{x}, \mathbf{y}) = \bar{\mathbf{x}}^T\mathbf{y} = 0$.

THEOREM 10.3.3. *Any hermitian (real symmetric) matrix with distinct characteristic roots is unitarily (orthogonally) similar to a diagonal matrix.*

Let $\lambda_1,\ldots,\lambda_n$ be the (distinct) characteristic roots of a hermitian or real symmetric matrix $\mathbf{A}$, and denote by $\mathbf{x}_1,\ldots,\mathbf{x}_n$ a set of unit characteristic vectors of $\mathbf{A}$, associated respectively with $\lambda_1,\ldots,\lambda_n$. When $\mathbf{A}$ is real and symmetric, real vectors can, of course, be chosen. Now, by Theorem 10.3.2, $\mathbf{x}_1,\ldots,\mathbf{x}_n$ constitute an orthonormal set, and the assertion therefore follows by Theorem 10.3.1.

We now possess an actual procedure for transforming to diagonal form certain hermitian (real symmetric) matrices by means of a unitary (orthogonal) similarity transformation. Consider, for example, the real symmetric matrix

$$\mathbf{A} = \begin{pmatrix} 1 & 0 & -4 \\ 0 & 5 & 4 \\ -4 & 4 & 3 \end{pmatrix}.$$

Its characteristic polynomial is

$$\begin{vmatrix} \lambda-1 & 0 & 4 \\ 0 & \lambda-5 & -4 \\ 4 & -4 & \lambda-3 \end{vmatrix} = \lambda^3-9\lambda^2-9\lambda+81 = (\lambda-3)(\lambda+3)(\lambda-9),$$

and the characteristic roots are thus 3, -3, 9. Now if $(x, y, z)^T$ is a characteristic vector associated with the characteristic root λ, then

$$\begin{pmatrix} 1 & 0 & -4 \\ 0 & 5 & 4 \\ -4 & 4 & 3 \end{pmatrix}\begin{pmatrix} x \\ y \\ z \end{pmatrix} = \lambda\begin{pmatrix} x \\ y \\ z \end{pmatrix},$$

i.e.

$$x-4z = \lambda x,$$

$$5y+4z = \lambda y,$$

$$-4x+4y+3z = \lambda z.$$

From this it is easily found that unit characteristic vectors associated with the characteristic roots 3, -3, 9 may be taken as

$$(\tfrac{2}{3}, \tfrac{2}{3}, -\tfrac{1}{3})^T, \quad (\tfrac{2}{3}, -\tfrac{1}{3}, \tfrac{2}{3})^T, \quad (\tfrac{1}{3}, -\tfrac{2}{3}, -\tfrac{2}{3})^T$$

respectively. Consequently the orthogonal matrix

$$\mathbf{S} = \begin{pmatrix} \frac{2}{3} & \frac{2}{3} & \frac{1}{3} \\ \frac{2}{3} & -\frac{1}{3} & -\frac{2}{3} \\ -\frac{1}{3} & \frac{2}{3} & -\frac{2}{3} \end{pmatrix}$$

has the property that $\mathbf{S}^{-1}\mathbf{A}\mathbf{S} = \mathbf{dg}(3, -3, 9)$.

10.3.2. We next turn to the more general problem in which the distinctness of characteristic roots is not presupposed. For the sake of simplicity we first deal with the real case, and in the proof of the next theorem we employ a type of argument that is found particularly effective in the discussion of canonical forms.† This argument will be used again for proving a number of subsequent results.

Theorem 10.3.4. (Orthogonal reduction of real symmetric matrices)

Every real symmetric matrix is orthogonally similar to a diagonal matrix.

This theorem contains, of course, Theorem 10.3.3 in so far as the latter refers to real symmetric matrices. We shall first give an existence proof of Theorem 10.3.4 and then describe the process for constructing the transforming matrix.

The proof is by induction with respect to n. For $n = 1$ the assertion is true trivially. Assume, then, that it is true for $n-1$, where $n \geqslant 2$. We shall show that it is then true for n.

If λ_1 is a characteristic root (necessarily real) of $\mathbf{A}$, then there exists a real unit vector $\mathbf{x}_1$ such that

$$\mathbf{A}\mathbf{x}_1 = \lambda_1 \mathbf{x}_1. \tag{10.3.1}$$

Let $\mathbf{S}$ be an orthogonal matrix with $\mathbf{x}_1$ as its first column.‡ Then, using (10.3.1), we obtain for $r = 1,..., n$,

$$\begin{aligned}(\mathbf{S}^{-1}\mathbf{A}\mathbf{S})_{r1} &= (\mathbf{S}^{-1})_{r*}\mathbf{A}\mathbf{S}_{*1} = (\mathbf{S}^{-1})_{r*}\mathbf{A}\mathbf{x}_1 \\ &= \lambda_1(\mathbf{S}^{-1})_{r*}\mathbf{x}_1 = \lambda_1(\mathbf{S}^{-1})_{r*}\mathbf{S}_{*1} \\ &= \lambda_1(\mathbf{S}^{-1}\mathbf{S})_{r1} = \lambda_1 \mathbf{I}_{r1} = \lambda_1 \delta_{r1}.\end{aligned}$$

† A similar idea was used in § 6.4.

‡ Such a matrix exists by Theorem 8.1.3 (p. 223).

Moreover, since $\mathbf{A}$ is symmetric, so is $\mathbf{S}^{-1}\mathbf{A}\mathbf{S} = \mathbf{S}^T\mathbf{A}\mathbf{S}$. Hence

$$(\mathbf{S}^{-1}\mathbf{A}\mathbf{S})_{1r} = \lambda_1 \delta_{r1} \quad (r = 1,..., n).$$

The matrix $\mathbf{B} = \mathbf{S}^{-1}\mathbf{A}\mathbf{S}$ thus has the form

$$\mathbf{B} = \begin{pmatrix} \lambda_1 & \mathbf{O}_1^{n-1} \\ \mathbf{O}_{n-1}^1 & \mathbf{B}_1 \end{pmatrix},$$

where $\mathbf{B}_1$ is a real symmetric matrix of order $n-1$. Now, by the induction hypothesis, there exist an orthogonal matrix $\mathbf{C}_1$ and a diagonal matrix $\mathbf{\Lambda}_1$, both of order $n-1$, such that $\mathbf{B}_1\mathbf{C}_1 = \mathbf{C}_1\mathbf{\Lambda}_1$. Hence

$$\begin{pmatrix} \lambda_1 & \mathbf{0} \\ \mathbf{0} & \mathbf{B}_1 \end{pmatrix}\begin{pmatrix} 1 & \mathbf{0} \\ \mathbf{0} & \mathbf{C}_1 \end{pmatrix} = \begin{pmatrix} 1 & \mathbf{0} \\ \mathbf{0} & \mathbf{C}_1 \end{pmatrix}\begin{pmatrix} \lambda_1 & \mathbf{0} \\ \mathbf{0} & \mathbf{\Lambda}_1 \end{pmatrix},$$

or say $\mathbf{BC} = \mathbf{C\Lambda}$. Here $\mathbf{\Lambda}$ and $\mathbf{C}$ are both of order n; $\mathbf{\Lambda}$ is obviously diagonal and $\mathbf{C}$ is orthogonal by Exercise 8.2.2 (p. 230). We thus have

$$\mathbf{S}^{-1}\mathbf{A}\mathbf{S}.\mathbf{C} = \mathbf{C\Lambda},$$

$$(\mathbf{SC})^{-1}\mathbf{A}(\mathbf{SC}) = \mathbf{\Lambda};$$

and since $\mathbf{SC}$ is orthogonal, the required result is established.

Although Theorem 10.3.4 deals with a rather restricted class of matrices it will be seen, in § 12.2, to be the appropriate instrument for the study of quadratic forms, where it is precisely transformations of real symmetric matrices that are relevant.

Let us now consider how an orthogonal matrix $\mathbf{S}$ may be constructed such that $\mathbf{S}^{-1}\mathbf{A}\mathbf{S}$ is diagonal. Denote the distinct characteristic roots of $\mathbf{A}$ by $\lambda_1,..., \lambda_k$ and let $m_1,..., m_k$ be their multiplicities, so that $m_1+...+m_k = n$. Since $\mathbf{A}$ is real and symmetric it is, by Theorem 10.3.4, similar to a diagonal matrix. Hence, by Theorem 10.2.3 (p. 294), the real vectors $\mathbf{x}$ satisfying the equation $\mathbf{Ax} = \lambda_i \mathbf{x}$ constitute a vector space of dimensionality m_i. Denote by $\mathfrak{B}_i$ an orthonormal basis of this space. If $i \neq j$, then, by Theorem 10.3.2, any vector in $\mathfrak{B}_i$ is orthogonal to any vector in $\mathfrak{B}_j$. Thus the n vectors contained in $\mathfrak{B}_1,..., \mathfrak{B}_k$ form a (real) orthonormal set. Hence, by Theorem 10.3.1, the orthogonal matrix $\mathbf{S}$ having these vectors as its columns possesses the required property.†

The preceding argument establishes, in particular, the following result.

† For a numerical illustration of the procedure described here, see § 12.2.2.

THEOREM 10.3.5. *If* $\mathbf{A}$ *is a real symmetric matrix having characteristic roots* $\lambda_1,\ldots,\lambda_n$, *then there exist characteristic vectors* $\mathbf{x}_1,\ldots,\mathbf{x}_n$ *of* $\mathbf{A}$ *which are associated with* $\lambda_1,\ldots,\lambda_n$ *respectively and which constitute an orthonormal set.*

Alternatively, we may infer this result from Theorem 10.3.4 by observing that if $\mathbf{S}^{-1}\mathbf{AS}$ is diagonal, then the columns of $\mathbf{S}$ are characteristic vectors of $\mathbf{A}$ associated with $\lambda_1,\ldots,\lambda_n$ respectively.

We also note that the converse of Theorem 10.3.4 is true. This may be expressed more precisely as follows.

THEOREM 10.3.6. *If a real matrix is orthogonally similar to a diagonal matrix, then it is symmetric.*

For, if $\mathbf{A}$ is real, $\mathbf{S}$ orthogonal, $\mathbf{\Lambda}$ diagonal, and $\mathbf{S}^{-1}\mathbf{AS} = \mathbf{\Lambda}$, then

$$\mathbf{A} = \mathbf{S\Lambda S}^{-1} = \mathbf{S\Lambda S}^T,$$

and so $\mathbf{A}^T = \mathbf{A}$, as asserted.†

We next consider complex matrices. A complex symmetric matrix need not be similar to a diagonal matrix. Thus, if the matrix

$$\mathbf{A} = \begin{pmatrix} 1 & i \\ i & -1 \end{pmatrix}$$

(whose characteristic roots are 0, 0) were similar to a diagonal matrix, we would have $\mathbf{S}^{-1}\mathbf{AS} = \mathbf{O}$, i.e. $\mathbf{A} = \mathbf{O}$. However, a valid analogue of Theorem 10.3.4 for complex matrices is not difficult to obtain.

THEOREM 10.3.7. *Every hermitian matrix is unitarily similar to a diagonal matrix.*

This result contains, of course, Theorem 10.3.3 in so far as that theorem relates to hermitian matrices, but it does not contain Theorem 10.3.4. However, the proof is virtually the same as that of Theorem 10.3.4, and we shall indicate it rather briefly.

Let λ_1 be a characteristic root of the hermitian matrix $\mathbf{A}$, and let $\mathbf{x}_1$ be a unit (complex) characteristic vector associated with λ_1. If $\mathbf{U}$ is a unitary matrix with $\mathbf{x}_1$ as its first column, then

$$(\mathbf{U}^{-1}\mathbf{AU})_{r1} = \lambda_1 \delta_{r1} \qquad (r = 1,\ldots,n).$$

Since, by Exercise 8.2.1 (p. 229), $\mathbf{U}^{-1}\mathbf{AU}$ is hermitian, it follows that

$$(\mathbf{U}^{-1}\mathbf{AU})_{1r} = \lambda_1 \delta_{r1} \qquad (r = 1,\ldots,n).$$

† The condition that $\mathbf{A}$ is real has not, in fact, been used and can be omitted from Theorem 10.3.6.

Thus
$$\mathbf{U}^{-1}\mathbf{A}\mathbf{U} = \begin{pmatrix} \lambda_1 & \mathbf{0} \\ \mathbf{0} & \mathbf{B}_1 \end{pmatrix},$$
where $\mathbf{B}_1$ is a hermitian matrix of order $n-1$. The proof is now completed by induction with respect to n.

Exercise 10.3.2. Let $\omega_1,..., \omega_k$ be the distinct characteristic roots of the hermitian $n \times n$ matrix **A**. Show that the minimum polynomial of **A** is $(\lambda-\omega_1)...(\lambda-\omega_k)$.

If **A** is a given hermitian matrix, then the construction of a unitary matrix **S** such that $\mathbf{S}^{-1}\mathbf{A}\mathbf{S}$ is diagonal proceeds by virtually the same steps as those described on p. 303. In the present case we again obtain a set of n orthonormal characteristic vectors of **A**, and the matrix having these vectors as its columns is the required matrix **S**.

Exercise 10.3.3. State and prove the analogue of Theorem 10.3.5 for hermitian matrices.

10.3.3. It is possible to give a far-reaching generalization of Theorem 10.3.7. With this end in view we introduce the notion of a normal matrix.

Definition 10.3.1. *A (complex) matrix* **A** *is* normal *if*
$$\bar{\mathbf{A}}^T\mathbf{A} = \mathbf{A}\bar{\mathbf{A}}^T.$$

Various types of matrices with which we are already familiar are, in fact, normal. Among these are diagonal matrices, unitary matrices, hermitian matrices, and skew-hermitian matrices.

Exercise 10.3.4. Verify that the property of normality is invariant under unitary similarity transformations, i.e. if **A** is normal and **U** is unitary, then $\mathbf{U}^{-1}\mathbf{A}\mathbf{U}$ is normal.

The following necessary and sufficient condition for the existence of diagonal canonical forms under the unitary similarity group was found by Schur and Toeplitz in 1910.

Theorem 10.3.8. *A matrix is unitarily similar to a diagonal matrix if and only if it is normal.*

This result contains Theorem 10.3.7 (but not Theorem 10.3.4) as a special case.

If **A** is a given matrix and $\mathbf{U}^{-1}\mathbf{A}\mathbf{U} = \mathbf{\Lambda}$, where **U** is unitary and $\mathbf{\Lambda}$ diagonal, then, by Exercise 10.3.4, it follows that **A** is normal.

Suppose, on the other hand, that **A** is normal. Let λ_1 be a characteristic root of **A**, $\mathbf{x}_1$ a unit characteristic vector associated with λ_1, and **U** any unitary matrix having $\mathbf{x}_1$ as its first column. By precisely the same argument as that employed in the proof of Theorem 10.3.4, we have

$$(\mathbf{U}^{-1}\mathbf{A}\mathbf{U})_{r1} = \lambda_1 \delta_{r1} \qquad (r = 1,..., n).$$

Hence $\mathbf{V} = \mathbf{U}^{-1}\mathbf{A}\mathbf{U}$ may be written in the form

$$\mathbf{V} = \begin{pmatrix} \lambda_1 & \mathbf{y}^T \\ \mathbf{O}^1_{n-1} & \mathbf{B}_1 \end{pmatrix},$$

where $\mathbf{y}$ is a (column) vector of order $n-1$, and $\mathbf{B}_1$ a square matrix of order $n-1$. Since

$$\overline{\mathbf{V}}^T = \begin{pmatrix} \bar{\lambda}_1 & \mathbf{O}^{n-1}_1 \\ \bar{\mathbf{y}} & \overline{\mathbf{B}}^T_1 \end{pmatrix},$$

we have

$$\overline{\mathbf{V}}^T\mathbf{V} = \begin{pmatrix} \bar{\lambda}_1 \lambda_1 & \bar{\lambda}_1 \mathbf{y}^T \\ \lambda_1 \bar{\mathbf{y}} & \bar{\mathbf{y}}\mathbf{y}^T + \overline{\mathbf{B}}^T_1 \mathbf{B}_1 \end{pmatrix}, \qquad \mathbf{V}\overline{\mathbf{V}}^T = \begin{pmatrix} \lambda_1 \bar{\lambda}_1 + \mathbf{y}^T\bar{\mathbf{y}} & \mathbf{y}^T\overline{\mathbf{B}}^T_1 \\ \mathbf{B}_1 \bar{\mathbf{y}} & \mathbf{B}_1 \overline{\mathbf{B}}^T_1 \end{pmatrix}.$$

But $\mathbf{V}$ is normal; hence $\bar{\lambda}_1 \lambda_1 = \lambda_1 \bar{\lambda}_1 + \mathbf{y}^T\bar{\mathbf{y}}$ and so $\mathbf{y} = \mathbf{O}^1_{n-1}$. Therefore $\overline{\mathbf{B}}^T_1 \mathbf{B}_1 = \mathbf{B}_1 \overline{\mathbf{B}}^T$ and, furthermore, $\mathbf{V}$ may be written in the form

$$\mathbf{V} = \begin{pmatrix} \lambda_1 & \mathbf{0} \\ \mathbf{0} & \mathbf{B}_1 \end{pmatrix},$$

where $\mathbf{B}_1$ is a normal matrix of order $n-1$.

The proof is now completed by induction with respect to n. If the theorem holds for $n-1$, then there exists a unitary matrix $\mathbf{C}_1$ and a diagonal matrix $\mathbf{\Lambda}_1$, both of order $n-1$, such that $\mathbf{B}_1 \mathbf{C}_1 = \mathbf{C}_1 \mathbf{\Lambda}_1$. Hence

$$\begin{pmatrix} \lambda_1 & \mathbf{0} \\ \mathbf{0} & \mathbf{B}_1 \end{pmatrix}\begin{pmatrix} 1 & \mathbf{0} \\ \mathbf{0} & \mathbf{C}_1 \end{pmatrix} = \begin{pmatrix} 1 & \mathbf{0} \\ \mathbf{0} & \mathbf{C}_1 \end{pmatrix}\begin{pmatrix} \lambda_1 & \mathbf{0} \\ \mathbf{0} & \mathbf{\Lambda}_1 \end{pmatrix},$$

or say $\mathbf{V}\mathbf{C} = \mathbf{C}\mathbf{\Lambda}$, where $\mathbf{\Lambda}$ is diagonal, $\mathbf{C}$ is unitary, and both are of order n. Thus

$$(\mathbf{U}\mathbf{C})^{-1}\mathbf{A}(\mathbf{U}\mathbf{C}) = \mathbf{\Lambda},$$

and the proof is complete since $\mathbf{U}\mathbf{C}$ is unitary.†

EXERCISE 10.3.5. (i) Show that all characteristic roots of a normal matrix are regular. (ii) Show that the characteristic vectors of a normal $n \times n$ matrix span the total vector space $\mathfrak{B}_n$.

10.4. Triangular canonical forms

10.4.1. We have so far confined our attention to diagonal canonical forms. In general, however, a matrix does not possess such a form. On the other hand, we may, in every case, assert the

† For an alternative proof see Perlis, **10**, 194.

existence of a triangular canonical form. The initial steps towards the proof of this result were taken by Jacobi in a paper published posthumously in 1857. The theorem to be proved next is due, in its present form, to Schur (1909).

Theorem 10.4.1. *Every matrix is unitarily similar to a triangular matrix.*

This statement is equally valid whether the term 'triangular' is taken to mean 'upper triangular' or 'lower triangular'. We shall prove it in the first instance for the former interpretation and then deduce at once that it continues to hold for the latter.

The argument is by induction with respect to n and is of the type with which we are now familiar from the proofs of Theorems 10.3.4, 10.3.7, and 10.3.8. Let us assume that the assertion is true for $n-1$, where $n \geqslant 2$. Let λ_1 be a characteristic root of $\mathbf{A}$, $\mathbf{x}_1$ a unit characteristic vector (generally complex) of $\mathbf{A}$ associated with λ_1, $\mathbf{U}$ a unitary matrix with $\mathbf{x}_1$ as its first column, and $\mathbf{V} = \mathbf{U}^{-1}\mathbf{A}\mathbf{U}$. Then $\mathbf{V}_{r1} = \lambda_1 \delta_{r1}$ $(r = 1,..., n)$, and so

$$\mathbf{V} = \begin{pmatrix} \lambda_1 & \mathbf{y}^T \\ \mathbf{0} & \mathbf{A}_1 \end{pmatrix},$$

where $\mathbf{y}$ is a vector of order $n-1$ and $\mathbf{A}_1$ a square matrix of order $n-1$. By the induction hypothesis there exists a unitary matrix $\mathbf{C}_1$ and an upper triangular matrix $\mathbf{\Delta}_1$, both of order $n-1$, such that $\mathbf{A}_1\mathbf{C}_1 = \mathbf{C}_1\mathbf{\Delta}_1$. Hence

$$\begin{pmatrix} \lambda_1 & \mathbf{y}^T \\ \mathbf{0} & \mathbf{A}_1 \end{pmatrix}\begin{pmatrix} 1 & \mathbf{0} \\ \mathbf{0} & \mathbf{C}_1 \end{pmatrix} = \begin{pmatrix} 1 & \mathbf{0} \\ \mathbf{0} & \mathbf{C}_1 \end{pmatrix}\begin{pmatrix} \lambda_1 & \mathbf{y}^T\mathbf{C}_1 \\ \mathbf{0} & \mathbf{\Delta}_1 \end{pmatrix}$$

and therefore $\mathbf{U}^{-1}\mathbf{A}\mathbf{U}.\mathbf{C} = \mathbf{C}\mathbf{\Delta}$, where $\mathbf{C}$ is unitary and $\mathbf{\Delta}$ upper triangular. The theorem now follows at once for upper triangular canonical forms.

The case of lower triangular canonical forms is an easy corollary. For if $\mathbf{A}$ is any given matrix there exists, as we have just shown, a unitary matrix $\mathbf{W}$ and an upper triangular matrix $\mathbf{T}$ such that $\mathbf{W}^{-1}\mathbf{A}^T\mathbf{W} = \mathbf{T}$. Hence $\mathbf{U}^{-1}\mathbf{A}\mathbf{U} = \mathbf{\Delta}$, where $\mathbf{U} = \overline{\mathbf{W}}$ is unitary and $\mathbf{\Delta} = \mathbf{T}^T$ is lower triangular.

It is instructive to consider an alternative proof of Theorem 10.4.1 which uses only invariant notions. We assert, in the first place, that *if L is a linear transformation of $\mathfrak{V}_n$ into itself, then there exists a basis $\{\mathbf{y}_1,..., \mathbf{y}_n\}$ of $\mathfrak{V}_n$ such that, for $k = 1,..., n$, $L\mathbf{y}_k$ is a linear combination of* $\mathbf{y}_1,..., \mathbf{y}_k$.

To prove this, let λ_1 be a characteristic root of the matrix representing L with respect to $\mathfrak{E}$, and let $\mathbf{y}_1$ be any characteristic vector associated with λ_1. Then

$$L\mathbf{y}_1 = \lambda_1 \mathbf{y}_1. \tag{10.4.1}$$

Denote by $\mathfrak{U}$ the space consisting of all scalar multiples of $\mathbf{y}_1$, and let $\mathfrak{U}'$ be any complement of $\mathfrak{U}$. Then, given any $\mathbf{y} \in \mathfrak{V}_n$, there exists a unique scalar $\alpha(\mathbf{y})$ such that

$$L\mathbf{y} - \alpha(\mathbf{y})\mathbf{y}_1 \in \mathfrak{U}'.$$

It is evident that

$$\alpha(t\mathbf{y}) = t.\alpha(\mathbf{y}), \qquad \alpha(\mathbf{y}+\mathbf{y}') = \alpha(\mathbf{y})+\alpha(\mathbf{y}'),$$

and therefore the mapping M of $\mathfrak{U}'$ into itself, defined by the formula

$$M\mathbf{y} = L\mathbf{y} - \alpha(\mathbf{y})\mathbf{y}_1 \qquad (\mathbf{y} \in \mathfrak{U}') \tag{10.4.2}$$

is linear. Since $\mathfrak{U}'$ is a complement of $\mathfrak{U}$, we have $d(\mathfrak{U}') = n-1$, and so $\mathfrak{U}'$ is isomorphic to $\mathfrak{V}_{n-1}$. We define a mapping M^* of $\mathfrak{V}_{n-1}$ into itself by the requirement that, if $\mathbf{y} \in \mathfrak{U}'$, $\boldsymbol{\eta} \in \mathfrak{V}_{n-1}$, and $\mathbf{y} \leftrightarrow \boldsymbol{\eta}$, then $M\mathbf{y} \leftrightarrow M^*\boldsymbol{\eta}$. Then M^* is evidently linear.

We now argue by induction with respect to n. The assertion is true trivially for $n = 1$. Assume that it is true for $n \geqslant 2$. Then $\mathfrak{V}_{n-1}$ possesses a basis $\{\boldsymbol{\eta}_2,\ldots,\boldsymbol{\eta}_n\}$ such that, for $k = 2,\ldots,n$, $M^*\boldsymbol{\eta}_k$ is a linear combination of $\boldsymbol{\eta}_2,\ldots,\boldsymbol{\eta}_n$, say

$$M^*\boldsymbol{\eta}_k = \lambda_{k2}\boldsymbol{\eta}_2+\ldots+\lambda_{kk}\boldsymbol{\eta}_k \qquad (k = 2,\ldots,n).$$

Denote by $\mathbf{y}_2,\ldots,\mathbf{y}_n$ the vectors in $\mathfrak{U}'$ corresponding to $\boldsymbol{\eta}_2,\ldots,\boldsymbol{\eta}_n$ respectively. Then $\{\mathbf{y}_2,\ldots,\mathbf{y}_n\}$ is a basis of $\mathfrak{U}'$ and

$$M\mathbf{y}_k = \lambda_{k2}\mathbf{y}_2+\ldots+\lambda_{kk}\mathbf{y}_k \qquad (k = 2,\ldots,n). \tag{10.4.3}$$

Now $\{\mathbf{y}_1,\mathbf{y}_2,\ldots,\mathbf{y}_n\}$ is a basis of $\mathfrak{V}_n$ and in view of (10.4.1), (10.4.2), and (10.4.3) it satisfies the required condition. Thus our assertion is proved.

We know by Schmidt's orthogonalization process (Theorem 2.5.6, p. 67) that $\mathfrak{V}_n$ possesses an orthonormal basis $\mathfrak{X} = \{\mathbf{x}_1,\ldots,\mathbf{x}_n\}$ such that

$$\mathbf{x}_k = c_{k1}\mathbf{y}_1+\ldots+c_{kk}\mathbf{y}_k, \quad c_{kk} \neq 0 \qquad (k = 1,\ldots,n).$$

In view of the result just proved it follows that, for $k = 1,\ldots,n$, $L\mathbf{x}_k$ is a linear combination of $\mathbf{x}_1,\ldots,\mathbf{x}_k$.

Let, now, $\mathbf{A}$ be a given matrix and let the linear transformation L be defined by the equation $\mathbf{A} = \mathscr{R}(L;\mathfrak{E})$. If $\mathbf{X}$ is the (unitary) matrix having $\mathbf{x}_1,\ldots,\mathbf{x}_n$ as its columns, then, by Theorem 4.2.8 (p. 121), $\mathbf{X}^{-1}\mathbf{A}\mathbf{X} = \mathscr{R}(L;\mathfrak{X})$. Writing $\mathbf{X}^{-1}\mathbf{A}\mathbf{X} = (b_{ij})$, we then have

$$L\mathbf{x}_k = b_{1k}\mathbf{x}_1+\ldots+b_{nk}\mathbf{x}_n \qquad (k = 1,\ldots,n).$$

But $L\mathbf{x}_k$ is a linear combination of $\mathbf{x}_1,\ldots,\mathbf{x}_k$ only. Hence $b_{ik} = 0$ $(i > k)$, and so $\mathbf{X}^{-1}\mathbf{A}\mathbf{X}$ is triangular. Theorem 10.4.1 is therefore proved once again.

Even if $\mathbf{A}$ is real its characteristic roots may be complex, and it is therefore not generally possible to find a real unitary (i.e. an orthogonal) matrix $\mathbf{U}$ such that $\mathbf{U}^{-1}\mathbf{A}\mathbf{U}$ is triangular. However, by adapting the argument used in proving Theorem 10.4.1, we obtain without difficulty the following companion result.

THEOREM 10.4.2. *A real matrix is orthogonally similar to a triangular matrix if and only if all its characteristic roots are real.*

EXERCISE 10.4.1. Write out a proof of this result.

It is interesting to observe that the main theorem on diagonal canonical forms previously established in § 10.3.3 follows once again from results on triangular canonical forms. To demonstrate this we require a simple preliminary result.

THEOREM 10.4.3. *A triangular matrix is normal if and only if it is diagonal.*

Assume, without loss of generality, that the given normal matrix $\mathbf{A} = (a_{ij})$ is of the upper triangular type. If $\mathbf{A}$ is non-diagonal, let r be the suffix of the first row in $\mathbf{A}$ which contains an element $a_{rs} \neq 0$ such that $r < s$. Then $a_{kr} = 0$ for $k \neq r$ and we have

$$(\bar{\mathbf{A}}^T\mathbf{A})_{rr} = \sum_{k=1}^{n} |a_{kr}|^2 = |a_{rr}|^2,$$

$$(\mathbf{A}\bar{\mathbf{A}}^T)_{rr} = \sum_{k=1}^{n} |a_{rk}|^2 \geqslant |a_{rr}|^2 + |a_{rs}|^2 > |a_{rr}|^2.$$

Hence $\bar{\mathbf{A}}^T\mathbf{A} \neq \mathbf{A}\bar{\mathbf{A}}^T$, and our assertion is therefore proved.

Theorem 10.3.8 can now be derived as follows. Let $\mathbf{A}$ be a normal matrix. By Theorem 10.4.1 there exists a unitary matrix $\mathbf{U}$ and a triangular matrix $\mathbf{\Delta}$ such that $\mathbf{U}^{-1}\mathbf{A}\mathbf{U} = \mathbf{\Delta}$. Hence $\mathbf{\Delta}$ is normal, and therefore diagonal by Theorem 10.4.3. Thus any normal matrix is unitarily similar to a diagonal matrix; and Theorem 10.3.8 follows.

10.4.2. Schur, in 1909, used the triangular canonical form to derive a number of interesting inequalities.

Theorem 10.4.4. *If* $\mathbf{A} = (a_{rs})$ *is any complex* $n \times n$ *matrix and* $\lambda_1, \ldots, \lambda_n$ *are its characteristic roots, then*

$$\sum_{r=1}^{n} |\lambda_r|^2 \leqslant \sum_{r,s=1}^{n} |a_{rs}|^2, \tag{10.4.4}$$

$$\sum_{r=1}^{n} |\mathfrak{Rl}\,\lambda_r|^2 \leqslant \sum_{r,s=1}^{n} \left|\frac{a_{rs}+\bar{a}_{sr}}{2}\right|^2, \tag{10.4.5}$$

$$\sum_{r=1}^{n} |\mathfrak{Im}\,\lambda_r|^2 \leqslant \sum_{r,s=1}^{n} \left|\frac{a_{rs}-\bar{a}_{sr}}{2}\right|^2. \tag{10.4.6}$$

Equality in any one of these relations implies equality in all three and occurs if and only if $\mathbf{A}$ *is normal.*

We note, in the first place, that if $\mathbf{X}$ is any matrix, then

$$\mathrm{tr}(\overline{\mathbf{X}}^T\mathbf{X}) = \sum_{s=1}^{n} (\overline{\mathbf{X}}^T\mathbf{X})_{ss} = \sum_{s=1}^{n}\sum_{r=1}^{n} (\overline{\mathbf{X}}^T)_{sr}\mathbf{X}_{rs} = \sum_{r,s=1}^{n} \overline{\mathbf{X}}_{rs}\mathbf{X}_{rs},$$

i.e.

$$\mathrm{tr}(\overline{\mathbf{X}}^T\mathbf{X}) = \sum_{r,s=1}^{n} |\mathbf{X}_{rs}|^2. \tag{10.4.7}$$

Next, suppose that $\mathbf{C}$, $\mathbf{K}$, $\mathbf{U}$ are matrices such that

$$\mathbf{C} = \mathbf{UKU}^{-1}, \quad \mathbf{U} \text{ unitary}. \tag{10.4.8}$$

Then $\overline{\mathbf{C}}^T\mathbf{C} = \mathbf{U}\overline{\mathbf{K}}^T\mathbf{KU}^{-1}$ and so $\mathrm{tr}(\overline{\mathbf{C}}^T\mathbf{C}) = \mathrm{tr}(\overline{\mathbf{K}}^T\mathbf{K})$. Hence, by (10.4.7),

$$\sum_{r,s=1}^{n} |\mathbf{C}_{rs}|^2 = \sum_{r,s=1}^{n} |\mathbf{K}_{rs}|^2. \tag{10.4.9}$$

Now, by Theorem 10.4.1, there exists a unitary matrix $\mathbf{U}$ and an upper triangular matrix $\mathbf{\Delta} = (d_{rs})$ such that

$$\mathbf{A} = \mathbf{U\Delta U}^{-1}. \tag{10.4.10}$$

Here evidently $d_{rr} = \lambda_r$ $(r = 1,..., n)$. Since (10.4.8) implies (10.4.9), we have

$$\sum_{r,s=1}^{n} |a_{rs}|^2 = \sum_{r,s=1}^{n} |d_{rs}|^2 = \sum_{r=1}^{n} |\lambda_r|^2 + \sum_{r<s} |d_{rs}|^2 \geqslant \sum_{r=1}^{n} |\lambda_r|^2.$$

Again, by (10.4.10),

$$\tfrac{1}{2}(\mathbf{A}\pm\overline{\mathbf{A}}^T) = \mathbf{U}.\tfrac{1}{2}(\mathbf{\Delta}\pm\overline{\mathbf{\Delta}}^T).\mathbf{U}^{-1},$$

and hence, by (10.4.9),

$$\sum_{r,s=1}^{n} \left|\frac{a_{rs}\pm\bar{a}_{sr}}{2}\right|^2 = \sum_{r,s=1}^{n} \left|\frac{d_{rs}\pm\bar{d}_{sr}}{2}\right|^2 = \sum_{r=1}^{n} \left|\frac{\lambda_r\pm\bar{\lambda}_r}{2}\right|^2 + \sum_{r\neq s} \left|\frac{d_{rs}\pm\bar{d}_{sr}}{2}\right|^2$$

$$\geqslant \sum_{r=1}^{n} \left|\frac{\lambda_r\pm\bar{\lambda}_r}{2}\right|^2.$$

The three inequalities are therefore established.

Equality in (10.4.4) evidently occurs if and only if $d_{rs} = 0$ $(r < s)$; it occurs in (10.4.5) if and only if $d_{rs}+\bar{d}_{sr} = 0$ $(r \neq s)$, and in (10.4.6) if and only if $d_{rs}-\bar{d}_{sr} = 0$ $(r \neq s)$. Now since $d_{rs} = 0$ for $r > s$, it follows that each of the above conditions is equivalent to the requirement that $\mathbf{\Delta}$ should be diagonal. The proof is therefore complete in view of Theorem 10.3.8.

Schur's inequalities imply a number of further results with most of which we are already familiar.

(i) The inequalities of Theorem 10.4.4 obviously imply the weaker inequalities of Theorem 7.5.3 (p. 211).

(ii) Theorem 7.5.2 may be deduced from (10.4.6) as follows. If $\mathbf{A} = (a_{rs})$ is real, then

$$\sum_{r=1}^{n} |\mathfrak{Im}\,\lambda_r|^2 \leqslant \sum_{\substack{r,s=1\\ r\neq s}}^{n} \left|\frac{a_{rs}-a_{sr}}{2}\right|^2 \leqslant \alpha^2 n(n-1), \tag{10.4.11}$$

where
$$\alpha = \max_{1\leqslant r,s\leqslant n} \tfrac{1}{2}|a_{rs}-a_{sr}|.$$

Let λ be any characteristic root of $\mathbf{A}$. Since the non-real characteristic roots of a real matrix occur in conjugate pairs, it follows that every non-vanishing value of $|\mathfrak{Im}\,\lambda_r|^2$ occurs at least twice on the left-hand side of (10.4.11). Hence
$$2|\mathfrak{Im}\,\lambda|^2 \leqslant \alpha^2 n(n-1),$$
and Theorem 7.5.2 follows.

(iii) Using the inequality of the arithmetic and geometric means and (10.4.4), we obtain

$$|\det \mathbf{A}|^2 = |\lambda_1|^2 ... |\lambda_n|^2 \leqslant \left\{\frac{|\lambda_1|^2+...+|\lambda_n|^2}{n}\right\}^n \leqslant \left\{\frac{1}{n}\sum_{r,s=1}^{n} |a_{rs}|^2\right\}^n.$$

If $\max_{1\leqslant r,s\leqslant n} |a_{rs}| = \rho$, then

$$|\det \mathbf{A}|^2 \leqslant \left(\frac{1}{n}\,.\,n^2\rho^2\right)^n,$$
$$|\det \mathbf{A}| \leqslant n^{\frac{1}{2}n}\rho^n.$$

An alternative derivation of this inequality will be given in § 13.5.

Exercise 10.4.2. Show that the weaker inequality $|\det \mathbf{A}| \leqslant n!\,\rho^n$ is trivial.

(iv) Let $\mathbf{A} = (a_{rs})$ be a unitary matrix. Then, by (10.4.7),

$$\sum_{r,s=1}^{n} |a_{rs}|^2 = \operatorname{tr}(\overline{\mathbf{A}}^T\mathbf{A}) = \operatorname{tr}\mathbf{I} = n,$$

and so, by (10.4.4),
$$\frac{1}{n}\sum_{r=1}^{n} |\lambda_r|^2 \leqslant 1.$$

Now the determinant of $\mathbf{A}$ has modulus 1, and therefore

$$1 = |\lambda_1 ... \lambda_n| = \{|\lambda_1|^2 ... |\lambda_n|^2\}^{1/n} \leqslant \frac{1}{n}\{|\lambda_1|^2+...+|\lambda_n|^2\} \leqslant 1.$$

Hence
$$\{|\lambda_1|^2 ... |\lambda_n|^2\}^{1/n} = \frac{1}{n}\{|\lambda_1|^2+...+|\lambda_n|^2\}.$$

But the inequality of the arithmetic and geometric means reduces to an equality if and only if the numbers involved are all equal. Hence
$$|\lambda_1| = ... = |\lambda_n| = 1.$$

Thus every characteristic root of a unitary matrix is of modulus 1, a result with which we are, of course, already familiar. Moreover, we have

$$\sum_{r=1}^{n} |\lambda_r|^2 = n = \sum_{r,s=1}^{n} |a_{rs}|^2,$$

and thus there is equality in (10.4.4). Hence any unitary matrix is unitarily similar to a diagonal matrix. This result (which is a special case of Theorem 10.3.8) was first discovered by Frobenius in 1883.

10.5. An intermediate canonical form

As we know, not every matrix is similar to a diagonal matrix. At the same time the knowledge of triangular canonical forms does not provide us with an adequate foundation for the study of the deeper properties of matrices, and some more special canonical form is therefore needed. It is, in point of fact, possible to show that every matrix $\mathbf{A}$ is similar to an 'almost diagonal' matrix $\mathbf{C}$. This means, to be rather more precise, that the diagonal elements of $\mathbf{C}$ are equal to the characteristic roots of $\mathbf{A}$, the elements immediately above the diagonal are equal to 1 or 0, and all remaining elements are equal to 0. The matrix $\mathbf{C}$ is known as the *classical canonical form* of $\mathbf{A}$, but the proof that every $\mathbf{A}$ is similar to a matrix $\mathbf{C}$ of the type just specified cannot be given here.† We propose, however, to derive a canonical form which is intermediate in type between the triangular canonical form (under the similarity group) and the classical canonical form.

Throughout the discussion below, $\mathbf{A}$ denotes a given $n \times n$ matrix, of which $\lambda_1,\ldots,\lambda_k$ are the distinct characteristic roots occurring with respective multiplicities $m_1,\ldots,m_k$. Furthermore, $\mathfrak{U}_\lambda$ denotes the vector space of vectors $\mathbf{x}$ such that

$$(\lambda\mathbf{I}-\mathbf{A})^n\mathbf{x} = \mathbf{0}.$$

We require a number of preliminary results.

THEOREM 10.5.1. *For every number λ we have*

$$d(\mathfrak{U}_\lambda) = m_\lambda(\mathbf{A}).$$

From the discussion on p. 279 we know that $(\lambda\mathbf{I}-\mathbf{A})^n$ is a group matrix. Hence, by Theorem 9.4.2,‡

$$R\{(\lambda\mathbf{I}-\mathbf{A})^n\} = n-m_0\{(\lambda\mathbf{I}-\mathbf{A})^n\} = n-m_0(\lambda\mathbf{I}-\mathbf{A}) = n-m_\lambda(\mathbf{A}),$$

and therefore

$$d(\mathfrak{U}_\lambda) = n-R\{(\lambda\mathbf{I}-\mathbf{A})^n\} = m_\lambda(\mathbf{A}).$$

COROLLARY. $\quad d(\mathfrak{U}_{\lambda_1})+\ldots+d(\mathfrak{U}_{\lambda_k}) = n.$

THEOREM 10.5.2. *Let $\mathbf{x}$ be a non-zero vector. Then there exists a unique non-constant monic polynomial ϕ having minimum degree and such that $\phi(\mathbf{A})\mathbf{x} = \mathbf{0}$.*

† Classical canonical forms and related questions constitute one of the central topics in the theory of matrices and are discussed fully in many textbooks. For references see the footnote on p. 290.

‡ The appeal to the results of Chapter IX is not essential since it is not very difficult to show directly that, for every $n \times n$ matrix $\mathbf{B}$, $R(\mathbf{B}^n) = n-m_0(\mathbf{B})$.

The $n+1$ vectors $\mathbf{x}$, $\mathbf{Ax}$,..., $\mathbf{A}^n\mathbf{x}$ are linearly dependent and there exist, therefore, scalars $c_0, c_1,..., c_n$, not all 0, such that

$$c_0\mathbf{x}+c_1\mathbf{Ax}+...+c_n\mathbf{A}^n\mathbf{x} = \mathbf{0}.$$

In fact, since $\mathbf{x} \neq \mathbf{0}$, not all scalars $c_1,..., c_n$ are 0, and thus the non-constant polynomial $f(t) = c_0+c_1 t+...+c_n t^n$ has the property that $f(\mathbf{A})\mathbf{x} = \mathbf{0}$. Among all such polynomials there is at least one, say ϕ, which is monic and has minimum degree. Then ϕ is unique, for if $\phi_1 \neq \phi$ is another polynomial having these properties, then $\{\phi(\mathbf{A})-\phi_1(\mathbf{A})\}\mathbf{x} = \mathbf{0}$. But $\phi-\phi_1$ is of lower degree than ϕ, and we therefore have a contradiction.

DEFINITION 10.5.1. (i) *Let* $\mathbf{x} \neq \mathbf{0}$. *Any non-constant polynomial* f *such that* $f(\mathbf{A})\mathbf{x} = \mathbf{0}$ *is an* ANNIHILATING POLYNOMIAL *of* $\mathbf{x}$. (ii) *The unique polynomial* ϕ *of Theorem* 10.5.2 *is the* MINIMUM ANNIHILATING POLYNOMIAL *of* $\mathbf{x}$.

THEOREM 10.5.3. *Let* $\mathbf{x} \neq \mathbf{0}$. *Then the minimum annihilating polynomial of* $\mathbf{x}$ *divides every annihilating polynomial of* $\mathbf{x}$.

Let ϕ be the minimum annihilating polynomial of $\mathbf{x}$, and f any annihilating polynomial of $\mathbf{x}$. Then there exist polynomials g, h such that $f = g\phi+h$, where h either vanishes identically or else is of lower degree than ϕ. We then have

$$\mathbf{0} = f(\mathbf{A})\mathbf{x} = \{g(\mathbf{A})\phi(\mathbf{A})+h(\mathbf{A})\}\mathbf{x} = h(\mathbf{A})\mathbf{x}.$$

Thus we cannot have $h \neq 0$, for this would be in contradiction to the minimal definition of ϕ. Hence $h = 0, f = g\phi$, and the theorem is proved.†

THEOREM 10.5.4. *Let* $\mathfrak{B}_1,..., \mathfrak{B}_k$ *be any bases of* $\mathfrak{U}_{\lambda_1},..., \mathfrak{U}_{\lambda_k}$ *respectively. Then the n vectors contained in these bases constitute a basis of* $\mathfrak{V}_n$.

If $k = 1$, then $\mathfrak{U}_{\lambda_1} = \mathfrak{V}_n$ by Theorem 10.5.1, and the theorem is therefore true. Next, assume that $k > 1$, and write

$$\mathfrak{B}_i = \{\mathbf{x}_1^{(i)},..., \mathbf{x}_{m_i}^{(i)}\} \qquad (i = 1,..., k).$$

Let α_{ij} be scalars such that

$$\sum_{i,j} \alpha_{ij}\mathbf{x}_j^{(i)} = \mathbf{0},$$

† It will be recalled that precisely the same reasoning was used in the proof of Theorem 7.4.2 (p. 204).

where i ranges over the values $1,..., k$ and, for each i, j ranges over the values $1,..., m_i$. Write

$$\mathbf{y}_i = \sum_{j=1}^{m_i} \alpha_{ij} \mathbf{x}_j^{(i)} \qquad (i = 1,..., k).$$

Then $\mathbf{y}_i \in \mathfrak{U}_{\lambda_i}$ $(i = 1,..., k)$, and

$$\mathbf{y}_1 + ... + \mathbf{y}_k = \mathbf{0}. \tag{10.5.1}$$

It is sufficient to show that $\mathbf{y}_1 = ... = \mathbf{y}_k = \mathbf{0}$, for this implies that all α's vanish and the theorem then follows at once.

Premultiplying (10.5.1) by $(\lambda_2 \mathbf{I} - \mathbf{A})^n ... (\lambda_k \mathbf{I} - \mathbf{A})^n$, we obtain

$$(\lambda_2 \mathbf{I} - \mathbf{A})^n ... (\lambda_k \mathbf{I} - \mathbf{A})^n \mathbf{y}_1 + (\lambda_2 \mathbf{I} - \mathbf{A})^n ... (\lambda_k \mathbf{I} - \mathbf{A})^n \mathbf{y}_2 + ... = \mathbf{0}.$$

The second term on the left is equal to

$$(\lambda_3 \mathbf{I} - \mathbf{A})^n ... (\lambda_k \mathbf{I} - \mathbf{A})^n (\lambda_2 \mathbf{I} - \mathbf{A})^n \mathbf{y}_2,$$

and this vanishes since $\mathbf{y}_2 \in \mathfrak{U}_{\lambda_2}$. Similarly the third,..., kth terms also vanish, and therefore

$$(\lambda_2 \mathbf{I} - \mathbf{A})^n ... (\lambda_k \mathbf{I} - \mathbf{A})^n \mathbf{y}_1 = \mathbf{0}. \tag{10.5.2}$$

Suppose that $\mathbf{y}_1 \neq \mathbf{0}$, and denote by ϕ the minimum annihilating polynomial of $\mathbf{y}_1$. Then, by (10.5.2) and Theorem 10.5.3,

$$\phi(t) \text{ divides } (\lambda_2 - t)^n ... (\lambda_k - t)^n. \tag{10.5.3}$$

Again, $\mathbf{y}_1 \in \mathfrak{U}_{\lambda_1}$, i.e. $(\lambda_1 \mathbf{I} - \mathbf{A})^n \mathbf{y}_1 = \mathbf{0}$; and therefore

$$\phi(t) \text{ divides } (\lambda_1 - t)^n. \tag{10.5.4}$$

Since, however, $\lambda_1,..., \lambda_k$ are distinct by hypothesis, there exists no non-constant polynomial ϕ satisfying both (10.5.3) and (10.5.4). We thus arrive at a contradiction, and it follows that $\mathbf{y}_1 = \mathbf{0}$. By symmetry $\mathbf{y}_2 = ... = \mathbf{y}_k = \mathbf{0}$, and the proof of the theorem is therefore complete.

In the statement and proof of the main theorem below we shall make use of the following notation. If $\mathbf{C}_1,..., \mathbf{C}_k$ are square matrices of order $m_1,..., m_k$ respectively, then we denote by

$$\mathbf{dg}(\mathbf{C}_1,..., \mathbf{C}_k)$$

the square matrix $\mathbf{C}$, of order $m_1 + ... + m_k$, of which $\mathbf{C}_1,..., \mathbf{C}_k$ are submatrices such that the diagonal of $\mathbf{C}$ consists of the diagonals of $\mathbf{C}_1,..., \mathbf{C}_k$ (in that order), while every element of $\mathbf{C}$ which does not belong to any of $\mathbf{C}_1,..., \mathbf{C}_k$ is equal to 0. Thus, for example, if

$$\mathbf{D} = \begin{pmatrix} d_{11} & d_{12} \\ d_{21} & d_{22} \end{pmatrix}, \qquad \mathbf{E} = \begin{pmatrix} e_{11} & e_{12} & e_{13} \\ e_{21} & e_{22} & e_{23} \\ e_{31} & e_{32} & e_{33} \end{pmatrix},$$

then
$$\mathbf{dg}(\mathbf{D}, \mathbf{E}) = \begin{pmatrix} d_{11} & d_{12} & 0 & 0 & 0 \\ d_{21} & d_{22} & 0 & 0 & 0 \\ 0 & 0 & e_{11} & e_{12} & e_{13} \\ 0 & 0 & e_{21} & e_{22} & e_{23} \\ 0 & 0 & e_{31} & e_{32} & e_{33} \end{pmatrix}.$$

If $\mathbf{C}_1,..., \mathbf{C}_k$ are all 1×1 matrices, then $\mathbf{dg}(\mathbf{C}_1,..., \mathbf{C}_k)$ is simply a diagonal matrix, and the notation in this case reduces to one that is already familiar.

We may note that if, for $i = 1,..., k$, the order of $\mathbf{C}_i$ is the same as that of $\mathbf{C}'_i$, then

$$\mathbf{dg}(\mathbf{C}_1,..., \mathbf{C}_k).\mathbf{dg}(\mathbf{C}'_1,..., \mathbf{C}'_k) = \mathbf{dg}(\mathbf{C}_1 \mathbf{C}'_1,..., \mathbf{C}_k \mathbf{C}'_k).$$

Theorem 10.5.5. *If the characteristic roots of the $n \times n$ matrix $\mathbf{A}$ are $\lambda_1,..., \lambda_k$, with multiplicities $m_1,..., m_k$ respectively, then there exists a (non-singular) matrix $\mathbf{S}$ and upper triangular matrices $\mathbf{\Delta}_1,..., \mathbf{\Delta}_k$, of order $m_1,..., m_k$ respectively, such that*

$$\mathbf{S}^{-1}\mathbf{A}\mathbf{S} = \mathbf{dg}(\mathbf{\Delta}_1,..., \mathbf{\Delta}_k).$$

In this statement 'lower triangular' can, of course, be substituted for 'upper triangular'.

Denote by L the linear transformation of $\mathfrak{V}_n$ into itself such that $\mathbf{A} = \mathscr{R}(L; \mathfrak{E})$. If $\mathbf{x} \in \mathfrak{U}_{\lambda_i}$, then

$$(\lambda_i \mathbf{I} - \mathbf{A})^n(L\mathbf{x}) = (\lambda_i \mathbf{I} - \mathbf{A})^n \mathbf{A}\mathbf{x} = \mathbf{A}(\lambda_i \mathbf{I} - \mathbf{A})^n \mathbf{x} = \mathbf{A}\mathbf{0} = \mathbf{0}.$$

Thus $\mathbf{x} \in \mathfrak{U}_{\lambda_i}$ implies $L\mathbf{x} \in \mathfrak{U}_{\lambda_i}$.

By Theorem 10.5.4 we know that a basis $\mathfrak{B}$ of $\mathfrak{V}_n$ can be selected having its first m_1 vectors of $\mathfrak{U}_{\lambda_1}$, the next m_2 vectors in $\mathfrak{U}_{\lambda_2}$, and so on. In view of the result just proved and (4.2.13) it follows that

$$\mathscr{R}(L; \mathfrak{B}) = \mathbf{dg}(\mathbf{A}_1,..., \mathbf{A}_k),$$

where $\mathbf{A}_1,..., \mathbf{A}_k$ are of order $m_1,..., m_k$ respectively. Thus, for some matrix $\mathbf{P}$,

$$\mathbf{P}^{-1}\mathbf{A}\mathbf{P} = \mathbf{dg}(\mathbf{A}_1,..., \mathbf{A}_k).$$

We now complete the proof by applying Theorem 10.4.1 separately to the matrices $\mathbf{A}_1,\ldots,\mathbf{A}_k$. There exist non-singular matrices $\mathbf{Q}_1,\ldots,\mathbf{Q}_k$, of order $m_1,\ldots,m_k$ respectively, such that

$$\mathbf{Q}_i^{-1}\mathbf{A}_i\mathbf{Q}_i = \mathbf{\Delta}_i \qquad (i = 1,\ldots,k),$$

where $\mathbf{\Delta}_i$ is an upper triangular matrix of order m_i. Writing

$$\mathbf{Q} = \mathbf{dg}(\mathbf{Q}_1,\ldots,\mathbf{Q}_k), \qquad \mathbf{S} = \mathbf{PQ},$$

we obtain

$$\begin{aligned}\mathbf{S}^{-1}\mathbf{AS} &= \mathbf{Q}^{-1}\mathbf{P}^{-1}\mathbf{APQ} \\ &= \mathbf{dg}(\mathbf{Q}_1^{-1},\ldots,\mathbf{Q}_k^{-1}).\mathbf{dg}(\mathbf{A}_1,\ldots,\mathbf{A}_k).\mathbf{dg}(\mathbf{Q}_1,\ldots,\mathbf{Q}_k) \\ &= \mathbf{dg}(\mathbf{Q}_1^{-1}\mathbf{A}_1\mathbf{Q}_1,\ldots,\mathbf{Q}_k^{-1}\mathbf{A}_k\mathbf{Q}_k) = \mathbf{dg}(\mathbf{\Delta}_1,\ldots,\mathbf{\Delta}_k).\end{aligned}$$

10.6. Simultaneous similarity transformations

In the preceding sections we have been investigating canonical forms of individual matrices. It is, however, of interest to inquire whether several matrices can be reduced to their canonical forms by the *same* transformation.

DEFINITION 10.6.1. *The matrices* $\mathbf{A}, \mathbf{B}, \mathbf{C},\ldots$ *are* SIMULTANEOUSLY SIMILAR *to diagonal (triangular) matrices if there exists a matrix* $\mathbf{S}$ *such that* $\mathbf{S}^{-1}\mathbf{AS}$, $\mathbf{S}^{-1}\mathbf{BS}$, $\mathbf{S}^{-1}\mathbf{CS},\ldots$ *are all diagonal (triangular). If, in addition,* $\mathbf{S}$ *is unitary, then* $\mathbf{A}, \mathbf{B}, \mathbf{C},\ldots$ *are said to be* SIMULTANEOUSLY UNITARILY SIMILAR *to diagonal (triangular) matrices.*†

The problem is to determine under what conditions the given matrices are simultaneously (unitarily) similar to diagonal or triangular matrices. As we shall see below, commutativity in pairs of the given matrices is the essential condition in each case. To simplify the argument we shall restrict ourselves, in the first place, to the case of two matrices only.

10.6.1. We require, to begin with, some preliminary results on matrices of a special class.

DEFINITION 10.6.2. *A matrix of* TYPE $(r_1,\ldots,r_k)$ *is a matrix of order* $r_1+\ldots+r_k$ *having the form* $\mathbf{dg}(\mathbf{A}_1,\ldots,\mathbf{A}_k)$, *where* $\mathbf{A}_1,\ldots,\mathbf{A}_k$ *are of order* $r_1,\ldots,r_k$ *respectively.*

The type of a matrix is not defined uniquely. Thus the matrix

$$\begin{pmatrix} a_{11} & a_{12} & 0 \\ a_{21} & a_{22} & 0 \\ 0 & 0 & a_{33} \end{pmatrix}$$

† When we speak of triangular matrices we must, in any one problem, restrict our attention either to upper triangular matrices or to lower triangular matrices.

may equally well be said to be of type (2, 1) or of type (3). However, the question whether a given matrix is of a given type can always be decided unambiguously.

THEOREM 10.6.1. *Let* $\lambda_1,..., \lambda_k$ *be distinct numbers, and write* $r_1+...+r_k = n$. *Then an* $n \times n$ *matrix commutes with*

$$\mathbf{D} = \mathbf{dg}(\lambda_1 \mathbf{I}_{r_1},..., \lambda_k \mathbf{I}_{r_k})$$

if and only if it is of type $(r_1,..., r_k)$.

Let $\mathbf{AD} = \mathbf{DA}$ and write $\mathbf{A}$ in the partitioned form

$$\mathbf{A} = \begin{pmatrix} \mathbf{A}^{(11)} & . & . & . & \mathbf{A}^{(1k)} \\ . & . & . & . & . \\ \mathbf{A}^{(k1)} & . & . & . & \mathbf{A}^{(kk)} \end{pmatrix},$$

where $\mathbf{A}^{(ij)}$ is an $r_i \times r_j$ matrix. Then

$$\mathbf{AD} = \begin{pmatrix} \lambda_1 \mathbf{A}^{(11)} & . & . & . & \lambda_k \mathbf{A}^{(1k)} \\ . & . & . & . & . \\ \lambda_1 \mathbf{A}^{(k1)} & . & . & . & \lambda_k \mathbf{A}^{(kk)} \end{pmatrix},$$

$$\mathbf{DA} = \begin{pmatrix} \lambda_1 \mathbf{A}^{(11)} & . & . & . & \lambda_1 \mathbf{A}^{(1k)} \\ . & . & . & . & . \\ \lambda_k \mathbf{A}^{(k1)} & . & . & . & \lambda_k \mathbf{A}^{(kk)} \end{pmatrix},$$

and therefore $\lambda_i \mathbf{A}^{(ij)} = \lambda_j \mathbf{A}^{(ij)}$ $(i, j = 1,..., k)$. This implies that $\mathbf{A}^{(ij)} = \mathbf{O}$ when $i \neq j$, and thus $\mathbf{A}$ is of type $(r_1,..., r_k)$. Conversely, if $\mathbf{A}$ is of type $(r_1,..., r_k)$, then it obviously commutes with $\mathbf{D}$.

EXERCISE 10.6.1. Deduce from Theorem 10.6.1 that, if $\omega_1,..., \omega_n$ are distinct, then a matrix commutes with $\mathbf{dg}(\omega_1,..., \omega_n)$ if and only if it is diagonal.

THEOREM 10.6.2. *If a matrix* $\mathbf{A}$ *of type* $(r_1,..., r_k)$ *is similar to a diagonal matrix, then there exists a matrix* $\mathbf{S}$ *of type* $(r_1,..., r_k)$ *such that* $\mathbf{S}^{-1}\mathbf{AS}$ *is diagonal.*

Write $r_1+...+r_k = n$ and $\mathbf{A} = \mathbf{dg}(\mathbf{A}_1,..., \mathbf{A}_k)$, where $\mathbf{A}_1,..., \mathbf{A}_k$ are of order $r_1,..., r_k$ respectively. If $\mathbf{X}$ is any $p \times p$ matrix and ω any number, put

$$f_\omega(\mathbf{X}) = R(\omega \mathbf{I}_p - \mathbf{X}) + m_\omega(\mathbf{X}) - p.$$

By the rank-multiplicity theorem (p. 214) we have, for all $\mathbf{X}$ and all ω,

$$f_\omega(\mathbf{X}) \geqslant 0. \tag{10.6.1}$$

Moreover, by Theorem 10.2.3 (p. 294),

$$f_\omega(\mathbf{X}) = 0 \quad \text{(for all } \omega)$$

if and only if $\mathbf{X}$ is similar to a diagonal matrix. Now clearly

$$f_\omega(\mathbf{A}) = \sum_{i=1}^{k} f_\omega(\mathbf{A}_i),$$

and so

$$\sum_{i=1}^{k} f_\omega(\mathbf{A}_i) = 0$$

for all ω. Hence, by (10.6.1), $f_\omega(\mathbf{A}_i) = 0$ for $i = 1,\ldots,k$ and all ω. Each matrix $\mathbf{A}_i$ is therefore similar to a diagonal matrix. For $i = 1,\ldots,k$ let $\mathbf{S}_i$ be a non-singular matrix and $\mathbf{D}_i$ a diagonal matrix, both of order r_i, such that $\mathbf{S}_i^{-1}\mathbf{A}_i\mathbf{S}_i = \mathbf{D}_i$. Writing $\mathbf{S} = \mathbf{dg}(\mathbf{S}_1,\ldots,\mathbf{S}_k)$, we obtain at once

$$\mathbf{S}^{-1}\mathbf{A}\mathbf{S} = \mathbf{dg}(\mathbf{D}_1,\ldots,\mathbf{D}_k),$$

and the theorem is therefore proved.

THEOREM 10.6.3. *Two matrices are simultaneously similar to diagonal matrices if and only if they commute and each is similar to a diagonal matrix.*

Let $\mathbf{A}$, $\mathbf{B}$ be given matrices. If there exists a matrix $\mathbf{S}$ such that $\mathbf{S}^{-1}\mathbf{A}\mathbf{S}$, $\mathbf{S}^{-1}\mathbf{B}\mathbf{S}$ are both diagonal, then these two matrices commute and therefore $\mathbf{A}$ and $\mathbf{B}$ commute.

Suppose, on the other hand, that $\mathbf{AB} = \mathbf{BA}$ and that $\mathbf{A}$ and $\mathbf{B}$ are both similar to diagonal matrices. Let $\lambda_1,\ldots,\lambda_k$ be the distinct characteristic roots of $\mathbf{A}$ and let their multiplicities be $r_1,\ldots,r_k$ respectively. There exists, then, a matrix $\mathbf{P}$ such that

$$\mathbf{P}^{-1}\mathbf{A}\mathbf{P} = \mathbf{dg}(\lambda_1\mathbf{I}_{r_1},\ldots,\lambda_k\mathbf{I}_{r_k}).$$

Now, in view of our hypothesis, $\mathbf{P}^{-1}\mathbf{A}\mathbf{P}$ commutes with $\mathbf{P}^{-1}\mathbf{B}\mathbf{P}$ and hence, by Theorem 10.6.1, $\mathbf{P}^{-1}\mathbf{B}\mathbf{P}$ is of type $(r_1,\ldots,r_k)$. Since $\mathbf{B}$ is similar to a diagonal matrix, so is $\mathbf{P}^{-1}\mathbf{B}\mathbf{P}$; therefore, by Theorem 10.6.2, there exists a matrix $\mathbf{Q}$, of type $(r_1,\ldots,r_k)$, such that $\mathbf{Q}^{-1}\mathbf{P}^{-1}\mathbf{B}\mathbf{P}\mathbf{Q}$ is diagonal. Moreover, again by Theorem 10.6.1, $\mathbf{Q}$ commutes with $\mathbf{P}^{-1}\mathbf{A}\mathbf{P}$, and therefore

$$\mathbf{Q}^{-1}\mathbf{P}^{-1}\mathbf{A}\mathbf{P}\mathbf{Q} = \mathbf{dg}(\lambda_1\mathbf{I}_{r_1},\ldots,\lambda_k\mathbf{I}_{r_k}).$$

Thus $(\mathbf{PQ})^{-1}\mathbf{A}(\mathbf{PQ})$ and $(\mathbf{PQ})^{-1}\mathbf{B}(\mathbf{PQ})$ are both diagonal, and the theorem is proved.

EXERCISE 10.6.2. Write out the proof of Theorem 10.6.3 for the case when the characteristic roots of one of the two given matrices are distinct.

It is worth mentioning that Theorem 10.6.3 can be appreciably extended. It can, in fact, be shown that the matrices $\mathbf{A}_1,\ldots,\mathbf{A}_m$

are simultaneously similar to diagonal matrices if and only if they commute in pairs and each is similar to a diagonal matrix.†

10.6.2. To deal with the case of unitary similarity transformations, we shall need the following result.

THEOREM 10.6.4. *Two commuting matrices possess a common characteristic vector.*

Suppose that $\mathbf{AB} = \mathbf{BA}$. Let λ be a characteristic root of $\mathbf{A}$ and $\{\mathbf{x}_1,\dots,\mathbf{x}_k\}$ a basis of the vector space $\mathfrak{U}$ of vectors $\mathbf{x}$ satisfying the equation $\mathbf{Ax} = \lambda\mathbf{x}$. Since $\mathbf{Ax}_i = \lambda\mathbf{x}_i$ $(i = 1,\dots, k)$ we obtain

$$\mathbf{A}(\mathbf{Bx}_i) = \mathbf{B}(\mathbf{Ax}_i) = \lambda(\mathbf{Bx}_i) \qquad (i = 1,\dots, k).$$

Thus $\mathbf{Bx}_i \in \mathfrak{U}$ $(i = 1,\dots, k)$ and there exist, therefore, scalars α_{ij} $(i, j = 1,\dots, k)$ such that

$$\mathbf{Bx}_j = \sum_{i=1}^{k} \alpha_{ij}\mathbf{x}_i \qquad (j = 1,\dots, k).$$

If μ is any characteristic root of the $k\times k$ matrix (α_{ij}), and $(t_1,\dots, t_k)^T$ is any characteristic vector of (α_{ij}) associated with μ, then

$$\mathbf{B}\Big(\sum_{j=1}^{k} t_j\mathbf{x}_j\Big) = \sum_{j=1}^{k} t_j \sum_{i=1}^{k} \alpha_{ij}\mathbf{x}_i = \sum_{i=1}^{k} \mathbf{x}_i \sum_{j=1}^{k} \alpha_{ij} t_j = \mu \sum_{i=1}^{k} t_i\mathbf{x}_i.$$

Thus $\mathbf{x} = \sum_{i=1}^{k} t_i\mathbf{x}_i$ is a common characteristic vector of $\mathbf{A}$ and $\mathbf{B}$.

THEOREM 10.6.5. *If two matrices commute, then they are simultaneously unitarily similar to triangular matrices.*

If $\mathbf{AB} = \mathbf{BA}$, then, by Theorem 10.6.4, there exists a unit vector $\mathbf{x}$ and scalars α, β such that $\mathbf{Ax} = \alpha\mathbf{x}$, $\mathbf{Bx} = \beta\mathbf{x}$. Let $\mathbf{U}$ be any unitary matrix having $\mathbf{x}$ as its first column. Then

$$\mathbf{U}^{-1}\mathbf{AU} = \begin{pmatrix} \alpha & \mathbf{p} \\ \mathbf{0} & \mathbf{A}_1 \end{pmatrix}, \qquad \mathbf{U}^{-1}\mathbf{BU} = \begin{pmatrix} \beta & \mathbf{q} \\ \mathbf{0} & \mathbf{B}_1 \end{pmatrix},$$

where $\mathbf{A}_1$, $\mathbf{B}_1$ are square matrices of order $n-1$ and $\mathbf{p}$, $\mathbf{q}$ row vectors of order $n-1$. Hence

$$\mathbf{U}^{-1}\mathbf{ABU} = \begin{pmatrix} \alpha\beta & \alpha\mathbf{q}+\mathbf{pB}_1 \\ \mathbf{0} & \mathbf{A}_1\mathbf{B}_1 \end{pmatrix}, \qquad \mathbf{U}^{-1}\mathbf{BAU} = \begin{pmatrix} \alpha\beta & \beta\mathbf{p}+\mathbf{qA}_1 \\ \mathbf{0} & \mathbf{B}_1\mathbf{A}_1 \end{pmatrix}.$$

Since $\mathbf{AB} = \mathbf{BA}$ it follows that $\mathbf{A}_1\mathbf{B}_1 = \mathbf{B}_1\mathbf{A}_1$; and we now argue by induction with respect to n. Suppose that any two commuting

† For a proof of this theorem and for many related results on simultaneous similarity transformations see M. P. Drazin, J. W. Dungey, and K. W. Gruenberg, *J. London Math. Soc.* 26 (1951), 221–8.

matrices of order $n-1$ are simultaneously unitarily similar to triangular matrices. Then there exists a unitary matrix $\mathbf{V}_1$, of order $n-1$, such that $\mathbf{V}_1^{-1}\mathbf{A}_1\mathbf{V}_1$ and $\mathbf{V}_1^{-1}\mathbf{B}_1\mathbf{V}_1$ are both triangular. Writing

$$\mathbf{V} = \begin{pmatrix} 1 & \mathbf{0} \\ \mathbf{0} & \mathbf{V}_1 \end{pmatrix}$$

we see that $\mathbf{UV}$ is unitary, and $(\mathbf{UV})^{-1}\mathbf{A}(\mathbf{UV})$, $(\mathbf{UV})^{-1}\mathbf{B}(\mathbf{UV})$ are both triangular. The proof is therefore complete.

EXERCISE 10.6.3. Deduce Theorem 10.3.8 from Theorem 10.6.5 without making use of Theorem 10.4.3.

EXERCISE 10.6.4. Show that the converse of Theorem 10.6.5 is false.

10.6.3. We shall next demonstrate that the restriction of Theorem 10.6.5 to the case of *two* matrices is not essential. We begin with a few obvious remarks.

When we say that the matrices $\mathbf{A}_1,..., \mathbf{A}_k$ are linearly independent we mean, of course, that the relation

$$\alpha_1 \mathbf{A}_1 + ... + \alpha_k \mathbf{A}_k = \mathbf{O}$$

implies $\alpha_1 = ... = \alpha_k = 0$. A set of $n \times n$ matrices contains at most n^2 linearly independent ones. If k is the maximum number of linearly independent matrices in the set and if $\mathbf{A}_1,..., \mathbf{A}_k$ are linearly independent matrices of the set, then every matrix of the set can be expressed as a linear combination of $\mathbf{A}_1,..., \mathbf{A}_k$.

THEOREM 10.6.6. *Let $\mathfrak{S}$ be a (finite or infinite) set of matrices which commute in pairs. Then all matrices in $\mathfrak{S}$ possess a common characteristic vector.*

Let k be the maximum number of linearly independent matrices in $\mathfrak{S}$ and let $\mathbf{A}_1,..., \mathbf{A}_k \in \mathfrak{S}$ be linearly independent. It is then clearly sufficient to show that $\mathbf{A}_1,..., \mathbf{A}_k$ possess a common characteristic vector.

Let $1 \leqslant r < k$ and assume that $\mathbf{A}_1,..., \mathbf{A}_r$ possess a common characteristic vector, say $\mathbf{x}_r$. Then there exist numbers $\alpha_1,..., \alpha_r$ such that

$$\mathbf{A}_i \mathbf{x}_r = \alpha_i \mathbf{x}_r \qquad (i = 1,..., r).$$

Let $\mathfrak{J}$ denote the set of all vectors $\mathbf{y}$ satisfying the r conditions

$$\mathbf{A}_i \mathbf{y} = \alpha_i \mathbf{y} \qquad (i = 1,..., r).$$

Then $\mathfrak{J}$ is not empty since $\mathbf{x}_r \in \mathfrak{J}$. Let s be the maximum number of linearly independent vectors in $\mathfrak{J}$, and let $\mathbf{y}_1,..., \mathbf{y}_s \in \mathfrak{J}$ be linearly independent. Then

$$\mathbf{A}_i \mathbf{y}_j = \alpha_i \mathbf{y}_j \qquad (i = 1,..., r;\ j = 1,..., s), \tag{10.6.2}$$

and we have, for $1 \leqslant i \leqslant r$, $1 \leqslant j \leqslant s$,

$$\mathbf{A}_i(\mathbf{A}_{r+1}\mathbf{y}_j) = \mathbf{A}_{r+1}(\mathbf{A}_i \mathbf{y}_j) = \alpha_i(\mathbf{A}_{r+1}\mathbf{y}_j).$$

Hence, for $j = 1,..., s$, $\mathbf{A}_{r+1}\mathbf{y}_j \in \mathfrak{J}$, and so there exist numbers c_{ij} such that

$$\mathbf{A}_{r+1}\mathbf{y}_j = \sum_{i=1}^{s} c_{ij}\mathbf{y}_i \qquad (j = 1,..., s). \tag{10.6.3}$$

There exist a number μ and numbers $t_1,..., t_s$, not all zero, such that†

$$\sum_{j=1}^{s} c_{ij} t_j = \mu t_i \qquad (i = 1,..., s). \tag{10.6.4}$$

Write
$$\mathbf{x}_{r+1} = \sum_{j=1}^{s} t_j \mathbf{y}_j.$$

Since $t_1,..., t_s$ are not all zero and $\mathbf{y}_1,..., \mathbf{y}_s$ are linearly independent, it follows that $\mathbf{x}_{r+1} \neq \mathbf{0}$. Moreover, for $1 \leqslant i \leqslant r$ we have by (10.6.2),

$$\mathbf{A}_i \mathbf{x}_{r+1} = \sum_{j=1}^{s} t_j \mathbf{A}_i \mathbf{y}_j = \sum_{j=1}^{s} t_j \alpha_i \mathbf{y}_j = \alpha_i \mathbf{x}_{r+1}.$$

Again, using (10.6.3) and (10.6.4), we obtain

$$\mathbf{A}_{r+1}\mathbf{x}_{r+1} = \sum_{j=1}^{s} t_j \mathbf{A}_{r+1}\mathbf{y}_j = \sum_{j=1}^{s} t_j \sum_{i=1}^{s} c_{ij}\mathbf{y}_i = \sum_{i=1}^{s} \mathbf{y}_i \sum_{j=1}^{s} c_{ij} t_j$$
$$= \mu \sum_{i=1}^{s} t_i \mathbf{y}_i = \mu \mathbf{x}_{r+1}.$$

We see, then, that the matrices $\mathbf{A}_1,..., \mathbf{A}_r, \mathbf{A}_{r+1}$ possess a common characteristic vector $\mathbf{x}_{r+1}$. Hence, by induction with respect to r, it follows that $\mathbf{A}_1,..., \mathbf{A}_k$ possess a common characteristic vector. The theorem is therefore established.‡

We are now able to prove an important generalization, due to Frobenius, of Theorem 10.6.5.

Theorem 10.6.7. *Let $\mathfrak{S}$ be a set of matrices which commute in pairs. Then the matrices of $\mathfrak{S}$ are simultaneously unitarily similar to triangular matrices.*

By Theorem 10.6.6, all matrices in $\mathfrak{S}$ possess a common characteristic vector, say $\mathbf{x}$, which may be assumed to be normalized. If

† In fact, μ is a characteristic root and $(t_1,..., t_s)^T$ an associated characteristic vector of the $s \times s$ matrix (c_{ij}).

‡ I owe this proof to Professor R. Rado.

$\mathbf{U}$ is a unitary matrix with $\mathbf{x}$ as its first column, then, for every $\mathbf{A} \in \mathfrak{S}$, we have

$$\mathbf{U}^{-1}\mathbf{A}\mathbf{U} = \begin{pmatrix} \alpha & \mathbf{y}^T \\ \mathbf{O}_{n-1}^1 & \mathbf{A}_1 \end{pmatrix}.$$

Here the scalar α, the vector $\mathbf{y}$ (of order $n-1$), and the square matrix $\mathbf{A}_1$ (of order $n-1$) depend on $\mathbf{A}$. Let $\mathfrak{S}_1$ be the set of matrices $\mathbf{A}_1$ corresponding to $\mathbf{A} \in \mathfrak{S}$. Since the matrices of $\mathfrak{S}$ commute in pairs, so do the matrices of $\mathfrak{S}_1$. The proof of the theorem can now be completed in the usual way by induction with respect to n.†

THEOREM 10.6.8. *The matrices of a set $\mathfrak{S}$ of normal matrices are simultaneously unitarily similar to diagonal matrices if and only if they commute in pairs.*

If the matrices of a set $\mathfrak{S}$ of normal matrices are simultaneously unitarily similar to diagonal matrices, then they obviously commute in pairs. If, on the other hand, they commute in pairs, then by Theorem 10.6.7 they are simultaneously unitarily similar to triangular matrices. Each of these triangular matrices is again normal and so, by Theorem 10.4.3 (p. 309), diagonal. Hence the matrices of $\mathfrak{S}$ are simultaneously unitarily similar to diagonal matrices.

PROBLEMS ON CHAPTER X

1. Let $\mathbf{A}$ be any given matrix. Show that there exist unique matrices $\mathbf{P}$, $\mathbf{Q}$ such that: (i) $\mathbf{A} = \mathbf{P}+\mathbf{Q}$, (ii) $\mathbf{P}$ is hermitian and $\mathbf{Q}$ skew-hermitian. Show further that $\mathbf{A}$ is normal if and only if $\mathbf{PQ} = \mathbf{QP}$.

2. Show that a matrix $\mathbf{A}$ is normal if and only if $|\mathbf{Ax}| = |\bar{\mathbf{A}}^T\mathbf{x}|$ for all vectors $\mathbf{x}$.

3. Let $\mathbf{A}$ be a normal matrix. Show that (i) if all characteristic roots of $\mathbf{A}$ are real, then $\mathbf{A}$ is hermitian; (ii) if all characteristic roots of $\mathbf{A}$ are of modulus 1, then $\mathbf{A}$ is unitary.

4. Suppose that $\mathbf{A}$ is similar to a diagonal matrix and that a positive power of $\mathbf{A}$ is equal to $\mathbf{O}$. Show that $\mathbf{A} = \mathbf{O}$.

5. Show that an orthogonal matrix all of whose characteristic roots are real is necessarily symmetric.

6. Find a similarity transformation which reduces the real matrix $\begin{pmatrix} a & b \\ -b & a \end{pmatrix}$ to diagonal form.

7. If $\mathbf{A}^2 = \mathbf{I}$, show that $\operatorname{tr}\mathbf{A}$ is an integer. Show also that, if $\mathbf{A}$ is not a scalar matrix, then $|\operatorname{tr}\mathbf{A}| < n$.

† Cf. the proof of Theorem 10.6.5.

8. Express the matrix

$$\frac{1}{9}\begin{pmatrix} 1 & -14 & -10 \\ -14 & -2 & -4 \\ -10 & -4 & 10 \end{pmatrix}$$

in the form $\mathbf{XKX}^T$, where $\mathbf{X}$ is orthogonal and $\mathbf{K}$ diagonal.

9. The following problem was considered in Chapter VII. 'If all elements of $\mathbf{A}^m$ are bounded as $m \to \infty$, show that all characteristic roots of $\mathbf{A}$ are, in modulus, less than or equal to 1.' Prove this result by the use of triangular canonical forms.

10. Let $\lambda_1,\ldots,\lambda_n$ be the characteristic roots of $\mathbf{A}$ and $\mu_1,\ldots,\mu_n$ the (necessarily real) characteristic roots of $\bar{\mathbf{A}}^T\mathbf{A}$. Show that

$$|\lambda_1|^2+\ldots+|\lambda_n|^2 \leqslant \mu_1+\ldots+\mu_n,$$

with equality if and only if $\mathbf{A}$ is normal.

11. Prove the result of Problem VI, 15 by the methods of the present chapter.

12. Show that every real symmetric 3×3 matrix $\mathbf{A}$ whose characteristic roots are $1, -1, -1$ can be written in the form $\mathbf{A} = 2\mathbf{l}\mathbf{l}^T - \mathbf{I}_3$, where $\mathbf{l}$ is a suitable unit vector.

13. Find orthogonal reductions for each of the matrices

(i) $\begin{pmatrix} 0 & 1 & 0 \\ 1 & 0 & 0 \\ 0 & 0 & 2 \end{pmatrix}$; (ii) $\begin{pmatrix} 0 & 1 & 0 \\ 1 & -3/2 & 0 \\ 0 & 0 & 1 \end{pmatrix}$; (iii) $\begin{pmatrix} 2 & 3\sqrt{2} & 3\sqrt{2} \\ 3\sqrt{2} & 7 & -5 \\ 3\sqrt{2} & -5 & 7 \end{pmatrix}$.

14. Find all solutions of the equation

$$\mathbf{X}^2 = \begin{pmatrix} 4 & 4 & 4 \\ 6 & 7 & 6 \\ -6 & -7 & -6 \end{pmatrix}.$$

15. Find a matrix $\mathbf{X}$ such that $2\mathbf{X}^2-5\mathbf{X} = \mathbf{A}$, where

$$\mathbf{A} = \begin{pmatrix} -1 & 0 & 2 \\ 0 & 1 & 2 \\ 2 & 2 & 0 \end{pmatrix}.$$

16. Suppose that the characteristic roots of $\mathbf{A}$ are distinct and that $\mathbf{B}$ has the same characteristic roots as $\mathbf{A}$. Show that there exist matrices $\mathbf{Q}$ and $\mathbf{R}$, at least one of them non-singular, such that $\mathbf{A} = \mathbf{QR}$, $\mathbf{B} = \mathbf{RQ}$.

17. Let $\mathbf{A} = (a_{rs})$ be an arbitrary $n\times n$ complex matrix and write $\rho = \max_{1\leqslant r,s\leqslant n} |a_{rs}|$. Show that $|\det \mathbf{A}| = n^{\frac{1}{2}n}\rho^n$ if and only if the following three conditions are satisfied. (i) $\mathbf{A}$ is normal; (ii) $|a_{rs}| = \rho$ $(1 \leqslant r,s \leqslant n)$; (iii) all characteristic roots of $\mathbf{A}$ are equal in modulus.

18. Show that $\mathbf{A}$ is normal if and only if the characteristic roots of $\bar{\mathbf{A}}^T\mathbf{A}$ are equal to the squares of the moduli of the characteristic roots of $\mathbf{A}$.

19. Show that an $n\times n$ matrix is normal if and only if it possesses an orthonormal set of n characteristic vectors.

20. Show that, if $f(\mathbf{A},\mathbf{B}) = \mathrm{tr}(\bar{\mathbf{A}}^T\mathbf{B})$, then

$$|f(\mathbf{A},\mathbf{B})|^2 \leqslant f(\mathbf{A},\mathbf{A}).f(\mathbf{B},\mathbf{B}).$$

Deduce that $$N(\mathbf{A}+\mathbf{B}) \leqslant N(\mathbf{A})+N(\mathbf{B}),$$

where $N(\mathbf{A}) = \{f(\mathbf{A},\mathbf{A})\}^{\frac{1}{2}}$.

21. Deduce from the inequality (10.4.4), p. 309, that all characteristic roots of a hermitian matrix are real.

22. The matrix $\mathbf{A}$ has distinct characteristic roots $\omega_1,\ldots,\omega_n$. If $\phi(t)$ is any function, let $\phi(\mathbf{A})$ be defined by the equation $\phi(\mathbf{A}) = f(\mathbf{A})$, where $f(t)$ is a polynomial such that $f(\omega_\nu) = \phi(\omega_\nu)$ $(\nu = 1,\ldots,n)$. Show that $\phi(\mathbf{A})$ depends on $\mathbf{A}$ and $\phi(t)$ only, but not on the choice of $f(t)$.

23. Use the result of Problem VII, 39 to show that, if $\mathbf{A}$ satisfies the equation $\mathbf{A}^2 = \mathbf{I}$, then it is similar to a diagonal matrix.

24. Show that a matrix which is similar to a diagonal matrix possesses a critical principal minor. Deduce that the rank of a skew-hermitian matrix is even.

25. Let $\mathbf{A}$ be a unitary matrix, $\mathbf{B}$ a normal matrix, and denote by β, β^* the characteristic roots of $\mathbf{B}$ of least and greatest modulus respectively. By using the result of Problem VIII, 19 show that if ω is any number which satisfies $|\mathbf{A}-\omega\mathbf{B}| = 0$, then $|\beta| \leqslant |\omega|^{-1} \leqslant |\beta^*|$.

26. Find necessary and sufficient conditions for the matrix $\begin{pmatrix} a & b \\ c & d \end{pmatrix}$ to be similar to a diagonal matrix.

27. Explain how to obtain all $n\times n$ matrices $\mathbf{X}$ which satisfy the equation $f(\mathbf{X}) = \mathbf{O}$, where f is a polynomial with distinct zeros.

Solve the equations

$$\text{(i)}\quad \mathbf{X}^2 = \mathbf{X}; \qquad \text{(ii)}\quad \mathbf{X}^2+4\mathbf{X}+3\mathbf{I} = \mathbf{O}.$$

28. $\mathbf{A}$ and $\mathbf{B}$ are matrices such that (i) the characteristic roots of $\mathbf{A}$ are distinct; (ii) every characteristic vector of $\mathbf{A}$ is also a characteristic vector of $\mathbf{B}$. Show that $\mathbf{AB} = \mathbf{BA}$. Show also that if $\mathbf{AB} = \mathbf{BA}$ and (i) is given, then (ii) is valid.

29. Show that a 2×2 matrix which is not similar to any diagonal matrix is similar to some matrix of the form $\begin{pmatrix} \lambda & 1 \\ 0 & \lambda \end{pmatrix}$.

Let $\mathfrak{S}$ be the set of all 2×2 matrices $\mathbf{A}$ having the property that any two matrices which commute with $\mathbf{A}$ commute with each other. If $\mathbf{A} \in \mathfrak{S}$ and $\mathbf{P}$ is non-singular, show that $\mathbf{P}^{-1}\mathbf{AP} \in \mathfrak{S}$; and deduce that a matrix belongs to $\mathfrak{S}$ if and only if it is non-scalar.

30. Show that $\mathbf{A}$ is similar to a diagonal matrix if and only if, for every number ω,
$$(\omega\mathbf{I}-\mathbf{A})^2\mathbf{x} = \mathbf{0} \quad \text{implies} \quad (\omega\mathbf{I}-\mathbf{A})\mathbf{x} = \mathbf{0}.$$

31. If the $n\times n$ matrix $\mathbf{A}$ has distinct characteristic roots and $\mathbf{AB} = \mathbf{BA}$, show that $\mathbf{B} = f(\mathbf{A})$, where f is a polynomial of degree $\leqslant n-1$.

32. Show that $\bar{\mathbf{A}}^T$ is equal to a polynomial in $\mathbf{A}$ if and only if $\mathbf{A}$ is normal.

33. Show, by using either the preceding question or Problem VII, 25 that $\mathbf{AB} = \mathbf{BA}$ implies $\bar{\mathbf{A}}^T\mathbf{B} = \mathbf{B}\bar{\mathbf{A}}^T$ for every $\mathbf{B}$ if and only if $\mathbf{A}$ is normal.

34. Suppose that the matrix $\mathbf{A}$ satisfies the equation

$$(\mathbf{A}-\omega_1\mathbf{I})...(\mathbf{A}-\omega_k\mathbf{I}) = \mathbf{O},$$

where $\omega_1,...,\omega_k$ are distinct numbers. If, for any $\mathbf{x} \in \mathfrak{V}_n$,

$$\mathbf{y}_i = \prod_{\substack{\kappa=1 \\ \kappa\neq i}}^{k} \left(\frac{\mathbf{A}-\omega_\kappa\mathbf{I}}{\omega_i-\omega_\kappa}\right).\mathbf{x} \qquad (i = 1,...,k),$$

show that $\mathbf{A}\mathbf{y}_i = \omega_i\mathbf{y}_i$ $(i = 1,...,k)$, and that $\mathbf{x} = \mathbf{y}_1+...+\mathbf{y}_k$. Using Problem II, 22, deduce that $\mathbf{A}$ possesses n linearly independent characteristic vectors. Hence show that a matrix, whose minimum polynomial is the product of distinct linear factors, is similar to a diagonal matrix.

35. Let $\omega_1,...,\omega_k$ be the distinct values of the characteristic roots of $\mathbf{A}$, and let, for $1 \leqslant i \leqslant k$, $\mathfrak{U}_i$ denote the space of vectors $\mathbf{x}$ such that $\mathbf{A}\mathbf{x} = \omega_i\mathbf{x}$. Show that, if the minimum polynomial of $\mathbf{A}$ is the product of distinct linear factors, then $d(\mathfrak{U}_1)+...+d(\mathfrak{U}_k) = n$.

36. Let $\mathbf{A}$ be a given matrix and let $\mathfrak{U}_1,..., \mathfrak{U}_k$ be defined as in the preceding question. Using Problem VII, 40, show that, if $d(\mathfrak{U}_1)+...+d(\mathfrak{U}_k) = n$, then $\mathbf{A}$ is similar to a diagonal matrix.

Deduce Theorem 10.2.3 (p. 294) from this result, and also establish the converse inference.

37. $\mathbf{A}$ and $\mathbf{B}$ are commuting 2×2 matrices. Prove that there exists a matrix $\mathbf{C}$ and polynomials f, g such that $\mathbf{A} = f(\mathbf{C})$, $\mathbf{B} = g(\mathbf{C})$. Show also that this assertion becomes false for matrices of order greater than 2.

38. Show that, if $\mathbf{A}$ and $\mathbf{B}$ are 2×2 matrices with the same characteristic vectors, then $\mathbf{AB} = \mathbf{BA}$. Show also that this inference is false for matrices of order greater than 2.

39. Let $\mathfrak{S}$ be a set of $n\times n$ matrices such that, if $\mathbf{A}, \mathbf{B} \in \mathfrak{S}$, then $\mathbf{A}^2 = \mathbf{A}$, $\mathbf{AB} = \mathbf{O}$. Prove, by induction with respect to the number of matrices in $\mathfrak{S}$, that the matrices in $\mathfrak{S}$ are simultaneously similar to diagonal matrices.

Deduce that, if $\mathbf{A}_1,..., \mathbf{A}_k \in \mathfrak{S}$, then

$$R(\mathbf{A}_1+...+\mathbf{A}_k) = R(\mathbf{A}_1)+...+R(\mathbf{A}_k),$$

and hence show that $\mathfrak{S}$ contains at most n non-zero matrices.

40. Let $\mathbf{A}$ be a matrix which is similar to a diagonal matrix, and put

$$\mathbf{E}_i = \prod_{\substack{1\leqslant\nu\leqslant k \\ \nu\neq i}} \left(\frac{\mathbf{A}-\alpha_\nu\mathbf{I}}{\alpha_i-\alpha_\nu}\right) \qquad (i = 1,...,k),$$

where $\alpha_1,...,\alpha_k$ are the distinct values of the characteristic roots of $\mathbf{A}$. Show that

(i) $\mathbf{E}_i\mathbf{E}_j = \mathbf{O}$ $(i,j = 1,...,k;\ i \neq j)$;

(ii) $\mathbf{E}_1+...+\mathbf{E}_k = \mathbf{I}$;

(iii) $\mathbf{E}_i^2 = \mathbf{E}_i$ $(i = 1,...,k)$;

(iv) $\mathbf{A} = \alpha_1\mathbf{E}_1+...+\alpha_k\mathbf{E}_k$.

41. Let $\mathfrak{S}$ be a set of matrices which commute in pairs and are similar to diagonal matrices. By considering the $\mathbf{E}$-matrices (defined in the preceding question) associated with every matrix of $\mathfrak{S}$, show that the matrices of $\mathfrak{S}$ are simultaneously similar to diagonal matrices.

42. Show that the matrices of a set $\mathfrak{S}$ are simultaneously similar to triangular matrices if and only if there exist linearly independent vectors $\mathbf{x}_1,\ldots,\mathbf{x}_n$ such that, for any $\mathbf{A} \in \mathfrak{S}$ and any k in the range $1 \leqslant k \leqslant n$, $\mathbf{A}\mathbf{x}_k$ is a linear combination of $\mathbf{x}_1,\ldots,\mathbf{x}_k$.

Deduce that, if the matrices of $\mathfrak{S}$ are simultaneously similar to triangular matrices, then they are simultaneously *unitarily* similar to triangular matrices.

43. Show that an orthogonal matrix all of whose characteristic roots are real is orthogonally similar to a diagonal matrix.

44. Let $\mathbf{A}$ be an orthogonal matrix, $e^{-i\alpha}$ a non-real characteristic root of $\mathbf{A}$, and $\mathbf{x}$ a vector such that $\mathbf{A}\mathbf{x} = e^{-i\alpha}\mathbf{x}$, $\bar{\mathbf{x}}^T\mathbf{x} = 2$. Verify that $\mathbf{x}^T\mathbf{x} = 0$, and prove that the vectors $\boldsymbol{\xi} = \frac{1}{2}(\mathbf{x}+\bar{\mathbf{x}})$, $\boldsymbol{\eta} = \frac{1}{2i}(\mathbf{x}-\bar{\mathbf{x}})$ satisfy the relations

$$\boldsymbol{\xi}^T\boldsymbol{\eta} = 0, \qquad \boldsymbol{\xi}^T\boldsymbol{\xi} = 1, \qquad \boldsymbol{\eta}^T\boldsymbol{\eta} = 1,$$

$$\mathbf{A}\boldsymbol{\xi} = \boldsymbol{\xi}\cos\alpha+\boldsymbol{\eta}\sin\alpha, \qquad \mathbf{A}\boldsymbol{\eta} = -\boldsymbol{\xi}\sin\alpha+\boldsymbol{\eta}\cos\alpha.$$

Hence show that if $\mathbf{S}$ is any orthogonal matrix with $\boldsymbol{\xi},\boldsymbol{\eta}$ as its first two columns, then

$$\mathbf{S}^{-1}\mathbf{A}\mathbf{S} = \begin{pmatrix} \cos\alpha & -\sin\alpha & 0 & . & . & . & 0 \\ \sin\alpha & \cos\alpha & 0 & . & . & . & 0 \\ 0 & 0 & & & & & \\ . & . & . & . & & \mathbf{A}_1 & \\ 0 & 0 & & & & & \end{pmatrix},$$

where $\mathbf{A}_1$ is a square matrix of order $n-2$. Deduce that every orthogonal matrix is orthogonally similar to a matrix of the form

$$\mathbf{dg}(\mathbf{C}_1,\ldots,\mathbf{C}_m,\mathbf{D}),$$

where $\quad \mathbf{C}_\nu = \begin{pmatrix} \cos\alpha_\nu & -\sin\alpha_\nu \\ \sin\alpha_\nu & \cos\alpha_\nu \end{pmatrix} \quad (\nu = 1,\ldots,m;\ \alpha_\nu \text{ real})$

and $\mathbf{D}$ is a diagonal matrix all of whose diagonal elements are ± 1.

45. Show that, if $\mathbf{A}$, $\mathbf{B}$, and $\mathbf{AB}$ are normal, then so is $\mathbf{BA}$.

46. If d denotes the discriminant of the characteristic polynomial of the $n\times n$ matrix $\mathbf{A} = (a_{rs})$, show that

$$|d|^{2/\{n(n-1)\}} \leqslant \frac{2}{n(n-1)} \sum_{r<s} |a_{rr}-a_{ss}|^2 + \frac{2}{n-1} \sum_{r\neq s} |a_{rs}|^2.$$

47. Show that, given any square matrix $\mathbf{A}$, there exist unitary matrices $\mathbf{U}$, $\mathbf{V}$ such that $\mathbf{UAV}$ is diagonal.

XI

MATRIX ANALYSIS

In our discussion so far we have confined ourselves to purely algebraic methods—that is, methods based on the four rational operations—and have avoided reference to limiting processes.† It is interesting, however, to remove this restriction in order to see how an *analysis* of matrices might be initiated. In the first three sections of the present chapter we propose to develop the more elementary parts of such a theory, but we shall be able to do no more than touch the fringe of an extensive subject.‡ In the last section we shall approach the question from the opposite point of view and use the theory of matrices to solve a problem in classical analysis. The distinctive common feature of both lines of inquiry is, of course, the fusion of algebraic and analytic ideas. The results obtained in this chapter will not be needed in Part III.

11.1. Convergent matrix sequences

11.1.1. An analytic theory of matrices must naturally be founded on the notion of convergence or some equivalent notion.

DEFINITION 11.1.1. *The sequence* $\{\mathbf{A}_m\}$ *of* $n \times n$ *matrices* CONVERGES (*or* TENDS TO) $\mathbf{A}$ (*in symbols*: $\mathbf{A}_m \to \mathbf{A}$ *as* $m \to \infty$ *or* $\lim_{m\to\infty} \mathbf{A}_m = \mathbf{A}$) *if*

$$(\mathbf{A}_m)_{rs} \to \mathbf{A}_{rs} \text{ as } m \to \infty \qquad (r, s = 1,..., n).$$

A sequence which does not converge is said to DIVERGE.

In other words, $\{\mathbf{A}_m\}$ converges to $\mathbf{A}$ if each element of $\mathbf{A}_m$ converges, as $m \to \infty$, to the corresponding element of $\mathbf{A}$.

In many ways the behaviour of sequences of matrices resembles that of sequences of numbers. Thus, if $\mathbf{A}_m \to \mathbf{A}$, $\mathbf{B}_m \to \mathbf{B}$ $(m \to \infty)$, then

$$\alpha\mathbf{A}_m + \beta\mathbf{B}_m \to \alpha\mathbf{A} + \beta\mathbf{B}$$

and

$$\mathbf{A}_m \mathbf{B}_m \to \mathbf{AB}.$$

† The only notable deviation from this course consisted in the appeal to the fundamental theorem of algebra, which was needed to establish the existence of characteristic roots of every matrix.

‡ For a more adequate treatment see Ferrar, **21**, chap. v; Frazer, Duncan, and Collar, **22**, chap. ii; and Wedderburn, **39**, chap. viii.

The last statement implies, in particular, that if **P**, **Q** are any matrices, then

$$\mathbf{P}\mathbf{A}_m\mathbf{Q} \to \mathbf{P}\mathbf{A}\mathbf{Q}.$$

EXERCISE 11.1.1. Prove these results.

DEFINITION 11.1.2. *Let* $\mathbf{A} = (a_{rs})$ *be a complex matrix and* $\mathbf{B} = (b_{rs})$ *a non-negative matrix (i.e. a matrix whose elements are real non-negative numbers.) Then* **B** *is said to* MAJORIZE **A** *(in symbols:* $\mathbf{A} \ll \mathbf{B}$*) if*

$$|a_{rs}| \leqslant b_{rs} \qquad (r, s = 1,..., n).$$

EXERCISE 11.1.2. Let **A** be a complex matrix, **B** a non-negative matrix, and m a positive integer. Show that $\mathbf{B}^m$ is non-negative and that, if $\mathbf{A} \ll \mathbf{B}$, then $\mathbf{A}^m \ll \mathbf{B}^m$.

EXERCISE 11.1.3. Let $\{\mathbf{A}_m\}$ be a sequence of complex matrices, $\{\mathbf{B}_m\}$ a sequence of non-negative matrices, and suppose that $\mathbf{A}_m \ll \mathbf{B}_m$ and $\mathbf{B}_m \to \mathbf{O}$. Show that $\mathbf{A}_m \to \mathbf{O}$.

DEFINITION 11.1.3. *For any matrix* $\mathbf{A} = (a_{rs})$ *we write*

$$||\mathbf{A}|| = \max_{1 \leqslant r,s \leqslant n} |a_{rs}|.$$

EXERCISE 11.1.4. Let **A** be a complex matrix and **B** a non-negative matrix, and suppose that $\mathbf{A} \ll \mathbf{B}$. Show that $||\mathbf{A}|| \leqslant ||\mathbf{B}||$.

EXERCISE 11.1.5. Show that $||\mathbf{ABC}|| \leqslant n^2\, ||\mathbf{A}|| \,.\, ||\mathbf{B}|| \,.\, ||\mathbf{C}||$.

EXERCISE 11.1.6. Show that $\mathbf{A}_m \to \mathbf{O}$ if and only if $||\mathbf{A}_m|| \to 0$.

The most natural sequences are those formed by powers of a single matrix, and some conditions governing the convergence of such sequences are given in the next theorem and in Exercise 11.1.8.

THEOREM 11.1.1. *A necessary and sufficient condition for the relation* $\mathbf{A}^m \to \mathbf{O}$ *to hold is that the moduli of all the characteristic roots of* **A** *should be less than* 1.

Let $\lambda_1,..., \lambda_n$ denote the characteristic roots of **A**. By Theorem 10.4.1 (p. 307), there exists a non-singular matrix **U** and an upper triangular matrix $\mathbf{\Delta}$ (with diagonal elements $\lambda_1,..., \lambda_n$) such that $\mathbf{U}^{-1}\mathbf{A}\mathbf{U} = \mathbf{\Delta}$.

To prove the necessity of the stated condition we note that

$$\mathbf{U}^{-1}\mathbf{A}^m\mathbf{U} = \mathbf{\Delta}^m.$$

Hence $\mathbf{A}^m \to \mathbf{O}$ implies $\mathbf{\Delta}^m \to \mathbf{O}$, and since the diagonal elements of $\mathbf{\Delta}^m$ are $\lambda_1^m,..., \lambda_n^m$, it follows that

$$|\lambda_1| < 1, \quad ..., \quad |\lambda_n| < 1. \tag{11.1.1}$$

Next, we establish the sufficiency of (11.1.1). We can choose *distinct* real numbers $\omega_1,\ldots,\omega_n$ and a real number q such that

$$|\lambda_i| \leqslant \omega_i \leqslant q < 1 \qquad (i = 1,\ldots,k).$$

Let $\mathbf{T}$ be any upper triangular non-negative matrix majorizing $\mathbf{\Delta}$ and having $\omega_1,\ldots,\omega_n$ as its diagonal elements. Since the characteristic roots $\omega_1,\ldots,\omega_n$ of $\mathbf{T}$ are distinct, there exists a non-singular matrix $\mathbf{S}$ such that

$$\mathbf{S}^{-1}\mathbf{T}\mathbf{S} = \mathbf{dg}(\omega_1,\ldots,\omega_n) = \mathbf{\Lambda}, \text{ say.}$$

Now $\mathbf{U}^{-1}\mathbf{A}\mathbf{U} \leqslant \mathbf{T}$ and so, by Exercise 11.1.2,

$$\mathbf{U}^{-1}\mathbf{A}^m\mathbf{U} \leqslant \mathbf{T}^m = \mathbf{S}\mathbf{\Lambda}^m\mathbf{S}^{-1}.$$

Hence, by Exercises 11.1.4 and 11.1.5,

$$\|\mathbf{U}^{-1}\mathbf{A}^m\mathbf{U}\| \leqslant \|\mathbf{S}\mathbf{\Lambda}^m\mathbf{S}^{-1}\| \leqslant n^2\|\mathbf{S}\|.\|\mathbf{\Lambda}^m\|.\|\mathbf{S}^{-1}\|$$
$$\leqslant n^2\|\mathbf{S}\|.\|\mathbf{S}^{-1}\|.q^m \to 0.$$

Consequently, by Exercise 11.1.6, $\mathbf{U}^{-1}\mathbf{A}^m\mathbf{U} \to \mathbf{O}$, and so $\mathbf{A}^m \to \mathbf{O}$.

EXERCISE 11.1.7. Interpret Theorem 11.1.1 for the case $n = 1$.

EXERCISE 11.1.8. Show that if the modulus of at least one characteristic root of $\mathbf{A}$ exceeds 1, then the sequence $\{\mathbf{A}^m\}$ diverges.

11.1.2. As an application of the result just proved we shall establish a theorem due to Frobenius (1908) on the location of characteristic roots of matrices. If the characteristic roots of $\mathbf{A}$ are $\lambda_1,\ldots,\lambda_n$, we shall write

$$\rho(\mathbf{A}) = \max_{1\leqslant r\leqslant n} |\lambda_r|.$$

THEOREM 11.1.2. *If* $\mathbf{A}$ *is a complex matrix,* $\mathbf{B}$ *a non-negative matrix, and* $\mathbf{A} \leqslant \mathbf{B}$, *then* $\rho(\mathbf{A}) \leqslant \rho(\mathbf{B})$.

Let λ be any characteristic root of $\mathbf{A}$, and let μ_0 be a characteristic root of $\mathbf{B}$ having maximum modulus. Let $\epsilon > 0$, and write

$$\mathbf{A}' = \frac{1}{|\mu_0|+\epsilon}\mathbf{A}, \qquad \mathbf{B}' = \frac{1}{|\mu_0|+\epsilon}\mathbf{B}.$$

All characteristic roots of $\mathbf{B}'$ are less than 1 in modulus and so, by the sufficiency part of Theorem 11.1.1, $\mathbf{B}'^m \to \mathbf{O}$. But $\mathbf{A}' \leqslant \mathbf{B}'$ and so, by Exercise 11.1.2, $\mathbf{A}'^m \leqslant \mathbf{B}'^m$. Hence, by Exercise 11.1.3, $\mathbf{A}'^m \to \mathbf{O}$ and so, by the necessity part of Theorem 11.1.1, all characteristic roots of $\mathbf{A}'$ are less than 1 in modulus. Thus $|\lambda|/(|\mu_0|+\epsilon) < 1$, and since ϵ can be chosen arbitrarily small, this implies that $|\lambda| \leqslant |\mu_0|$, which is equivalent to the assertion.

COROLLARY. *If* $\mathbf{A} = (a_{rs})$, $\mathbf{A}' = (|a_{rs}|)$, *then* $\rho(\mathbf{A}) \leqslant \rho(\mathbf{A}')$.

THEOREM 11.1.3. *If* $\mathbf{B}$ *is any non-negative matrix obtained from a non-negative matrix* $\mathbf{A}$ *when the latter is 'bordered' by rows and columns, then* $\rho(\mathbf{A}) \leqslant \rho(\mathbf{B})$.

Let $\mathbf{A}$ be of order n and $\mathbf{B}$ of order N $(> n)$. Write

$$\mathbf{A}' = \begin{pmatrix} \mathbf{A} & \mathbf{O}_n^{N-n} \\ \mathbf{O}_{N-n}^n & \mathbf{O}_{N-n}^{N-n} \end{pmatrix}.$$

The characteristic roots of $\mathbf{A}'$ are the characteristic roots of $\mathbf{A}$ together with $N-n$ zeros. Hence $\rho(\mathbf{A}) = \rho(\mathbf{A}')$. Moreover, by Theorem 11.1.2, $\rho(\mathbf{A}') \leqslant \rho(\mathbf{B})$, and the assertion therefore follows.

11.2. Power series and matrix functions

In the analysis of functions of a scalar variable power series play a prominent part. It is therefore natural that in seeking to develop an analysis of matrices we should again make extensive use of this notion. Most theorems derived below are concerned with power series, but some of the results they contain are necessarily fragmentary since a satisfactory discussion of the problems involved can only be based on the classical canonical form. Here, on the other hand, we must content ourselves with making what use we can of triangular and diagonal canonical forms.

11.2.1. In ordinary analysis convergence of infinite series is defined in terms of the behaviour of their partial sums. The extension of this idea to matrices is immediate.

DEFINITION 11.2.1. *The series of matrices*

$$\sum_{m=0}^{\infty} \mathbf{A}_m \tag{11.2.1}$$

is said to CONVERGE *to, or to have the* SUM, $\mathbf{S}$ *if the sequence of partial sums*

$$\mathbf{S}_N = \sum_{m=0}^{N} \mathbf{A}_m$$

converges to $\mathbf{S}$ *as* $N \to \infty$. *A series which does not converge is said to* DIVERGE.

The statement $\sum_{m=0}^{\infty} \mathbf{A}_m = \mathbf{S}$ thus means that

$$\sum_{m=0}^{\infty} (\mathbf{A}_m)_{rs} = \mathbf{S}_{rs} \qquad (r, s = 1,..., n). \tag{11.2.2}$$

EXERCISE 11.2.1. Show that if $\sum_{m=0}^{\infty} \mathbf{A}_m$ is a convergent matrix series, then $\sum_{m=0}^{\infty} \mathbf{P}\mathbf{A}_m \mathbf{Q}$ is also convergent, and

$$\sum_{m=0}^{\infty} \mathbf{P}\mathbf{A}_m \mathbf{Q} = \mathbf{P}\Big(\sum_{m=0}^{\infty} \mathbf{A}_m\Big)\mathbf{Q}.$$

DEFINITION 11.2.2. *The series* (11.2.1) *is* ABSOLUTELY CONVERGENT *if each of the series on the left-hand side of* (11.2.2) *is absolutely convergent.*

It follows at once from corresponding results in scalar analysis that, if the series (11.2.1) is absolutely convergent, then it is convergent and its terms may be rearranged in any manner without alteration of the sum.

THEOREM 11.2.1. *The series $\sum_{m=0}^{\infty} \mathbf{A}_m$ is absolutely convergent if and only if $\sum_{m=0}^{\infty} \|\mathbf{A}_m\|$ is convergent.*

Suppose that $\sum_{m=0}^{\infty} \mathbf{A}_m$ is absolutely convergent. Then there exists a number K, independent of N, r, s such that

$$\sum_{m=0}^{N} |(\mathbf{A}_m)_{rs}| < K \qquad (N \geqslant 0;\ r, s = 1,..., n).$$

Hence
$$\sum_{m=0}^{N} \|\mathbf{A}_m\| \leqslant \sum_{m=0}^{N} \sum_{r,s=1}^{n} |(\mathbf{A}_m)_{rs}| < n^2 K,$$

and $\sum_{m=0}^{\infty} \|\mathbf{A}_m\|$ is therefore convergent.

On the other hand, if $\sum_{m=0}^{\infty} \|\mathbf{A}_m\|$ is convergent then, since

$$|(\mathbf{A}_m)_{rs}| \leqslant \|\mathbf{A}_m\| \qquad (r, s = 1,..., n),$$

it follows that each of the series on the left-hand side of (11.2.2) is absolutely convergent. The theorem is therefore proved.

THEOREM 11.2.2. *If $\sum_{m=0}^{\infty} \mathbf{A}_m$ is absolutely convergent, then so is $\sum_{m=0}^{\infty} \mathbf{P}\mathbf{A}_m \mathbf{Q}$.*

By Exercise 11.1.5 we have

$$\|\mathbf{P}\mathbf{A}_m \mathbf{Q}\| \leqslant n^2 \|\mathbf{P}\| . \|\mathbf{A}_m\| . \|\mathbf{Q}\| \leqslant K\|\mathbf{A}_m\|,$$

where K is independent of m. Now, by Theorem 11.2.1, $\sum_{m=0}^{\infty} \|\mathbf{A}_m\|$

is convergent. Hence $\sum_{m=0}^{\infty} \|\mathbf{PA}_m\mathbf{Q}\|$ is convergent; and therefore, again by Theorem 11.2.1, $\sum_{m=0}^{\infty} \mathbf{PA}_m\mathbf{Q}$ is absolutely convergent.

11.2.2. After the preliminary remarks of § 11.2.1. we turn to the discussion of power series and begin with an easy result on infinite geometric progressions.

THEOREM 11.2.3. *If the moduli of all characteristic roots of* $\mathbf{A}$ *are less than* 1, *then* $\mathbf{I}-\mathbf{A}$ *is non-singular, and the series*

$$\mathbf{I}+\mathbf{A}+\mathbf{A}^2+\mathbf{A}^3+...$$

converges to $(\mathbf{I}-\mathbf{A})^{-1}$.

The first assertion is obvious. Writing

$$\mathbf{S}_N = \mathbf{I}+\mathbf{A}+...+\mathbf{A}^N,$$

we obtain

$$\mathbf{S}_N(\mathbf{I}-\mathbf{A}) = \mathbf{I}-\mathbf{A}^{N+1}.$$

Hence, by Theorem 11.1.1, $\mathbf{S}_N(\mathbf{I}-\mathbf{A})\rightarrow\mathbf{I}$, and the required conclusion follows.

More generally, it is possible to obtain a striking relation between the matrix power series $\sum_{m=0}^{\infty} c_m\mathbf{A}^m$ and the corresponding scalar power series $\sum_{m=0}^{\infty} c_m z^m$.

Theorem 11.2.4. (i) *If all characteristic roots of* $\mathbf{A}$ *lie in the interior of the circle of convergence of the power series*

$$\phi(z) = \sum_{m=0}^{\infty} c_m z^m, \tag{11.2.3}$$

then the matrix power series

$$\sum_{m=0}^{\infty} c_m\mathbf{A}^m \tag{11.2.4}$$

converges absolutely. (ii) *If at least one characteristic root of* $\mathbf{A}$ *lies outside the circle of convergence of* (11.2.3), *then* (11.2.4) *diverges.*

This theorem is due to Weyr (1887). A more complete result was found in 1926 by Hensel who dealt fully with the critical cases when some of the characteristic roots of $\mathbf{A}$ lie *on* the circle of convergence. Hensel showed, in fact, that the matrix power series (11.2.4) converges if and only if all characteristic roots of $\mathbf{A}$ lie within or on the circle of convergence of (11.2.3) and satisfy the further condition that, for every k-fold characteristic root λ on the circle of convergence, the power series for $\phi^{(k-1)}(\lambda)$ is convergent.

To establish Theorem 11.2.4 we argue in much the same way as in the proof of Theorem 11.1.1. Let R be the radius of convergence of the power series (11.2.3) and denote by $\lambda_1,\ldots,\lambda_n$ the characteristic roots of $\mathbf{A}$. Let $\mathbf{U}$ be a non-singular matrix and $\boldsymbol{\Delta}$ an upper triangular matrix, with diagonal elements $\lambda_1,\ldots,\lambda_n$, such that $\mathbf{U}^{-1}\mathbf{A}\mathbf{U} = \boldsymbol{\Delta}$. Then the diagonal elements of $\sum_{m=0}^{N} c_m \boldsymbol{\Delta}^m$ are

$$\sum_{m=0}^{N} c_m \lambda_i^m \qquad (i = 1,\ldots,n);$$

and if, for at least one i, $|\lambda_i| > R$, then the series

$$\sum_{m=0}^{\infty} c_m \boldsymbol{\Delta}^m \tag{11.2.5}$$

diverges; in that case (11.2.4) also diverges in view of Exercise 11.2.1.

Suppose, on the other hand, that $|\lambda_1| < R,\ldots, |\lambda_n| < R$; and let $\omega_1,\ldots,\omega_n$ be *distinct* real numbers and q a real number such that

$$|\lambda_i| \leqslant \omega_i \leqslant q < R \qquad (i = 1,\ldots,n).$$

Let $\mathbf{T}$ be any upper triangular matrix majorizing $\boldsymbol{\Delta}$ and having $\omega_1,\ldots,\omega_n$ as its diagonal elements, and let $\mathbf{S}$ be a non-singular matrix such that

$$\mathbf{S}^{-1}\mathbf{T}\mathbf{S} = \mathbf{dg}(\omega_1,\ldots,\omega_n) = \boldsymbol{\Lambda}, \text{ say.}$$

Then $\boldsymbol{\Delta}^m \ll \mathbf{S}\boldsymbol{\Lambda}^m\mathbf{S}^{-1}$, and so

$$c_m \boldsymbol{\Delta}^m \ll |c_m|\mathbf{S}\boldsymbol{\Lambda}^m\mathbf{S}^{-1} \qquad (m \geqslant 0).$$

Hence, by Exercises 11.1.4 and 11.1.5,

$$\|c_m \boldsymbol{\Delta}^m\| \leqslant \|\, |c_m|\mathbf{S}\boldsymbol{\Lambda}^m\mathbf{S}^{-1}\| \leqslant |c_m| n^2 \|\mathbf{S}\| . \|\boldsymbol{\Lambda}^m\| . \|\mathbf{S}^{-1}\| \leqslant K|c_m| q^m,$$

where K is independent of m. Since $0 \leqslant q < R$, it follows that $\sum_{m=0}^{\infty} \|c_m \boldsymbol{\Delta}^m\|$ converges. Hence, by Theorem 11.2.1, the series (11.2.5) converges absolutely; and therefore, by Theorem 11.2.2, so does the series (11.2.4).

It is worth pointing out that for two special cases the proof of Theorem 11.2.4 can be simplified still further. In the first instance assume that **A** is similar to a diagonal matrix, say

$$\mathbf{S}^{-1}\mathbf{AS} = \mathbf{dg}(\lambda_1,..., \lambda_n),$$

where $\lambda_1,..., \lambda_n$ are, of course, the characteristic roots of **A**. Writing

$$\phi_N(z) = \sum_{m=0}^{N} c_m z^m$$

we obtain $$\phi_N(\mathbf{A}) = \mathbf{S}.\mathbf{dg}\{\phi_N(\lambda_1),..., \phi_N(\lambda_n)\}.\mathbf{S}^{-1}.$$

Now if some λ_i lies outside the circle of convergence of (11.2.3), then the sequence $\phi_N(\lambda_i)$ does not converge, and so (11.2.4) diverges. If, on the other hand, all λ_i lie within the circle of convergence of (11.2.3), then

$$\phi_N(\lambda_i) \to \phi(\lambda_i) \quad \text{as } N \to \infty \qquad (i = 1,..., n)$$

and therefore the series (11.2.4) converges (absolutely) to

$$\mathbf{S}.\mathbf{dg}\{\phi(\lambda_1),..., \phi(\lambda_n)\}.\mathbf{S}^{-1}. \tag{11.2.6}$$

In this case, then, we not only recognize the fact of convergence but also obtain an explicit formula for the sum of the matrix power series (11.2.4).

We state the second special case in the form of a corollary.

COROLLARY. *If the power series* (11.2.3) *converges in the whole complex plane, then the matrix power series* (11.2.4) *converges absolutely for every matrix* **A**.

An independent proof of this result can be given very easily. We can show at once by induction with respect to m that, for $m \geqslant 0$,

$$\|\mathbf{A}^m\| \leqslant (n\|\mathbf{A}\|)^m.$$

Hence $$\sum_{m=0}^{N} \|c_m \mathbf{A}^m\| \leqslant \sum_{m=0}^{N} |c_m|(n\|\mathbf{A}\|)^m < K,$$

where K is independent of N. The required conclusion therefore follows at once by Theorem 11.2.1.

11.2.3. Theorem 11.2.4 suggests a method of defining new classes of functions of a matrix variable. Before we can adopt this method, however, we need to consider briefly the case of power series whose sums are rational functions.

THEOREM 11.2.5. *Let $\phi(z)$ be a rational function and suppose that, for $|z| < R$,*

$$\sum_{m=0}^{\infty} c_m z^m = \phi(z).$$

If **A** *is any matrix whose characteristic roots are, in modulus, less than R, then $\phi(\mathbf{A})$ exists and is equal to the sum of the series $\sum_{m=0}^{\infty} c_m \mathbf{A}^m$.*

It is essential to emphasize that $\phi(\mathbf{A})$ is not *defined* as the sum of the power series $\sum_{m=0}^{\infty} c_m \mathbf{A}^m$; it is defined by Definition 3.7.2 (p. 99), and the fact that it is equal to the sum of the power series is precisely what Theorem 11.2.5 asserts.

Write $\phi(z) = f(z)/g(z)$, where $f(z)$, $g(z)$ are polynomials and, by hypothesis, $g(z) \neq 0$ for $|z| < R$. If $\lambda_1,..., \lambda_n$ are the characteristic roots of **A**, then, by Theorem 7.3.1 (p. 201),

$$|g(\mathbf{A})| = g(\lambda_1) \dots g(\lambda_n) \neq 0.$$

Hence $\phi(\mathbf{A}) = \{g(\mathbf{A})\}^{-1} f(\mathbf{A})$ exists.

Put
$$g(z) = \sum_{m=0}^{k} p_m z^m, \qquad f(z) = \sum_{m=0}^{l} q_m z^m.$$

We are given that, for $|z| < R$,

$$g(z) \sum_{m=0}^{\infty} c_m z^m = f(z),$$

and this implies that

$$p_0 c_m + p_1 c_{m-1} + \dots + p_k c_{m-k} = \begin{cases} q_m & (m = 0, 1,..., l), \\ 0 & (m > l), \end{cases}$$

where $c_r = 0$ for $r < 0$. Hence

$$\begin{aligned} g(\mathbf{A}) \sum_{m=0}^{\infty} c_m \mathbf{A}^m &= (p_0 \mathbf{I} + p_1 \mathbf{A} + \dots + p_k \mathbf{A}^k) \sum_{m=0}^{\infty} c_m \mathbf{A}^m \\ &= \sum_{m=0}^{\infty} p_0 c_m \mathbf{A}^m + \sum_{m=0}^{\infty} p_1 c_m \mathbf{A}^{m+1} + \dots + \sum_{m=0}^{\infty} p_k c_m \mathbf{A}^{m+k} \\ &= \sum_{m=0}^{\infty} p_0 c_m \mathbf{A}^m + \sum_{m=0}^{\infty} p_1 c_{m-1} \mathbf{A}^m + \dots + \sum_{m=0}^{\infty} p_k c_{m-k} \mathbf{A}^m \\ &= \sum_{m=0}^{\infty} (p_0 c_m + p_1 c_{m-1} + \dots + p_k c_{m-k}) \mathbf{A}^m \\ &= \sum_{m=0}^{l} q_m \mathbf{A}^m = f(\mathbf{A}), \end{aligned}$$

and therefore
$$\sum_{m=0}^{\infty} c_m \mathbf{A}^m = \{g(\mathbf{A})\}^{-1} f(\mathbf{A}) = \phi(\mathbf{A}).$$

EXERCISE 11.2.2. Interpret Theorem 11.2.3 in the light of Theorems 11.2.4 and 11.2.5.

We are now able to extend our definition of functions of matrices.

DEFINITION 11.2.3. *If all characteristic roots of* $\mathbf{A}$ *lie in the interior of the circle of convergence of the power series* (11.2.3), *then* $\phi(\mathbf{A})$ *is defined as the sum of the series* (11.2.4).

If $\phi(z)$ is a rational function, then $\phi(\mathbf{A})$ is defined in two ways—by Definitions 3.7.2 and 11.2.3. However, the preceding theorem shows that the two definitions of $\phi(\mathbf{A})$ are consistent.

In future, when we use the term 'matrix function' we shall mean the sum of a power series in $\mathbf{A}$ and assume that all characteristic roots of $\mathbf{A}$ lie in the interior of the circle of convergence of this power series.

11.2.4. Since the power series for $\exp z$, $\cos z$, $\sin z$ converge everywhere, it follows that the matrix functions

$$\exp \mathbf{A} = \mathbf{I} + \frac{1}{1!}\mathbf{A} + \frac{1}{2!}\mathbf{A}^2 + \frac{1}{3!}\mathbf{A}^3 + \ldots,$$

$$\cos \mathbf{A} = \mathbf{I} - \frac{1}{2!}\mathbf{A}^2 + \frac{1}{4!}\mathbf{A}^4 - \ldots,$$

$$\sin \mathbf{A} = \mathbf{A} - \frac{1}{3!}\mathbf{A}^3 + \frac{1}{5!}\mathbf{A}^5 - \ldots$$

are defined for every matrix $\mathbf{A}$. From this it follows at once that

$$\exp(i\mathbf{A}) = \cos \mathbf{A} + i \sin \mathbf{A};$$

and this, in turn, leads to the identities

$$\left.\begin{aligned} \cos \mathbf{A} &= \tfrac{1}{2}\{\exp(i\mathbf{A}) + \exp(-i\mathbf{A})\} \\ \sin \mathbf{A} &= \frac{1}{2i}\{\exp(i\mathbf{A}) - \exp(-i\mathbf{A})\} \end{aligned}\right\}. \qquad (11.2.7)$$

EXERCISE 11.2.3. Define the hyperbolic functions of $\mathbf{A}$ and derive the basic identities connecting them.

The actual evaluation of a function $\phi(\mathbf{A})$ may be carried out in different ways. We can, for instance, make use of the diagonal canonical form—if $\mathbf{A}$ possesses such a form—and obtain $\phi(\mathbf{A})$ by means of the expression (11.2.6). Alternatively we can sum the power series defining $\phi(\mathbf{A})$ by appealing to the theorem of Cayley and Hamilton. As a simple example of the latter procedure consider a matrix $\mathbf{A}$ of order 4 whose characteristic roots are π, $-\pi$, 0, 0.

Then **A** satisfies the equation $\mathbf{A}^4 - \pi^2\mathbf{A}^2 = \mathbf{O}$, and we have

$$\begin{aligned}\sin\mathbf{A} &= \mathbf{A} - \frac{1}{3!}\mathbf{A}^3 + \frac{1}{5!}\mathbf{A}^5 - \frac{1}{7!}\mathbf{A}^7 + \dots \\ &= \mathbf{A} - \frac{1}{3!}\mathbf{A}^3 + \frac{1}{5!}\pi^2\mathbf{A}^3 - \frac{1}{7!}\pi^4\mathbf{A}^3 + \dots \\ &= \mathbf{A} + \left(-\frac{1}{3!} + \frac{\pi^2}{5!} - \frac{\pi^4}{7!} + \dots\right)\mathbf{A}^3 = \mathbf{A} + \frac{\sin\pi - \pi}{\pi^3}\mathbf{A}^3 \\ &= \mathbf{A} - \pi^{-2}\mathbf{A}^3.\end{aligned}$$

In this case, then, $\sin\mathbf{A}$ is equal to a polynomial in **A**. We shall see in § 11.3 that this is not due to the specific circumstances of the problem but illustrates a general result on matrix functions.

The question to which we must now turn is whether functional equations of scalar analysis continue to remain valid when the scalar variable z is replaced by a matrix **A**. The next theorem furnishes us with most of the information we need.

Theorem 11.2.6. *Let $\phi(z)$, $\psi(z)$, $\chi(z)$ be sums of power series convergent for $|z| < R$ and suppose that*

$$\phi(z)\psi(z) = \chi(z) \qquad (|z| < R).$$

If all characteristic roots of **A** *are less than R in modulus, then*

$$\phi(\mathbf{A})\psi(\mathbf{A}) = \chi(\mathbf{A}).$$

Write

$$\phi(z) = \sum_{m=0}^{\infty} a_m z^m, \qquad \psi(z) = \sum_{m=0}^{\infty} b_m z^m, \qquad \chi(z) = \sum_{m=0}^{\infty} c_m z^m,$$

so that

$$c_m = a_0 b_m + a_1 b_{m-1} + \dots + a_m b_0 \qquad (m = 0, 1, 2,\dots).$$

Let

$$d_m = \sum_{\nu=0}^{m} |a_\nu|\,|b_{m-\nu}|.$$

Then $\sum_{m=0}^{\infty} d_m z^m$ is convergent for $|z| < R$ and therefore, by Theorem 11.2.4 (i), $\sum_{m=0}^{\infty} d_m \mathbf{A}^m$ is absolutely convergent. Hence, by Theorem 11.2.1,

$$\sum_{m=0}^{\infty}\left(\sum_{\nu=0}^{m} |a_\nu|\,|b_{m-\nu}|\right)\|\mathbf{A}^m\| = \sum_{m=0}^{\infty} \|d_m \mathbf{A}^m\|$$

is convergent. Since this is a series of non-negative terms, we see that

$$\sum_{m=0}^{\infty}\sum_{\nu=0}^{m} |a_\nu|\,|b_{m-\nu}|\,\|\mathbf{A}^m\| = \sum_{m=0}^{\infty}\sum_{\nu=0}^{m} \|a_\nu b_{m-\nu}\mathbf{A}^m\|$$

is convergent. Hence, again by Theorem 11.2.1, the matrix series

$$\sum_{m=0}^{\infty} \sum_{\nu=0}^{m} a_\nu b_{m-\nu} \mathbf{A}^m$$

is absolutely convergent and may therefore be rearranged. Accordingly, we have

$$\chi(\mathbf{A}) = \sum_{m=0}^{\infty} c_m \mathbf{A}^m = \sum_{m=0}^{\infty} \Big(\sum_{\nu=0}^{m} a_\nu b_{m-\nu}\Big)\mathbf{A}^m = \sum_{m=0}^{\infty} \sum_{\nu=0}^{m} a_\nu b_{m-\nu} \mathbf{A}^m$$

$$= \sum_{\nu=0}^{\infty} a_\nu \sum_{m=\nu}^{\infty} b_{m-\nu} \mathbf{A}^m = \sum_{\nu=0}^{\infty} a_\nu \sum_{\mu=0}^{\infty} b_\mu \mathbf{A}^{\nu+\mu} = \sum_{\nu=0}^{\infty} a_\nu \mathbf{A}^\nu \sum_{\mu=0}^{\infty} b_\mu \mathbf{A}^\mu$$

$$= \phi(\mathbf{A})\psi(\mathbf{A}).$$

EXERCISE 11.2.4. Prove Theorem 11.2.6 for the case of a matrix **A** which is similar to a diagonal matrix.

EXERCISE 11.2.5. Let $\phi(z)$, $\psi(z)$, $\chi(z)$ be sums of power series convergent for $|z| < R$, and suppose that $\phi(z)+\psi(z) = \chi(z)$. Show that if all characteristic roots of **A** are less than R in modulus, then $\phi(\mathbf{A})+\psi(\mathbf{A}) = \chi(\mathbf{A})$.

By means of Theorem 11.2.6 and Exercise 11.2.5 we can derive functional equations for matrices from corresponding scalar results. Thus, since $\exp z.\exp(-z) = 1$, it follows that, for all **A**,

$$\exp \mathbf{A}.\exp(-\mathbf{A}) = \mathbf{I}.$$

This equation shows that $\exp \mathbf{A}$ is non-singular for all **A**, and that

$$(\exp \mathbf{A})^{-1} = \exp(-\mathbf{A}). \qquad (11.2.8)$$

Again, we have $\cos 2z = \cos^2 z - \sin^2 z$. Hence, for all **A**,

$$\cos 2\mathbf{A} = (\cos \mathbf{A})^2 - (\sin \mathbf{A})^2,$$

and it is clear how other relations of the same type may be obtained.

The principal result expressing the behaviour of the scalar exponential function is the addition theorem

$$\exp x.\exp y = \exp(x+y).$$

We are naturally interested to know whether the analogous result is true for matrices, i.e. whether the identity

$$\exp \mathbf{A}.\exp \mathbf{B} = \exp(\mathbf{A}+\mathbf{B}) \qquad (11.2.9)$$

is valid. It is easy to see that this is not the case. Indeed, the validity of (11.2.9) for all **A** and **B** would imply that $\exp \mathbf{A}$ and $\exp \mathbf{B}$ always commute, which in general they do not. Consider, for example, the matrices

$$\mathbf{A} = \begin{pmatrix} 1 & 1 \\ 0 & 0 \end{pmatrix}, \qquad \mathbf{B} = \begin{pmatrix} 1 & -1 \\ 0 & 0 \end{pmatrix}.$$

By the theorem of Cayley and Hamilton we have $\mathbf{A}^2 = \mathbf{A}$, $\mathbf{B}^2 = \mathbf{B}$, and it is therefore easily verified that

$$\exp \mathbf{A} = \mathbf{I}+(e-1)\mathbf{A} = \begin{pmatrix} e & e-1 \\ 0 & 1 \end{pmatrix},$$

$$\exp \mathbf{B} = \mathbf{I}+(e-1)\mathbf{B} = \begin{pmatrix} e & 1-e \\ 0 & 1 \end{pmatrix}.$$

Hence

$$\exp \mathbf{A}.\exp \mathbf{B} = \begin{pmatrix} e^2 & -(e-1)^2 \\ 0 & 1 \end{pmatrix}, \qquad \exp \mathbf{B}.\exp \mathbf{A} = \begin{pmatrix} e^2 & (e-1)^2 \\ 0 & 1 \end{pmatrix},$$

and this suffices to show that (11.2.9) is not generally valid. In the particular case under consideration we have, moreover,

$$\mathbf{A}+\mathbf{B} = \begin{pmatrix} 2 & 0 \\ 0 & 0 \end{pmatrix},$$

so that $(\mathbf{A}+\mathbf{B})^2 = 2(\mathbf{A}+\mathbf{B})$. Therefore

$$\exp(\mathbf{A}+\mathbf{B}) = \mathbf{I}+\tfrac{1}{2}(e^2-1)(\mathbf{A}+\mathbf{B}) = \begin{pmatrix} e^2 & 0 \\ 0 & 1 \end{pmatrix}.$$

The fact that so fundamental a law as the addition theorem for the exponential function is not valid for matrices shows that matrix analysis does not follow a course analogous to that of scalar analysis but diverges from it almost at the start.

There exists, however, a valid modification of the addition theorem. To obtain it, we shall consider rectangular matrices whose elements are functions of a single variable, say t. We shall denote matrices of this type by symbols such as $\mathbf{A}(t) = (a_{ij}(t))$.

DEFINITION 11.2.4. *The matrix* $\mathbf{A}(t) = (a_{ij}(t))$ *is said to be differentiable if all its elements* $a_{ij}(t)$ *are differentiable. Its derivative is then defined by the formula*

$$\dot{\mathbf{A}}(t) = \frac{d}{dt}\mathbf{A}(t) = \left(\frac{d}{dt}a_{ij}(t)\right).$$

EXERCISE 11.2.6. Let $\mathbf{A}(t)$, $\mathbf{B}(t)$ be rectangular differentiable matrices, for which the product $\mathbf{A}(t)\mathbf{B}(t)$ is defined. Show that

$$\frac{d}{dt}\{\mathbf{A}(t)\mathbf{B}(t)\} = \dot{\mathbf{A}}(t)\mathbf{B}(t)+\mathbf{A}(t)\dot{\mathbf{B}}(t).$$

THEOREM 11.2.7. *For any matrix* $\mathbf{A}$ *we have*

$$\frac{d}{dt}\exp(t\mathbf{A}) = \mathbf{A}\exp(t\mathbf{A}).$$

To prove this identity, we observe that

$$\{\exp(t\mathbf{A})\}_{ij} = \sum_{m=0}^{\infty} \frac{1}{m!} t^m (\mathbf{A}^m)_{ij}.$$

Now the power series in t on the right-hand side converges for all values of t. It may, therefore, be differentiated term by term, and so

$$\frac{d}{dt}[\{\exp(t\mathbf{A})\}_{ij}] = \sum_{m=1}^{\infty} \frac{1}{(m-1)!} t^{m-1} (\mathbf{A}^m)_{ij}.$$

Hence

$$\frac{d}{dt}\exp(t\mathbf{A}) = \sum_{m=1}^{\infty} \frac{1}{(m-1)!} t^{m-1}\mathbf{A}^m$$

$$= \mathbf{A} \sum_{m=1}^{\infty} \frac{1}{(m-1)!} t^{m-1}\mathbf{A}^{m-1} = \mathbf{A}\exp(t\mathbf{A}).$$

THEOREM 11.2.8. *If* $\mathbf{A}$ *and* $\mathbf{B}$ *commute, then*

$$\exp\mathbf{A}.\exp\mathbf{B} = \exp\mathbf{B}.\exp\mathbf{A} = \exp(\mathbf{A}+\mathbf{B}).$$

We first observe that if $\mathbf{A}$ and $\mathbf{B}$ commute, then $\exp(t\mathbf{A})$ and $\mathbf{B}$ commute also. For, using Exercise 11.2.1 (p. 331), we have

$$\exp(t\mathbf{A}).\mathbf{B} = \left(\sum_{m=0}^{\infty} \frac{t^m}{m!}\mathbf{A}^m\right)\mathbf{B} = \sum_{m=0}^{\infty} \frac{t^m}{m!}\mathbf{A}^m\mathbf{B} = \sum_{m=0}^{\infty} \frac{t^m}{m!}\mathbf{B}\mathbf{A}^m$$

$$= \mathbf{B}\sum_{m=0}^{\infty} \frac{t^m}{m!}\mathbf{A}^m = \mathbf{B}.\exp(t\mathbf{A}).$$

We next employ an argument which is modelled on the proof of the scalar functional equation $\exp x.\exp y = \exp(x+y)$. Writing

$$\mathbf{C}(t) = \exp\{t(\mathbf{A}+\mathbf{B})\}.\exp(-t\mathbf{A}).\exp(-t\mathbf{B})$$

and using Exercise 11.2.6 and Theorem 11.2.7, we obtain

$$\dot{\mathbf{C}}(t) = (\mathbf{A}+\mathbf{B})\exp\{t(\mathbf{A}+\mathbf{B})\}.\exp(-t\mathbf{A}).\exp(-t\mathbf{B})+$$
$$+\exp\{t(\mathbf{A}+\mathbf{B})\}.(-\mathbf{A}).\exp(-t\mathbf{A}).\exp(-t\mathbf{B})+$$
$$+\exp\{t(\mathbf{A}+\mathbf{B})\}.\exp(-t\mathbf{A}).(-\mathbf{B}).\exp(-t\mathbf{B}).$$

Hence, in view of the preceding remarks, $\dot{\mathbf{C}}(t) = \mathbf{O}$, and so $\mathbf{C}(t)$ is independent of t. Thus $\mathbf{C}(1) = \mathbf{C}(0) = \mathbf{I}$, i.e.

$$\exp(\mathbf{A}+\mathbf{B}).\exp(-\mathbf{A}).\exp(-\mathbf{B}) = \mathbf{I},$$

and so, by (11.2.8),

$$\exp(\mathbf{A}+\mathbf{B}) = \exp\mathbf{B}.\exp\mathbf{A}.$$

Hence, by symmetry,

$$\exp(\mathbf{A}+\mathbf{B}) = \exp\mathbf{A}.\exp\mathbf{B}.$$

From Theorem 11.2.8 and the identities (11.2.7) on p. 336, we can easily deduce the addition theorems for trigonometric functions of a matrix variable. Thus, if $\mathbf{A}$ and $\mathbf{B}$ commute, then

$$\cos(\mathbf{A}+\mathbf{B}) = \cos\mathbf{A}\cos\mathbf{B}-\sin\mathbf{A}\sin\mathbf{B},$$

and so on. Broadly speaking, we may say that functional equations in a single scalar variable retain their validity for matrices, but functional equations in two variables remain true in matrix algebra only for pairs of *commuting* matrices.

11.3. The relation between matrix functions and matrix polynomials

Every term of a power series in $\mathbf{A}$ is equal to a polynomial in $\mathbf{A}$ whose degree is less than that of the minimum polynomial of $\mathbf{A}$. This suggests that the sum of every power series in $\mathbf{A}$ is also equal to a polynomial in $\mathbf{A}$. To prove this is the object of the present section.

THEOREM 11.3.1. *Let* $\mathbf{C}_1,..., \mathbf{C}_k$ *be linearly independent matrices,* t_{mi} $(m \geqslant 1;\ i = 1,..., k)$ *given numbers, and*

$$\mathbf{\Gamma}_m = \sum_{i=1}^{k} t_{mi}\mathbf{C}_i \qquad (m \geqslant 1). \tag{11.3.1}$$

If $\lim_{m\to\infty}\mathbf{\Gamma}_m$ *exists, then* $\lim_{m\to\infty} t_{mi}$ *exists for every* i *in the range* $1 \leqslant i \leqslant k$.

For $r, s = 1,..., n$, we denote by $\mathbf{I}^{(r,s)}$ the matrix whose (r, s)th element is equal to 1 and all of whose remaining elements are equal to 0. Writing

$$(\mathbf{\Gamma}_m)_{rs} = \gamma_{rsm} \qquad (r, s = 1,..., n;\ m \geqslant 1)$$

we have

$$\mathbf{\Gamma}_m = \sum_{r,s=1}^{n} \gamma_{rsm}\mathbf{I}^{(r,s)} \qquad (m \geqslant 1). \tag{11.3.2}$$

The existence of $\lim_{m\to\infty}\mathbf{\Gamma}_m$ means that

$$\lim_{m\to\infty}\gamma_{rsm} \text{ exists for all } r, s = 1,..., n. \tag{11.3.3}$$

By Theorem 2.3.5 (p. 54) we know that matrices $\mathbf{C}_{k+1},..., \mathbf{C}_{n^2}$ can be found such that $\mathbf{C}_1,..., \mathbf{C}_{n^2}$ constitute a basis of the linear manifold of all $n \times n$ matrices. Every matrix can then be represented as a unique linear combination of $\mathbf{C}_1,..., \mathbf{C}_{n^2}$. In particular, we write

$$\mathbf{I}^{(r,s)} = \sum_{i=1}^{n^2} u_{rsi} \mathbf{C}_i \qquad (r, s = 1,..., n).$$

Hence, by (11.3.2),

$$\mathbf{\Gamma}_m = \sum_{i=1}^{n^2} \Big(\sum_{r,s=1}^{n} \gamma_{rsm} u_{rsi} \Big) \mathbf{C}_i \qquad (m \geqslant 1),$$

and therefore, by (11.3.1),

$$t_{mi} = \sum_{r,s=1}^{n} \gamma_{rsm} u_{rsi} \qquad (m \geqslant 1; i = 1,..., k).$$

The assertion now follows at once by (11.3.3).

Theorem 11.3.2. *If $\phi(\mathbf{A})$ is a function of the matrix $\mathbf{A}$, then there exists a polynomial $p(x)$, whose degree is less than that of the minimum polynomial of $\mathbf{A}$ and whose coefficients depend on ϕ and $\mathbf{A}$, such that $\phi(\mathbf{A}) = p(\mathbf{A})$.*

It should be stressed that, if $\mathbf{B}$ is any other matrix, then, in general, $\phi(\mathbf{B}) \neq p(\mathbf{B})$. Hence the theorem must not be taken to imply that the terms 'matrix function' and 'matrix polynomial' are synonymous.†

To prove the theorem we write

$$\phi(\mathbf{A}) = \sum_{\nu=0}^{\infty} c_\nu \mathbf{A}^\nu, \qquad \phi_m(\mathbf{A}) = \sum_{\nu=0}^{m} c_\nu \mathbf{A}^\nu.$$

If k is the degree of the minimum polynomial of $\mathbf{A}$, then every polynomial in $\mathbf{A}$ is equal to a suitable polynomial in $\mathbf{A}$ of degree not exceeding $k-1$; and in particular we can write

$$\mathbf{A}^\nu = \sum_{i=0}^{k-1} p_{\nu i} \mathbf{A}^i \qquad (\nu \geqslant 0).$$

Putting

$$t_{mi} = \sum_{\nu=0}^{m} c_\nu p_{\nu i} \qquad (m \geqslant 0; i = 0, 1,..., k-1)$$

we obtain

$$\phi_m(\mathbf{A}) = \sum_{i=0}^{k-1} t_{mi} \mathbf{A}^i.$$

Now $\lim_{m\to\infty} \phi_m(\mathbf{A}) = \phi(\mathbf{A})$; furthermore, the matrices $\mathbf{I}, \mathbf{A},..., \mathbf{A}^{k-1}$ are linearly independent by the definition of k. Hence, by Theorem 11.3.1, there exist numbers $t_0,..., t_{k-1}$ such that

$$\lim_{m\to\infty} t_{mi} = t_i \qquad (i = 0, 1,..., k-1),$$

and therefore

$$\phi(\mathbf{A}) = \sum_{i=0}^{k-1} t_i \mathbf{A}^i.$$

† Cf. Theorem 7.4.3 (p. 204), of which Theorem 11.3.2 is a generalization.

The assertion is therefore established, and it is interesting to note that we have actually proved a little more than Theorem 11.3.2. For in the argument above we used only the fact that $\phi(\mathbf{A})$ is the sum of a convergent power series without relying on the stronger assumption that $\phi(\mathbf{A})$ is a matrix function in the sense of Definition 11.2.3.

11.4. Systems of linear differential equations

11.4.1. In the present section we shall be concerned, as we were in the latter part of § 11.2.4, with rectangular matrices whose elements are functions of a single variable. We begin with a few definitions.

The derivative of $\mathbf{A}(t) = (a_{ij}(t))$ has already been defined. Analogously, we denote by $\int_p^q \mathbf{A}(t)\,dt$ the matrix whose (i,j)th element is $\int_p^q a_{ij}(t)\,dt$. The matrix $\mathbf{A}(t)$ is called integrable, continuous, or bounded respectively if all its elements have the property in question.

In conformity with Definition 11.1.3 (p. 328), we write

$$\|\mathbf{A}(t)\| = \max_{1 \leqslant i,j \leqslant n} |\{\mathbf{A}(t)\}_{ij}|.$$

If $\mathbf{A}(t)$ is continuous, then $\|\mathbf{A}(t)\|$ is continuous and therefore integrable.

Exercise 11.4.1. Let $f(u)$, $g(u)$ be continuous functions such that $f(u) \leqslant g(u)$ for $0 \leqslant |u| \leqslant |t|$. Show that

$$\left|\int_0^t f(u)\,du\right| \leqslant \left|\int_0^t g(u)\,du\right|,$$

and deduce that, if $\mathbf{C}(t)$ is continuous, then

$$\left\|\int_0^t \mathbf{C}(u)\,du\right\| \leqslant \left|\int_0^t \|\mathbf{C}(u)\|\,du\right|.$$

For $r \geqslant 0$ we define

$$m(r; \mathbf{A}) = \overline{\mathrm{bd}}_{0 \leqslant |t| \leqslant r} \|\mathbf{A}(t)\|.$$

If $\mathbf{P}(t)$, $\mathbf{Q}(t)$ are $n \times n$ matrices, then obviously

$$\|\mathbf{P}(t)\mathbf{Q}(t)\| \leqslant n\|\mathbf{P}(t)\| . \|\mathbf{Q}(t)\|,$$

and therefore

$$\|\mathbf{P}(u)\mathbf{Q}(u)\| \leqslant n\|\mathbf{P}(u)\| . m(r; \mathbf{Q}) \quad (0 \leqslant |u| \leqslant r). \quad (11.4.1)$$

It will be recalled that the series

$$\sum_{k=0}^{\infty} \mathbf{A}_k(t) \tag{11.4.2}$$

is said to converge if each of the series

$$\sum_{k=0}^{\infty} \{\mathbf{A}_k(t)\}_{ij} \qquad (i, j = 1, \ldots, n)$$

is convergent. If, in addition, all these series converge uniformly in $t_1 \leqslant t \leqslant t_2$, we say that the series (11.4.2) converges uniformly in that interval. This is certainly the case if, for $k \geqslant 0$,

$$\|\mathbf{A}_k(t)\| \leqslant c_k \quad (t_1 \leqslant t \leqslant t_2)$$

and $\sum_{k=0}^{\infty} c_k$ is a convergent series.

11.4.2. We propose to discuss systems of linear differential equations† of the form

$$\left.\begin{aligned} \frac{dx_1}{dt} &= a_{11}(t)x_1 + \ldots + a_{1n}(t)x_n + b_1(t) \\ &\cdots\cdots\cdots\cdots \\ \frac{dx_n}{dt} &= a_{n1}(t)x_1 + \ldots + a_{nn}(t)x_n + b_n(t) \end{aligned}\right\}. \tag{11.4.3}$$

Here the $a_{ij}(t)$ and $b_i(t)$ are given functions and we wish to determine whether there exist functions $x_1(t), \ldots, x_n(t)$ satisfying (11.4.3) and also some set of initial conditions and, if so, whether these functions are unique. The result we shall establish is as follows.

THEOREM 11.4.1. *If $a_{ij}(t)$, $b_i(t)$ $(i, j = 1, \ldots, n)$ are continuous functions of t, and $c_1, \ldots, c_n$ are any constants, then there exists one and only one set of functions $x_1(t), \ldots, x_n(t)$ satisfying the system* (11.4.3) *with the initial conditions*

$$x_1(0) = c_1, \quad \ldots, \quad x_n(0) = c_n. \tag{11.4.4}$$

We begin by restating the problem in terms of matrices.‡ Let

$$\mathbf{A}(t) = (a_{ij}(t)), \qquad \mathbf{b}(t) = (b_1(t), \ldots, b_n(t))^T,$$

$$\mathbf{x} = \mathbf{x}(t) = (x_1(t), \ldots, x_n(t))^T.$$

† We intend here to do no more than consider an isolated problem. For a systematic treatment of differential equations by the methods of matrix algebra see Frazer, Duncan, and Collar, **22**, chaps v–vii, Gantmacher, **29**, ii. chap. xiv, and Lefschetz, *Lectures on Differential Equations*. The reader's attention is also drawn to the remarks in Perlis, **10**, 138–42 and 165–6.

‡ I am indebted to Professor H. A. Heilbronn for the proof given below.

With this notation (11.4.3) and (11.4.4) become respectively

$$\dot{\mathbf{x}} = \mathbf{A}(t)\mathbf{x}+\mathbf{b}(t), \qquad (11.4.5)$$

$$\mathbf{x}(0) = (c_1,..., c_n)^T. \qquad (11.4.6)$$

We have, therefore, to show that there exists a unique vector $\mathbf{x}(t)$ which satisfies (11.4.5) and for which $\mathbf{x}(0)$ is prescribed.

Assume, for the moment, that there exists an $n\times n$ matrix $\mathbf{M}(t)$ such that

$$\dot{\mathbf{M}}(t) = -\mathbf{M}(t)\mathbf{A}(t) \quad (\text{all } t), \qquad (11.4.7)$$

$$|\mathbf{M}(t)| \neq 0 \quad (\text{all } t). \qquad (11.4.8)$$

If (11.4.5) possesses a solution $\mathbf{x}$, then

$$\frac{d}{dt}\{\mathbf{M}(t)\mathbf{x}\} = \dot{\mathbf{M}}(t)\mathbf{x}+\mathbf{M}(t)\dot{\mathbf{x}}$$
$$= -\mathbf{M}(t)\mathbf{A}(t)\mathbf{x}+\mathbf{M}(t)\{\mathbf{A}(t)\mathbf{x}+\mathbf{b}(t)\} = \mathbf{M}(t)\mathbf{b}(t),$$

and therefore

$$\mathbf{M}(t)\mathbf{x} = \int_0^t \mathbf{M}(u)\mathbf{b}(u)\,du + \mathbf{M}(0)\mathbf{x}(0),$$

$$\mathbf{x}(t) = \{\mathbf{M}(t)\}^{-1}\left\{\int_0^t \mathbf{M}(u)\mathbf{b}(u)\,du + \mathbf{M}(0)\mathbf{x}(0)\right\}. \qquad (11.4.9)$$

Thus, if (11.4.5) is soluble subject to (11.4.6), then the solution is given by (11.4.9) and so is unique. On the other hand, it can be verified at once that the vector $\mathbf{x}(t)$ defined by (11.4.9) satisfies (11.4.5) and (11.4.6). The proof of the theorem will therefore be complete if we can establish the existence of a matrix $\mathbf{M}(t)$ satisfying (11.4.7) and (11.4.8). We recognize, of course, that $\mathbf{M}(t)$ plays the part of an 'integrating factor' of the equation (11.4.5).

We write

$$\mathbf{M}_0(t) = \mathbf{I}; \qquad \mathbf{M}_{k+1}(t) = \int_0^t \mathbf{M}_k(u)\mathbf{A}(u)\,du \quad (k = 0, 1, 2,...).$$

These recurrence relations define a sequence $\{\mathbf{M}_k(t)\}$ of matrices. Each of these matrices is continuous and, indeed, differentiable; and we have

$$\dot{\mathbf{M}}_{k+1}(t) = \mathbf{M}_k(t)\mathbf{A}(t) \quad (k = 0, 1, 2,...). \qquad (11.4.10)$$

We now assert, for all $r \geqslant 0$, the inequality

$$\|\mathbf{M}_k(t)\| \leqslant \frac{1}{k!}\{n.\,|t|\,.\,m(r;\mathbf{A})\}^k \quad (|t| \leqslant r;\ k = 0, 1, 2,...). \qquad (11.4.11)$$

For $k = 0$ this is true trivially. Assume that the inequality holds for some $k \geqslant 0$. Using Exercise 11.4.1, we have

$$\|\mathbf{M}_{k+1}(t)\| \leqslant \left|\int_0^t \|\mathbf{M}_k(u)\mathbf{A}(u)\|\, du\right|.$$

Now in the integral on the right-hand side $0 \leqslant |u| \leqslant |t| \leqslant r$ and so, by (11.4.1),

$$\|\mathbf{M}_{k+1}(t)\| \leqslant \left|\int_0^t n\|\mathbf{M}_k(u)\|m(r;\, \mathbf{A})\, du\right|.$$

Hence, by the induction hypothesis,

$$\|\mathbf{M}_{k+1}(t)\| \leqslant \frac{1}{k!}\{n\,.\,m(r;\, \mathbf{A})\}^{k+1}\left|\int_0^t |u|^k\, du\right|$$

$$= \frac{1}{(k+1)!}\{n\,.\,|t|\,.\,m(r;\, \mathbf{A})\}^{k+1}.$$

The inequality (11.4.11) is therefore proved, and it follows that

$$\|\mathbf{M}_k(t)\| \leqslant c_k \quad (|t| \leqslant r;\ k = 0, 1, 2,...),$$

where
$$c_k = \frac{1}{k!}\{nr\,.\,m(r;\, \mathbf{A})\}^k.$$

Now $\sum_{k=0}^{\infty} c_k$ is an exponential series and so is convergent. Therefore, for any $r \geqslant 0$, the series

$$\mathbf{M}_0(t) - \mathbf{M}_1(t) + \mathbf{M}_2(t) - \mathbf{M}_3(t) + ...$$

converges uniformly in the interval $|t| \leqslant r$. Denoting the sum of this series by $\mathbf{M}(t)$, differentiating formally term by term, and using (11.4.10), we obtain

$$\dot{\mathbf{M}}(t) = -\{\mathbf{M}_0(t) - \mathbf{M}_1(t) + \mathbf{M}_2(t) - ...\}\mathbf{A}(t). \qquad (11.4.12)$$

Now since $\mathbf{A}(t)$ is bounded for $|t| \leqslant r$, it follows that the series (11.4.12) is uniformly convergent in that interval. Hence $\mathbf{M}(t)$ is differentiable for $|t| < r$, and its derivative is given by (11.4.12). But r can be chosen arbitrarily large; hence $\mathbf{M}(t)$ is everywhere differentiable and satisfies (11.4.7).

We next consider the sequence $\{\mathbf{N}_k(t)\}$ defined by the relations

$$\mathbf{N}_0(t) = \mathbf{I}; \qquad \mathbf{N}_{k+1}(t) = \int_0^t \mathbf{A}(u)\mathbf{N}_k(u)\, du \quad (k = 0, 1, 2,...). \qquad (11.4.13)$$

By an argument analogous to that just used it follows that the matrix

$$\mathbf{N}(t) = \mathbf{N}_0(t)+\mathbf{N}_1(t)+\mathbf{N}_2(t)+... \qquad (11.4.14)$$

satisfies the relation

$$\dot{\mathbf{N}}(t) = \mathbf{A}(t)\mathbf{N}(t), \qquad \mathbf{N}(0) = \mathbf{I}.$$

Hence

$$\frac{d}{dt}\{\mathbf{M}(t)\mathbf{N}(t)\} = \dot{\mathbf{M}}(t)\mathbf{N}(t)+\mathbf{M}(t)\dot{\mathbf{N}}(t)$$

$$= -\mathbf{M}(t)\mathbf{A}(t).\mathbf{N}(t)+\mathbf{M}(t).\mathbf{A}(t)\mathbf{N}(t) = \mathbf{O},$$

and therefore $\quad \mathbf{M}(t)\mathbf{N}(t) = \mathbf{M}(0)\mathbf{N}(0) = \mathbf{I}.$

Thus $\mathbf{M}(t)$ possesses an inverse for all values of t, and (11.4.8) is satisfied. The theorem is therefore proved.

The solution of the system of differential equations (11.4.3) reduces to a particularly simple form when all $a_{ij}(t)$ are constants and all $b_i(t)$ vanish. In this case we have the following result.

THEOREM 11.4.2. *The system of differential equations*

$$\left.\begin{aligned} \frac{dx_1}{dt} &= a_{11}x_1+...+a_{1n}x_n \\ &\cdots\cdots \\ \frac{dx_n}{dt} &= a_{n1}x_1+...+a_{nn}x_n \end{aligned}\right\}, \qquad (11.4.15)$$

subject to the initial conditions (11.4.4), *possesses a unique solution given by*

$$\mathbf{x}(t) = \exp(t\mathbf{A}).\mathbf{x}(0), \qquad (11.4.16)$$

where $\mathbf{A} = (a_{ij}), \quad \mathbf{x}(t) = (x_1(t),..., x_n(t))^T, \quad \mathbf{x}(0) = (c_1,..., c_n)^T.$

The uniqueness of the solution is guaranteed by Theorem 11.4.1. Moreover, we infer from (11.4.9) that this solution is given by

$$\mathbf{x}(t) = \{\mathbf{M}(t)\}^{-1}\mathbf{x}(0) = \mathbf{N}(t)\mathbf{x}(0),$$

where $\mathbf{N}(t)$ is defined by (11.4.14). From (11.4.13) we obtain at once $\mathbf{N}_k(t) = t^k\mathbf{A}^k/k!$ $(k = 0, 1, 2,...)$. Hence $\mathbf{N}(t) = \exp(t\mathbf{A})$, and (11.4.16) follows.

It is even easier to verify directly that $\mathbf{x}(t)$, as defined by (11.4.16), satisfies (11.4.15). For we have, by Theorem 11.2.7,

$$\dot{\mathbf{x}}(t) = \mathbf{A}\exp(t\mathbf{A})\mathbf{x}(0) = \mathbf{A}.\mathbf{x}(t),$$

and this is equivalent to (11.4.15).

EXERCISE 11.4.2. State the case $n = 1$ of Theorem 11.4.2.

EXERCISE 11.4.3. Obtain a proof of Theorem 11.4.2, independent of Theorem 11.4.1, for the case when $\mathbf{A}$ is similar to a diagonal matrix.

PROBLEMS ON CHAPTER XI

1. Let $\phi(\mathbf{A})$ be a matrix function. Show that $\phi(\mathbf{A}^T) = \{\phi(\mathbf{A})\}^T$.

2. Show that, for every matrix $\mathbf{A}$, $|\exp \mathbf{A}| = \exp(\operatorname{tr} \mathbf{A})$. Deduce that, for a skew-symmetric matrix $\mathbf{A}$, $|\exp \mathbf{A}| = 1$.

3. Let $\lambda_1,\ldots,\lambda_n$ be the characteristic roots of $\mathbf{A}$, and let $\phi(\mathbf{A})$ be a function of $\mathbf{A}$. Show that the characteristic roots of $\phi(\mathbf{A})$ are $\phi(\lambda_1),\ldots,\phi(\lambda_n)$.

4. If

$$\mathbf{A} = \begin{pmatrix} 0 & 1 & 0 \\ 0 & 0 & 1 \\ 1 & 0 & 0 \end{pmatrix},$$

show that

$$\exp \mathbf{A} = \tfrac{1}{3}\begin{pmatrix} l & m & n \\ n & l & m \\ m & n & l \end{pmatrix},$$

where $l = e+e^{\rho}+e^{\rho^2}$, $m = e+\rho^2 e^{\rho}+\rho e^{\rho^2}$, $n = e+\rho e^{\rho}+\rho^2 e^{\rho^2}$ and $\rho = e^{2\pi i/3}$.

5. If

$$\mathbf{A}(x) = \begin{pmatrix} x & x-1 \\ 0 & 1 \end{pmatrix},$$

where x is any complex number, show that

$$\exp\{\mathbf{A}(x)\} = e\mathbf{A}(e^{x-1}).$$

6. If

$$\mathbf{A} = \begin{pmatrix} 0 & 1 & 0 \\ 2 & 0 & 2 \\ 0 & 1 & 0 \end{pmatrix},$$

show that

$$\exp \mathbf{A} = \begin{pmatrix} a^2 & ab & b^2 \\ 2ab & a^2+b^2 & 2ab \\ b^2 & ab & a^2 \end{pmatrix},$$

where $a = \cosh 1$, $b = \sinh 1$.

7. Show that if $\mathbf{A}$ is a real skew-symmetric matrix (skew-hermitian matrix), then $\exp \mathbf{A}$ is orthogonal (unitary).

8. Prove that if $\mathbf{A}$ is hermitian, then $\exp(i\mathbf{A})$ is unitary; and interpret this result for the case $n = 1$.

9. Show that, in general, the relation

$$\frac{d}{dt}\{\mathbf{A}(t)\}^m = m\{\mathbf{A}(t)\}^{m-1}\dot{\mathbf{A}}(t)$$

is not valid. Under what conditions is it valid?

10. Show that every unitary matrix $\mathbf{U}$ can be represented in the form $\mathbf{U} = \exp(i\mathbf{H})$, where $\mathbf{H}$ is hermitian.

11. For any complex matrix $\mathbf{A} = (a_{rs})$, let $N(\mathbf{A})$ denote the positive square root of $\sum_{r,s=1}^{n} |a_{rs}|^2$. Show that the matrix series $\sum_{m=0}^{\infty} \mathbf{A}_m$ converges absolutely if and only if $\sum_{m=0}^{\infty} N(\mathbf{A}_m)$ converges.

12. By differentiating the matrix $\exp(t\mathbf{A})\exp(-t\mathbf{A})$ with respect to t establish the identity $(\exp \mathbf{A})^{-1} = \exp(-\mathbf{A})$.

13. Let $\mathbf{A} = (a_{rs})$, $\mathbf{B} = (b_{rs})$ be $n \times n$ matrices. Show that the inequalities $|a_{rs}| \leqslant |b_{rs}|$ $(r, s = 1,..., n)$ do not imply that $\rho(\mathbf{A}) \leqslant \rho(\mathbf{B})$.

14. Let $\phi(z)$ be a power series and $\mathbf{A}$ the $n \times n$ matrix

$$\begin{pmatrix} \lambda & 1 & 0 & 0 & . & . & . & 0 \\ 0 & \lambda & 1 & 0 & . & . & . & 0 \\ 0 & 0 & \lambda & 1 & . & . & . & 0 \\ . & . & . & . & . & . & . & . \\ 0 & 0 & 0 & 0 & . & . & . & \lambda \end{pmatrix}.$$

Show that the matrix power series $\phi(\mathbf{A})$ converges if and only if the scalar power series $\phi(z), \phi'(z),..., \phi^{(n-1)}(z)$ all converge for $z = \lambda$.

15. Evaluate $\exp \mathbf{A}$, when $\mathbf{A}$ is the matrix given in the preceding question.

16. Show that, if

$$\mathbf{S} = \begin{pmatrix} 0 & \nu & -\mu \\ -\nu & 0 & \lambda \\ \mu & -\lambda & 0 \end{pmatrix},$$

then

$$\exp \mathbf{S} = \mathbf{I} + \frac{\sin\omega}{\omega}\mathbf{S} + \frac{1-\cos\omega}{\omega^2}\mathbf{S}^2$$

and

$$\sin \mathbf{S} = \mathbf{I} + \left(\frac{\sinh\omega}{\omega} - 1\right)\mathbf{S},$$

where $\omega^2 = \lambda^2 + \mu^2 + \nu^2$. How are these results to be interpreted when $\omega = 0$?

17. Assuming that $\mathbf{A}_m \to \mathbf{A}$ $(m \to \infty)$, where $|\mathbf{A}| \neq 0$, show that $\mathbf{A}_m$ is non-singular for all sufficiently large values of m and that $(\mathbf{A}_m)^{-1} \to \mathbf{A}^{-1}$.

18. The characteristic roots of a matrix $\mathbf{A}$ of order 3 are $\frac{1}{2}\pi$, π, $-\pi$. Prove, by induction with respect to n, that

$$\mathbf{A}^{2n+2} = \tfrac{1}{3}\pi^{2n}(4 - 2^{-2n})\mathbf{A}^2 - \tfrac{1}{3}\pi^{2n+2}(1 - 2^{-2n})\mathbf{I}.$$

Deduce that

$$\sin \mathbf{A} = \frac{4}{3}\mathbf{I} - \frac{4}{3\pi^2}\mathbf{A}^2.$$

19. If the characteristic roots of $\mathbf{A}$ are the nth roots of unity, prove that

$$(\mathbf{A}^2 - 5\mathbf{A} + 6\mathbf{I})^{-1} = \sum_{k=0}^{n-1} \frac{2^{n-k-1}(3^n - 1) - 3^{n-k-1}(2^n - 1)}{(2^n - 1)(3^n - 1)}\mathbf{A}^k.$$

20. The characteristic roots of a matrix $\mathbf{A}$ are the nth roots of unity. Prove that, for $|c| < 1$,

$$(\mathbf{I} - c\mathbf{A})^{-2} = (1 - c^n)^{-2}\sum_{k=0}^{n-1} c^k\{(k+1) + (n-k-1)c^n\}\mathbf{A}^k.$$

21. Deduce the first inequality in Theorem 7.5.3 (p. 211) from Theorem 11.1.2 (p. 329).

22. Show that if $\lim_{N\to\infty} \mathbf{A}^N$ exists, then it is equal to a polynomial in $\mathbf{A}$.

23. Show that, if k is a positive integer and $\mathbf{A}$ is a matrix all of whose characteristic roots are less than 1 in modulus, then

$$(\mathbf{I} - \mathbf{A})^{-k} = \sum_{\nu=0}^{\infty} \binom{k+\nu-1}{k-1}\mathbf{A}^\nu.$$

24. All elements above the diagonal of the matrix $\mathbf{A}$ are equal to 1 and all other elements are equal to 0. Show that

$$(\mathbf{A}^k)_{rs} = \binom{s-r-1}{k-1},$$

where $\binom{p}{q}$ is interpreted as 0 when $p < q$.

25. Determine all skew-symmetric matrices $\mathbf{S}$ which satisfy the equation $\exp \mathbf{S} = \mathbf{A}$, where $\mathbf{A}$ is a given proper orthogonal 2×2 matrix.

26. Show that, if $\mathbf{A}$ is a proper orthogonal 3×3 matrix, then there exists a skew-symmetric matrix $\mathbf{B}$ such that $\mathbf{A} = \exp \mathbf{B}$.

27. Let $f_1(x),..., f_n(x)$, $g(x)$ be continuous functions. Show that there exists one and only one function y of x, with prescribed values for

$$y, \frac{dy}{dx},..., \frac{d^{n-1}y}{dx^{n-1}}$$

when $x = 0$, and satisfying the differential equation

$$\frac{d^ny}{dx^n}+f_1(x)\frac{d^{n-1}y}{dx^{n-1}}+...+f_{n-1}(x)\frac{dy}{dx}+f_n(x)y = g(x).$$

28. Show that the solution of the differential equation

$$\frac{d^ny}{dx^n}+a_1\frac{d^{n-1}y}{dx^{n-1}}+...+a_{n-1}\frac{dy}{dx}+a_n y = Q(x)$$

($a_1,..., a_n$ constants; $Q(x)$ a continuous function), subject to the initial conditions

$$y = \frac{dy}{dx} = ... = \frac{d^{n-1}y}{dx^{n-1}} = 0 \qquad (x = 0),$$

can be written in the form

$$y = \int_0^x \{\exp(t\mathbf{A})\}_{1n}\, Q(x-t)\, dt,$$

where

$$\mathbf{A} = \begin{pmatrix} 0 & 1 & 0 & 0 & . & . & . & 0 \\ 0 & 0 & 1 & 0 & . & . & . & 0 \\ . & . & . & . & . & . & . & . \\ 0 & 0 & 0 & 0 & . & . & . & 1 \\ -a_n & -a_{n-1} & -a_{n-2} & -a_{n-3} & . & . & . & -a_1 \end{pmatrix}.$$

29. Let a, b, c, d be real numbers, and suppose that b, c are not both zero. Prove that all solutions of the system

$$\ddot{x} = ax+by, \qquad \ddot{y} = cx+dy$$

are bounded if and only if the characteristic roots of $\begin{pmatrix} a & b \\ c & d \end{pmatrix}$ are negative and distinct.

30. Show that there exists one and only one set of functions $x_1(t),..., x_n(t)$ satisfying the system of differential equations

$$\ddot{x}_i(t) = \sum_{j=1}^{n} a_{ij}\, x_j(t) \qquad (i = 1,..., n)$$

and having prescribed values for $x_1(0),..., x_n(0)$, $\dot{x}_1(0),..., \dot{x}_n(0)$. Show further that these functions are specified by the identity

$$\mathbf{x}(t) = \phi(t^2\mathbf{A})\mathbf{x}(0)+t\psi(t^2\mathbf{A})\dot{\mathbf{x}}(0),$$

where $\mathbf{x}(t) = (x_1(t),..., x_n(t))^T$, $\mathbf{A} = (a_{ij})$,

$$\phi(z) = \sum_{m=0}^{\infty} \frac{z^m}{(2m)!}, \qquad \psi(z) = \sum_{m=0}^{\infty} \frac{z^m}{(2m+1)!}.$$

If the characteristic roots of $\mathbf{A}$ are distinct negative numbers, prove that all solutions of the given system of differential equations are bounded.

31. Use the preceding question to find the solution of the equations

$$\frac{d^2u}{dt^2}+2u+v+w = 0, \qquad \frac{d^2v}{dt^2}+u+2v+w = 0, \qquad \frac{d^2w}{dt^2}+u+v+2w = 0,$$

such that $u = v = w = 0$, $\dfrac{du}{dt} = a$, $\dfrac{dv}{dt} = b$, $\dfrac{dw}{dt} = c$ when $t = 0$.

32. Show that the general solution of the equation $\exp \mathbf{X} = \mathbf{I}_2$ can be written in the form

$$\mathbf{X} = \mathbf{S}.\mathbf{dg}(2\pi ki,\ 2\pi li)\mathbf{S}^{-1},$$

where $\mathbf{S}$ is an arbitrary non-singular 2×2 matrix and k, l are arbitrary integers.

33. Let $a_1,..., a_n$ be complex and $b_1,..., b_n$ real non-negative numbers, and suppose that $|a_k| \leqslant b_k$ $(k = 1,..., n)$. Using Theorem 11.1.2 (p. 329) and Problem VII, 32 show that, if λ denotes a root of greatest modulus of the equation $x^n+a_1 x^{n-1}+...+a_n = 0$ and μ denotes a root of greatest modulus of the equation $x^n-b_1 x^{n-1}-...-b_n = 0$, then $|\lambda| \leqslant |\mu|$.

34. Let $f(\mathbf{A}) = \sum_{m=0}^{\infty} c_m \mathbf{A}^m$ be a convergent power series in the matrix $\mathbf{A}$. Show that, for $0 < t < 1$,

$$f(\mathbf{A};\, t) = \sum_{m=0}^{\infty} c_m t^m \mathbf{A}^m$$

is again a convergent power series in $\mathbf{A}$, and that

$$\lim_{t\to 1-0} f(\mathbf{A};\, t) = f(\mathbf{A}).$$

35. $\mathbf{A}$ is a matrix with positive elements and characteristic roots $1, \omega_2,..., \omega_n$, where $|\omega_2| < 1,..., |\omega_n| < 1$. Show that there exists a non-singular matrix $\mathbf{T}$ such that $\mathbf{T}^{-1}\mathbf{A}\mathbf{T}$ is of the form

$$\begin{pmatrix} 1 & \mathbf{b} \\ \mathbf{0} & \mathbf{B} \end{pmatrix},$$

where $\mathbf{b}$ is a row vector of order $n-1$ and $\mathbf{B}$ is an $(n-1)\times(n-1)$ matrix. Prove that, for $m \to \infty$,

$$\mathbf{A}^m \to \mathbf{T}\begin{pmatrix} 1 & \mathbf{b}(\mathbf{I}-\mathbf{B})^{-1} \\ \mathbf{0} & \mathbf{O} \end{pmatrix}\mathbf{T}^{-1}.$$

By considering the vector $\mathbf{x} = (\lim_{m\to\infty} \mathbf{A}^m)\mathbf{y}$, where $\mathbf{y}$ is any vector with positive components, show that $\mathbf{A}$ possesses a characteristic vector with positive components associated with the characteristic root 1.

36. The characteristic roots of the matrix $\mathbf{A}$ are $\omega_1, \omega_2, \ldots, \omega_n$, and $|\omega_1| > |\omega_k|$ $(k = 2, \ldots, n)$. Show that $\lim_{m \to \infty} (\omega_1^{-m} \mathbf{A}^m)$ exists and is non-zero; and deduce that, for a suitable vector $\mathbf{x}$, $\lim_{m \to \infty} (\omega_1^{-m} \mathbf{A}^m \mathbf{x})$ is a characteristic vector of $\mathbf{A}$ associated with ω_1.

37. $\mathbf{A}$ is a real matrix with positive determinant. Show that there exists a matrix $\mathbf{A}(t)$, continuous for $0 \leqslant t \leqslant 1$, and such that

$$\mathbf{A}(0) = \mathbf{A}, \qquad \mathbf{A}(1) = \mathbf{I}, \qquad |\mathbf{A}(t)| \neq 0 \quad (0 \leqslant t \leqslant 1).$$

Show further that the matrix $\mathbf{A}(t) = (1-t)\mathbf{A} + t\mathbf{I}$ has the required properties if and only if no characteristic root of $\mathbf{A}$ is real and negative.

38. Let $\mathbf{A}$ be a proper orthogonal matrix. Show, by using Problem 25 and Problem X, 44, that there exists a real skew-symmetric matrix $\mathbf{S}$ such that $\mathbf{A} = \exp \mathbf{S}$.

39. Let $k \geqslant 1$ and suppose that the characteristic roots $\omega_1, \ldots, \omega_n$ of a matrix $\mathbf{A}$ satisfy the conditions

$$|\omega_i| = 1 \quad (i = 1, \ldots, k), \qquad |\omega_i| < 1 \quad (i = k+1, \ldots, n).$$

Show, by using Problem IX, 32, that, as $m \to \infty$, all elements of $\mathbf{A}^m$ are of the form $O(m^{k-1})$.

40. If $\{\mathbf{A}_m\}$ is a sequence of square matrices and $\lim_{m \to \infty} (\mathbf{A}_m \mathbf{x})$ exists for every vector $\mathbf{x}$, show that $\lim_{m \to \infty} \mathbf{A}_m$ also exists.

PART III

QUADRATIC FORMS

XII

BILINEAR, QUADRATIC, AND HERMITIAN FORMS

WE have now carried the discussion of matrices as far as we intend to, and in this chapter and the next we shall apply the results previously obtained to the study of quadratic forms and related topics.

12.1. Operators and forms of the bilinear and quadratic types

12.1.1. It is necessary at this stage to make use of some of the properties of linear manifolds derived in Chapters II and IV.

DEFINITION 12.1.1. *Let $\mathfrak{M}$ and $\mathfrak{N}$ be two linear manifolds over a field $\mathfrak{F}$.*† *Let $\phi(X,Y)$ be a function of two variables X, Y which are elements of $\mathfrak{M}$, $\mathfrak{N}$ respectively, and suppose that the functional values of $\phi(X,Y)$ are numbers of $\mathfrak{F}$. If $\phi(X,Y)$ is linear in both variables, i.e. if for all $\alpha \in \mathfrak{F}$, $X, X' \in \mathfrak{M}$, $Y, Y' \in \mathfrak{N}$ we have*

$$\left.\begin{aligned} \phi(\alpha X, Y) &= \alpha\phi(X,Y) = \phi(X,\alpha Y) \\ \phi(X+X',Y) &= \phi(X,Y)+\phi(X',Y) \\ \phi(X,Y+Y') &= \phi(X,Y)+\phi(X,Y') \end{aligned}\right\}, \tag{12.1.1}$$

then $\phi(X, Y)$ is called a BILINEAR OPERATOR *on $\mathfrak{M}$ and $\mathfrak{N}$.*

EXERCISE 12.1.1. Show that the relation

$$\phi(\alpha X+\alpha' X', \beta Y+\beta' Y')$$
$$= \alpha\beta\phi(X,Y)+\alpha\beta'\phi(X,Y')+\alpha'\beta\phi(X',Y)+\alpha'\beta'\phi(X',Y')$$

holds for all $\alpha,\alpha',\beta,\beta' \in \mathfrak{F}$, $X,X' \in \mathfrak{M}$, $Y,Y' \in \mathfrak{N}$ if and only if (12.1.1) is satisfied.

In the discussion below we shall write

$$d(\mathfrak{M}) = m, \qquad d(\mathfrak{N}) = n.$$

Since linear operators on linear manifolds possess matrix representations, it is natural to seek some such representation for bilinear operators.

† The possibility that $\mathfrak{M}$ and $\mathfrak{N}$ are identical is not, of course, excluded.

THEOREM 12.1.1. *Let* $\mathfrak{B} = \{E_1,..., E_m\}$, $\mathfrak{C} = \{F_1,..., F_n\}$ *be bases of* $\mathfrak{M}$, $\mathfrak{N}$ *respectively. If* $\phi(X, Y)$ *is a bilinear operator on* $\mathfrak{M}$ *and* $\mathfrak{N}$, *and if*

$$\mathbf{x} = \mathscr{R}(X; \mathfrak{B}), \qquad \mathbf{y} = \mathscr{R}(Y; \mathfrak{C}),$$

then

$$\phi(X, Y) = \mathbf{x}^T\mathbf{A}\mathbf{y},$$

where the $m \times n$ *matrix* $\mathbf{A} = (a_{rs})$ *is defined by the relations*

$$a_{rs} = \phi(E_r, F_s) \qquad (r = 1,..., m;\ s = 1,..., n). \tag{12.1.2}$$

The assertion is virtually obvious. Writing

$$\mathbf{x} = (x_1,..., x_m)^T, \qquad \mathbf{y} = (y_1,..., y_n)^T$$

we have

$$X = \sum_{r=1}^{m} x_r E_r, \qquad Y = \sum_{s=1}^{n} y_s F_s,$$

and so, by (12.1.1),

$$\phi(X, Y) = \sum_{r=1}^{m} \sum_{s=1}^{n} x_r y_s \phi(E_r, F_s) = \sum_{r=1}^{m} \sum_{s=1}^{n} a_{rs} x_r y_s = \mathbf{x}^T\mathbf{A}\mathbf{y}.$$

Matrix products of the type $\mathbf{x}^T\mathbf{A}\mathbf{y}$ will clearly play a prominent role in the theory of bilinear operators.

DEFINITION 12.1.2. *Any polynomial*

$$\sum_{r=1}^{m} \sum_{s=1}^{n} a_{rs} x_r y_s = \mathbf{x}^T\mathbf{A}\mathbf{y}, \tag{12.1.3}$$

where $\mathbf{A} = (a_{rs})$, $\mathbf{x} = (x_1,..., x_m)^T$, $\mathbf{y} = (y_1,..., y_n)^T$ *is called a* BILINEAR FORM *in the two sets of variables* $x_1,..., x_m$ *and* $y_1,..., y_n$. *The* $m \times n$ *matrix* $\mathbf{A}$ *is called the matrix of this bilinear form.*

There is thus clearly a biunique correspondence between rectangular matrices and bilinear forms.

DEFINITION 12.1.3. *If, with the notation of Theorem* 12.1.1, $\phi(X, Y) = \mathbf{x}^T\mathbf{A}\mathbf{y}$, *then the bilinear form* $\mathbf{x}^T\mathbf{A}\mathbf{y}$ (*or alternatively the matrix* $\mathbf{A}$) *is said to* REPRESENT *the bilinear operator* $\phi(X, Y)$ *with respect to the bases* $\mathfrak{B}$, $\mathfrak{C}$.

If $\mathfrak{M}$ and $\mathfrak{N}$ are the same manifold and $\mathfrak{B}$ and $\mathfrak{C}$ the same basis, we say that $\mathbf{x}^T\mathbf{A}\mathbf{y}$ represents $\phi(X, Y)$ with respect to $\mathfrak{B}$.

THEOREM 12.1.2. *For any fixed choice of bases* $\mathfrak{B}$, $\mathfrak{C}$ *in* $\mathfrak{M}$, $\mathfrak{N}$ *respectively, there is a biunique correspondence between bilinear operators* (*on* $\mathfrak{M}$ *and* $\mathfrak{N}$) *and their representing matrices.*

This result is implicit in our previous remarks. For to a given bilinear operator $\phi(X, Y)$ corresponds the representing matrix $\mathbf{A}$

defined by (12.1.2). On the other hand, if $\mathbf{A}$ is given, then the bilinear operator ϕ which it represents with respect to $\mathfrak{B}$, $\mathfrak{C}$ is uniquely determined by (12.1.2) and (12.1.1).

The representation of a bilinear operator by a bilinear form (or by a matrix) depends, of course, on the arbitrary choice of bases in $\mathfrak{M}$ and $\mathfrak{N}$. The relation between bilinear operators and bilinear forms is in this respect analogous to that between the elements of a linear manifold and vectors, or that between linear operators and matrices.

THEOREM 12.1.3. *The $m \times n$ matrices $\mathbf{A}$ and $\mathbf{B}$ represent the same bilinear operator with respect to suitable pairs of bases if and only if there exist non-singular matrices $\mathbf{P}$, $\mathbf{Q}$, of order m, n respectively, such that*

$$\mathbf{B} = \mathbf{P}^T\mathbf{A}\mathbf{Q}. \tag{12.1.4}$$

Suppose, in the first place, that $\mathbf{A}$ represents the bilinear operator ϕ with respect to $\mathfrak{B}$, $\mathfrak{C}$ and $\mathbf{B}$ represents the same operator with respect to $\mathfrak{B}'$, $\mathfrak{C}'$. Then

$$\phi(X, Y) = \mathbf{x}^T\mathbf{A}\mathbf{y}, \tag{12.1.5}$$

where
$$\mathbf{x} = \mathscr{R}(X;\, \mathfrak{B}), \qquad \mathbf{y} = \mathscr{R}(Y;\, \mathfrak{C}). \tag{12.1.6}$$

Write
$$\mathbf{x}' = \mathscr{R}(X;\, \mathfrak{B}'), \qquad \mathbf{y}' = \mathscr{R}(Y;\, \mathfrak{C}'). \tag{12.1.7}$$

Then, by Theorem 4.1.2 (i) (p. 112), there exist non-singular matrices $\mathbf{P}$, $\mathbf{Q}$, of order m, n respectively, and independent of X, Y, such that

$$\mathbf{x} = \mathbf{P}\mathbf{x}', \qquad \mathbf{y} = \mathbf{Q}\mathbf{y}'. \tag{12.1.8}$$

Hence, by (12.1.5),

$$\phi(X, Y) = (\mathbf{P}\mathbf{x}')^T\mathbf{A}(\mathbf{Q}\mathbf{y}') = \mathbf{x}'^T(\mathbf{P}^T\mathbf{A}\mathbf{Q})\mathbf{y}',$$

and so $\mathbf{P}^T\mathbf{A}\mathbf{Q}$ represents ϕ with respect to $\mathfrak{B}'$, $\mathfrak{C}'$. Hence (12.1.4) follows.

On the other hand, let $\mathbf{A}$ and $\mathbf{B}$ be given matrices connected by the relation (12.1.4), where $\mathbf{P}$ and $\mathbf{Q}$ are non-singular. Let $\mathfrak{B}$, $\mathfrak{C}$ be any bases in $\mathfrak{M}$, $\mathfrak{N}$ respectively. By Theorem 4.1.2 (ii), there exist bases $\mathfrak{B}'$, $\mathfrak{C}'$ in $\mathfrak{M}$, $\mathfrak{N}$ respectively such that, if $\mathbf{x}$, $\mathbf{y}$ are defined by (12.1.6) and $\mathbf{x}'$, $\mathbf{y}'$ by (12.1.7), then (12.1.8) is valid. Let now ϕ be the (uniquely determined) bilinear operator represented by $\mathbf{A}$ with respect to $\mathfrak{B}$, $\mathfrak{C}$. Then, by (12.1.8),

$$\phi(X, Y) = \mathbf{x}^T\mathbf{A}\mathbf{y} = \mathbf{x}'^T\mathbf{B}\mathbf{y}',$$

and therefore $\mathbf{B}$ represents the same operator ϕ with respect to $\mathfrak{B}', \mathfrak{C}'$.

DEFINITION 12.1.4. *If* $\mathbf{P}, \mathbf{Q}$ *are non-singular matrices, then the substitutions* $\mathbf{x} = \mathbf{P}\mathbf{x}'$, $\mathbf{y} = \mathbf{Q}\mathbf{y}'$ *for the variables in the bilinear form* (12.1.3) *are jointly called a non-singular linear transformation of the bilinear form.*

EXERCISE 12.1.2. Show that, with the obvious definition of the rule of composition, the non-singular linear transformations of a bilinear form constitute a group.

Our preceding remarks show that if ϕ is a given bilinear operator on $\mathfrak{M}$ and $\mathfrak{N}$, then a change of bases in $\mathfrak{M}, \mathfrak{N}$ induces a non-singular linear transformation of the representing bilinear form, and conversely.

For many purposes the notion of a bilinear operator is best set aside in favour of the notion of a bilinear form, since matrix theory provides us with a ready technique for handling such forms. It should, nevertheless, be realized that our concern is not so much with bilinear forms as with the underlying bilinear operators. For this reason properties common to all bilinear forms representing the same bilinear operator are of particular interest since they express intrinsic features of that operator. One such property is that of rank.

DEFINITION 12.1.5. *The* RANK *of the bilinear form* $\mathbf{x}^T\mathbf{A}\mathbf{y}$ *is the rank of* $\mathbf{A}$.

By Theorems 12.1.3 and 5.6.3 (p. 160) it follows that any two bilinear forms representing the same bilinear operator have equal rank.† We may therefore introduce the following definition.

DEFINITION 12.1.6. *The* RANK *of a bilinear operator is the rank of any bilinear form which represents it.*‡

12.1.2. The most important subclass of bilinear operators is that of quadratic operators.

† By virtue of the theorem on the equivalence of matrices (Theorem 6.2.3, p. 176), the converse is also true, and two $m \times n$ matrices of equal rank represent the same bilinear operator with respect to suitable pairs of bases.

‡ For further discussion of bilinear operators and bilinear forms, see Jacobson, **28**, chap. v, and Julia, **34**, 110–16.

DEFINITION 12.1.7. *Let $\phi(X, Y)$ be a bilinear operator in which both variables belong to the same linear manifold $\mathfrak{M}$. If the variables are taken equal to each other, then the resulting function $\phi(X, X)$ is called a* QUADRATIC OPERATOR *on $\mathfrak{M}$.*

The dimensionality of $\mathfrak{M}$ will be denoted by n. As an immediate consequence of Theorem 12.1.1 we see that if $\mathfrak{B} = \{E_1,..., E_n\}$ is a basis of $\mathfrak{M}$ and $\mathbf{x} = \mathscr{R}(X; \mathfrak{B})$, then

$$\phi(X, X) = \mathbf{x}^T\mathbf{A}\mathbf{x} = \sum_{r,s=1}^{n} a_{rs} x_r x_s,$$

where $\mathbf{x} = (x_1,..., x_n)^T$ and $\mathbf{A} = (a_{rs})$ is the $n \times n$ matrix defined by the equations $a_{rs} = \phi(E_r, E_s)$ $(r, s = 1,..., n)$.

If, by analogy with the usage of § 12.1.1, we were now to say that the quadratic function $\sum_{r,s=1}^{n} a_{rs} x_r x_s$ represents the quadratic operator $\phi(X, X)$ we would be setting up a correspondence between quadratic operators and quadratic functions which is not biunique (even for a fixed choice of basis in $\mathfrak{M}$), since if

$$b_{rs}+b_{sr} = a_{rs}+a_{sr} \qquad (r, s = 1,..., n)$$

and $\mathbf{B} = (b_{rs})$, then $\mathbf{x}^T\mathbf{A}\mathbf{x} = \mathbf{x}^T\mathbf{B}\mathbf{x}$, and so $\phi(X, X)$ can be represented equally well by $\mathbf{x}^T\mathbf{A}\mathbf{x}$ and $\mathbf{x}^T\mathbf{B}\mathbf{x}$. We can overcome the difficulty by insisting that the matrix $\mathbf{A}$ should be symmetric. There is no loss of generality in making this assumption, for if

$$a'_{rs} = \tfrac{1}{2}(a_{rs}+a_{sr}) \qquad (r, s = 1,..., n),$$

then $\mathbf{A}' = (a'_{rs})$ is symmetric and $\phi(X, X) = \mathbf{x}^T\mathbf{A}'\mathbf{x}$. We shall, accordingly, always represent quadratic operators by expressions such as $\mathbf{x}^T\mathbf{A}'\mathbf{x}$, where $\mathbf{A}'$ is symmetric. The insistence on symmetry enables us to set up a biunique correspondence between quadratic operators, quadratic forms (defined below), and symmetric matrices.

DEFINITION 12.1.8. *Any polynomial*

$$\begin{aligned}\phi(x_1,..., x_n) &= \sum_{r,s=1}^{n} a_{rs} x_r x_s \\ &= a_{11} x_1^2+...+a_{nn} x_n^2+2a_{12} x_1 x_2+...+2a_{n-1,n} x_{n-1} x_n \\ &= \mathbf{x}^T\mathbf{A}\mathbf{x},\end{aligned}$$

where $\mathbf{x} = (x_1,..., x_n)^T$ and $\mathbf{A} = (a_{rs})$ is a symmetric matrix, is called a QUADRATIC FORM *in the variables $x_1,..., x_n$. $\mathbf{A}$ is the matrix of this form and the numbers a_{rs} are its coefficients.*

It is clear that there is a biunique correspondence between quadratic forms and their matrices.

Quadratic forms in $2, 3, 4, \ldots, n$ variables are called *binary, ternary, quaternary,..., n-ary* respectively.

DEFINITION 12.1.9. *If $\phi(X, X)$ is a quadratic operator on $\mathfrak{M}$ and $\phi(X, X) = \mathbf{x}^T\mathbf{A}\mathbf{x}$, where $\mathbf{A}$ is symmetric and $\mathbf{x}$ is the vector representing X with respect to some basis $\mathfrak{B}$ of $\mathfrak{M}$, then the quadratic form $\mathbf{x}^T\mathbf{A}\mathbf{x}$ (or, alternatively, the matrix $\mathbf{A}$) is said to* REPRESENT *the quadratic operator $\phi(X, X)$ with respect to $\mathfrak{B}$.*†

We leave it to the reader to prove the next statement.

THEOREM 12.1.4. *For every given basis $\mathfrak{B}$ of $\mathfrak{M}$ there is a biunique correspondence between quadratic operators on $\mathfrak{M}$ and the quadratic forms representing them.*

The problem concerning the relation between different representations arises here as it does in the case of linear and bilinear operators.

THEOREM 12.1.5. *Two symmetric $n \times n$ matrices $\mathbf{A}$ and $\mathbf{B}$ represent the same quadratic operator with respect to suitable bases in $\mathfrak{M}$ if and only if they are congruent, i.e. $\mathbf{B} = \mathbf{P}^T\mathbf{A}\mathbf{P}$, where $\mathbf{P}$ is non-singular.*

Suppose that the matrices $\mathbf{A}$ and $\mathbf{B}$ represent the same quadratic operator $\phi(X, X)$ with respect to bases $\mathfrak{B}$ and $\mathfrak{B}'$ (of $\mathfrak{M}$) respectively. Thus, if

$$\mathbf{x} = \mathscr{R}(X; \mathfrak{B}), \qquad \mathbf{x}' = \mathscr{R}(X; \mathfrak{B}'),$$

then

$$\phi(X, X) = \mathbf{x}^T\mathbf{A}\mathbf{x} = \mathbf{x}'^T\mathbf{B}\mathbf{x}'.$$

But $\mathbf{x} = \mathbf{P}\mathbf{x}'$, where $\mathbf{P}$ is a non-singular $n \times n$ matrix, and so

$$\mathbf{x}^T\mathbf{A}\mathbf{x} = \mathbf{x}'^T(\mathbf{P}^T\mathbf{A}\mathbf{P})\mathbf{x}'.$$

Thus the symmetric matrix $\mathbf{P}^T\mathbf{A}\mathbf{P}$ represents $\phi(X, X)$ with respect to $\mathfrak{B}'$, i.e. $\mathbf{B} = \mathbf{P}^T\mathbf{A}\mathbf{P}$. We leave it to the reader to complete the proof of the theorem by establishing the converse.

Theorem 12.1.5 shows that when the basis in $\mathfrak{M}$ is changed, the matrix representing a *quadratic operator* is subjected to a *congruence transformation*. By way of contrast we recall that a change of

† It is a common practice to designate quadratic operators and quadratic forms (and similarly bilinear operators and bilinear forms) by the same term. However, the distinction of name may serve to remind us of the difference of status of the two concepts.

basis in $\mathfrak{M}$ induces a *similarity transformation* in matrices representing *linear operators*.

One of the main reasons for studying quadratic forms is their usefulness in the geometry of conics and quadrics; and the relation between quadratic operators and quadratic forms can, indeed, be aptly illustrated in this context. If Q is a given quadric, then, with respect to a system S of projective coordinates x_0, x_1, x_2, x_3, its equation assumes the form

$$\sum_{r,s=0}^{3} a_{rs}\, x_r\, x_s = \mathbf{x}^T\mathbf{A}\mathbf{x} = 0,$$

where $\mathbf{A} = \mathbf{A}^T = (a_{rs})$ and $\mathbf{x} = (x_0, x_1, x_2, x_3)^T$. If any new system S' of coordinates is taken, then vectors $\mathbf{x}$, $\mathbf{x}'$ representing the same point with respect to the two systems are connected by a linear relation $\mathbf{x} = \mathbf{P}\mathbf{x}'$, where $|\mathbf{P}| \neq 0$; and so the equation of Q, with respect to S', has the form

$$\mathbf{x}'^T(\mathbf{P}^T\mathbf{A}\mathbf{P})\mathbf{x}' = 0.$$

We may thus think of the quadric Q as being associated with a quadratic operator, and then its various equations (with respect to different systems of coordinates) will represent this operator in terms of quadratic forms.† We are, of course, interested primarily in the intrinsic (i.e. the geometrical) properties of Q rather than in the pecularities of any particular equation reresenting it. Similarly, in our discussion of quadratic forms in n variables we shall be primarily concerned with those properties which are common to all quadratic forms representing the same quadratic operator, i.e. properties invariant under congruence transformations.

DEFINITION 12.1.10. *A quadratic form is real* (*complex*) *if its coefficients belong to the real* (*complex*) *field.*

In future, unless the contrary is stated, all quadratic forms will be assumed to be real, and this will be taken to imply that the variables are also real-valued.

DEFINITION 12.1.11. *The determinant* $|\mathbf{A}|$ *of the matrix* $\mathbf{A}$ *associated with a quadratic form* ϕ *is called the determinant of* ϕ. *A quadratic form is said to be singular or non-singular according as its determinant vanishes or does not vanish.*

† This statement requires some qualification owing to the fact that, for any $k \neq 0$, the equations $\mathbf{x}^T\mathbf{A}\mathbf{x} = 0$ and $\mathbf{x}^T(k\mathbf{A})\mathbf{x} = 0$ represent the same quadric.

DEFINITION 12.1.12. (i) *The rank* $R(\phi)$ *of a quadratic form* ϕ *is the rank of the matrix of* ϕ. (ii) *The rank of a quadratic operator is the rank of any quadratic form representing that operator.*

In view of Theorem 12.1.5, the second part of this definition is unambiguous.

DEFINITION 12.1.13. *The substitution* $\mathbf{x}' = \mathbf{P}\mathbf{x}$ *for the variables of a quadratic form* $\mathbf{x}^T\mathbf{A}\mathbf{x}$ *is called a linear transformation of the quadratic form. This transformation is said to be real or complex according as the matrix* $\mathbf{P}$ *of the transformation is real or complex; it is said to be singular or non-singular according as the determinant* $|\mathbf{P}|$ *of the transformation vanishes or does not vanish.*

We shall, generally speaking, be concerned with real non-singular linear transformations.

EXERCISE 12.1.3. Show that a binary quadratic form in the variables x, y may be written as $ax^2+2bxy+cy^2$. What is the result of applying the substitutions $x = \alpha x'+\beta y'$, $y = \gamma x'+\delta y'$ to the variables?

We collect below a number of almost obvious results which are implicit in our previous discussion, the proof of which may be regarded as an exercise.

THEOREM 12.1.6. (i) *The substitution* $\mathbf{x} = \mathbf{P}\mathbf{x}'$ *changes the quadratic form associated with the matrix* $\mathbf{A}$ *into that associated with the matrix* $\mathbf{P}^T\mathbf{A}\mathbf{P}$.

(ii) *A linear transformation has the effect of multiplying the determinant of a quadratic form by the square of the determinant of the transformation.*

(iii) *The rank of a quadratic form is invariant under non-singular linear transformations.*

Finally, it may be observed that non-singular linear (real or complex) transformations of a quadratic form constitute a group.

EXERCISE 12.1.4. Show that if a quadratic form is subjected to a linear transformation specified by the matrix $\mathbf{P}$ and the resulting quadratic form is then subjected to a linear transformation specified by $\mathbf{Q}$, then the total effect of the two transformations is the same as if the original quadratic form were transformed by $\mathbf{PQ}$.

12.1.3. The algebraic treatment of the projective geometry of conics and quadrics, which is based ultimately on Joachimsthal's equation, makes extensive use of a procedure whereby every quadratic form gives rise unambiguously to a symmetric bilinear form, i.e. to a bilinear form whose matrix is symmetric. The necessary process of symmetrization was, in effect, carried out on p. 357, but the relation just mentioned between quadratic forms and the associated bilinear forms has an invariant character and is best described in terms of operators.

Let $\mathfrak{M}$ be a linear manifold of dimensionality n and let $\phi(X,X)$ be a quadratic operator on $\mathfrak{M}$. If $\phi(X,Y)$ is a bilinear operator which gives rise to the quadratic operator $\phi(X,X)$ when the substitution $Y = X$ is made, then, in general, $\phi(X,Y) \neq \phi(Y,X)$. It is, however, easy to symmetrize $\phi(X,Y)$. By definition 12.1.1 (p. 353) the operator $\psi(X,Y)$, defined by the equation

$$\psi(X,Y) = \tfrac{1}{2}\{\phi(X,Y)+\phi(Y,X)\}, \tag{12.1.9}$$

is again bilinear; and it evidently satisfies the relations

$$\psi(X,Y) = \psi(Y,X), \tag{12.1.10}$$

$$\psi(X,X) = \phi(X,X), \tag{12.1.11}$$

for all $X,Y \in \mathfrak{M}$. We express the identity (12.1.10) by calling $\psi(X,Y)$ a *symmetric* bilinear operator. It is, in fact, the only symmetric bilinear operator which gives rise to $\phi(X,X)$, i.e. which satisfies (12.1.11). For, by Definition 12.1.1 and (12.1.9), we at once obtain the identity

$$\phi(X+\lambda Y, X+\lambda Y) = \phi(X,X)+2\lambda\psi(X,Y)+\lambda^2\phi(Y,Y). \tag{12.1.12}$$

If, now, $\psi'(X,Y)$ denotes a bilinear operator such that

$$\psi'(X,Y) = \psi'(Y,X), \qquad \psi'(X,X) = \phi(X,X),$$

then, similarly,

$$\begin{aligned}\phi(X+\lambda Y, X+\lambda Y) &= \psi'(X+\lambda Y, X+\lambda Y)\\ &= \phi(X,X)+2\lambda\psi'(X,Y)+\lambda^2\phi(Y,Y),\end{aligned} \tag{12.1.13}$$

and it follows by (12.1.12) and (12.1.13) that $\psi'(X,Y) = \psi(X,Y)$.

The unique bilinear operator $\psi(X,Y)$ associated with the given quadratic form $\phi(X,X)$ is called the *polarized operator* of $\phi(X,X)$. Its representation by a bilinear form is easily obtained. Let the quadratic form $\mathbf{x}^T\mathbf{A}\mathbf{x}$ represent $\phi(X,X)$ with respect to a basis $\mathfrak{B}$ of $\mathfrak{M}$ and denote by $\chi(X,Y)$ the bilinear operator represented by the bilinear form $\mathbf{x}^T\mathbf{A}\mathbf{y}$ with respect to $\mathfrak{B}$. Then clearly

$$\chi(X,X) = \mathbf{x}^T\mathbf{A}\mathbf{x} = \phi(X,X).$$

Moreover, since $\mathbf{A}$ is symmetric,

$$\chi(X,Y) = \mathbf{x}^T\mathbf{A}\mathbf{y} = \mathbf{y}^T\mathbf{A}\mathbf{x} = \chi(Y,X).$$

Hence $\chi(X,Y)$ is, in fact, the polarized operator $\psi(X,Y)$ of $\phi(X,X)$. Thus, if the quadratic operator $\phi(X,X)$ is represented by the quadratic form $\mathbf{x}^T\mathbf{A}\mathbf{x}$, its polarized operator $\psi(X,Y)$ is represented, with respect to the same basis, by the bilinear form $\mathbf{x}^T\mathbf{A}\mathbf{y}$. The bilinear form $\mathbf{x}^T\mathbf{A}\mathbf{y}$ is called the *polarized form* of the quadratic form $\mathbf{x}^T\mathbf{A}\mathbf{x}$.

The geometrical significance of polarized operators depends primarily on the identity (12.1.12). In terms of the representing forms we have the corresponding identity

$$(\mathbf{x}+\lambda\mathbf{y})^T\mathbf{A}(\mathbf{x}+\lambda\mathbf{y}) = \mathbf{x}^T\mathbf{A}\mathbf{x}+2\lambda\mathbf{x}^T\mathbf{A}\mathbf{y}+\lambda^2\mathbf{y}^T\mathbf{A}\mathbf{y}. \qquad (12.1.14)$$

If the right-hand side of (12.1.14) is equated to zero we obtain a quadratic equation in λ, which is known as Joachimsthal's equation.†

12.2. Orthogonal reduction to diagonal form

12.2.1. In studying problems of euclidean geometry by algebraic methods, we normally employ a non-homogeneous, rectangular, cartesian frame of reference. It is often desirable to change this frame while preserving the origin. The resulting transformation of coordinates is then, as we know, represented by an orthogonal matrix. The purpose of changing the frame of reference is to simplify the equations of the curves or surfaces we are investigating. Since these are frequently conics or quadrics we are led to inquire in what way the quadratic form $\mathbf{x}^T\mathbf{A}\mathbf{x}$ can be simplified as the result of the substitution $\mathbf{x} = \mathbf{P}\mathbf{x}'$, where $\mathbf{P}$ is orthogonal. The answer to this question turns out to be particularly simple and satisfactory.

DEFINITION 12.2.1. A DIAGONAL (UNIT) QUADRATIC FORM *is a quadratic form associated with a diagonal* (*unit*) *matrix.*

Thus the quadratic form associated with the matrix $\mathbf{dg}(a_1,..., a_n)$ is the diagonal quadratic form $a_1 x_1^2+...+a_n x_n^2$. In particular, the n-ary unit quadratic form is $x_1^2+...+x_n^2 = \mathbf{x}^T\mathbf{x}$.

Before proceeding further let us agree on a useful convention. If the quadratic form $\mathbf{x}^T\mathbf{A}\mathbf{x}$ is subjected to the linear transformation $\mathbf{x} = \mathbf{P}\mathbf{x}'$, it becomes $\mathbf{x}'^T(\mathbf{P}^T\mathbf{A}\mathbf{P})\mathbf{x}' = \mathbf{x}'^T\mathbf{B}\mathbf{x}'$, say. Now generally it does not matter what symbols are used for the variables

† For its use in geometry see Semple and Kneebone, *Algebraic Projective Geometry*, 107.

of a quadratic form, since the quadratic form is completely characterized by its matrix. We may therefore say that the transformation $\mathbf{x} = \mathbf{P}\mathbf{x}'$ changes the quadratic form $\mathbf{x}^T\mathbf{A}\mathbf{x}$ into $\mathbf{x}^T\mathbf{B}\mathbf{x}$; and we shall employ this mode of expression whenever we find it convenient to do so.

EXERCISE 12.2.1. Show that the linear transformation $\mathbf{x} = \mathbf{P}\mathbf{x}'$ is orthogonal if and only if it transforms the unit quadratic form into itself.

In preparation for the next theorem let us note that if $\mathbf{A}$ is real and symmetric and $R(\mathbf{A}) = r$, then, by Theorem 10.3.4 (p. 302) and the corollary to Theorem 10.2.3 (p. 296), the number of non-vanishing characteristic roots of $\mathbf{A}$ is r.

Theorem 12.2.1. (Orthogonal reduction of quadratic forms) *Let $\mathbf{x}^T\mathbf{A}\mathbf{x}$ be a real quadratic form of rank r. Then there exists an orthogonal transformation which transforms it into the diagonal form*

$$\lambda_1 x_1^2 + ... + \lambda_r x_r^2,$$

where $\lambda_1,..., \lambda_r$ are the non-vanishing characteristic roots of $\mathbf{A}$.

This result is essentially a restatement of the theorem on the orthogonal reduction of real symmetric matrices (Theorem 10.3.4). For by virtue of that theorem there exists an orthogonal matrix $\mathbf{P}$ such that†

$$\mathbf{P}^T\mathbf{A}\mathbf{P} = \mathbf{P}^{-1}\mathbf{A}\mathbf{P} = \mathbf{dg}(\lambda_1,..., \lambda_r, 0,..., 0),$$

where the expression on the right-hand side contains $n-r$ zeros. Hence the orthogonal transformation $\mathbf{x} = \mathbf{P}\mathbf{x}'$ carries the given quadratic form $\mathbf{x}^T\mathbf{A}\mathbf{x}$ into

$$\mathbf{x}'(\mathbf{P}^T\mathbf{A}\mathbf{P})\mathbf{x}' = \mathbf{x}'^T.\mathbf{dg}(\lambda_1,..., \lambda_r, 0,..., 0).\mathbf{x}' = \lambda_1 x_1'^2 + ... + \lambda_r x_r'^2.$$

The process of transforming a given quadratic form into a diagonal quadratic form is known as *reduction to diagonal form*. If the transformation is orthogonal, we speak of an *orthogonal reduction*.

12.2.2. One of the most important applications of the transformation theory of quadratic forms occurs in the reduction of central quadrics (or conics) to principal axes.

† It should be remembered that 'orthogonal similarity transformation' and 'orthogonal congruence transformation' are identical notions.

THEOREM 12.2.2. *Let the equation of a central quadric Q, with its centre at the origin, be*

$$a_{11}x_1^2+a_{22}x_2^2+a_{33}x_3^2+2a_{23}x_2x_3+2a_{31}x_3x_1+2a_{12}x_1x_2 = 1,$$

where x_1, x_2, x_3 are non-homogeneous rectangular coordinates. Let $\lambda_1, \lambda_2, \lambda_3$ be the characteristic roots of the real symmetric 3×3 matrix $\mathbf{A} = (a_{rs})$ and let $\boldsymbol{\xi}_1, \boldsymbol{\xi}_2, \boldsymbol{\xi}_3$ be any set of orthonormal characteristic vectors of $\mathbf{A}$ associated with $\lambda_1, \lambda_2, \lambda_3$ respectively.† *Then the directions specified by $\boldsymbol{\xi}_1, \boldsymbol{\xi}_2, \boldsymbol{\xi}_3$*‡ *are the directions of a set of principal axes of Q; and with respect to these axes as coordinate axes the equation of Q assumes the form*

$$\lambda_1 x_1^2+\lambda_2 x_2^2+\lambda_3 x_3^2 = 1. \tag{12.2.1}$$

Let $\mathbf{P}$ denote the (orthogonal) matrix having $\boldsymbol{\xi}_1, \boldsymbol{\xi}_2, \boldsymbol{\xi}_3$ as its columns. Then, by Theorem 10.2.1 (p. 293),

$$\mathbf{P}^T\mathbf{A}\mathbf{P} = \mathbf{dg}(\lambda_1, \lambda_2, \lambda_3). \tag{12.2.2}$$

The original equation of Q is given by $\mathbf{x}^T\mathbf{A}\mathbf{x} = 1$, where $\mathbf{x} = (x_1, x_2, x_3)^T$. If S denotes the original system of coordinates and S' the system for which $\boldsymbol{\xi}_1, \boldsymbol{\xi}_2, \boldsymbol{\xi}_3$ specify the directions of the axes, then, by Theorem 8.4.4 (p. 238), $\mathbf{x} = \mathbf{P}\mathbf{x}'$, where $\mathbf{x}' = (x_1', x_2', x_3')^T$ and x_1', x_2', x_3' are the coordinates, with respect to S', of a point whose coordinates with respect to S are x_1, x_2, x_3. Hence, with respect to S', the equation of Q assumes the form

$$(\mathbf{P}\mathbf{x}')^T\mathbf{A}(\mathbf{P}\mathbf{x}) = 1,$$

and in view of (12.2.2) this is equivalent to

$$\lambda_1 x_1'^2+\lambda_2 x_2'^2+\lambda_3 x_3'^2 = 1.$$

It follows that $\boldsymbol{\xi}_1, \boldsymbol{\xi}_2, \boldsymbol{\xi}_3$ specify the directions of a set of principal axes of Q, and the theorem is therefore proved.

In examining the reduction of central quadrics to principal axes we must distinguish between three cases that may arise.

Case I. ($\lambda_1, \lambda_2, \lambda_3$ distinct.)

In view of the corollary to Theorem 7.6.1 (p. 215), each of the three vectors $\boldsymbol{\xi}_1, \boldsymbol{\xi}_2, \boldsymbol{\xi}_3$ is, in this case, uniquely determined to

† The existence of such a set of vectors is guaranteed by Theorem 10.3.5 (p. 304).

‡ If $\boldsymbol{\xi}$ is the vector (p, q, r) then by 'the direction specified by $\boldsymbol{\xi}$' we mean, of course, the direction of the straight line joining the origin to the point (p, q, r).

within a scalar factor ± 1. Hence Q has a unique set of principal axes. This conclusion is confirmed by the equation (12.2.1) which shows that Q is an ellipsoid or a hyperboloid, but not a quadric of revolution.

Case II. (Precisely two of λ_1, λ_2, λ_3 are equal.)

To fix our ideas, let us assume that $\lambda_1 \neq \lambda_2 = \lambda_3$. In that case $\boldsymbol{\xi}_1$ is determined to within a factor ± 1. Now $\boldsymbol{\xi}_1$, $\boldsymbol{\xi}_2$, $\boldsymbol{\xi}_3$ constitute a basis of $\mathfrak{V}_3$, and therefore any vector $\boldsymbol{\eta}$ may be expressed in the form

$$\boldsymbol{\eta} = t_1 \boldsymbol{\xi}_1 + t_2 \boldsymbol{\xi}_2 + t_3 \boldsymbol{\xi}_3.$$

If $\boldsymbol{\eta}$ is orthogonal to $\boldsymbol{\xi}_1$, then

$$0 = (\boldsymbol{\xi}_1, \boldsymbol{\eta}) = t_1(\boldsymbol{\xi}_1, \boldsymbol{\xi}_1) = t_1,$$

and therefore
$$\boldsymbol{\eta} = t_2 \boldsymbol{\xi}_2 + t_3 \boldsymbol{\xi}_3,$$

$$\mathbf{A}\boldsymbol{\eta} = t_2 \mathbf{A}\boldsymbol{\xi}_2 + t_3 \mathbf{A}\boldsymbol{\xi}_3 = t_2 \lambda_2 \boldsymbol{\xi}_2 + t_3 \lambda_2 \boldsymbol{\xi}_3 = \lambda_2 \boldsymbol{\eta}.$$

Hence any non-zero vector $\boldsymbol{\eta}$ orthogonal to $\boldsymbol{\xi}_1$ is necessarily a characteristic vector of $\mathbf{A}$ associated with λ_2. Thus to choose $\boldsymbol{\xi}_2$ we may take any unit vector in the plane through O perpendicular to $\boldsymbol{\xi}_1$. Finally $\boldsymbol{\xi}_3$ must be at right angles to both $\boldsymbol{\xi}_1$ and $\boldsymbol{\xi}_2$, and so, for any given $\boldsymbol{\xi}_1$ and $\boldsymbol{\xi}_2$, it is fixed to within a scalar factor ± 1.

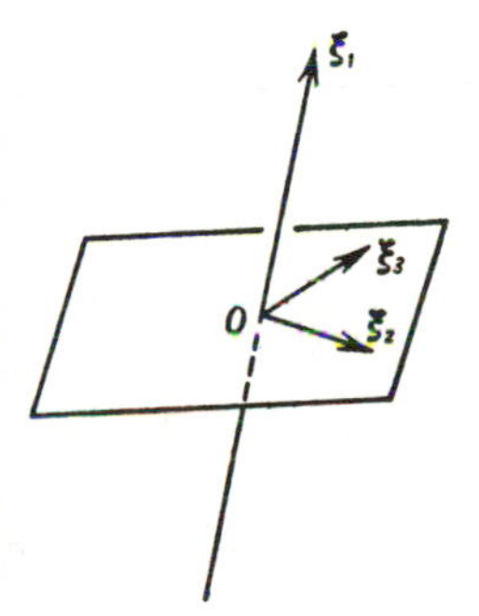

One principal axis of Q (that corresponding to $\boldsymbol{\xi}_1$) is therefore uniquely determined; the remaining two may be chosen arbitrarily (subject to their being at right angles) in the plane through O perpendicular to the first axis. This is borne out by the equation (12.2.1) from which we can infer that Q is an ellipsoid of revolution (but not a sphere) or a hyperboloid of revolution.

Case III. (λ_1, λ_2, λ_3 all equal.)

By (12.2.2) we have in this case

$$\mathbf{P}^T\mathbf{A}\mathbf{P} = \mathbf{dg}(\lambda_1, \lambda_1, \lambda_1) = \lambda_1 \mathbf{I},$$

and so $\mathbf{A} = \lambda_1 \mathbf{I}$. Since $(\lambda_1 \mathbf{I})\boldsymbol{\xi} = \lambda_1 \boldsymbol{\xi}$, we see that every non-zero vector is a characteristic vector of $\mathbf{A}$. It follows that any three mutually perpendicular lines through the origin may be taken as principal axes of Q. The equation (12.2.1) leads to the same conclusion since it tells us that Q is a sphere.

To carry out the reduction to principal axes in a numerical case we have to effect an orthogonal reduction of a symmetric matrix, and we do this by using the procedure described on p. 303. Consider the central quadric Q given by the equation

$$3y^2+3z^2-2yz-4zx-4xy = 1. \tag{12.2.3}$$

The associated matrix is

$$\mathbf{A} = \begin{pmatrix} 0 & -2 & -2 \\ -2 & 3 & -1 \\ -2 & -1 & 3 \end{pmatrix},$$

and the characteristic roots of $\mathbf{A}$ are found to be $-2, 4, 4$. Hence, referred to a set of principal axes, the equation of Q assumes the form

$$-2x'^2+4y'^2+4z'^2 = 1, \tag{12.2.4}$$

and Q is therefore seen to be a hyperboloid of revolution of one sheet. Let us next determine the formulae of transformation. If $(p,q,r)^T$ is a characteristic vector of $\mathbf{A}$ associated with the characteristic root λ, then

$$\left.\begin{aligned} -2q-2r &= \lambda p \\ -2p+3q-r &= \lambda q \\ -2p-q+3r &= \lambda r \end{aligned}\right\}. \tag{12.2.5}$$

When $\lambda = -2$ this system of equations gives us $p = 2q$, $r = q$, and therefore a unit characteristic vector associated with the characteristic root -2 may be taken as $(2/\sqrt{6},\ 1/\sqrt{6},\ 1/\sqrt{6})^T$. Again, when $\lambda = 4$ the system (12.2.5) reduces to the single equation $2p+q+r = 0$. We have to choose two unit characteristic vectors, say $(p,q,r)^T$ and $(p',q',r')^T$, both of which are associated with the characteristic root 4 and the same time are orthogonal to each other. Hence

$$2p+q+r = 0, \qquad 2p'+q'+r' = 0, \qquad pp'+qq'+rr' = 0.$$

The two vectors in question can be chosen in an infinity of ways and one possible choice is $(1/\sqrt{3},\ -1/\sqrt{3},\ -1/\sqrt{3})^T$, $(0,\ 1/\sqrt{2},\ -1/\sqrt{2})^T$.† The orthogonal matrix $\mathbf{P}$ whose columns are the three characteristic vectors just found is then given by

$$\mathbf{P} = \begin{pmatrix} 2/\sqrt{6} & 1/\sqrt{3} & 0 \\ 1/\sqrt{6} & -1/\sqrt{3} & 1/\sqrt{2} \\ 1/\sqrt{6} & -1/\sqrt{3} & -1/\sqrt{2} \end{pmatrix},$$

and we have $$\mathbf{P}^T\mathbf{A}\mathbf{P} = \mathbf{dg}(-2,4,4).$$

† The two vectors are, of course, automatically orthogonal to the vector $(2/\sqrt{6},\ 1/\sqrt{6},\ 1/\sqrt{6})^T$ chosen earlier.

The substitution $\mathbf{x} = \mathbf{P}\mathbf{x}'$, where $\mathbf{x} = (x, y, z)^T$, $\mathbf{x}' = (x', y', z')^T$, thus reduces the equation (12.2.3) to the form (12.2.4). This substitution can be written, more explicitly, as

$$x = \frac{2}{\sqrt{6}}x' + \frac{1}{\sqrt{3}}y',$$

$$y = \frac{1}{\sqrt{6}}x' - \frac{1}{\sqrt{3}}y' + \frac{1}{\sqrt{2}}z',$$

$$z = \frac{1}{\sqrt{6}}x' - \frac{1}{\sqrt{3}}y' - \frac{1}{\sqrt{2}}z'.$$

The three characteristic vectors specify the direction ratios of a set of principal axes. The equations of these axes are therefore given by

$$x = 2y = 2z; \qquad -x = y = z; \qquad x = 0,\ y = -z.$$

12.2.3. The relation between the principal axes of a quadric Q, given by the equation $\mathbf{x}^T\mathbf{A}\mathbf{x} = 1$, and the characteristic vectors of $\mathbf{A}$ was established above by means of the transformation theory of quadratic forms (i.e. of symmetric matrices.) A more direct derivation proceeds as follows. If $\mathbf{x}$ is the position vector of a point P, then the direction ratios of the normal to the polar plane π of P with respect to Q are given by the components of the vector $\mathbf{A}\mathbf{x}$. Hence the line OP is perpendicular to π if and only if $\mathbf{A}\mathbf{x} = \lambda\mathbf{x}$ for some (non-zero) value of λ. But OP is perpendicular to π if and only if P lies on one of the principal axes of Q. Thus each principal axis of Q is specified by a characteristic vector of $\mathbf{A}$.

12.3. General reduction to diagonal form

In the preceding section we studied the reduction of quadratic forms to diagonal form by means of orthogonal transformations. In the geometrical context the insistence on orthogonality is necessary only when we investigate metrical properties of figures and it loses its significance in affine and projective geometry. We propose, therefore, to consider next reductions to diagonal form effected by non-singular linear transformations which are not subject to further restrictions. An analogous problem arises for bilinear forms and we shall deal with it before turning again to quadratic forms.

12.3.1. A bilinear operator can be represented in many different ways by a bilinear form, and we are therefore confronted by the problem of determining representations which shall be as simple as possible. The next theorem shows that a 'diagonal representation' can always be found.

THEOREM 12.3.1. *Any bilinear form of rank r can be changed into the bilinear form*
$$x_1 y_1 + \ldots + x_r y_r \tag{12.3.1}$$
by means of a non-singular linear transformation.

This result states, in fact, that if ϕ is a bilinear operator, of rank r, on the linear manifolds $\mathfrak{M}$ and $\mathfrak{N}$, then bases in $\mathfrak{M}$ and $\mathfrak{N}$ can be found with respect to which ϕ is represented by the bilinear form (12.3.1).

To prove the theorem denote the given bilinear form by $\mathbf{x}^T\mathbf{A}\mathbf{y}$, where $\mathbf{A}$ is an $m \times n$ matrix and $R(\mathbf{A}) = r$. Then, by Theorem 6.2.3 (p. 176), there exist non-singular matrices $\mathbf{P}$, $\mathbf{Q}$ (of order m, n respectively) such that
$$\mathbf{P}^T\mathbf{A}\mathbf{Q} = \mathbf{N}_r^{(m,n)},$$
where $\mathbf{N}_r^{(m,n)}$ is the $m \times n$ matrix
$$\begin{pmatrix} \mathbf{I}_r & \mathbf{O} \\ \mathbf{O} & \mathbf{O} \end{pmatrix}.$$
Hence the non-singular linear transformation $\mathbf{x} = \mathbf{P}\boldsymbol{\xi}$, $\mathbf{y} = \mathbf{Q}\boldsymbol{\eta}$ changes the bilinear form $\mathbf{x}^T\mathbf{A}\mathbf{y}$ into
$$\boldsymbol{\xi}^T(\mathbf{P}^T\mathbf{A}\mathbf{Q})\boldsymbol{\eta} = \boldsymbol{\xi}^T\mathbf{N}_r^{(m,n)}\boldsymbol{\eta} = \xi_1\eta_1 + \ldots + \xi_r\eta_r,$$
where $\boldsymbol{\xi} = (\xi_1, \ldots, \xi_m)^T$, $\boldsymbol{\eta} = (\eta_1, \ldots, \eta_n)^T$. This proves the theorem. We may, at the same time, note that if the original bilinear form is real, then it can be reduced to (12.3.1) by a real transformation.

DEFINITION 12.3.1. *Two bilinear forms are* EQUIVALENT *if one of them can be transformed into the other by a non-singular linear transformation.*

Thus equivalence of bilinear forms means equivalence with respect to the group of non-singular linear transformations.

EXERCISE 12.3.1. Show that the relation between bilinear forms specified in Definition 12.3.1 is an equivalence relation in the sense of Definition 6.5.1 (p. 186).

THEOREM 12.3.2. *Each of the following three conditions relating to the bilinear forms ϕ, ψ implies the other two.*

(i) *ϕ and ψ are equivalent.*

(ii) *ϕ and ψ represent the same bilinear operator.*

(iii) *ϕ and ψ have the same rank.*

Let the matrices associated with ϕ, ψ be denoted by **A**, **B** respectively. Statement (i) then means simply that there exist non-singular matrices **P**, **Q** such that

$$\mathbf{B} = \mathbf{P}^T\mathbf{A}\mathbf{Q}, \tag{12.3.2}$$

and, by Theorem 12.1.3 (p. 355), this implies and is implied by (ii). Again, by Theorem 6.2.3, (12.3.2) implies and is implied by (iii). The assertion is therefore proved.

EXERCISE 12.3.2. Show that two bilinear forms are equivalent if and only if their matrices are equivalent in the sense of Definition 6.2.2 (p. 176).

12.3.2. Turning now to quadratic forms we observe at once the following result.

THEOREM 12.3.3. *A real quadratic form of rank r can be transformed by a real non-singular linear transformation into the diagonal form*

$$\alpha_1 x_1^2 + \ldots + \alpha_r x_r^2,$$

where $\alpha_1,\ldots,\alpha_r$ *are all non-zero.*

This is, of course, a weakened version of Theorem 12.2.1. Alternatively it follows immediately by Exercise 6.4.3 (p. 185) since non-singular linear transformations of quadratic forms correspond to real congruence transformations of the associated matrix.

Theorem 12.3.3 shows, in fact, that every quadratic operator on a linear manifold $\mathfrak{M}$ possesses a 'diagonal representation' for an appropriate choice of basis in $\mathfrak{M}$. The theorem is also important in projective geometry since it shows that, with respect to a suitable coordinate system, the equation of a conic or a quadric assumes a form involving only the squares of the coordinates.

EXERCISE 12.3.3. Show that the transformation in Theorem 12.3.3 can be chosen in such a way that each α_k has the value ± 1.

EXERCISE 12.3.4. Show that Theorem 12.3.3 remains valid for complex quadratic forms, provided that complex transformations are admitted. By using the additional transformation

$$x_k = \begin{cases} y_k/\sqrt{\alpha_k} & (k = 1,\ldots, r), \\ y_k & (k = r+1,\ldots, n), \end{cases}$$

show also that in this case all α's can be made equal to 1.

As a consequence of Theorem 12.3.3 we have the following corollary.

COROLLARY. *An n-ary real singular quadratic form can be reduced by a real non-singular linear transformation to the form*

$$\alpha_1 x_1^2 + \ldots + \alpha_{n-1} x_{n-1}^2.$$

In view of Theorem 12.3.3 it may be asked whether it is possible to reduce a quadratic form of rank r to diagonal form in which the number of non-vanishing coefficients is not equal to r. It is, however, almost obvious that such a reduction is not possible.

THEOREM 12.3.4. *If a quadratic form of rank r is reduced by a non-singular linear transformation to diagonal form, then the latter must have precisely r non-vanishing coefficients.*

For, by Theorem 12.1.6 (iii) (p. 360), the rank of a quadratic form is invariant under non-singular linear transformations, and the rank of a diagonal quadratic form is equal to the number of its non-vanishing coefficients.

The theory of linear transformations of quadratic forms may be stated in a language slightly different from that used so far. Consider, for instance, the binary quadratic form

$$ax^2+2bxy+cy^2, \tag{12.3.3}$$

and suppose that a change of variables, specified by the equations

$$x' = \alpha x+\beta y, \qquad y' = \gamma x+\delta y,$$

transforms it into

$$a'x'^2+2b'x'y'+c'y'^2. \tag{12.3.4}$$

Instead of regarding the transition from (12.3.3) to (12.3.4) as a result of the transformation of variables, we can suppress all reference to the introduction of new variables x', y' and exhibit the relation between the original and the resulting quadratic form as an *identity*, namely,

$$ax^2+2bxy+cy^2 = a'(\alpha x+\beta y)^2+2b'(\alpha x+\beta y)(\gamma x+\delta y)+c'(\gamma x+\delta y)^2.$$

By adopting, as we obviously can, the same point of view in the case of n-ary quadratic forms we are led to the following conclusion.

Theorem 12.3.5. *A quadratic form in $x_1,..., x_n$, of rank r, can be expressed as a linear combination, with non-zero coefficients, of the squares of r linearly independent linear forms in $x_1,..., x_n$.*

This result is implied by Theorem 12.3.3. Let

$$\phi(x_1,..., x_n) = \mathbf{x}^T\mathbf{A}\mathbf{x},$$

where $\mathbf{x} = (x_1,..., x_n)^T$, be the given quadratic form. Then a suitable non-singular linear transformation $\mathbf{x} = \mathbf{P}\mathbf{y}$ changes it to $\alpha_1 y_1^2+...+\alpha_r y_r^2$ $(\alpha_1,..., \alpha_r \neq 0)$, where $\mathbf{y} = (y_1,..., y_n)^T$. We may think of each y_i as a linear form in $x_1,..., x_n$ determined by the relation $\mathbf{y} = \mathbf{P}^{-1}\mathbf{x}$, and we then have, identically in $x_1,..., x_n$,

$$\phi(x_1,..., x_n) = \alpha_1 y_1^2 + ... + \alpha_r y_r^2.$$

Moreover, since $|\mathbf{P}^{-1}| \neq 0$, it follows by Theorem 5.5.2 (p. 153) that not only $y_1,..., y_r$ but, indeed, $y_1,..., y_n$ are linearly independent linear forms.

EXERCISE 12.3.5. Express $3x^2+2xy+7y^2$ as a linear combination of two squares.

12.3.3. Theorems 12.3.3 and 12.3.5 are pure existence theorems and do not provide a procedure for carrying out the reduction of a given quadratic form to diagonal form. Such a reduction can, of course, be always effected by an orthogonal transformation.† However, the amount of computation involved in this process is often prohibitive; and when orthogonality is not essential, Lagrange's method of reduction, explained below, is preferable. In addition to its practical utility this method (which was devised by Lagrange in 1759 and rediscovered by Gauss in 1823) provides, in conjunction with Theorem 12.3.4, a new proof of Theorems 12.3.3 and 12.3.5, and so makes them independent of the relatively difficult Theorems 6.4.1 and 10.3.4.

Lagrange's method is a process of successive reduction by means of which we eventually remove all mixed terms $a_{rs} x_r x_s$ ($r \neq s$) from the initial quadratic form

$$\phi = \phi(x_1,..., x_n) = \sum_{r,s=1}^{n} a_{rs} x_r x_s$$

$$= a_{11} x_1^2 + ... + a_{nn} x_n^2 + 2a_{12} x_1 x_2 + ... + 2a_{n-1,n} x_{n-1} x_n.$$

If at least one of $a_{11},..., a_{nn}$ is not zero we may assume, without loss of generality, that $a_{11} \neq 0$.‡ Then

$$\phi = a_{11} x_1^2 + 2a_{12} x_1 x_2 + ... + 2a_{1n} x_1 x_n + \sum_{r,s=2}^{n} a_{rs} x_r x_s$$

$$= a_{11}\left(x_1^2 + 2\frac{a_{12}}{a_{11}} x_1 x_2 + ... + 2\frac{a_{1n}}{a_{11}} x_1 x_n\right) + \phi_1(x_2,..., x_n), \quad \text{say}.$$

† Cf. the numerical example in § 12.2.2.

‡ For, if $a_{11} = 0$ and $a_{rr} \neq 0$ for some $r > 1$, we may first apply the non-singular linear transformation

$$x_1 = \xi_r, \quad x_r = \xi_1, \quad x_s = \xi_s \quad (s \neq 1, s \neq r).$$

Hence

$$\phi = a_{11}\left\{\left(x_1+\frac{a_{12}}{a_{11}}x_2+\dots+\frac{a_{1n}}{a_{11}}x_n\right)^2-\left(\frac{a_{12}}{a_{11}}\right)^2x_2^2-\dots-\left(\frac{a_{1n}}{a_{11}}\right)^2x_n^2-\right.$$
$$\left.-2\frac{a_{12}a_{13}}{a_{11}^2}x_2x_3-\dots\right\}+\phi_1(x_2,\dots,x_n)$$
$$= a_{11}\left(x_1+\frac{a_{12}}{a_{11}}x_2+\dots+\frac{a_{1n}}{a_{11}}x_n\right)^2+\phi_2(x_2,\dots,x_n),$$

say, where ϕ_2 is a quadratic form in $x_2,\dots, x_n$. We now see that the (obviously non-singular) linear transformation

$$y_1 = x_1+\frac{a_{12}}{a_{11}}x_2+\dots+\frac{a_{1n}}{a_{11}}x_n, \qquad y_2 = x_2, \quad \dots, \quad y_n = x_n$$

changes ϕ into the form

$$\phi = \alpha_1 y_1^2+\phi_2(y_2,\dots, y_n) \tag{12.3.5}$$

(where actually $\alpha_1 = a_{11}$).

However, the above procedure breaks down if $a_{11} = \dots = a_{nn} = 0$, and in that case an additional step is required. We still have $a_{rs} \neq 0$ for some r, s such that $r \neq s$, for otherwise the quadratic form would vanish identically. Assume, then, without loss of generality, that $a_{12} \neq 0$ and use the transformation

$$x_2 = \xi_1+\xi_2, \qquad x_k = \xi_k \quad (k \neq 2).$$

This transformation has determinant 1 and so is non-singular. It carries ϕ into a quadratic form in $\xi_1,\dots, \xi_n$ in which the term in ξ_1^2 is present; to this new quadratic form we now apply the method described earlier and obtain ϕ in the form (12.3.5).

It is thus in every case possible to change the given quadratic form ϕ into (12.3.5) by a non-singular linear transformation. The reduction to diagonal form may now be completed by repeating the process as often as it is necessary. Treating ϕ_2 in the same way as we treated ϕ above, we observe that there exists a real non-singular transformation of $y_2,\dots, y_n$ into $z_2,\dots, z_n$ such that

$$\phi_2 = \alpha_2 z_2^2+\phi_3(z_3,\dots, z_n),$$

where ϕ_3 is a quadratic form in $z_3,\dots, z_n$. If the equations connecting $y_2,\dots, y_n$ with $z_2,\dots, z_n$ are supplemented by the equation $y_1 = z_1$, then the transformation from $y_1,\dots, y_n$ to $z_1,\dots, z_n$ is again non-singular; and ϕ is seen to assume the form

$$\phi = \alpha_1 z_1^2+\alpha_2 z_2^2+\phi_3(z_3,\dots, z_n),$$

the change being effected by a non-singular transformation from $x_1,..., x_n$ to $z_1,..., z_n$.

We continue this process of reduction as long as any mixed terms are left, and ultimately obtain ϕ in the form

$$\phi = \alpha_1 w_1^2+...+\alpha_r w_r^2,$$

where $r = R(\phi)$. The reduction has been effected by a succession of real non-singular linear transformations whose resultant is therefore again a real non-singular linear transformation. It is clear, moreover, that the procedure applies equally well to complex quadratic forms. We conclude, therefore, that any quadratic form, real or complex, can be reduced to diagonal form by means of a non-singular linear transformation.

To illustrate Lagrange's method of reduction in a numerical case, let us consider the quaternary quadratic form

$$\phi = 2x_1 x_2-x_1 x_3+x_1 x_4-x_2 x_3+x_2 x_4-2x_3 x_4.$$

Putting

$$x_1 = y_1, \quad x_2 = y_1+y_2, \quad x_3 = y_3, \quad x_4 = y_4, \tag{12.3.6}$$

we obtain

$$\begin{aligned}\phi &= 2y_1^2+2y_1 y_2-2y_1 y_3+2y_1 y_4-y_2 y_3+y_2 y_4-2y_3 y_4 \\ &= 2(y_1+\tfrac{1}{2}y_2-\tfrac{1}{2}y_3+\tfrac{1}{2}y_4)^2-\tfrac{1}{2}y_2^2-\tfrac{1}{2}y_3^2-\tfrac{1}{2}y_4^2-y_3 y_4.\end{aligned}$$

Next, putting

$$z_1 = y_1+\tfrac{1}{2}y_2-\tfrac{1}{2}y_3+\tfrac{1}{2}y_4, \quad z_2 = y_2, \quad z_3 = y_3, \quad z_4 = y_4, \tag{12.3.7}$$

we obtain

$$\phi = 2z_1^2-\tfrac{1}{2}z_2^2-\tfrac{1}{2}z_3^2-\tfrac{1}{2}z_4^2-z_3 z_4 = 2z_1^2-\tfrac{1}{2}z_2^2-\tfrac{1}{2}(z_3+z_4)^2.$$

Hence, writing

$$w_1 = z_1, \quad w_2 = z_2, \quad w_3 = z_3+z_4, \quad w_4 = z_4, \tag{12.3.8}$$

we have

$$\phi = 2w_1^2-\tfrac{1}{2}w_2^2-\tfrac{1}{2}w_3^2. \tag{12.3.9}$$

Moreover, we see by (12.3.6), (12.3.7), and (12.3.8) that the transformation effecting the reduction of ϕ to the diagonal form (12.3.9) is given by

$$\begin{aligned}x_1 &= w_1-\tfrac{1}{2}w_2+\tfrac{1}{2}w_3-w_4, \\ x_2 &= w_1+\tfrac{1}{2}w_2+\tfrac{1}{2}w_3-w_4, \\ x_3 &= \qquad\qquad\quad\; w_3-w_4, \\ x_4 &= \qquad\qquad\qquad\quad\;\; w_4.\end{aligned}$$

Alternatively, we may write

$$\begin{aligned} w_1 &= \tfrac{1}{2}x_1+\tfrac{1}{2}x_2-\tfrac{1}{2}x_3+\tfrac{1}{2}x_4, \\ w_2 &= -x_1+x_2, \\ w_3 &= x_3+x_4, \\ w_4 &= x_4, \end{aligned}$$

and (12.3.9) may be expressed in the form of the identity

$$2x_1x_2-x_1x_3+x_1x_4-x_2x_3+x_2x_4-2x_3x_4$$
$$= 2(\tfrac{1}{2}x_1+\tfrac{1}{2}x_2-\tfrac{1}{2}x_3+\tfrac{1}{2}x_4)^2-\tfrac{1}{2}(-x_1+x_2)^2-\tfrac{1}{2}(x_3+x_4)^2.$$

We recognize, incidentally, that $R(\phi) = 3$.

12.3.4. The next theorem shows in what circumstances a given quadratic form can be expressed as a product of two linear forms.

THEOREM 12.3.6. *Let ϕ be a complex quadratic form.* (i) *ϕ is the square of a non-vanishing linear form if and only if $R(\phi) = 1$.* (ii) *ϕ is the product of two linearly independent linear forms if and only if $R(\phi) = 2$.*

If $\phi = (\alpha_1 x_1+...+\alpha_n x_n)^2$ and $\alpha_1 x_1+...+\alpha_n x_n$ does not vanish identically we may assume, without loss of generality, that $\alpha_1 \neq 0$. The non-singular linear transformation

$$y_1 = \alpha_1 x_1+...+\alpha_n x_n, \qquad y_2 = x_2, \quad ..., \quad y_n = x_n$$

carries ϕ into y_1^2. Hence $R(\phi) = 1$. On the other hand, if $R(\phi) = 1$, then, by virtue of Theorem 12.3.5 (as applied to complex quadratic forms), it is possible to write ϕ in the form $\phi = \alpha_1 y_1^2$, where $\alpha_1 \neq 0$ and y_1 is a (non-vanishing) linear form in $x_1,..., x_n$. Thus

$$\phi = \{\sqrt{\alpha_1}.(p_1 x_1+...+p_n x_n)\}^2,$$

and (i) is proved.

Next, let
$$\phi = (\alpha_1 x_1+...+\alpha_n x_n)(\beta_1 x_1+...+\beta_n x_n),$$

where the two linear forms on the right-hand side are linearly independent, i.e. are not multiples of each other. We may then assume that

$$\begin{vmatrix} \alpha_1 & \alpha_2 \\ \beta_1 & \beta_2 \end{vmatrix} \neq 0,$$

since at least one such minor is not zero. The non-singular linear transformation

$$\begin{aligned} y_1 &= \alpha_1 x_1 + \alpha_2 x_2 + \ldots + \alpha_n x_n \\ y_2 &= \beta_1 x_1 + \beta_2 x_2 + \ldots + \beta_n x_n \\ y_3 &= \qquad\qquad x_3 \\ &\cdot \quad \cdot \quad \cdot \quad \cdot \quad \cdot \quad \cdot \quad \cdot \quad \cdot \\ y_n &= \qquad\qquad\qquad x_n, \end{aligned}$$

carries ϕ into $y_1 y_2$, and the non-singular linear transformation

$$y_1 = z_1 + z_2, \quad y_2 = z_1 - z_2, \quad y_3 = z_3, \quad \ldots, \quad y_n = z_n$$

carries $y_1 y_2$ into $z_1^2 - z_2^2$. Hence $R(\phi) = 2$. On the other hand, if $R(\phi) = 2$, then, by Theorem 12.3.5, ϕ can be written as $\phi = \alpha_1 y_1^2 + \alpha_2 y_2^2$ ($\alpha_1 \neq 0$, $\alpha_2 \neq 0$), where y_1, y_2 are linearly independent linear forms in $x_1, \ldots, x_n$. Hence

$$\phi = (\sqrt{\alpha_1}\, y_1 + i\sqrt{\alpha_2}\, y_2)(\sqrt{\alpha_1}\, y_1 - i\sqrt{\alpha_2}\, y_2),$$

and the expressions in brackets are again linearly independent linear forms in $x_1, \ldots, x_n$. The proof is therefore complete.

COROLLARY. *A complex ternary quadratic form is the product of two linear factors if and only if it is singular.*

This corollary provides us with a criterion for the degeneracy of conics. If in homogeneous (cartesian, affine, or projective) coordinates x, y, z the equation of the conic C is of the form $\mathbf{x}^T\mathbf{A}\mathbf{x} = 0$, where $\mathbf{x} = (x, y, z)^T$, then C is degenerate (i.e. breaks up into two distinct lines or a repeated line) if and only if the determinant $|\mathbf{A}|$ vanishes.

12.4. The problem of equivalence. Rank and signature

12.4.1. Consider a system $S(x, y, z)$ of projective coordinates in the plane. With respect to this system any conic C will have an equation of the form

$$\phi(x, y, z) = 0, \tag{12.4.1}$$

where ϕ is a ternary quadratic form. If $S'(x', y', z')$ is a second system of projective coordinates, then the two systems are connected by a relation $(x', y', z')^T = \mathbf{A}(x, y, z)^T$, where $\mathbf{A}$ is a non-singular 3×3 matrix. This linear transformation changes the equation (12.4.1) of C into

$$\psi(x', y', z') = 0,$$

where ψ is again a ternary quadratic form. The new equation still represents the same conic; and consequently, from the point of view of geometrical interpretation, what is important is not an individual quadratic form ϕ but the entire class of quadratic forms that can be obtained from ϕ by non-singular linear transformations. All these quadratic forms will represent, when equated to 0, the same conic C, referred in each case to a suitable system of projective coordinates.

It is principally for this reason that we are interested in quadratic forms which can be obtained from each other by non-singular linear transformations. Accordingly we introduce the following definition which is, in point of fact, implicit in our previous terminology.

DEFINITION 12.4.1. *If a quadratic form ψ is obtained from the quadratic form ϕ by a complex (real) non-singular linear transformation, then ψ is said to be* EQUIVALENT *to ϕ under the group of complex (real) non-singular linear transformations.*

The two groups of transformations mentioned here will be denoted, for brevity, by T_C and T_R. The two relations between ϕ and ψ are clearly equivalence relations. We can express these relations in terms of the matrices $\mathbf{A}$, $\mathbf{B}$ of ϕ, ψ respectively by saying that there exists a non-singular matrix $\mathbf{P}$ such that $\mathbf{B} = \mathbf{P}^T\mathbf{A}\mathbf{P}$. Here $\mathbf{P}$ is complex in the case of equivalence with respect to T_C and real in the case of equivalence with respect to T_R.

Our problem is to determine all equivalence classes defined by the groups T_C and T_R; and for T_C the solution of this problem is very simple indeed.

THEOREM 12.4.1. *Two complex quadratic forms are equivalent with respect to the group of complex non-singular linear transformations if and only if they have the same rank.*†

If the two quadratic forms are equivalent (in the sense stated), then their ranks are obviously equal. Consider next two quadratic forms ϕ, ψ, both of rank r. In view of Exercise 12.3.4 (p. 369) both these forms are equivalent to $x_1^2+...+x_r^2$; hence they are equivalent to each other.

EXERCISE 12.4.1. Show that an n-ary quadratic form is non-singular if and only if it is equivalent, under the group T_C, to the n-ary unit form.

12.4.2. Theorem 12.4.1 settles the problem of equivalence with respect to T_C, since it shows that each equivalence class consists

† Cf. Theorem 12.3.2 (p. 368).

simply of all n-ary quadratic forms having the same rank. For real transformations the problem is more difficult and more interesting. The basis of our discussion is a result found by Sylvester in 1852.

Theorem 12.4.2. (Sylvester's law of inertia)

If a real quadratic form in $x_1,..., x_n$, of rank r, is reduced by two real non-singular linear transformations to the diagonal forms

$$\alpha_1 y_1^2+...+\alpha_r y_r^2, \tag{12.4.2}$$

$$\beta_1 z_1^2+...+\beta_r z_r^2, \tag{12.4.3}$$

respectively, then the number of positive α's is equal to the number of positive β's.†

Let the numbers of positive α's and positive β's be denoted by s, t respectively. The y's and z's may be assumed to be so numbered that the positive α's and positive β's come first. Since the forms (12.4.2) and (12.4.3) are obtained by transformations of the same initial quadratic form, we have

$$\alpha_1 y_1^2+...+\alpha_s y_s^2-|\alpha_{s+1}|y_{s+1}^2-...-|\alpha_r|y_r^2$$
$$= \beta_1 z_1^2+...+\beta_t z_t^2-|\beta_{t+1}|z_{t+1}^2-...-|\beta_r|z_r^2. \tag{12.4.4}$$

This relation is an identity in $x_1,..., x_n$, since the y's and z's are linear forms in the x's.

Suppose that $s > t$. Then the system of $n-s+t < n$ linear homogeneous equations

$$y_{s+1} = 0, \quad ..., \quad y_n = 0, \qquad z_1 = 0, \quad ..., \quad z_t = 0 \tag{12.4.5}$$

in the unknowns $x_1,..., x_n$ possesses a solution

$$x_1 = \xi_1, \quad ..., \quad x_n = \xi_n, \tag{12.4.6}$$

in which $\xi_1,..., \xi_n$ are not all zero.

Denote by y_k^*, z_k^* ($k = 1,..., n$) the *values* assumed by the linear forms y_k, z_k respectively when the substitution (12.4.6) has been made. Then, by (12.4.4) and (12.4.5),

$$\alpha_1 y_1^{*2}+...+\alpha_s y_s^{*2} = -|\beta_{t+1}|z_{t+1}^{*2}-...-|\beta_r|z_r^{*2},$$

and therefore

$$y_1^* = ... = y_s^* = z_{t+1}^* = ... = z_r^* = 0. \tag{12.4.7}$$

Hence, by (12.4.5) and (12.4.7),

$$y_1^* = ... = y_s^* = y_{s+1}^* = ... = y_n^* = 0,$$

† In view of Theorem 12.3.4 (p. 370) all α's and β's are non-zero.

and so the system of n linear homogeneous equations $y_1 = 0,..., y_n = 0$ in the unknowns $x_1,..., x_n$ possesses a non-trivial solution (12.4.6). Hence the determinant of the linear forms $y_1,..., y_n$ vanishes, and this is incompatible with our hypothesis that the original quadratic form is transformed into (12.4.2) by a *non-singular* linear transformation. Thus the assumption $s > t$ leads to a contradiction; and by symmetry so does the assumption $s < t$. Hence $s = t$, and the theorem is proved. It follows at once, of course, that the number of negative α's is equal to the number of negative β's.

It is convenient to introduce the following definition.

DEFINITION 12.4.2. *If a quadratic form ϕ is transformed by a real non-singular linear transformation into a diagonal quadratic form ϕ_1, then ϕ_1 is said to be a* CANONICAL FORM *of ϕ.*

EXERCISE 12.4.2. Show that two real quadratic forms are equivalent with respect to T_R if and only if they possess a common canonical form.

Sylvester's law of inertia asserts that if ϕ_1, ϕ_2 are any two canonical forms of a quadratic form ϕ, then the number of positive coefficients in ϕ_1 is equal to the number of positive coefficients in ϕ_2, and similarly for negative coefficients. We are therefore entitled to introduce the following notation.

DEFINITION 12.4.3. *The number of positive and negative coefficients in any canonical form of the quadratic form $\mathbf{x}^T\mathbf{A}\mathbf{x}$ will be denoted by $P = P(\mathbf{A})$ and $N = N(\mathbf{A})$ respectively.*

The numbers P and N so defined can easily be related to the characteristic roots of $\mathbf{A}$.

THEOREM 12.4.3. *If $\mathbf{A}$ is a real symmetric matrix, then $P(\mathbf{A})$ is equal to the number of positive and $N(\mathbf{A})$ to the number of negative characteristic roots of $\mathbf{A}$.*

If $R(\mathbf{A}) = r$ and $\lambda_1,..., \lambda_r$ are the non-zero characteristic roots of $\mathbf{A}$, then, by the theorem on orthogonal reduction of quadratic forms (p. 363), we know that $\mathbf{x}^T\mathbf{A}\mathbf{x}$ possesses the canonical form $\lambda_1 x_1^2 + ... + \lambda_r x_r^2$. The assertion therefore follows.

DEFINITION 12.4.4. *The number $s = P(\mathbf{A}) - N(\mathbf{A})$ is called the* SIGNATURE *of the quadratic form $\mathbf{x}^T\mathbf{A}\mathbf{x}$ (or of the symmetric matrix $\mathbf{A}$).*

Since $P+N = r$, $P-N = s$, any two of r, s, P, N determine the other two. The evaluation of r, s, P, N is carried out particularly easily by the use of Lagrange's method of reduction.

EXERCISE 12.4.3. Determine the values of P, N, r, s for the quaternary quadratic form $2x_1x_2-x_1x_3+x_1x_4-x_2x_3+x_2x_4-2x_3x_4$ considered on p. 373.

We are now able to formulate the solution of our principal problem.

Theorem 12.4.4. (Equivalence theorem for quadratic forms)

Two real quadratic forms are equivalent with respect to the group of real non-singular linear transformations if and only if they have the same rank and the same signature.

This result should be compared with Theorem 12.4.1 where a less restrictive condition (i.e. equality of rank) suffices to characterize a less restrictive definition of equivalence (i.e. equivalence with respect to T_C).

To prove Theorem 12.4.4 we first observe that if two quadratic forms are equivalent (with respect to T_R), then, by Exercise 12.4.2, they possess a common canonical form and so have the same rank and the same signature.

Next, let $\mathbf{x}^T\mathbf{A}\mathbf{x}$ be a quadratic form of rank r, and write $P(\mathbf{A}) = P$. We know that $\mathbf{x}^T\mathbf{A}\mathbf{x}$ is equivalent to

$$\alpha_1 x_1^2+\ldots+\alpha_P x_P^2-\alpha_{P+1} x_{P+1}^2-\ldots-\alpha_r x_r^2, \tag{12.4.8}$$

where all α's are positive. The additional non-singular linear transformation

$$x_k = \begin{cases} y_k/\sqrt{\alpha_k} & (k = 1,\ldots,r) \\ y_k & (k = r+1,\ldots,n) \end{cases}$$

carries (12.4.8) into

$$y_1^2+\ldots+y_P^2-y_{P+1}^2-\ldots-y_r^2, \tag{12.4.9}$$

and $\mathbf{x}^T\mathbf{A}\mathbf{x}$ is therefore equivalent to (12.4.9). If now a second quadratic form $\mathbf{x}^T\mathbf{B}\mathbf{x}$ has the same rank and the same signature as $\mathbf{x}^T\mathbf{A}\mathbf{x}$, then $P(\mathbf{B}) = P(\mathbf{A})$ and so $\mathbf{x}^T\mathbf{B}\mathbf{x}$ is also equivalent to (12.4.9). Hence $\mathbf{x}^T\mathbf{A}\mathbf{x}$ and $\mathbf{x}^T\mathbf{B}\mathbf{x}$ are equivalent to each other, and the proof is complete. Thus each equivalence class with respect to the group T_R consists of all n-ary quadratic forms having the same rank and the same signature.

EXERCISE 12.4.4. Show that there are altogether $n+1$ equivalence classes with respect to T_C and $\frac{1}{2}(n+1)(n+2)$ equivalence classes with respect to T_R.

EXERCISE 12.4.5. Show that an n-ary quadratic form is equivalent (with respect to T_R) to the n-ary unit form if and only if its rank and its signature are both equal to n.

12.4.3. In conclusion we mention, without proof, two further results in which rank and signature play a part. Let

$$x^n + a_1 x^{n-1} + \ldots + a_n = 0 \tag{12.4.10}$$

be an equation with complex coefficients, and let s_r denote the sum of the rth powers of its roots. Then the rank of the (symmetric) matrix

$$\mathbf{A} = \begin{pmatrix} s_{2n-2} & s_{2n-3} & . & . & . & s_{n-1} \\ s_{2n-3} & s_{2n-4} & . & . & . & s_{n-2} \\ . & . & . & . & . & . \\ s_{n-1} & s_{n-2} & . & . & . & s_0 \end{pmatrix}$$

is equal to the number of distinct roots of (12.4.10). This implies, in particular, that (12.4.10) has at least two coincident roots if and only if $\mathbf{A}$ is singular—a result with which we are already familiar from § 1.4.1. If the coefficients of (12.4.10) are real, then $\mathbf{A}$ is real and symmetric and its signature is equal to the number of distinct *real* roots of (12.4.10).†

12.5. Classification of quadrics

The equivalence theory of quadratic forms plays an important part in geometry and enables us, in particular, to classify quadrics with respect to various groups of transformations. We shall briefly discuss some of the principal modes of classification.

(i) *Projective classification*

Let the equation of a quadric be written as

$$\mathbf{x}^T \mathbf{A} \mathbf{x} = 0, \tag{12.5.1}$$

where $\mathbf{A}$ is a symmetric 4×4 matrix, $\mathbf{x} = (x_0, x_1, x_2, x_3)^T$, and x_0, x_1, x_2, x_3 are projective coordinates. Two quadrics are said to be *projectively equivalent* if the equation of one of them can be transformed into that of the other by a (*complex*) *projective collineation*, i.e. by a transformation

$$\mathbf{x}' = \mathbf{P}\mathbf{x}, \tag{12.5.2}$$

where $\mathbf{P}$ is a non-singular complex 4×4 matrix. Thus projective equivalence of quadrics means equivalence of the associated quadratic forms with respect to the group T_C. Two quadrics are therefore projectively equivalent if and only if the associated quadratic forms have the same rank. Accordingly there are just

† For a proof of these and related results see Perron, **18**, ii, 2–5.

four equivalence classes; and these are exhibited in the following table, where r denotes the rank of $\mathbf{A}$.

r	*Standard equation*	*Type of quadric*
4	$x_0^2+x_1^2+x_2^2+x_3^2 = 0$	Proper quadric
3	$x_0^2+x_1^2+x_2^2 = 0$	Quadric cone
2	$x_0^2+x_1^2 = 0$	Pair of distinct planes
1	$x_0^2 = 0$	Repeated plane

(ii) *Complex affine classification*

Suppose that (12.5.1) is again the equation of a quadric, but that now x_0, x_1, x_2, x_3 denote homogeneous affine coordinates, with $x_0 = 0$ as the equation of the plane at infinity. Write, moreover,

$$\mathbf{A} = \begin{pmatrix} a_{00} & a_{01} & a_{02} & a_{03} \\ a_{10} & a_{11} & a_{12} & a_{13} \\ a_{20} & a_{21} & a_{22} & a_{23} \\ a_{30} & a_{31} & a_{32} & a_{33} \end{pmatrix}, \qquad \mathbf{A}' = \begin{pmatrix} a_{11} & a_{12} & a_{13} \\ a_{21} & a_{22} & a_{23} \\ a_{31} & a_{32} & a_{33} \end{pmatrix},$$

so that $\mathbf{A}'$ is a matrix specifying the conic of intersection of the quadric (12.5.1) and the plane at infinity. A *complex affine collineation* is a complex non-singular linear transformation of coordinates which transforms the plane at infinity into itself, i.e. a non-singular transformation of the type

$$\left.\begin{aligned} x_0' &= p_{00}x_0 \\ x_1' &= p_{10}x_0+p_{11}x_1+p_{12}x_2+p_{13}x_3 \\ x_2' &= p_{20}x_0+p_{21}x_1+p_{22}x_2+p_{23}x_3 \\ x_3' &= p_{30}x_0+p_{31}x_1+p_{32}x_2+p_{33}x_3 \end{aligned}\right\}. \tag{12.5.3}$$

In matrix form this may be written as (12.5.2), where

$$\mathbf{P} = \begin{pmatrix} p_{00} & 0 & 0 & 0 \\ p_{10} & p_{11} & p_{12} & p_{13} \\ p_{20} & p_{21} & p_{22} & p_{23} \\ p_{30} & p_{31} & p_{32} & p_{33} \end{pmatrix}$$

is a complex non-singular matrix. We shall discuss equivalence with respect to the group of complex affine collineations, and shall determine all equivalence classes and the corresponding standard forms of the equations of quadrics.

Write $R(\mathbf{A}) = r$, $R(\mathbf{A}') = \rho$. We know, by virtue of Exercise 12.3.4 (p. 369), that there exists a non-singular transformation

$$x_0' = x_0, \qquad x_k' = c_{k1}x_1+c_{k2}x_2+c_{k3}x_3 \quad (k = 1, 2, 3)$$

which reduces the quadratic form $\mathbf{x}^T\mathbf{A}\mathbf{x}$ to

$$a_{00}x_0^2+2b_{01}x_0x_1+2b_{02}x_0x_2+2b_{03}x_0x_3+x_1^2+...+x_\rho^2,$$

and this quadratic form may be written as

$$(a_{00}-b_{01}^2-...-b_{0\rho}^2)x_0^2+2b_{0,\rho+1}x_0x_{\rho+1}+...+$$
$$+2b_{03}x_0x_3+(x_1+b_{01}x_0)^2+...+(x_\rho+b_{0\rho}x_0)^2.$$

Applying the transformation

$$x_k' = \begin{cases} x_k+b_{0k}x_0 & (k=1,...,\rho) \\ x_k & (k=0,\ \rho+1,...,3) \end{cases}$$

and writing $a_{00}-b_{01}^2-...-b_{0\rho}^2 = b$, we obtain the quadratic form

$$bx_0^2+2b_{0,\rho+1}x_0x_{\rho+1}+...+2b_{03}x_0x_3+x_1^2+...+x_\rho^2. \tag{12.5.4}$$

This last expression can be simplified still further; and we need to consider three cases, each of which gives rise to a distinct type of quadratic form.

(i) When $b_{0,\rho+1} = ... = b_{03} = 0$ and $b = 0$ (or when $\rho = 3$ and $b = 0$) (12.5.4) becomes

$$x_1^2+...+x_\rho^2.$$

(ii) When $b_{0,\rho+1} = ... = b_{03} = 0$ (or $\rho = 3$) and $b \neq 0$, then (12.5.4) can be transformed into

$$x_0^2+x_1^2+...+x_\rho^2.$$

(iii) When $b_{0,\rho+1},..., b_{03}$ are not all zero we may assume, without loss of generality, that $b_{0,\rho+1} \neq 0$. Putting

$$x_{\rho+1}' = bx_0+2b_{0,\rho+1}x_{\rho+1}+...+2b_{03}x_3$$

and leaving the other x's unchanged in (12.5.4), we obtain the quadratic form

$$x_0x_{\rho+1}+x_1^2+...+x_\rho^2.$$

Since all transformations concerned are complex affine collineations, it follows that the original quadratic form $\mathbf{x}^T\mathbf{A}\mathbf{x}$ is equivalent, with respect to the group of these collineations, to one or the other of the forms

$$x_1^2+...+x_\rho^2, \tag{12.5.5}$$

$$x_0^2+x_1^2+...+x_\rho^2, \tag{12.5.6}$$

$$x_0x_{\rho+1}+x_1^2+...+x_\rho^2. \tag{12.5.7}$$

Moreover, no two of these quadratic forms are equivalent to each other, for r is invariant under collineations and we have $r = \rho$, $r = \rho+1$, $r = \rho+2$ for the forms (12.5.5), (12.5.6), (12.5.7) respectively. It follows that two quadrics $\mathbf{x}^T\mathbf{A}\mathbf{x} = 0$ and $\mathbf{x}^T\mathbf{B}\mathbf{x} = 0$ are equivalent with respect to the group of complex affine collineations if and only if the values of r and ρ are the same for **A** as for **B**. The equivalence classes are thus completely specified by the values of r and ρ, and each class has a representative of the form (12.5.5), (12.5.6), or (12.5.7). These classes are exhibited in the following table.

r	ρ	*Standard equation*	*Type of quadric*
4	3	$x_0^2+x_1^2+x_2^2+x_3^2 = 0$	Central quadric
4	2	$x_1^2+x_2^2+x_0 x_3 = 0$	Paraboloid
3	3	$x_1^2+x_2^2+x_3^2 = 0$	Quadric cone
3	2	$x_0^2+x_1^2+x_2^2 = 0$	Central cylinder
3	1	$x_1^2+x_0 x_2 = 0$	Parabolic cylinder
2	2	$x_1^2+x_2^2 = 0$	Pair of intersecting planes
2	1	$x_0^2+x_1^2 = 0$	Pair of parallel planes
2	0	$x_0 x_1 = 0$	A plane together with the plane at infinity
1	1	$x_1^2 = 0$	Repeated plane
1	0	$x_0^2 = 0$	Plane at infinity repeated

(iii) *Real affine classification*

So far we have been considering complex quadrics, i.e. quadrics specified by complex matrices. We shall now confine our attention to real quadrics. We shall continue to write the equation of a quadric in the form (12.5.1), where **A** is a real 4×4 matrix and x_0, x_1, x_2, x_3 denote homogeneous affine coordinates, with $x_0 = 0$ again as the equation of the plane at infinity. A *real affine collineation* is a real non-singular linear transformation of the type (12.5.3). Our object is to find all equivalence classes with respect to the group of real affine collineations.

Projective classification and complex affine classification depend on complex transformations, and considerations of rank are sufficient for determining all equivalence classes. In the case of *real* affine classification, however, account must be taken of the signs of the various terms which appear in the equations of the quadrics; and use must be made, therefore, of the notion of signature. We shall denote by s, σ the moduli of the signatures of **A**, **A**$'$ respectively. The symbols r, ρ will have the same significance as before.

Arguing as in (ii), we see that any quadratic form $\mathbf{x}^T\mathbf{A}\mathbf{x}$ can be transformed, by a real affine collineation, into a quadratic form of one of the following types:

$$\epsilon_1 x_1^2+\dots+\epsilon_\rho x_\rho^2;$$
$$\epsilon_0 x_0^2+\epsilon_1 x_1^2+\dots+\epsilon_\rho x_\rho^2;$$
$$x_0 x_{\rho+1}+\epsilon_1 x_1^2+\dots+\epsilon_\rho x_\rho^2.$$

Here each ϵ has the value 1 or -1. We do not need to consider all possible combinations of values of the ϵ's, since our main concern is with the equation $\mathbf{x}^T\mathbf{A}\mathbf{x} = 0$ rather than with the quadratic form $\mathbf{x}^T\mathbf{A}\mathbf{x}$, and any choice of values of $\epsilon_0, \epsilon_1,..., \epsilon_\rho$ leads to the same equation as the choice $-\epsilon_0, -\epsilon_1,..., -\epsilon_\rho$. Bearing this in mind, we can write out the table of standard equations. In this table we note down in each case the values of r, ρ, s, and σ.

r	ρ	s	σ	*Standard equation*	*Type of quadric*
4	3	4	3	$x_0^2+x_1^2+x_2^2+x_3^2 = 0$	Virtual quadric
4	3	2	3	$-x_0^2+x_1^2+x_2^2+x_3^2 = 0$	Ellipsoid
4	3	2	1	$x_0^2+x_1^2+x_2^2-x_3^2 = 0$	Hyperboloid of two sheets
4	3	0	1	$-x_0^2+x_1^2+x_2^2-x_3^2 = 0$	Hyperboloid of one sheet
4	2	2	2	$x_1^2+x_2^2+x_0x_3 = 0$	Elliptic paraboloid
4	2	0	0	$x_1^2-x_2^2+x_0x_3 = 0$	Hyperbolic paraboloid
3	3	3	3	$x_1^2+x_2^2+x_3^2 = 0$	Virtual quadric with a single real point
3	3	1	1	$x_1^2+x_2^2-x_3^2 = 0$	Quadric cone
3	2	3	2	$x_0^2+x_1^2+x_2^2 = 0$	Virtual cylinder
3	2	1	2	$-x_0^2+x_1^2+x_2^2 = 0$	Elliptic cylinder
3	2	1	0	$x_0^2+x_1^2-x_2^2 = 0$	Hyperbolic cylinder
3	1	1	1	$x_1^2+x_0x_2 = 0$	Parabolic cylinder
2	2	2	2	$x_1^2+x_2^2 = 0$	Pair of virtual planes intersecting in a real line
2	2	0	0	$x_1^2-x_2^2 = 0$	Pair of intersecting planes
2	1	2	1	$x_0^2+x_1^2 = 0$	Pair of parallel virtual planes
2	1	0	1	$-x_0^2+x_1^2 = 0$	Pair of parallel planes
2	0	0	0	$x_0x_1 = 0$	A plane together with the plane at infinity
1	1	1	1	$x_1^2 = 0$	Repeated plane
1	0	1	0	$x_0^2 = 0$	Plane at infinity repeated

It is clear that the numbers r, ρ, s, σ are invariant under the group of real affine collineations. The above table therefore shows that no two quadrics having different standard equations are equivalent with respect to this group. Thus each equivalence class is specified by the set of values of r, ρ, s, σ; and the table gives one representative for each class.

The different types of equivalence we have discussed give rise to a hierarchy of classifications. Thus all proper quadrics are projectively equivalent, but with respect to the group of complex affine collineations they separate into central quadrics and paraboloids. Again, with respect to the group of real affine collineations, central quadrics separate into ellipsoids, hyperboloids of one sheet, and hyperboloids of two sheets, while paraboloids separate into those of the elliptic and those of the hyperbolic type.

EXERCISE 12.5.1. Discuss the projective and affine classification of conics.

EXERCISE 12.5.2. Extend to n-ary quadratic forms the reduction of quaternary forms carried out in (ii) and (iii).

(iv) *Metric classification*

By specializing further the group of transformations, we are led to refinements of the previous systems of classification. By far the most important system that arises in this way is that of 'metric classification'.

Let (12.5.1) again be the equation of the quadric, with x_0, x_1, x_2, x_3 now interpreted as homogeneous rectangular coordinates, and with $x_0 = 0$ as the equation of the plane at infinity. A *euclidean collineation* (i.e. a combination of translations, rotations, and reflections) can be shown to be a transformation of the type

$$x_0' = x_0, \qquad x_k' = p_{k0}x_0+p_{k1}x_1+p_{k2}x_2+p_{k3}x_3 \quad (k = 1, 2, 3),$$

where

$$\begin{pmatrix} p_{11} & p_{12} & p_{13} \\ p_{21} & p_{22} & p_{23} \\ p_{31} & p_{32} & p_{33} \end{pmatrix}$$

is an orthogonal matrix. Two quadrics are said to be *metrically equivalent* if they are equivalent with respect to the group of euclidean collineations. The number of equivalence classes is, of course, infinite, and the individual classes cannot now be specified by means of 'arithmetic' invariants, such as rank or signature. The problem of classification involves the study of 'algebraic' invariants, which are polynomials in the coefficients a_{ij} of the quadric $\mathbf{x}^T\mathbf{A}\mathbf{x} = 0$. Thus $|\mathbf{A}|$, $|\mathbf{A}'|$, $a_{11}+a_{22}+a_{33}$, and

$$a_{22}a_{33}+a_{33}a_{11}+a_{11}a_{22}-a_{23}a_{32}-a_{31}a_{13}-a_{12}a_{21}$$

are all invariant under the group of euclidean collineations. The systematic discussion of such invariants falls, however, outside the scope of the present treatment.†

12.6. Hermitian forms

12.6.1. The reason why in the preceding sections we confined ourselves largely to the discussion of *real* quadratic forms is that the most significant generalization of a real quadratic form is not a complex quadratic form but a 'hermitian form'. The theory of hermitian forms (initiated by Hermite in 1854) closely resembles

† For further information see Sommerville, *Analytical Geometry of Three Dimensions*, 171–3 and 321–2.

that of quadratic forms and we shall therefore deal with it rather summarily.

DEFINITION 12.6.1. *An n-ary* HERMITIAN FORM ϕ *in the (complex-valued) variables* $x_1,..., x_n$ *is a function of the type*

$$\phi = \phi(x_1,..., x_n) = \sum_{r,s=1}^{n} a_{rs}\bar{x}_r x_s = \bar{\mathbf{x}}^T\mathbf{A}\mathbf{x},$$

where $\mathbf{A} = (a_{rs})$ *is a hermitian matrix and* $\mathbf{x} = (x_1,..., x_n)^T$.

There is therefore a biunique correspondence between n-ary hermitian forms and $n \times n$ hermitian matrices.

The value which a hermitian form assumes for any values of its variables is necessarily real, for

$$(\overline{\bar{\mathbf{x}}^T\mathbf{A}\mathbf{x}}) = \mathbf{x}^T\bar{\mathbf{A}}\bar{\mathbf{x}} = (\mathbf{x}^T\bar{\mathbf{A}}\bar{\mathbf{x}})^T = \bar{\mathbf{x}}^T\bar{\mathbf{A}}^T\mathbf{x} = \bar{\mathbf{x}}^T\mathbf{A}\mathbf{x}.$$

If the matrix $\mathbf{A}$ of a hermitian form is real (and so symmetric), then $\bar{\mathbf{x}}^T\mathbf{A}\mathbf{x}$ is called a real hermitian form. If, in addition, its variables are restricted to the real field, then the hermitian form becomes, in fact, the quadratic form $\mathbf{x}^T\mathbf{A}\mathbf{x}$. A quadratic form may therefore be regarded as a special case of a hermitian form.

A substitution
$$\mathbf{x} = \mathbf{P}\mathbf{x}' \tag{12.6.1}$$

for the variables of the hermitian form $\phi = \bar{\mathbf{x}}^T\mathbf{A}\mathbf{x}$ is called a linear transformation of ϕ, and it is said to be singular or non-singular according as $|\mathbf{P}| = 0$ or $|\mathbf{P}| \neq 0$. If $\mathbf{P}$ is unitary, the substitution is called unitary. Unless the contrary is stated, all transformations of hermitian forms are understood to be complex. The transformation (12.6.1) changes the hermitian form $\bar{\mathbf{x}}^T\mathbf{A}\mathbf{x}$ to

$$(\overline{\mathbf{P}\mathbf{x}'})^T\mathbf{A}(\mathbf{P}\mathbf{x}') = \bar{\mathbf{x}}'^T(\bar{\mathbf{P}}^T\mathbf{A}\mathbf{P})\mathbf{x}'.$$

Since $\bar{\mathbf{P}}^T\mathbf{A}\mathbf{P}$ is a hermitian matrix, the new function is again a hermitian form. We observe that the matrix $\mathbf{A}$ associated with the original hermitian form is changed to

$$\mathbf{B} = \bar{\mathbf{P}}^T\mathbf{A}\mathbf{P}.$$

A matrix transformation of this type is known as a *conjunctive transformation* if $|\mathbf{P}| \neq 0$. When $\mathbf{P}$ is unitary this transformation becomes, of course, a unitary similarity transformation. Conjunctive transformations clearly form a group.

The rank of a hermitian form is defined as the rank of the associated (hermitian) matrix. It is plainly invariant with respect to non-singular linear transformations of the hermitian form.

A hermitian form of type $\alpha_1 \bar{x}_1 x_1 + ... + \alpha_n \bar{x}_n x_n$ is called diagonal. Here the coefficients $\alpha_1, ..., \alpha_n$ are necessarily real since they are the diagonal elements of a hermitian matrix. The form $\bar{x}_1 x_1 + ... + \bar{x}_n x_n$ is known as the unit n-ary hermitian form.

EXERCISE 12.6.1. Express Theorem 8.2.5 (p. 230) in terms of hermitian forms.

As in the theory of quadratic forms it is interesting to investigate reduction to diagonal form. Here the essential tool is Theorem 10.3.7 which states that every hermitian matrix is unitarily similar to a diagonal matrix. This leads at once to the following result which is analogous to the orthogonal reduction theorem for quadratic forms (Theorem 12.2.1).

THEOREM 12.6.1. (Unitary reduction of hermitian forms)

Let $\bar{\mathbf{x}}^T \mathbf{A} \mathbf{x}$ *be a hermitian form of rank* r. *Then there exists a unitary transformation which reduces* $\bar{\mathbf{x}}^T \mathbf{A} \mathbf{x}$ *to the diagonal form*

$$\lambda_1 \bar{x}_1 x_1 + ... + \lambda_r \bar{x}_r x_r,$$

where $\lambda_1, ..., \lambda_r$ *are the non-vanishing characteristic roots of* $\mathbf{A}$.†

If we do not insist that the reducing matrix should be unitary, then the coefficients in the resulting diagonal form will not necessarily be equal to the characteristic roots of $\mathbf{A}$, but the non-vanishing coefficients will still be r in number, and they will all be real. The question as to their sign therefore arises just as in the case of quadratic forms, and we are led to a result analogous to Sylvester's law of inertia (Theorem 12.4.2).

THEOREM 12.6.2. (Law of inertia for hermitian forms)

If a hermitian form of rank r *is reduced by two complex non-singular linear transformations to the diagonal forms*

$$\alpha_1 \bar{x}_1 x_1 + ... + \alpha_r \bar{x}_r x_r, \qquad \beta_1 \bar{x}_1 x_1 + ... + \beta_r \bar{x}_r x_r$$

respectively, then the number of positive α*'s is equal to the number of positive* β*'s.*

The proof is virtually the same as that of the law of inertia for quadratic forms, and the details may be left to the reader.

The signature of a hermitian form is defined in precisely the same way as the signature of a quadratic form. It is then seen that rank and signature characterize completely all equivalence classes of hermitian forms.

† The number of non-vanishing characteristic roots is equal to r by virtue of Theorem 10.3.7 and the corollary to Theorem 10.2.3 (p. 296).

THEOREM 12.6.3. (Equivalence theorem for hermitian forms) *Two hermitian forms are equivalent with respect to the group of complex non-singular linear transformations if and only if they have the same rank and the same signature.*

The connexion between rank, signature, and the signs of the characteristic roots of the associated matrix is the same for hermitian as for quadratic forms. We state the analogue of Theorem 12.4.3 (p. 378).

THEOREM 12.6.4. (i) *The rank of a hermitian form* $\phi = \bar{\mathbf{x}}^T\mathbf{A}\mathbf{x}$ *is equal to the number of non-vanishing characteristic roots of* $\mathbf{A}$. (ii) *The signature of* ϕ *is equal to the difference between the number of positive and the number of negative characteristic roots of* $\mathbf{A}$.

EXERCISE 12.6.2. Give detailed proofs of all assertions made above.

EXERCISE 12.6.3. Show that the following statement is false. 'Every hermitian form of rank r can be reduced to the diagonal form $\bar{x}_1 x_1 + \ldots + \bar{x}_r x_r$ by a suitable complex non-singular linear transformation.'

EXERCISE 12.6.4. Extend Lagrange's method of reduction to hermitian forms.

12.6.2. The theory of hermitian forms can be used to narrow the bounds, previously obtained in § 7.5, for the real and imaginary parts of the characteristic roots of an arbitrary matrix. We first deduce a preliminary inequality for the values assumed by hermitian forms.

THEOREM 12.6.5. *If* λ and Λ *are the least and greatest characteristic roots of the hermitian matrix* $\mathbf{A}$, *then, for all* $\mathbf{x}$,

$$\lambda \bar{\mathbf{x}}^T\mathbf{x} \leqslant \bar{\mathbf{x}}^T\mathbf{A}\mathbf{x} \leqslant \Lambda \bar{\mathbf{x}}^T\mathbf{x}.$$

Denote the characteristic roots of $\mathbf{A}$ by $\lambda_1,\ldots,\lambda_n$, and write $\mathbf{D} = \mathbf{dg}(\lambda_1,\ldots,\lambda_n)$. By Theorem 10.3.7 there exists a unitary matrix $\mathbf{U}$ such that $\bar{\mathbf{U}}^T\mathbf{A}\mathbf{U} = \mathbf{D}$. Let $\mathbf{x}$ be any (complex) vector and put $\mathbf{y} = \bar{\mathbf{U}}^T\mathbf{x}$. Since obviously

$$\lambda \bar{\mathbf{y}}^T\mathbf{y} \leqslant \bar{\mathbf{y}}^T\mathbf{D}\mathbf{y} \leqslant \Lambda \bar{\mathbf{y}}^T\mathbf{y},$$

the assertion follows at once.

From Theorem 12.6.5 we derive important inequalities due to Bromwich (1906).

THEOREM 12.6.6. *Let* $\mathbf{A}$ *be any complex matrix. Let* μ, M *be the least and greatest characteristic roots of the hermitian matrix* $\frac{1}{2}(\mathbf{A}+\overline{\mathbf{A}}^T)$ *and* ν, N *the least and greatest characteristic roots of the hermitian matrix* $\frac{1}{2i}(\mathbf{A}-\overline{\mathbf{A}}^T)$. *If* λ *is any characteristic root of* $\mathbf{A}$, *then*

$$\mu \leqslant \mathfrak{Rl}\,\lambda \leqslant M, \qquad \nu \leqslant \mathfrak{Im}\,\lambda \leqslant N.$$

Let $\mathbf{x}$ be a unit vector such that $\lambda\mathbf{x} = \mathbf{A}\mathbf{x}$. Then

$$\lambda = \lambda\bar{\mathbf{x}}^T\mathbf{x} = \bar{\mathbf{x}}^T\mathbf{A}\mathbf{x},$$

$$\bar{\lambda} = \mathbf{x}^T\overline{\mathbf{A}}\bar{\mathbf{x}} = (\mathbf{x}^T\overline{\mathbf{A}}\bar{\mathbf{x}})^T = \bar{\mathbf{x}}^T\overline{\mathbf{A}}^T\mathbf{x}.$$

Hence
$$\mathfrak{Rl}\,\lambda = \tfrac{1}{2}(\lambda+\bar{\lambda}) = \bar{\mathbf{x}}^T.\tfrac{1}{2}(\mathbf{A}+\overline{\mathbf{A}}^T).\mathbf{x},$$

$$\mathfrak{Im}\,\lambda = \frac{1}{2i}(\lambda-\bar{\lambda}) = \bar{\mathbf{x}}^T.\frac{1}{2i}(\mathbf{A}-\overline{\mathbf{A}}^T).\mathbf{x},$$

and the required inequalities follow by Theorem 12.6.5.

Bromwich's bounds for $\mathfrak{Rl}\,\lambda$ and $\mathfrak{Im}\,\lambda$ are narrower than those of Hirsch and of Bendixson given in Theorems 7.5.2 and 7.5.3 (pp. 210–11). For, applying the first inequality of Theorem 7.5.3 to the matrices $\frac{1}{2}(\mathbf{A}+\overline{\mathbf{A}}^T)$ and $\frac{1}{2i}(\mathbf{A}-\overline{\mathbf{A}}^T)$ respectively, we obtain

$$|\mu| \leqslant n\sigma, \qquad |M| \leqslant n\sigma, \qquad |\nu| \leqslant n\tau, \qquad |N| \leqslant n\tau.$$

Hence
$$-n\sigma \leqslant \mu \leqslant \mathfrak{Rl}\,\lambda \leqslant M \leqslant n\sigma,$$

$$-n\tau \leqslant \nu \leqslant \mathfrak{Im}\,\lambda \leqslant N \leqslant n\tau,$$

and Theorem 12.6.6 is therefore sharper than Theorem 7.5.3.

Next, let $\mathbf{A}$ be real and denote the characteristic roots of $\frac{1}{2i}(\mathbf{A}-\mathbf{A}^T)$ by $\nu_1,\ldots,\nu_n$. The characteristic roots of

$$\mathbf{B} = \tfrac{1}{2}(\mathbf{A}-\mathbf{A}^T) = (b_{rs})$$

and then $i\nu_1,\ldots, i\nu_n$, and since the non-vanishing ones among them occur in conjugate pairs, we have

$$\nu_1+\ldots+\nu_n = 0,$$

$$\nu_1^2+\ldots+\nu_n^2 = -2\sum_{1\leqslant r<s\leqslant n}\nu_r\nu_s.$$

Now, by Theorem 7.1.3 (p. 198),

$$\sum_{1\leqslant r<s\leqslant n} i\nu_r . i\nu_s = \sum_{1\leqslant r<s\leqslant n} \begin{vmatrix} b_{rr} & b_{rs} \\ b_{sr} & b_{ss} \end{vmatrix},$$

i.e.
$$-\sum_{1\leqslant r<s\leqslant n} \nu_r \nu_s = \sum_{1\leqslant r<s\leqslant n} \left(\frac{a_{rs}-a_{sr}}{2}\right)^2.$$

Hence
$$\nu_1^2+\ldots+\nu_n^2 = 2\sum_{1\leqslant r<s\leqslant n} \left(\frac{a_{rs}-a_{sr}}{2}\right)^2 \leqslant n(n-1)\alpha^2,$$

where
$$\alpha = \max_{1\leqslant r,s\leqslant n} \tfrac{1}{2}|a_{rs}-a_{sr}|.$$

Now the value of any non-vanishing ν_k^2 occurs at least twice among $\nu_1^2,\ldots,\nu_n^2$, and therefore

$$|\nu_k| \leqslant \alpha\sqrt{(n(n-1)/2)} \qquad (k = 1,\ldots,n).$$

Thus

$$-\alpha\sqrt{(n(n-1)/2)} \leqslant \nu \leqslant \mathfrak{Im}\,\lambda \leqslant N \leqslant \alpha\sqrt{(n(n-1)/2)},$$

and these inequalities demonstrate the superiority of Theorem 12.6.6 over Theorem 7.5.2. Nevertheless the bounds given by Theorems 7.5.2 and 7.5.3 are still useful, since μ, M, ν, N are not known directly.

PROBLEMS ON CHAPTER XII

1. Determine the rank and signature of the ternary quadratic forms (i) $2y^2-z^2+xy+xz$; (ii) $2xy-xz-yz$.

2. Determine the rank and signature of the following quaternary quadratic forms.

 (i) $yz+zx+xy+xt+yt+zt$;

 (ii) $x^2+4y^2+4z^2-t^2+4yz-2zx+2xy+2xt+2yt-2zt$;

 (iii) $3(x^2+y^2+z^2+t^2)+2xy+2xz+2xt-2yz-2yt-2zt$.

3. Suppose that $\phi(x_1,\ldots,x_n) = y_1^2+\ldots+y_r^2$, where $y_1,\ldots,y_r$ are linearly independent linear forms in $x_1,\ldots,x_n$. Show that $R(\phi) = r$.

4. Let ϕ be a quadratic form in $x_1,\ldots,x_m$ of rank r and signature s, and let ψ be a quadratic form in $x_{m+1},\ldots,x_{m+n}$ of rank r' and signature s'. Show that the $(m+n)$-ary quadratic form $\phi+\psi$ has rank $r+r'$ and signature $s+s'$.

5. Obtain an orthogonal reduction of the quadratic form

$$x^2+5y^2-z^2+4\sqrt{2}\,zx.$$

6. The equation, in rectangular non-homogeneous coordinates, of a central quadric Q is given by $4yz+4zx-2xy+z^2 = 1$. Find the equations of the principal axes and the equation of Q referred to these axes as coordinate axes.

7. Determine the principal axes of the quadric Q given by

$$2x^2+2y^2-\sqrt{2}\,yz+\sqrt{2}\,zx+4xy = 1,$$

and obtain the equation of Q referred to its principal axes as coordinate axes.

8. Let $\mathbf{A}$ be the matrix of the quadratic form $Q(x_1,\ldots,x_n)$, and let λ be a characteristic root of $\mathbf{A}$. Show that there exist values of $x_1,\ldots,x_n$, not all zero, which satisfy the equation $Q(x_1,\ldots,x_n) = \lambda(x_1^2+\ldots+x_n^2)$.

9. An n-ary quadratic form ϕ vanishes only when all its variables are zero. Show that the rank of ϕ is n and that its signature is n or $-n$.

10. Show that the maximum (minimum) value of the hermitian form $\bar{\mathbf{x}}^T\mathbf{A}\mathbf{x}$, subject to the condition $\bar{\mathbf{x}}^T\mathbf{x} = 1$, is equal to the greatest (least) characteristic root of $\mathbf{A}$. Obtain the analogous result for quadratic forms.

11. Find an orthogonal matrix $\mathbf{P}$ such that $\mathbf{P}^{-1}\mathbf{A}\mathbf{P}$ is diagonal, where

$$\mathbf{A} = \begin{pmatrix} 2 & 1 & 1 \\ 1 & 2 & 1 \\ 1 & 1 & 2 \end{pmatrix}.$$

What are the maximum and minimum values of $x^2+y^2+z^2+yz+zx+xy$, subject to the condition $x^2+y^2+z^2 = 1$?

12. Let $\mathbf{A} = \begin{pmatrix} 1 & 2 \\ 0 & 2 \end{pmatrix}$. Show that the least value assumed by the quadratic form $\mathbf{x}^T\mathbf{A}^T\mathbf{A}\mathbf{x}$, subject to the condition $\mathbf{x}^T\mathbf{x} = 1$, is $(9-\sqrt{65})/2$.

13. Let λ be a characteristic root of the complex matrix $\mathbf{A}$, and let M and m be the greatest and least characteristic roots of $\bar{\mathbf{A}}^T\mathbf{A}$. Show that

$$m^{\frac{1}{2}} \leqslant |\lambda| \leqslant M^{\frac{1}{2}}.$$

14. Show that, given a bilinear form ϕ in the variables $x_1,\ldots,x_n, y_1,\ldots,y_n$, substitutions of the type

$$\mathbf{x} = \bar{\mathbf{P}}\boldsymbol{\xi}, \qquad \mathbf{y} = \mathbf{P}\boldsymbol{\eta}, \qquad \mathbf{P} \text{ unitary},$$

can be found which turn ϕ into a bilinear form with a triangular matrix.

15. Two quadratic forms are said to be inverse if their matrices are inverses of each other. Show that if the inverse of a quadratic form ϕ exists, then it has the same signature as ϕ.

16. Deduce from Theorem 6.4.1 (p. 183) that a quadratic form of rank 2 is the product of two linearly independent linear forms.

17. Show that a non-vanishing bilinear form ψ in the two sets of variables $x_1,\ldots,x_m$ and $y_1,\ldots,y_n$ is equal to the product of a linear form in $x_1,\ldots,x_m$ and a linear form in $y_1,\ldots,y_n$ if and only if the rank of ψ is 1.

18. Let $\mathbf{A}$ be symmetric, $\mathbf{S}$ skew-symmetric, and $\mathbf{A}+\mathbf{S}$ non-singular. Prove that the matrix $\mathbf{T} = (\mathbf{A}+\mathbf{S})^{-1}(\mathbf{A}-\mathbf{T})$ satisfies the relations

$$\mathbf{T}^T(\mathbf{A}+\mathbf{S})\mathbf{T} = \mathbf{A}+\mathbf{S}, \qquad \mathbf{T}^T(\mathbf{A}-\mathbf{S})\mathbf{T} = \mathbf{A}-\mathbf{S}.$$

Deduce that the quadratic form $\mathbf{x}^T\mathbf{A}\mathbf{x}$ is left invariant by the transformation of variables $\mathbf{x} = \mathbf{T}\mathbf{y}$.

Prove that all real transformations of this type which leave invariant the form $x_1^2-x_2^2$ may be expressed as

$$x_1 = \lambda y_1 - k\lambda y_2, \qquad x_2 = -k\lambda y_1 + \lambda y_2,$$

where $\lambda = \pm 1/\sqrt{(1-k^2)}$.

19. Suppose that the principal axes of the quadric $\mathbf{x}^T\mathbf{A}\mathbf{x} = 1$ come into coincidence with the coordinate axes when the quadric is rotated about the line specified by the vector $\boldsymbol{\xi}$. Show that $\boldsymbol{\xi}^T\mathbf{B}\boldsymbol{\xi} = 0$, where

$$\mathbf{B} = \mathbf{A} - \mathbf{dg}(\lambda_1, \lambda_2, \lambda_3),$$

and λ_1, λ_2, λ_3 are the characteristic roots of $\mathbf{A}$.

20. Determine the rank and signature of the $(2n)$-ary quadratic form $x_1 x_2 + x_3 x_4 + ... + x_{2n-1} x_{2n}$.

21. Show that the rank and signature of the quadratic form

$$\sum_{r,s=1}^{n} (\lambda rs + r + s) x_r x_s$$

are independent of λ.

22. Determine the rank and signature of the ternary quadratic form $ayz + bzx + cxy$.

23. (i) Find the rank and signature of the quadratic form

$$a(x_1^2 + ... + x_n^2) + 2b(x_1 x_2 + x_1 x_3 + ... + x_{n-1} x_n).$$

(ii) Show that there exist real, linearly independent, linear forms $y_1,..., y_n$ in $x_1,..., x_n$ such that, identically,

$$x_1 x_2 + x_1 x_3 + ... + x_{n-1} x_n = y_1^2 - y_2^2 - ... - y_n^2.$$

Express this result in terms of matrices.

24. Determine the rank and signature of the n-ary quadratic form $\sum_{r \neq s} (x_r - x_s)^2$.

25. $\mathbf{A}$ is a real symmetric matrix with a negative determinant. Show that there exists a real vector $\mathbf{x}$ such that $\mathbf{x}^T\mathbf{A}\mathbf{x} < 0$.

Find real values of x, y, z for which

$$x^2 + 2y^2 + 3z^2 + 2yz - 2zx + 2xy < 0.$$

26. Show that, if $[\mathbf{A}] = \overline{\mathrm{bd}}_{|\mathbf{x}|=1} |\mathbf{A}\mathbf{x}|$, then $[\mathbf{A}]^2 = \lambda$ and $[\mathbf{A}] \geqslant |\omega|$, where λ is the greatest characteristic root of $\bar{\mathbf{A}}^T\mathbf{A}$ and ω is a characteristic root, greatest in modulus, of $\mathbf{A}$.

27. Write $m(\mathbf{A}) = \max_{\lambda \in \mathscr{A}} |\lambda|$, where $\mathscr{A}$ denotes the set of characteristic roots of $\mathbf{A}$. Show that, for every normal matrix $\mathbf{A}$,

$$m(\mathbf{A}) = \overline{\mathrm{bd}}_{|\mathbf{x}|=1} |\mathbf{A}\mathbf{x}|.$$

Deduce that, if $\mathbf{A}$, $\mathbf{B}$, and $\mathbf{A}+\mathbf{B}$ are normal, then

$$m(\mathbf{A}+\mathbf{B}) \leqslant m(\mathbf{A}) + m(\mathbf{B});$$

and that, if $\mathbf{A}, \mathbf{B}$, and $\mathbf{AB}$ are normal, then

$$m(\mathbf{AB}) \leqslant m(\mathbf{A}) m(\mathbf{B}).$$

28. Let $\mathbf{A} = (a_{rs})$ be a hermitian matrix of order n, and let λ and λ' be its greatest and least characteristic roots. Show that $\lambda' \leqslant a_{rr} \leqslant \lambda$ $(r = 1,..., n)$.

29. Let $\mathbf{A}$ be a hermitian $n \times n$ matrix, and let $\mathbf{B}$ be the submatrix obtained when the last row and last column of $\mathbf{A}$ are deleted. Show that, if μ denotes any characteristic root of $\mathbf{B}$ and λ, λ' denote the greatest and least characteristic roots of $\mathbf{A}$, then $\lambda' \leqslant \mu \leqslant \lambda$.

30. Let $\mathbf{A} = (a_{rs})$ be a hermitian matrix of order n; and let, for $1 \leqslant r \leqslant n$, λ_r and λ'_r denote the greatest and least characteristic roots of the submatrix

$$\begin{pmatrix} a_{11} & . & . & . & a_{1r} \\ . & . & . & . & . \\ a_{r1} & . & . & . & a_{rr} \end{pmatrix}.$$

Show that

$$\lambda_n \geqslant \lambda_{n-1} \geqslant \dots \geqslant \lambda_2 \geqslant \lambda_1 = \lambda'_1 \geqslant \lambda'_2 \geqslant \dots \geqslant \lambda'_{n-1} \geqslant \lambda'_n.$$

31. Let (a_{rs}) be a real symmetric $n \times n$ matrix, and write

$$m_k = \begin{vmatrix} a_{11} & . & . & . & a_{1k} \\ . & . & . & . & . \\ a_{k1} & . & . & . & a_{kk} \end{vmatrix} \qquad (k = 1,\dots, n).$$

Show that, if no m_k vanishes, then there exist linearly independent linear forms $y_1,\dots, y_n$ in $x_1,\dots, x_n$, such that

$$\sum_{r,s=1}^{n} a_{rs} x_r x_s = \frac{1}{m_1} y_1^2 + \frac{m_1}{m_2} y_2^2 + \dots + \frac{m_{n-1}}{m_n} y_n^2.$$

Deduce that the signature of (a_{rs}) is equal to $n-2t$, where t is the number of changes of sign in the sequence 1, m_1, $m_2,\dots, m_n$.

32. Let $L_1,\dots, L_k$, $\Lambda_1,\dots, \Lambda_k$ be linear forms in $x_1,\dots, x_n$, and let

$$f(x_1,\dots, x_n) = L_1 \Lambda_1 + \dots + L_k \Lambda_k$$

be a non-singular n-ary quadratic form. Show that among $L_1,\dots, L_k$, $\Lambda_1,\dots, \Lambda_k$ there must be a set of n linearly independent linear forms.

33. The real non-singular quadratic form $f(x_1,\dots, x_n)$ in n ($\geqslant 2k$) variables vanishes if $x_{k+1} = x_{k+2} = \dots = x_n = 0$. Show that, by a real non-singular linear transformation, f can be reduced to the form

$$y_1 y_{k+1} + y_2 y_{k+2} + \dots + y_k y_{2k} + \phi(y_{2k+1},\dots, y_n),$$

where ϕ is non-singular; and deduce that the signature s of f satisfies the inequality $|s| \leqslant n-2k$.

34. Show that, if $\mathbf{x}^T\mathbf{A}\mathbf{x}$ is a quadratic form of rank r and signature s, then there exists a subspace $\mathfrak{U}$ of $\mathfrak{V}_n$ such that $d(\mathfrak{U}) = \frac{1}{2}(r-s)$ and $\mathbf{x}^T\mathbf{A}\mathbf{x} < 0$ for every non-zero vector $\mathbf{x} \in \mathfrak{U}$.

35. Let s_k denote the sum of kth powers of the roots of the equation $x^n + a_1 x^{n-1} + \dots + a_n = 0$ having complex coefficients. Show that the number of distinct roots is equal to the rank of the matrix

$$\begin{pmatrix} s_{2n-2} & s_{2n-3} & . & . & . & s_{n-1} \\ s_{2n-3} & s_{2n-4} & . & . & . & s_{n-2} \\ . & . & . & . & . & . \\ s_{n-1} & s_{n-2} & . & . & . & s_0 \end{pmatrix}.$$

36. Show that a real n-ary quadratic form is equal to the product of two real, linearly independent, linear forms if and only if its rank is 2 and its signature 0.

XIII

DEFINITE AND INDEFINITE FORMS

In the previous chapter quadratic and hermitian forms were classified according to rank and signature. We shall now introduce another classification, which is based on the values that the forms in question are capable of assuming.† Though this new classification is cruder than the former (in the sense that each class will now consist of one or more of the former classes) it is, for some purposes, of even greater importance.

13.1. The value classes

13.1.1. Definition 13.1.1. *Let ϕ be a hermitian or a quadratic form in the variables $x_1,..., x_n$.*

(i) *ϕ is* POSITIVE DEFINITE (NEGATIVE DEFINITE) *if $\phi > 0$ ($\phi < 0$) except when $x_1 = ... = x_n = 0$. A form which is positive definite or negative definite is called* DEFINITE.

(ii) *ϕ is* POSITIVE SEMI-DEFINITE (NEGATIVE SEMI-DEFINITE) *if $\phi \geqslant 0$ ($\phi \leqslant 0$) for all values of $x_1,..., x_n$ and $\phi = 0$ for some values of $x_1,..., x_n$, not all zero. A form which is positive semi-definite or negative semi-definite is called* SEMI-DEFINITE.

(iii) *ϕ is* INDEFINITE *if it is capable of assuming both positive and negative values.*

Definition 13.1.2. *The five classes of hermitian (or quadratic) forms specified in the preceding definition will be called the* VALUE CLASSES.‡

It is evident that the subdivision of hermitian or quadratic forms into the five value classes is exhaustive and also (if the trivial case of the identically vanishing form is ignored) exclusive.

The following binary quadratic forms illustrate the five possible cases.

† In speaking of the values of a quadratic (hermitian) form we mean, of course, the values it assumes when the variables range over all real (complex) numbers. We may also remind the reader that a hermitian form assumes real values only.

‡ The separation of quadratic forms into value classes has an important application in the study of maxima and minima of functions of several variables. See Stoll, **11**, 127, Ex. 5.7.

Positive definite: x^2+y^2 Positive semi-definite: $(x+y)^2$

Negative definite: $-x^2-y^2$ Negative semi-definite: $-(x+y)^2$

Indefinite: x^2-y^2

If ϕ is positive definite (positive semi-definite), then $-\phi$ is negative definite (negative semi-definite). If ϕ is indefinite, then so is $-\phi$.

A real symmetric matrix, say $\mathbf{A}$, is associated with two forms—the hermitian form $\bar{\mathbf{x}}^T\mathbf{A}\mathbf{x}$ and the quadratic form $\mathbf{x}^T\mathbf{A}\mathbf{x}$. Each of these forms belongs, of course, to a certain value class, but there is fortunately no possibility of confusion since the two forms belong, in fact, to the same value class.

Theorem 13.1.1. *If* $\mathbf{A}$ *is a real symmetric matrix, then the hermitian form* $\phi = \bar{\mathbf{x}}^T\mathbf{A}\mathbf{x}$ *and the quadratic form* $\psi = \mathbf{x}^T\mathbf{A}\mathbf{x}$ *belong to the same value class.*

If ϕ is positive definite, then (since the set of values assumed by ψ is contained in the set of values assumed by ϕ) ψ is also positive definite. If, on the other hand, ψ is positive definite, we write $\mathbf{x} = \boldsymbol{\xi}+i\boldsymbol{\eta}$ (where $\boldsymbol{\xi}$, $\boldsymbol{\eta}$ are real) and obtain

$$\bar{\mathbf{x}}^T\mathbf{A}\mathbf{x} = \boldsymbol{\xi}^T\mathbf{A}\boldsymbol{\xi}+\boldsymbol{\eta}^T\mathbf{A}\boldsymbol{\eta}. \tag{13.1.1}$$

Hence $\bar{\mathbf{x}}^T\mathbf{A}\mathbf{x} > 0$ except for $\boldsymbol{\xi} = \boldsymbol{\eta} = \mathbf{0}$, i.e. except for $\mathbf{x} = \mathbf{0}$. Thus ϕ is positive definite. In the same way it is shown that ϕ and ψ are negative definite together.

Next, let ψ be positive semi-definite. Then, by (13.1.1), $\bar{\mathbf{x}}^T\mathbf{A}\mathbf{x} \geqslant 0$ for all $\mathbf{x}$. Now, in view of the previous result, ϕ cannot be positive definite; hence it is positive semi-definite. Again, if ϕ is positive semi-definite, then *a fortiori* $\mathbf{x}^T\mathbf{A}\mathbf{x} \geqslant 0$ for all real $\mathbf{x}$, and so ψ is positive semi-definite. Negative semi-definiteness is, of course, dealt with in exactly the same way.

Again, if ψ is indefinite, then ϕ is obviously indefinite, too. On the other hand, if ϕ is indefinite, then so is ψ in view of the previous results.

It is sometimes convenient to transform the terms 'positive definite' and so on from forms to matrices.

Definition 13.1.3. *A hermitian matrix* $\mathbf{A}$ *is said to belong to the same value class as the associated hermitian form.*

Thus we may speak of positive definite matrices, indefinite matrices, and so on. When $\mathbf{A}$ is real and symmetric, the terms 'hermitian matrix' and 'hermitian form' in the above definition may, in view of Theorem 13.1.1, be replaced by 'real symmetric matrix' and 'quadratic form'.

EXERCISE 13.1.1. Let (a_{rs}) be a positive definite hermitian matrix and let $b_1,..., b_n$ be real non-zero numbers. Show that the (hermitian) matrix $(a_{rs} b_r b_s)$ is again positive definite.

13.1.2. In order to devise tests for distinguishing between hermitian (or quadratic) forms belonging to different value classes it is useful to note in the first place that non-singular linear transformations do not affect the set of values assumed by a form. This is expressed more precisely in the next theorem.

THEOREM 13.1.2. *Let ϕ, ψ be two hermitian (quadratic) forms such that one can be transformed into the other by a complex (real) non-singular linear transformation. If $\mathfrak{S}(\phi)$, $\mathfrak{S}(\psi)$ denote the sets of values assumed by ϕ, ψ respectively as their variables take all complex (real) values, not all zero, then $\mathfrak{S}(\phi) = \mathfrak{S}(\psi)$.*

We shall state the proof in the language of hermitian forms. Let

$$\phi = \bar{\mathbf{x}}^T\mathbf{A}\mathbf{x}, \qquad \psi = \bar{\mathbf{x}}^T\mathbf{B}\mathbf{x}.$$

By hypothesis there exists a non-singular matrix $\mathbf{P}$ such that $\overline{\mathbf{P}}^T\mathbf{A}\mathbf{P} = \mathbf{B}$. Suppose that $\alpha \in \mathfrak{S}(\psi)$, i.e. for some complex vector $\boldsymbol{\xi} \neq \mathbf{0}$, $\bar{\boldsymbol{\xi}}^T\mathbf{B}\boldsymbol{\xi} = \alpha$. Hence $(\overline{\mathbf{P}\boldsymbol{\xi}})^T\mathbf{A}(\mathbf{P}\boldsymbol{\xi}) = \alpha$, and since $\mathbf{P}\boldsymbol{\xi} \neq \mathbf{0}$ it follows that $\alpha \in \mathfrak{S}(\phi)$. Thus $\mathfrak{S}(\psi) \subset \mathfrak{S}(\phi)$. By symmetry $\mathfrak{S}(\phi) \subset \mathfrak{S}(\psi)$, and the theorem is therefore proved.

THEOREM 13.1.3. *The value class of a hermitian (quadratic) form is invariant under complex (real) non-singular linear transformations.*

This result follows at once from Theorem 13.1.2 since a hermitian, or quadratic, form ϕ is positive definite (negative definite) precisely when $\mathfrak{S}(\phi)$ contains only positive (negative) numbers, positive semi-definite (negative semi-definite) precisely when $\mathfrak{S}(\phi)$ contains only positive (negative) numbers and 0, and indefinite precisely when $\mathfrak{S}(\phi)$ contains both positive and negative numbers.

Alternatively, we may express Theorem 13.1.3 by saying that two hermitian (quadratic) forms which are equivalent with respect to the group of complex (real) non-singular linear transformations belong to the same value class. The converse need not be true, and this fact bears out our earlier remark about the relative crudity of the distribution of hermitian forms into value classes, as compared with their classification by rank and signature.

We shall next assign each hermitian or quadratic form to its value class by means of the characteristic roots of the associated matrix.

THEOREM 13.1.4. *Let ϕ be the hermitian form associated with the (hermitian) matrix* $\mathbf{A}$.

(i) *ϕ is positive definite (negative definite) if and only if all characteristic roots of* $\mathbf{A}$ *are positive (negative).*

(ii) *ϕ is positive semi-definite (negative semi-definite) if and only if all characteristic roots of* $\mathbf{A}$ *are non-negative (non-positive) and at least one root is equal to zero.*

(iii) *ϕ is indefinite if and only if* $\mathbf{A}$ *has at least one positive and at least one negative characteristic root.*

When $\mathbf{A}$ is real and symmetric this result enables us (in view of Theorem 13.1.1) to distinguish between value classes of quadratic forms.

By Theorems 12.6.1 (p. 387) and Theorem 13.1.3 we see that it suffices to consider the hermitian form

$$\phi = \lambda_1 \bar{x}_1 x_1 + \ldots + \lambda_n \bar{x}_n x_n,$$

where $\lambda_1,\ldots,\lambda_n$ are the characteristic roots of $\mathbf{A}$. If $\lambda_1,\ldots,\lambda_n > 0$, then $\phi > 0$ except when $x_1 = \ldots = x_n = 0$, and so ϕ is positive definite. Next, let ϕ be positive definite and assume that $\lambda_1,\ldots,\lambda_n$ are not all positive, say $\lambda_1 \leqslant 0$. Then, for $x_1 = 1, x_2 = \ldots = x_n = 0$, we have $\phi = \lambda_1 \leqslant 0$; and this contradicts our hypothesis. Hence $\lambda_1,\ldots,\lambda_n > 0$, and (i) is established, since the case of negative definite forms is treated similarly.

Again, suppose that all λ_i are non-negative and that at least one of them vanishes, say $\lambda_1 = 0$; $\lambda_2,\ldots,\lambda_n \geqslant 0$. Then $\phi \geqslant 0$ for all $x_1,\ldots,x_n$, and $\phi = 0$ for $x_1 = 1$, $x_2 = \ldots = x_n = 0$. Hence ϕ is positive semi-definite. If, on the other hand, ϕ is known to be positive semi-definite, then $\lambda_1,\ldots,\lambda_n \geqslant 0$, for otherwise ϕ would be capable of assuming negative values. If now we had $\lambda_1,\ldots,\lambda_n > 0$, then ϕ would be positive definite; hence at least one λ_i must vanish. This proves (ii), since the case of negative semi-definite forms is treated similarly.

In view of (i) and (ii), the third part of the assertion now follows automatically.

EXERCISE 13.1.2. Show that a definite form is non-singular, a semi-definite form singular, and that an indefinite form can be either.

Theorem 13.1.4 can easily be restated in terms of rank and signature.

THEOREM 13.1.5. *Let ϕ be an n-ary hermitian form of rank r and signature s.*

(i) *ϕ is positive definite (negative definite) if and only if $r = s = n$ $(r = -s = n)$.*

(ii) *ϕ is positive semi-definite (negative semi-definite) if and only if $r = s < n$ $(r = -s < n)$.*

(iii) *ϕ is indefinite if and only if $|s| < r$.*

This result follows as an immediate consequence of Theorems 12.6.4 (p. 388) and 13.1.4. It is of greater practical importance than Theorem 13.1.4, for while the location of characteristic roots may be troublesome, rank and signature can be computed fairly easily by means of Lagrange's method of reduction.

EXERCISE 13.1.3. Show that two quadratic forms which belong to the same value class need not be equivalent (with respect to T_R) unless both are positive definite or both negative definite.

13.2. Transformations of positive definite forms

We know by Theorem 12.6.1 that every hermitian matrix can be reduced to diagonal form. The proof of this fact depends on the comparatively difficult Theorem 10.3.7 (p. 304), but for the special case of positive definite forms a particularly simple proof independent of earlier results can be given. The proof below not only establishes the existence of a canonical form for every positive definite hermitian form but also enables us to obtain more precise information about the reducing matrix than could be extracted from earlier, and more general, arguments.

THEOREM 13.2.1. *If $\phi = \phi(x_1,..., x_n)$ is a positive definite hermitian (quadratic) form, then there exists a complex (real) linear transformation of determinant* 1, *specified by the equations*

$$\left.\begin{array}{rl} y_1 = & x_1+\alpha_{12}x_2+\alpha_{13}x_3+\ldots+\alpha_{1n}x_n \\ y_2 = & x_2+\alpha_{23}x_3+\ldots+\alpha_{2n}x_n \\ \cdots & \cdots\cdots\cdots\cdots\cdots \\ y_{n-1} = & x_{n-1}+\alpha_{n-1,n}x_n \\ y_n = & x_n \end{array}\right\}, \qquad (13.2.1)$$

which changes ϕ *into the diagonal form*

$$c_1 \bar{y}_1 y_1 + \dots + c_n \bar{y}_n y_n \quad (c_1 y_1^2 + \dots + c_n y_n^2), \tag{13.2.2}$$

where $c_1, \dots, c_n > 0$.

We state the proof for hermitian forms. For quadratic forms the reasoning is exactly the same except that each symbol $\bar{t}t$ must be replaced by t^2. The argument below depends effectively on Lagrange's method of reduction.

Write $$\phi = \phi(x_1, \dots, x_n) = \sum_{r,s=1}^{n} a_{rs} \bar{x}_r x_s.$$

Then $$a_{11} = \phi(1, 0, \dots, 0) > 0,$$

and so

$$\phi(x_1, \dots, x_n) = a_{11} \bar{x}_1 x_1 + \sum_{r=2}^{n} (a_{1r} \bar{x}_1 x_r + a_{r1} \bar{x}_r x_1) + \sum_{r,s=2}^{n} a_{rs} \bar{x}_r x_s$$

$$= a_{11} \left\{ \left(\bar{x}_1 + \frac{a_{21}}{a_{11}} \bar{x}_2 + \dots + \frac{a_{n1}}{a_{11}} \bar{x}_n \right) \left(x_1 + \frac{a_{12}}{a_{11}} x_2 + \dots + \frac{a_{1n}}{a_{11}} x_n \right) - \right.$$

$$\left. - \sum_{r,s=2}^{n} \frac{a_{r1}}{a_{11}} \bar{x}_r \cdot \frac{a_{1s}}{a_{11}} x_s \right\} + \sum_{r,s=2}^{n} a_{rs} \bar{x}_r x_s$$

$$= a_{11} \left| x_1 + \frac{a_{12}}{a_{11}} x_2 + \dots + \frac{a_{1n}}{a_{11}} x_n \right|^2 + \sum_{r,s=2}^{n} \left(a_{rs} - \frac{a_{r1} a_{1s}}{a_{11}} \right) \bar{x}_r x_s \tag{13.2.3}$$

$$= c_1 |x_1 + \alpha_{12} x_2 + \dots + \alpha_{1n} x_n|^2 + \psi(x_2, \dots, x_n),$$

say, where $c_1 = a_{11} > 0$ and $\psi(x_2, \dots, x_n)$ is a hermitian form since

$$\overline{a_{rs} - a_{r1} a_{1s}/a_{11}} = a_{sr} - a_{s1} a_{1r}/a_{11}.$$

Moreover, $\psi(x_2, \dots, x_n)$ is positive definite. For otherwise there exist numbers $\xi_2, \dots, \xi_n$, not all zero, such that $\psi(\xi_2, \dots, \xi_n) \leqslant 0$. In that case let ξ_1 be defined by the equation

$$\xi_1 + \alpha_{12} \xi_2 + \dots + \alpha_{1n} \xi_n = 0.$$

Then $\phi(\xi_1, \xi_2, \dots, \xi_n) = \psi(\xi_2, \dots, \xi_n) \leqslant 0$, and this contradicts the hypothesis that ϕ is positive definite.

Since, then, $\psi(x_2, \dots, x_n)$ is positive definite we may apply to it the same process that has just been applied to ϕ. Carrying out this process repeatedly, we ultimately obtain

$$\phi(x_1, \dots, x_n) = c_1 |x_1 + \alpha_{12} x_2 + \dots + \alpha_{1n} x_n|^2 +$$

$$+ c_2 |x_2 + \alpha_{23} x_3 + \dots + \alpha_{2n} x_n|^2 + \dots + c_{n-1} |x_{n-1} + \alpha_{n-1,n} x_n|^2 + c_n |x_n|^2,$$

where $c_1, \dots, c_n > 0$. The transformation specified by (13.2.1) is now seen to change ϕ into the diagonal form (13.2.2).

As immediate consequences of Theorem 13.2.1 we have the following results which are also implicit in our earlier discussion.

COROLLARY 1. (i) *Every positive definite n-ary hermitian form can be changed into* $\bar{x}_1 x_1 + \ldots + \bar{x}_n x_n$ *by a complex non-singular linear transformation.*

(ii) *Any two positive definite n-ary hermitian forms are equivalent with respect to the group of complex non-singular linear transformations.*

COROLLARY 2. (i) *Every positive definite n-ary quadratic form can be changed into* $x_1^2 + \ldots + x_n^2$ *by a real non-singular linear transformation.*

(ii) *Any two positive definite n-ary quadratic forms are equivalent with respect to the group of real non-singular linear transformations.*

A further useful result which can be derived from the above discussion is as follows.

THEOREM 13.2.2. *The determinant of any positive definite hermitian matrix is positive.*

Let $\mathbf{A}$ be the matrix associated with a positive definite hermitian form. Then, by Corollary 1 (i), we know that there exists a matrix $\mathbf{P}$ such that $\overline{\mathbf{P}}^T\mathbf{A}\mathbf{P} = \mathbf{I}$. Hence $|\overline{\mathbf{P}}||\mathbf{P}||\mathbf{A}| = 1$, and so $|\mathbf{A}| > 0$.

13.3. Determinantal criteria

13.3.1. In § 13.1.2 the value classes were discussed in relation to the characteristic roots and to rank and signature. We shall now approach the question from a different point of view and derive methods for deciding the value class of a hermitian or quadratic form in terms of the principal minors of its matrix. Our first and most fundamental result, which was discovered by Frobenius in 1894, relates to positive definite forms.

Theorem 13.3.1. *A necessary and sufficient condition for the hermitian form*

$$\phi = \phi(x_1,\ldots, x_n) = \sum_{r,s=1}^{n} a_{rs}\bar{x}_r x_s$$

to be positive definite is that

$$a_{11} > 0, \quad \begin{vmatrix} a_{11} & a_{12} \\ a_{21} & a_{22} \end{vmatrix} > 0, \quad \ldots, \quad \begin{vmatrix} a_{11} & . & . & . & a_{1n} \\ . & . & . & . & . \\ a_{n1} & . & . & . & a_{nn} \end{vmatrix} > 0. \tag{13.3.1}$$

To prove the necessity of (13.3.1), suppose that $\phi(x_1,..., x_n)$ is positive definite and consider, for $1 \leqslant m \leqslant n$, the hermitian form

$$\phi(x_1,..., x_m, 0,..., 0) = \sum_{r,s=1}^{m} a_{rs}\bar{x}_r x_s$$

in the variables $x_1,..., x_m$. This form is positive definite, for otherwise we should have $\phi(x_1,..., x_m, 0,..., 0) \leqslant 0$ for some values of $x_1,..., x_m$, not all 0; and this would be contrary to the hypothesis that $\phi(x_1,..., x_n)$ is positive definite. Hence, by Theorem 13.2.2,

$$\begin{vmatrix} a_{11} & . & . & . & a_{1m} \\ . & . & . & . & . \\ a_{m1} & . & . & . & a_{mm} \end{vmatrix} > 0.$$

This holds for $m = 1,..., n$, and (13.3.1) is therefore satisfied.

Next, we use induction with respect to n to prove the sufficiency of (13.3.1). For $n = 1$ the assertion is trivial. Assume that it holds for $n-1$, where $n \geqslant 2$. We have to show that if (13.3.1) is satisfied, then ϕ is positive definite.

Since $a_{11} > 0$ we can, by virtue of (13.2.3), write

$$\phi(x_1,..., x_n) = \frac{1}{a_{11}}|a_{11}x_1+...+a_{1n}x_n|^2+\psi(x_2,..., x_n), \quad (13.3.2)$$

where
$$\psi(x_2,..., x_n) = \sum_{r,s=2}^{n} b_{rs}\bar{x}_r x_s$$
is a hermitian form and

$$b_{rs} = a_{rs}-\frac{a_{r1}a_{1s}}{a_{11}} \quad (r, s = 2,..., n).$$

Now, for $m = 2,..., n$, we have

$$a_{11}\begin{vmatrix} b_{22} & . & . & . & b_{2m} \\ . & . & . & . & . \\ b_{m2} & . & . & . & b_{mm} \end{vmatrix} = \begin{vmatrix} a_{11} & . & . & . & a_{1m} \\ . & . & . & . & . \\ a_{m1} & . & . & . & a_{mm} \end{vmatrix}.$$

This identity follows immediately on subtracting (for $s = 2,..., m$) a_{1s}/a_{11} times the first column from the sth column in the determinant on the right-hand side. Hence, by (13.3.1),

$$\begin{vmatrix} b_{22} & . & . & . & b_{2m} \\ . & . & . & . & . \\ b_{m2} & . & . & . & b_{mm} \end{vmatrix} > 0 \qquad (m = 2,..., n),$$

and so, by the induction hypothesis, $\psi(x_2,..., x_n)$ is positive definite.

Suppose now that, for some values $x_1,..., x_n$, we have

$$\phi(x_1,..., x_n) \leqslant 0.$$

Then, using (13.3.2), the fact that $\psi(x_2,..., x_n)$ is positive definite, and the inequality $a_{11} > 0$, we infer the relations

$$\psi(x_2,..., x_n) = 0, \tag{13.3.3}$$

$$a_{11}x_1+...+a_{1n}x_n = 0, \tag{13.3.4}$$

for these values of $x_1,..., x_n$. Now (13.3.3) implies $x_2 = ... = x_n = 0$. Hence, by (13.3.4), $x_1 = 0$. It follows that $\phi(x_1,..., x_n)$ is positive definite. The proof of the theorem is therefore complete.

In the formulation of the theorem just proved we considered the variables in the order of their suffixes, and our selection of minors in **A** was dependent on that order. But the value class of a hermitian or quadratic form is obviously unaffected by the labelling of the variables, and the theorem will therefore continue to hold if the variables are relabelled in any manner. Thus, for example, we know by Theorem 13.3.1 that the ternary hermitian form

$$\phi = \sum_{r,s=1}^{3} a_{rs}\bar{x}_r x_s$$

is positive definite if and only if

$$a_{11} > 0, \qquad \begin{vmatrix} a_{11} & a_{12} \\ a_{21} & a_{22} \end{vmatrix} > 0, \qquad \begin{vmatrix} a_{11} & a_{12} & a_{13} \\ a_{21} & a_{22} & a_{23} \\ a_{31} & a_{32} & a_{33} \end{vmatrix} > 0.$$

Let us now write y_2, y_3, y_1 for x_1, x_2, x_3 respectively. Then ϕ assumes the form

$$\begin{aligned} \psi = {} & a_{33}\bar{y}_1 y_1+a_{31}\bar{y}_1 y_2+a_{32}\bar{y}_1 y_3 \\ & +a_{13}\bar{y}_2 y_1+a_{11}\bar{y}_2 y_2+a_{12}\bar{y}_2 y_3 \\ & +a_{23}\bar{y}_3 y_1+a_{21}\bar{y}_3 y_2+a_{22}\bar{y}_3 y_3. \end{aligned}$$

The matrix associated with ψ is therefore given by

$$\begin{pmatrix} a_{33} & a_{31} & a_{32} \\ a_{13} & a_{11} & a_{12} \\ a_{23} & a_{21} & a_{22} \end{pmatrix}.$$

Hence ψ, and so ϕ, is positive definite if and only if

$$a_{33} > 0, \qquad \begin{vmatrix} a_{33} & a_{31} \\ a_{13} & a_{11} \end{vmatrix} > 0, \qquad \begin{vmatrix} a_{33} & a_{31} & a_{32} \\ a_{13} & a_{11} & a_{12} \\ a_{23} & a_{21} & a_{22} \end{vmatrix} > 0,$$

i.e. if and only if

$$a_{33} > 0, \qquad \begin{vmatrix} a_{11} & a_{13} \\ a_{31} & a_{33} \end{vmatrix} > 0, \qquad \begin{vmatrix} a_{11} & a_{12} & a_{13} \\ a_{21} & a_{22} & a_{23} \\ a_{31} & a_{32} & a_{33} \end{vmatrix} > 0.$$

In the same way we obtain other sets of conditions necessary and sufficient to ensure that ϕ should be positive definite, and we can readily satisfy ourselves that ϕ is positive definite if and only if *all* principal minors of the matrix of ϕ are positive. In the general case the situation is analogous, as is shown by the next theorem which, for convenience, we state in terms of matrices.

THEOREM 13.3.2. *A hermitian matrix is positive definite if and only if all its principal minors are positive.*

The proof of this result is implicit in the preceding illustration. Let the matrix in question be $\mathbf{A} = (a_{rs})$. If all its principal minors are positive, then (13.3.1) is satisfied *a fortiori* and $\mathbf{A}$ is then positive definite by Theorem 13.3.1.

Assume next that $\mathbf{A}$ is positive definite and consider the principal minor

$$\Delta = \begin{vmatrix} a_{k_1 k_1} & . & . & . & a_{k_1 k_m} \\ . & . & . & . & . \\ a_{k_m k_1} & . & . & . & a_{k_m k_m} \end{vmatrix},$$

where $1 \leqslant m \leqslant n$, $1 \leqslant k_1 < ... < k_m \leqslant n$. Let the numbers $k_{m+1},..., k_n$ be defined by the conditions

$$1 \leqslant k_{m+1} < ... < k_n \leqslant n, \qquad (k_1,..., k_n) = \mathscr{A}(1,..., n).$$

The hermitian form $\phi = \sum_{r,s=1}^{n} a_{rs} \bar{x}_r x_s$ is positive definite by hypothesis. Changing the order of its terms we can rewrite this form as

$$\phi = \sum_{r,s=1}^{n} a_{k_r k_s} \bar{x}_{k_r} x_{k_s}.$$

Now let the variables $x_{k_1},..., x_{k_n}$ be replaced by $y_1,..., y_n$ respectively. The resulting hermitian form

$$\psi = \sum_{r,s=1}^{n} a_{k_r k_s} \bar{y}_r y_s$$

is again positive definite and therefore, by Theorem 13.3.1, $\Delta > 0$. The proof is thus complete.

EXERCISE 13.3.1. Prove Theorem 13.3.2 by equating to zero selected x's in the form $\sum_{r,s=1}^{n} a_{rs}\bar{x}_r x_s$.

EXERCISE 13.3.2. Show that the hermitian $n \times n$ matrix (a_{rs}) is positive definite if and only if

$$a_{nn} > 0, \quad \begin{vmatrix} a_{n-1,n-1} & a_{n-1,n} \\ a_{n,n-1} & a_{nn} \end{vmatrix} > 0, \quad ..., \quad \begin{vmatrix} a_{11} & . & . & . & a_{1n} \\ . & . & . & . & . \\ a_{n1} & . & . & . & a_{nn} \end{vmatrix} > 0.$$

13.3.2. From Frobenius's basic result on positive definite forms it is easy to deduce determinantal criteria for the remaining value classes. We note in the first place that since a hermitian form ϕ is negative definite if and only if $-\phi$ is positive definite, Theorem 13.3.1 leads at once to conditions for negative definiteness.

THEOREM 13.3.3. *A necessary and sufficient condition for the hermitian form* $\sum_{r,s=1}^{n} a_{rs}\bar{x}_r x_s$ *to be negative definite is that*

$$a_{11} < 0, \quad \begin{vmatrix} a_{11} & a_{12} \\ a_{21} & a_{22} \end{vmatrix} > 0, \quad \begin{vmatrix} a_{11} & a_{12} & a_{13} \\ a_{21} & a_{22} & a_{23} \\ a_{31} & a_{32} & a_{33} \end{vmatrix} < 0, \quad ....$$

As in the case of positive definite forms, this result can be reformulated in such a way as to become independent of any particular labelling of the variables. Indeed, Theorem 13.3.2 implies immediately our next theorem.

THEOREM 13.3.4. *A hermitian matrix is negative definite if and only if all its principal minors of even order are positive and all its principal minors of odd order are negative.*

One of the advantages of the determinantal criteria is the ease with which they can be applied. To illustrate their use consider the ternary quadratic form

$$\phi = 10x^2 - 2y^2 + 3z^2 + 4zx + 4xy,$$

whose matrix is
$$\begin{pmatrix} 10 & 2 & 2 \\ 2 & -2 & 0 \\ 2 & 0 & 3 \end{pmatrix}.$$

We have

$$10 > 0; \quad \begin{vmatrix} 10 & 2 \\ 2 & -2 \end{vmatrix} = -24 < 0; \quad \begin{vmatrix} 10 & 2 & 2 \\ 2 & -2 & 0 \\ 2 & 0 & 3 \end{vmatrix} = -64 < 0.$$

Hence ϕ is neither positive definite nor negative definite. Moreover, it is non-singular and so cannot be semi-definite. Hence ϕ is indefinite.

We next turn to semi-definite forms and matrices.

THEOREM 13.3.5. *A hermitian matrix is positive semi-definite if and only if it is singular and all its principal minors are non-negative.*

Denote the matrix in question by $\mathbf{A} = (a_{rs})$ and suppose that it is positive semi-definite. Then, by Theorem 13.1.4 (ii) (p. 397), at least one of its characteristic roots vanishes and the matrix is therefore singular. Moreover, by hypothesis,

$$\phi(x_1,..., x_n) = \bar{\mathbf{x}}^T\mathbf{A}\mathbf{x} \geqslant 0$$

for all $\mathbf{x}$. Let $1 \leqslant m < n$, $1 \leqslant k_1 < ... < k_m \leqslant n$ and denote by $\psi(x_{k_1},..., x_{k_m})$ the m-ary hermitian form obtained from $\phi(x_1,..., x_n)$ on equating to zero every x_i whose suffix is not equal to one of the k_j. Then clearly $\psi(x_{k_1},..., x_{k_m}) \geqslant 0$ for all $x_{k_1},..., x_{k_m}$, and so the matrix

$$\begin{pmatrix} a_{k_1k_1} & . & . & . & a_{k_1k_m} \\ . & . & . & . & . \\ a_{k_mk_1} & . & . & . & a_{k_mk_m} \end{pmatrix}$$

is either positive definite or positive semi-definite. Its characteristic roots are therefore all non-negative and consequently its determinant is also non-negative. Thus all principal minors of $\mathbf{A}$ are non-negative.

Assume next that $\mathbf{A}$ is singular and that all its principal minors are non-negative. Writing

$$\mathbf{A}_m = \begin{pmatrix} a_{11} & . & . & . & a_{1m} \\ . & . & . & . & . \\ a_{m1} & . & . & . & a_{mm} \end{pmatrix} \quad (m = 1,..., n) \qquad (13.3.5)$$

we have

$$|t\mathbf{I}_m + \mathbf{A}_m| = t^m + p_1 t^{m-1} + ... + p_{m-1} t + p_m,$$

where, by Theorem 7.1.2 (p. 197), p_i is equal to the sum of all i-rowed principal minors of $\mathbf{A}_m$. Hence, by hypothesis,

$$|t\mathbf{I}_m + \mathbf{A}_m| > 0 \qquad (m = 1,..., n)$$

when $t > 0$. It follows by Theorem 13.3.1 that $t\mathbf{I}+\mathbf{A}$ is positive definite for every $t > 0$. Now assume that, for some $\mathbf{x} \neq \mathbf{0}$ and some $\tau > 0$,

$$\bar{\mathbf{x}}^T\mathbf{A}\mathbf{x} = -\tau.$$

Writing $t = \tau/\bar{\mathbf{x}}^T\mathbf{x}$, we then have $\bar{\mathbf{x}}^T(t\mathbf{I}+\mathbf{A})\mathbf{x} = 0$, and this is not possible in view of our earlier conclusion. Hence $\bar{\mathbf{x}}^T\mathbf{A}\mathbf{x} \geqslant 0$ for all $\mathbf{x}$, and since $\mathbf{A}$ is singular it must be positive semi-definite.

As an immediate consequence of the theorem just proved we have the following result for negative semi-definite matrices.

THEOREM 13.3.6. *A hermitian matrix is negative semi-definite if and only if it is singular and all its principal minors of even order are non-negative, while all those of odd order are non-positive.*

In view of Theorem 13.3.1 it is plausible to conjecture that the hermitian matrix $\mathbf{A} = (a_{rs})$ is positive semi-definite if and only if $|\mathbf{A}| = 0$ and $|\mathbf{A}_m| \geqslant 0$ $(m = 1,..., n-1)$, where $\mathbf{A}_m$ is defined by (13.3.5). This conjecture is, however, false as is demonstrated, for example, by the matrix $\begin{pmatrix} 0 & 0 \\ 0 & -1 \end{pmatrix}$. We have, on the other hand, the following correct (though incomplete) analogy with Theorem 13.3.1.

THEOREM 13.3.7. *Let* $\mathbf{A} = (a_{rs})$ *be a hermitian matrix. If*

$$|\mathbf{A}_1| > 0, \quad |\mathbf{A}_2| > 0, \quad ..., \quad |\mathbf{A}_{n-1}| > 0, \quad |\mathbf{A}_n| = 0,$$

where $\mathbf{A}_m$ *is defined by* (13.3.5), *then* $\mathbf{A}$ *is positive semi-definite.*

Since $\mathbf{A}$ is singular it cannot be positive definite and it suffices, therefore, to show that $\phi(x_1,..., x_n) = \bar{\mathbf{x}}^T\mathbf{A}\mathbf{x}$ is non-negative for all $\mathbf{x}$. Assume, on the contrary, that for some $\mathbf{x} \neq \mathbf{0}$ and some $\tau > 0$ we have $\bar{\mathbf{x}}^T\mathbf{A}\mathbf{x} = -\tau$. Then $x_n \neq 0$, for

$$\phi(x_1,..., x_{n-1}, 0) = \bar{\mathbf{y}}^T\mathbf{A}_{n-1}\,\mathbf{y},$$

where $\mathbf{y} = (x_1,..., x_{n-1})^T$, is a positive definite form in $x_1,..., x_{n-1}$ by virtue of Theorem 13.3.1. Writing $t = \tau/\bar{x}_n x_n$, we therefore have $\bar{\mathbf{x}}^T\mathbf{A}\mathbf{x}+t\bar{x}_n x_n = 0$, i.e.

$$\bar{\mathbf{x}}^T\mathbf{B}\mathbf{x} = 0, \tag{13.3.6}$$

where $\mathbf{x} \neq \mathbf{0}$ and

$$\mathbf{B} = \begin{pmatrix} a_{11} & . & . & . & a_{1n} \\ . & . & . & . & . \\ a_{n1} & . & . & . & a_{nn}+t \end{pmatrix}.$$

Now $t > 0$, and therefore

$$|\mathbf{B}| = |\mathbf{A}|+t|\mathbf{A}_{n-1}| > 0.$$

Hence, again by Theorem 13.3.1, **B** is a positive definite matrix and therefore (13.3.6) implies $\mathbf{x} = \mathbf{0}$. We thus arrive at a contradiction, and it follows that $\bar{\mathbf{x}}^T\mathbf{A}\mathbf{x} \geqslant 0$ for all $\mathbf{x}$.

EXERCISE 13.3.3. Suppose that **A** is a singular hermitian matrix and that

$$\begin{vmatrix} a_{mm} & . & . & . & a_{mn} \\ . & . & . & . & . \\ a_{nm} & . & . & . & a_{nn} \end{vmatrix} > 0 \qquad (m = 2,\ldots, n).$$

Show that **A** is positive semi-definite.

A criterion for indefinite matrices can be obtained without difficulty from Theorems 13.3.2, 13.3.4, 13.3.5, and 13.3.6.

THEOREM 13.3.8. *A hermitian matrix* **A** *is indefinite if and only if it satisfies at least one of the following two conditions.*

(i) **A** *possesses a negative principal minor of even order.*

(ii) **A** *possesses a positive principal minor of odd order and also a negative principal minor of odd order.*

It is, in the first place, obvious that if **A** satisfies either of the two stated conditions, then it is indefinite. It remains, therefore, to show that if **A** is indefinite and does not satisfy (i), it must satisfy (ii). Since **A** is indefinite it possesses at least one negative principal minor, and by hypothesis all its negative principal minors are of odd order. Now if *all* principal minors of odd order were non-positive, then **A** would be negative definite or negative semi-definite Hence at least one principal minor of odd order is positive and (ii) is therefore satisfied.

It should be noted that conditions (i) and (ii) are not incompatible but that neither of them implies the other. Thus, of the three indefinite real symmetric matrices

$$\begin{pmatrix} 1 & 0 & 1 \\ 0 & -1 & 1 \\ 1 & 1 & 1 \end{pmatrix}, \quad \begin{pmatrix} 2 & 4 & 2 \\ 4 & 2 & 2 \\ 2 & 2 & 1 \end{pmatrix}, \quad \begin{pmatrix} 2 & 1 & 1 \\ 1 & 2 & -1 \\ 1 & -1 & 1 \end{pmatrix},$$

the first satisfies both (i) and (ii), the second (i) only, and the third (ii) only.

EXERCISE 13.3.4. Use Theorem 13.3.8 to show that the ternary quadratic form mentioned on p. 404 is indefinite.

13.4. Simultaneous reduction of two quadratic forms

13.4.1. In § 12.2 and § 12.3 we considered the reduction of a given quadratic form to diagonal form. There are occasions, however, when we wish to go a step further and to effect a *simultaneous* reduction of two quadratic forms. This notion is explained more precisely in the next definition.

DEFINITION 13.4.1. *The quadratic forms* $\mathbf{x}^T\mathbf{A}\mathbf{x}$, $\mathbf{x}^T\mathbf{B}\mathbf{x}$ *are said to be* SIMULTANEOUSLY REDUCIBLE (*to diagonal forms*) *if there exists a non-singular linear substitution* $\mathbf{x} = \mathbf{U}\mathbf{y}$ *which transforms both the given quadratic forms into diagonal forms.*

The need for simultaneous reduction arises, for example, in geometrical problems when we attempt to choose a system of coordinates in such a way that the equations of two given conics both assume diagonal form. Now it is not possible to carry out in every case a simultaneous reduction of two quadratic forms, and the aim of the present section is to obtain certain conditions which ensure the existence of such a reduction. Of the two cases discussed below the first might, in fact, have been disposed of in Chapter XII, but it is more convenient to treat both cases together.†

If $\mathbf{A}$ and $\mathbf{B}$ are given matrices we shall need to consider the equation $|\mathbf{A}-\lambda\mathbf{B}| = 0$ in the unknown λ. If $\mathbf{B}$ is non-singular, this equation is of degree n and so has n roots. When $\mathbf{B} = \mathbf{I}$ the equation becomes simply the characteristic equation of $\mathbf{A}$.

13.4.2. It does not matter greatly whether the discussion in this paragraph is concerned with real or with complex matrices. We choose the latter alternative since it enables us to formulate the main result (Theorem 13.4.2) in a slightly neater form.

THEOREM 13.4.1. *If* $\mathbf{A}, \mathbf{B}$ *are complex symmetric matrices,* μ *and* ν *distinct roots of the equation* $|\mathbf{A}-\lambda\mathbf{B}| = 0$, *and* $\mathbf{x}$, $\mathbf{y}$ *vectors such that*

$$(\mathbf{A}-\mu\mathbf{B})\mathbf{x} = \mathbf{0}, \qquad (\mathbf{A}-\nu\mathbf{B})\mathbf{y} = \mathbf{0},$$

then

$$\mathbf{x}^T\mathbf{A}\mathbf{y} = 0, \qquad \mathbf{x}^T\mathbf{B}\mathbf{y} = 0.$$

We have

$$\mu\mathbf{x}^T\mathbf{B}\mathbf{y} = \mu\mathbf{y}^T\mathbf{B}\mathbf{x} = \mathbf{y}^T\mathbf{A}\mathbf{x} = \mathbf{x}^T\mathbf{A}\mathbf{y} = \nu\mathbf{x}^T\mathbf{B}\mathbf{y}.$$

Since $\mu \neq \nu$, this implies $\mathbf{x}^T\mathbf{B}\mathbf{y} = 0$, and hence $\mathbf{x}^T\mathbf{A}\mathbf{y} = \nu\mathbf{x}^T\mathbf{B}\mathbf{y} = 0$.

† For a treatment of simultaneous reductions of two quadratic forms which makes no use of matrix algebra see Bôcher, **16**, 167–73.

THEOREM 13.4.2. *If* $\mathbf{A}$, $\mathbf{B}$ *are complex symmetric matrices,* $\mathbf{B}$ *is non-singular, and the roots* $\lambda_1,\ldots,\lambda_n$ *of the equation* $|\mathbf{A}-\lambda\mathbf{B}| = 0$ *are distinct, then the quadratic forms* $\mathbf{x}^T\mathbf{A}\mathbf{x}$, $\mathbf{x}^T\mathbf{B}\mathbf{x}$ *can be reduced simultaneously to the diagonal forms*

$$\lambda_1 y_1^2+\ldots+\lambda_n y_n^2, \qquad y_1^2+\ldots+y_n^2$$

respectively.

Let $\mathbf{x}_1,\ldots,\mathbf{x}_n$ be any non-zero (complex) vectors such that, for $r = 1,\ldots,n$, $(\mathbf{A}-\lambda_r\mathbf{B})\mathbf{x}_r = \mathbf{0}$, i.e.

$$\mathbf{A}\mathbf{x}_r = \lambda_r\mathbf{B}\mathbf{x}_r \qquad (r = 1,\ldots,n). \tag{13.4.1}$$

Assume that $\mathbf{x}_1,\ldots,\mathbf{x}_n$ are linearly dependent. Then there exists a number k in the range $2 \leqslant k \leqslant n$ such that $\mathbf{x}_1,\ldots,\mathbf{x}_{k-1}$ are linearly independent while $\mathbf{x}_1,\ldots,\mathbf{x}_{k-1},\mathbf{x}_k$ are linearly dependent. There exist, therefore, numbers $\alpha_1,\ldots,\alpha_k$, not all zero, such that

$$\mathbf{0} = \alpha_1\mathbf{x}_1+\ldots+\alpha_k\mathbf{x}_k. \tag{13.4.2}$$

Now $\alpha_1,\ldots,\alpha_{k-1}$ cannot all be zero, and we may assume without loss of generality that $\alpha_1 \neq 0$. Using (13.4.1) we obtain

$$\mathbf{0} = \alpha_1\mathbf{A}\mathbf{x}_1+\ldots+\alpha_k\mathbf{A}\mathbf{x}_k = \mathbf{B}(\alpha_1\lambda_1\mathbf{x}_1+\ldots+\alpha_k\lambda_k\mathbf{x}_k).$$

But $|\mathbf{B}| \neq 0$, and so

$$\mathbf{0} = \alpha_1\lambda_1\mathbf{x}_1+\ldots+\alpha_k\lambda_k\mathbf{x}_k. \tag{13.4.3}$$

By (13.4.2) and 13.4.3) we have

$$\mathbf{0} = \alpha_1(\lambda_1-\lambda_k)\mathbf{x}_1+\ldots+\alpha_{k-1}(\lambda_{k-1}-\lambda_k)\mathbf{x}_{k-1},$$

and since $\alpha_1 \neq 0$ and all λ_i are distinct, this implies that, contrary to our assumption, $\mathbf{x}_1,\ldots,\mathbf{x}_{k-1}$ are linearly dependent. Thus we arrive at a contradiction and it follows that $\mathbf{x}_1,\ldots,\mathbf{x}_n$ are, in fact, linearly independent. Hence the matrix $\mathbf{U}$ defined by the relations $\mathbf{U}_{*r} = \mathbf{x}_r$ $(r = 1,\ldots,n)$ is non-singular. Writing

$$\mathbf{x}_r^T\mathbf{B}\mathbf{x}_r = \mu_r \qquad (r = 1,\ldots,n),$$

we have

$$\mathbf{x}_r^T\mathbf{A}\mathbf{x}_r = \lambda_r\mu_r \qquad (r = 1,\ldots,n).$$

Hence, by Theorem 13.4.1,

$$\mathbf{x}_r^T\mathbf{A}\mathbf{x}_s = \delta_{rs}\lambda_r\mu_r, \qquad \mathbf{x}_r^T\mathbf{B}\mathbf{x}_s = \delta_{rs}\mu_r \qquad (r,s = 1,\ldots,n),$$

and therefore

$$(\mathbf{U}^T\mathbf{A}\mathbf{U})_{rs} = (\mathbf{U}^T)_{r*}\mathbf{A}\mathbf{U}_{*s} = (\mathbf{U}_{*r})^T\mathbf{A}\mathbf{U}_{*s} = \mathbf{x}_r^T\mathbf{A}\mathbf{x}_s = \delta_{rs}\lambda_r\mu_r.$$

Thus

$$\mathbf{U}^T\mathbf{A}\mathbf{U} = \mathbf{dg}(\lambda_1\mu_1,\ldots,\lambda_n\mu_n),$$

and similarly

$$\mathbf{U}^T\mathbf{B}\mathbf{U} = \mathbf{dg}(\mu_1,\ldots,\mu_n).$$

Since $\mathbf{B}$ and $\mathbf{U}$ are both non-singular, it follows from the last equation that $\mu_1 \neq 0,..., \mu_n \neq 0$. Hence $\mathbf{x}_1,..., \mathbf{x}_n$ can be chosen in such a way that $\mu_1 = ... = \mu_n = 1$, and we then have

$$\mathbf{U}^T\mathbf{A}\mathbf{U} = \mathbf{dg}(\lambda_1,..., \lambda_n), \qquad \mathbf{U}^T\mathbf{B}\mathbf{U} = \mathbf{I}.$$

This proves our assertion.

EXERCISE 13.4.1. Interpret Theorems 13.4.1 and 13.4.2 for the case $\mathbf{B} = \mathbf{I}$.

A result analogous to Theorem 13.4.2 can, of course, be established for real quadratic forms and real transformations by precisely the same argument as used above. Indeed, if $\mathbf{x}^T\mathbf{A}\mathbf{x}$, $\mathbf{x}^T\mathbf{B}\mathbf{x}$ are real quadratic forms, $\mathbf{B}$ is non-singular, and all roots of $|\mathbf{A}-\lambda\mathbf{B}| = 0$ are real and distinct, then there exists a real non-singular linear transformation $\mathbf{x} = \mathbf{U}\mathbf{y}$ which carries the two given quadratic forms into $\alpha_1 y_1^2+...+\alpha_n y_n^2$ and $\beta_1 y_1^2+...+\beta_n y_n^2$ respectively.

13.4.3. We shall now assume that all matrices and quadratic forms under consideration are real and that one of the two given quadratic forms, say $\mathbf{x}^T\mathbf{B}\mathbf{x}$, is positive definite. Two vectors $\mathbf{x}$, $\mathbf{y}$ will be said to be $\mathbf{B}$-*orthogonal* if

$$\mathbf{x}^T\mathbf{B}\mathbf{y} = 0.$$

EXERCISE 13.4.2. What is the meaning of the term '$\mathbf{I}$-orthogonal'?

THEOREM 13.4.3. *If* $\mathbf{A}$, $\mathbf{B}$ *are real symmetric matrices and* $\mathbf{B}$ *is positive definite, then there exists a real non-singular linear transformation* $\mathbf{x} = \mathbf{U}\mathbf{y}$ *which changes the quadratic forms* $\mathbf{x}^T\mathbf{A}\mathbf{x}$, $\mathbf{x}^T\mathbf{B}\mathbf{x}$ *into*

$$\lambda_1 y_1^2+...+\lambda_n y_n^2, \qquad y_1^2+...+y_n^2$$

respectively, where $\lambda_1,..., \lambda_n$ *are the roots of the equation* $|\mathbf{A}-\lambda\mathbf{B}| = 0$.

This theorem is due to Weierstrass (1858). The special case when $\lambda_1,..., \lambda_n$ are distinct had been discussed earlier by Cauchy and by Jacobi.

We apply, in succession, two real non-singular linear transformations to $\mathbf{x}^T\mathbf{A}\mathbf{x}$ and $\mathbf{x}^T\mathbf{B}\mathbf{x}$. The first is chosen so as to carry $\mathbf{x}^T\mathbf{B}\mathbf{x}$ into the unit form.† It carries $\mathbf{x}^T\mathbf{A}\mathbf{x}$ into some new quadratic form which we call ϕ. The second transformation is orthogonal and chosen so as to carry ϕ into a diagonal form;‡ it will, of course, leave the unit form unchanged. The product of the two transforma-

† This is possible by virtue of Corollary 2 (i) to Theorem 13.2.1 (p. 400).

‡ This is possible by virtue of the theorem on the orthogonal reduction of quadratic forms (p. 363).

tions is a real non-singular transformation $\mathbf{x} = \mathbf{U}\mathbf{y}$ which carries $\mathbf{x}^T\mathbf{B}\mathbf{x}$ into $y_1^2+...+y_n^2$ and $\mathbf{x}^T\mathbf{A}\mathbf{x}$ into some diagonal quadratic form, say $\omega_1 y_1^2+...+\omega_n y_n^2$. Then

$$\mathbf{U}^T\mathbf{A}\mathbf{U} = \mathbf{dg}(\omega_1,..., \omega_n), \qquad \mathbf{U}^T\mathbf{B}\mathbf{U} = \mathbf{I},$$

and so
$$\mathbf{U}^T(\mathbf{A}-\lambda\mathbf{B})\mathbf{U} = \mathbf{dg}(\omega_1-\lambda,..., \omega_n-\lambda).$$

Hence
$$|\mathbf{U}^T(\mathbf{A}-\lambda\mathbf{B})\mathbf{U}| = (\omega_1-\lambda) ... (\omega_n-\lambda),$$

and
$$|\mathbf{A}-\lambda\mathbf{B}| = c(\lambda-\omega_1) ... (\lambda-\omega_n) \qquad (c \neq 0).$$

Hence $\omega_1,..., \omega_n$ are the roots of $|\mathbf{A}-\lambda\mathbf{B}| = 0$, and the theorem is proved.

Theorem 13.4.3 is mainly valuable as an existence theorem, for the actual construction of the reducing matrix $\mathbf{U}$ as described in the proof above is extremely tedious in all but the simplest numerical cases. The reason for this is that to obtain the required transformation we have first to determine the two auxiliary transformations and then to compute their product. It is therefore desirable to devise a method which effects the simultaneous reduction in a single step. Such a method will be found in Theorem 13.4.7, but before we can prove this result we shall have to consider certain preliminary questions.

In each of the next three theorems it is assumed that $\mathbf{A}$, $\mathbf{B}$ are real and symmetric and that $\mathbf{B}$ is positive definite.

THEOREM 13.4.4. *All roots of the equation* $|\mathbf{A}-\lambda\mathbf{B}| = 0$ *are real.*

This result, which is a generalization of the real case of Theorem 7.5.1 (p. 209), is almost obvious. For Theorem 13.4.3 establishes the existence of a real matrix $\mathbf{U}$ such that $\mathbf{U}^T\mathbf{A}\mathbf{U} = \mathbf{dg}(\lambda_1,..., \lambda_n)$, where $\lambda_1,..., \lambda_n$ are the roots of $|\mathbf{A}-\lambda\mathbf{B}| = 0$. Hence all these roots are real.

We may also note an alternative proof depending on the same idea as that used in the proof of Theorem 7.5.1. Let ω be any root of $|\mathbf{A}-\lambda\mathbf{B}| = 0$ and let $\mathbf{x}$ be any (possibly complex) non-zero vector satisfying

$$\mathbf{A}\mathbf{x} = \omega\mathbf{B}\mathbf{x}. \qquad (13.4.4)$$

The quadratic form $\mathbf{x}^T\mathbf{B}\mathbf{x}$ and the hermitian form $\bar{\mathbf{x}}^T\mathbf{B}\mathbf{x}$ belong, by Theorem 13.1.1 (p. 395), to the same value class. Hence the hermitian form $\bar{\mathbf{x}}^T\mathbf{B}\mathbf{x}$ is positive definite, and so the vector $\mathbf{x}$ in (13.4.4) may be chosen so as to satisfy the additional relation $\bar{\mathbf{x}}^T\mathbf{B}\mathbf{x} = 1$. We then have

$$\omega = \omega\bar{\mathbf{x}}^T\mathbf{B}\mathbf{x} = \bar{\mathbf{x}}^T\mathbf{A}\mathbf{x},$$

$$\bar{\omega} = \mathbf{x}^T\mathbf{A}\bar{\mathbf{x}} = (\mathbf{x}^T\mathbf{A}\bar{\mathbf{x}})^T = \bar{\mathbf{x}}^T\mathbf{A}\mathbf{x} = \omega;$$

hence $\omega = \bar{\omega}$, and the theorem is proved once again.

THEOREM 13.4.5. *If ω is an m-fold root of the equation $|\mathbf{A}-\lambda\mathbf{B}| = 0$, then the vector space of vectors $\mathbf{x}$ such that $(\mathbf{A}-\omega\mathbf{B})\mathbf{x} = \mathbf{0}$ has dimensionality m.*

Let $\lambda_1,\dots,\lambda_n$ be the roots of $|\mathbf{A}-\lambda\mathbf{B}| = 0$, and write

$$\mathbf{\Lambda} = \mathbf{dg}(\lambda_1,\dots,\lambda_n).$$

The value ω occurs exactly m times among $\lambda_1,\dots,\lambda_n$, and therefore

$$R(\mathbf{\Lambda}-\omega\mathbf{I}) = n-m.$$

By Theorem 13.4.3 there exists a non-singular matrix $\mathbf{U}$ such that

$$\mathbf{U}^T\mathbf{A}\mathbf{U} = \mathbf{\Lambda}, \qquad \mathbf{U}^T\mathbf{B}\mathbf{U} = \mathbf{I}.$$

Hence

$$R(\mathbf{A}-\omega\mathbf{B}) = R\{\mathbf{U}^T(\mathbf{A}-\omega\mathbf{B})\mathbf{U}\} = R(\mathbf{\Lambda}-\omega\mathbf{I}) = n-m,$$

and the theorem follows.

THEOREM 13.4.6. *Let $m > 1$. If ω is an m-fold root of the equation $|\mathbf{A}-\lambda\mathbf{B}| = 0$, then there exist m real non-zero vectors which satisfy $(\mathbf{A}-\omega\mathbf{B})\mathbf{x} = \mathbf{0}$ and are $\mathbf{B}$-orthogonal in pairs.*

Let $\mathbf{x}_1$ be any non-zero vector satisfying

$$(\mathbf{A}-\omega\mathbf{B})\mathbf{x}_1 = \mathbf{0}.$$

Next choose, in succession, non-zero vectors $\mathbf{x}_2,\dots,\mathbf{x}_m$ such that, for $k = 2,\dots,m$,

$$(\mathbf{A}-\omega\mathbf{B})\mathbf{x}_k = \mathbf{0}, \tag{13.4.5}$$

$$\mathbf{x}_k^T\mathbf{B}\mathbf{x}_1 = 0, \quad \dots, \quad \mathbf{x}_k^T\mathbf{B}\mathbf{x}_{k-1} = 0. \tag{13.4.6}$$

This choice is possible for every value of k. For suppose that $\mathbf{x}_1,\mathbf{x}_2,\dots,\mathbf{x}_{k-1}$ have been already chosen. By Theorem 13.4.5 the space of vectors $\mathbf{x}$ satisfying $(\mathbf{A}-\omega\mathbf{B})\mathbf{x} = \mathbf{0}$ has dimensionality m. Let the vectors $\boldsymbol{\xi}_1,\dots,\boldsymbol{\xi}_m$ constitute a basis of this space. Then $\mathbf{x}_k$ satisfies (13.4.5) if and only if it is of the form

$$\mathbf{x}_k = \alpha_1\boldsymbol{\xi}_1+\dots+\alpha_m\boldsymbol{\xi}_m.$$

Writing, for brevity, $\mathbf{B}\mathbf{x}_i = \mathbf{y}_i$ $(i = 1,\dots,k-1)$ we see that (13.4.6) means that $(\mathbf{x}_k, \mathbf{y}_i) = 0$ for $i = 1,\dots,k-1$, i.e.

$$\alpha_1(\boldsymbol{\xi}_1, \mathbf{y}_i)+\dots+\alpha_m(\boldsymbol{\xi}_m, \mathbf{y}_i) = 0 \qquad (i = 1,\dots,k-1).$$

This is a system of $k-1 \leqslant m-1 < m$ linear homogeneous equations in the m unknowns $\alpha_1,\dots,\alpha_m$. Hence values of $\alpha_1,\dots,\alpha_m$, not all zero, can be found satisfying these equations. There exists, therefore, a vector $\mathbf{x}_k \neq \mathbf{0}$ satisfying (13.4.5) and (13.4.6). The theorem now follows by induction with respect to k.

Theorem 13.4.7. *Let* **A, B** *be real symmetric* $n \times n$ *matrices, and let* **B** *be positive definite. Let, moreover,* $\omega_1,..., \omega_k$ *be the distinct values of the roots of the equation* $|\mathbf{A}-\lambda\mathbf{B}| = 0$ *and denote their multiplicities by* $m_1,..., m_k$ *respectively (so that* $m_1+...+m_k = n$*). Corresponding to each root* ω_r *let* m_r *real non-zero vectors be chosen which satisfy the equation* $(\mathbf{A}-\omega_r \mathbf{B})\mathbf{x} = \mathbf{0}$ *and are* **B**-*orthogonal in pairs.*† *The vectors* $\mathbf{x}_1,..., \mathbf{x}_n$ *so obtained may, furthermore, be assumed to satisfy the relations*

$$\mathbf{x}_r^T \mathbf{B}\mathbf{x}_r = 1 \qquad (r = 1,..., n).‡$$

The (real) matrix **U** *having* $\mathbf{x}_1,..., \mathbf{x}_n$ *as its columns is then non-singular, and the substitution* $\mathbf{x} = \mathbf{U}\mathbf{y}$ *transforms* $\mathbf{x}^T\mathbf{A}\mathbf{x}$, $\mathbf{x}^T\mathbf{B}\mathbf{x}$ *into*

$$\lambda_1 y_1^2+...+\lambda_n y_n^2, \qquad y_1^2+...+y_n^2$$

respectively, where $\lambda_1,..., \lambda_n$ *are the roots of* $|\mathbf{A}-\lambda\mathbf{B}| = 0$ *(with their correct multiplicities).*§

We note that, for r, $s = 1,..., n$,

$$\mathbf{x}_r^T \mathbf{B}\mathbf{x}_s = \delta_{rs}. \tag{13.4.7}$$

For, when $r \neq s$ but $\mathbf{x}_r$, $\mathbf{x}_s$ are associated with the same root of $|\mathbf{A}-\lambda\mathbf{B}| = 0$, then (13.4.7) is satisfied by construction; while when $\mathbf{x}_r$, $\mathbf{x}_s$ are associated with different roots, (13.4.7) is satisfied by virtue of Theorem 13.4.1. The proof is now completed in precisely the same way as the proof of Theorem 13.4.2.

EXERCISE 13.4.3. Supply the details of the proof given above in outline.

The simultaneous transformation described by Theorem 13.4.7 is effected in a single step and its matrix can therefore be computed without too much trouble in numerical cases. Consider, for example, the two quadratic forms

$$y^2+2z^2-2yz+2zx-2xy, \qquad 2x^2+9y^2+2z^2-2yz+2zx+6xy.$$

Their matrices are

$$\mathbf{A} = \begin{pmatrix} 0 & -1 & 1 \\ -1 & 1 & -1 \\ 1 & -1 & 2 \end{pmatrix}, \qquad \mathbf{B} = \begin{pmatrix} 2 & 3 & 1 \\ 3 & 9 & -1 \\ 1 & -1 & 2 \end{pmatrix},$$

and it is easily verified (say by Theorem 13.3.1) that **B** is positive definite.

† When $m_r > 1$, this is possible by Theorem 13.4.6; and when $m_r = 1$ the condition of **B**-orthogonality is, of course, inoperative.

‡ This is possible since **B** is positive definite.

§ The construction of the matrix **U** should be compared with the construction of a matrix effecting the orthogonal reduction of a *single* symmetric matrix (p. 303).

Now

$$\mathbf{A}-\lambda\mathbf{B} = \begin{pmatrix} -2\lambda & -1-3\lambda & 1-\lambda \\ -1-3\lambda & 1-9\lambda & -1+\lambda \\ 1-\lambda & -1+\lambda & 2-2\lambda \end{pmatrix}$$

and the roots of $|\mathbf{A}-\lambda\mathbf{B}| = 0$ are found to be $\lambda = -1, 1, 1$. Let $(p, q, r)^T$ be a vector associated with the root λ, so that

$$\begin{pmatrix} -2\lambda & -1-3\lambda & 1-\lambda \\ -1-3\lambda & 1-9\lambda & -1+\lambda \\ 1-\lambda & -1+\lambda & 2-2\lambda \end{pmatrix}\begin{pmatrix} p \\ q \\ r \end{pmatrix} = \mathbf{0}.$$

Simplifying this we obtain

$$\left.\begin{aligned} 2\lambda p+(3\lambda+1)q+(\lambda-1)r &= 0 \\ (3\lambda+1)p+(9\lambda-1)q-(\lambda-1)r &= 0 \\ (1-\lambda)(p-q+2r) &= 0 \end{aligned}\right\}. \tag{13.4.8}$$

For $\lambda = -1$ this reduces to

$$p+q+r = 0, \qquad p+5q-r = 0, \qquad p-q+2r = 0,$$

and so $$(p,q,r) = (3a, -a, -2a).$$

For $\lambda = 1$ (13.4.8) reduces to $p+2q = 0$. We need to find two vectors, say $(p',q',r')^T$, $(p'',q'',r'')^T$, satisfying this equation and $\mathbf{B}$-orthogonal to each other. If we take

$$(p',q',r') = (2b, -b, 0), \tag{13.4.9}$$

then $$p''+2q'' = 0, \qquad (p'',q'',r'')\mathbf{B}(p',q',r')^T = 0,$$

and in view of (13.4.9) this means

$$p''+2q'' = 0, \qquad p''-3q''+3r'' = 0.$$

Hence $$(p'',q'',r'') = (-6c, 3c, 5c).$$

The three vectors now chosen must each be made to satisfy the condition

$$(p,q,r)\mathbf{B}(p,q,r)^T = 1,$$

i.e. $$2p^2+9q^2+2r^2-2qr+2rp+6pq = 1.$$

This gives $a = 1$, $b = 1/\sqrt{5}$, $c = 1/\sqrt{5}$, and the required reducing matrix is therefore given by

$$\mathbf{U} = \begin{pmatrix} 3 & 2/\sqrt{5} & -6/\sqrt{5} \\ -1 & -1/\sqrt{5} & 3/\sqrt{5} \\ -2 & 0 & \sqrt{5} \end{pmatrix}.$$

The substitution

$$x = 3\xi + \frac{2}{\sqrt{5}}\eta - \frac{6}{\sqrt{5}}\zeta,$$

$$y = -\xi - \frac{1}{\sqrt{5}}\eta + \frac{3}{\sqrt{5}}\zeta,$$

$$z = -2\xi \qquad + \sqrt{5}\,\zeta,$$

therefore reduces the two given quadratic forms to $-\xi^2+\eta^2+\zeta^2$, $\xi^2+\eta^2+\zeta^2$ respectively.

When hermitian forms are considered in place of quadratic forms the theory of simultaneous reduction remains virtually unchanged, and we leave it to the reader to supply the details.

Exercise 13.4.4. Let **A, B** be hermitian matrices and suppose that **B** is positive definite. Show that all roots of the equation $|\mathbf{A}-\lambda\mathbf{B}| = 0$ are real.

Exercise 13.4.5. Let $\bar{\mathbf{x}}^T\mathbf{A}\mathbf{x}$, $\bar{\mathbf{x}}^T\mathbf{B}\mathbf{x}$ be hermitian forms of which the second is positive definite. Show that there exists a non-singular matrix **U** such that the substitution $\mathbf{x} = \mathbf{U}\mathbf{y}$ transforms the given hermitian forms into $\lambda_1 \bar{y}_1 y_1 + \ldots + \lambda_n \bar{y}_n y_n$, $\bar{y}_1 y_1 + \ldots + \bar{y}_n y_n$ respectively, where $\lambda_1,\ldots, \lambda_n$ are the roots of $|\mathbf{A}-\lambda\mathbf{B}| = 0$.

Deduce that if $\bar{\mathbf{x}}^T\mathbf{A}\mathbf{x}$ is also positive definite, then all roots of $|\mathbf{A}-\lambda\mathbf{B}| = 0$ are real and positive.

Exercise 13.4.6. Let $\bar{\mathbf{x}}^T\mathbf{A}\mathbf{x}$, $\bar{\mathbf{x}}^T\mathbf{B}\mathbf{x}$ be hermitian forms of which the second is positive definite. Show that there exists a matrix **P**, satisfying $|\det \mathbf{P}| = 1$, such that the substitution $\mathbf{x} = \mathbf{P}\mathbf{y}$ transforms both the given hermitian forms into diagonal hermitian forms.

13.4.4. One of the most important applications of the theory of simultaneous reduction of quadratic forms arises in the study of small vibrations, and was noted by Weierstrass in 1858. The configuration of a dynamical system can be specified by means of a set of Lagrangian coordinates $q_1,\ldots, q_n$; and if the q's are measured from zero at a certain configuration of stable equilibrium, then (to the first order) the kinetic energy T is a quadratic form in $\dot{q}_1,\ldots, \dot{q}_n$, which is necessarily positive definite; and the potential energy V is a quadratic form (also positive definite) in $q_1,\ldots, q_n$. Now, by applying a suitable non-singular linear transformation

$$u_r = \sum_{s=1}^{n} c_{rs} q_s \qquad (r = 1,\ldots, n), \tag{13.4.10}$$

which implies, of course, that

$$\dot{u}_r = \sum_{s=1}^{n} c_{rs} \dot{q}_s \qquad (r = 1,\ldots, n),$$

we can reduce both T and V to diagonal forms. Suppose, therefore, that

$$2T = \sum_{r=1}^{n} \alpha_r \dot{u}_r^2, \qquad 2V = \sum_{r=1}^{n} \beta_r u_r^2,$$

where all coefficients are positive. Lagrange's equations

$$\frac{d}{dt}\left(\frac{\partial T}{\partial \dot{u}_r}\right) - \frac{\partial T}{\partial u_r} + \frac{\partial V}{\partial u_r} = 0 \qquad (r = 1,..., n)$$

then take the simple form

$$\alpha_r \ddot{u}_r + \beta_r u_r = 0 \qquad (r = 1,..., n),$$

each equation involving only one of the coordinates. These equations for $u_1,..., u_n$ can be solved at once in terms of trigonometric functions and the original coordinates $q_1,..., q_n$ then obtained from (13.4.10).†

13.5. The inequalities of Hadamard, Minkowski, Fischer, and Oppenheim

The theory of positive definite hermitian forms gives rise to a number of striking inequalities, some of which we propose to discuss below. In our discussion we follow substantially the treatment given by A. Oppenheim.‡

If $\mathbf{A} = (a_{rs})$ is a given $n \times n$ matrix then A_{11} will denote, as usual, the determinant of the matrix obtained from $\mathbf{A}$ by deleting the first row and the first column. When $\mathbf{A}$ is a positive definite hermitian matrix, then $A_{11} > 0$ and we define

$$\alpha = |\mathbf{A}|/A_{11}.$$

If $\mathbf{A}$ is positive semi-definite, we put $\alpha = 0$; so that in either case

$$|\mathbf{A}| = \alpha A_{11}. \qquad (13.5.1)$$

The matrix $\mathbf{A}'$ is defined as

$$\mathbf{A}' = \begin{pmatrix} a_{11}-\alpha & . & . & . & a_{1n} \\ . & . & . & . & . \\ a_{n1} & . & . & . & a_{nn} \end{pmatrix}.$$

Our subsequent discussion rests on the following preliminary remark.

THEOREM 13.5.1. *If $\mathbf{A}$ is positive definite or positive semi-definite, then $\mathbf{A}'$ is positive semi-definite.*

† For further discussion of the theory see Whittaker, *Analytical Dynamics* (3rd edition), chap. vii.

‡ *J. London Math. Soc.* 5 (1930), 114–19.

When $\mathbf{A}$ is positive semi-definite, then $\mathbf{A}' = \mathbf{A}$ and there is nothing to prove. When $\mathbf{A}$ is positive definite we have

$$|\mathbf{A}'| = |\mathbf{A}| - \alpha A_{11} = 0.$$

Moreover, we know by Theorem 13.3.2 (p. 403), that

$$\begin{vmatrix} a_{mm} & . & . & . & a_{mn} \\ . & . & . & . & . \\ a_{nm} & . & . & . & a_{nn} \end{vmatrix} > 0 \qquad (m = 2,\ldots, n).$$

The assertion now follows by Exercise 13.3.3 (p. 407).

THEOREM 13.5.2. *If* $\mathbf{A} = (a_{rs})$ *is a positive definite hermitian* $n \times n$ *matrix, then*

$$|\mathbf{A}| \leqslant a_{11} a_{22} \ldots a_{nn}, \tag{13.5.2}$$

with equality in (13.5.2) *if and only if* $\mathbf{A}$ *is diagonal.*

The proof is by induction with respect to n. For $n = 1$ the theorem is true trivially. Assume, then, that it holds for $n-1$, where $n \geqslant 2$. Since $\mathbf{A}$ is positive definite we know, by Theorem 13.5.1, that $\mathbf{A}'$ is positive semi-definite. Hence, by Theorem 13.3.5 (p. 405), all diagonal elements of $\mathbf{A}'$ are non-negative, and so $\alpha \leqslant a_{11}$. Therefore, by (13.5.1) and Theorem 13.3.5,

$$|\mathbf{A}| \leqslant a_{11} A_{11}. \tag{13.5.3}$$

Now A_{11} is the determinant of a hermitian positive definite matrix of order $n-1$. Hence, by the induction hypothesis,

$$A_{11} \leqslant a_{22} \ldots a_{nn}, \tag{13.5.4}$$

and (13.5.2) follows from (13.5.3) and (13.5.4).

Next, suppose that

$$|\mathbf{A}| = a_{11} a_{22} \ldots a_{nn}. \tag{13.5.5}$$

Then, by (13.5.3) and (13.5.4), $A_{11} = a_{22} \ldots a_{nn}$, and by virtue of the induction hypothesis this implies that $a_{rs} = 0$ for $r, s = 2,\ldots, n$, $r \neq s$. Hence, by (13.5.5),

$$\begin{vmatrix} a_{11} & a_{12} & a_{13} & . & . & . & a_{1n} \\ a_{21} & a_{22} & 0 & . & . & . & 0 \\ a_{31} & 0 & a_{33} & . & . & . & 0 \\ . & . & . & . & . & . & . \\ a_{n1} & 0 & 0 & . & . & . & a_{nn} \end{vmatrix} = a_{11} a_{22} \ldots a_{nn}.$$

Subtracting, for $r = 2,..., n$, a_{1r}/a_{rr} times the rth row from the first row we obtain

$$\left(a_{11}-\frac{|a_{12}|^2}{a_{22}}-...-\frac{|a_{1n}|^2}{a_{nn}}\right)a_{22}\,...\,a_{nn} = a_{11}\,a_{22}\,...\,a_{nn}.$$

This implies that $a_{12} = ... = a_{1n} = 0$. The matrix $\mathbf{A}$ is therefore diagonal.

An alternative proof can be made to depend on the inequality of the arithmetic and geometric means.† The characteristic roots, say $\lambda_1,..., \lambda_n$, of a positive definite hermitian matrix $\mathbf{C}$ are positive and satisfy, therefore, the relation

$$\lambda_1\,...\,\lambda_n \leqslant \left(\frac{\lambda_1+...+\lambda_n}{n}\right)^n,$$

i.e.

$$|\mathbf{C}| \leqslant \left(\frac{1}{n}\operatorname{tr}\mathbf{C}\right)^n. \tag{13.5.6}$$

Here the sign of equality holds if and only if all λ's are equal; by Exercise 10.1.1 (p. 292) this occurs if and only if $\mathbf{C}$ is a scalar matrix. If $\mathbf{A} = (a_{rs})$ is positive definite, then

$$\mathbf{C} = \left(\frac{a_{rs}}{\sqrt{(a_{rr}\,a_{ss})}}\right) \tag{13.5.7}$$

is also positive definite by Exercise 13.1.1 (p. 396); and for this matrix

$$|\mathbf{C}| = |\mathbf{A}|(a_{11}\,...\,a_{nn})^{-1}, \qquad \operatorname{tr}\mathbf{C} = n.$$

Hence (13.5.2) follows by (13.5.6). Moreover, we have equality in (13.5.2) if and only if the matrix $\mathbf{C}$, given by (13.5.7), is scalar, i.e. if and only if $\mathbf{A}$ is diagonal.‡

Theorem 13.5.2 enables us to derive a celebrated result proved by Hadamard in 1893.

Theorem 13.5.3. (Hadamard's inequality)

If $\mathbf{A} = (a_{rs})$ *is any complex non-singular* $n \times n$ *matrix, then*§

$$|\det\mathbf{A}|^2 \leqslant \prod_{r=1}^{n}\{|a_{1r}|^2+...+|a_{nr}|^2\}. \tag{13.5.8}$$

To prove the theorem we consider the Gram matrix $\mathbf{B} = \bar{\mathbf{A}}^T\mathbf{A} = (b_{rs})$ of $\mathbf{A}$. Since $\bar{\mathbf{x}}^T\mathbf{B}\mathbf{x} = |\mathbf{A}\mathbf{x}|^2$, it follows that $\bar{\mathbf{x}}^T\mathbf{B}\mathbf{x} \geqslant 0$ for all $\mathbf{x}$ and $\bar{\mathbf{x}}^T\mathbf{B}\mathbf{x} = 0$ if and only if $\mathbf{A}\mathbf{x} = \mathbf{0}$, i.e. if

† This proof is taken from Hardy, Littlewood, and Pólya, *Inequalities*, 34–35.

‡ For yet another proof of Theorem 13.5.2 see Schwerdtfeger, **9**, 149–50.

§ For a geometrical interpretation of this inequality see Hardy, Littlewood, and Pólya, op. cit. 34.

and only if $\mathbf{x} = \mathbf{0}$. The hermitian matrix $\mathbf{B}$ is therefore positive definite, and applying to it the inequality (13.5.2) we obtain

$$|\bar{\mathbf{A}}^T\mathbf{A}| \leqslant b_{11} \dots b_{nn},$$

where, for $r = 1,\dots, n$,

$$b_{rr} = (\bar{\mathbf{A}}^T\mathbf{A})_{rr} = \sum_{k=1}^{n} |a_{kr}|^2.$$

The assertion therefore follows.

The inequality (13.5.8) may clearly be restated in the form

$$|\det \mathbf{A}| \leqslant |\mathbf{A}_{*1}| \dots |\mathbf{A}_{*n}|,$$

and in view of the symmetry between rows and columns this implies

$$|\det \mathbf{A}| \leqslant |\mathbf{A}_{1*}| \dots |\mathbf{A}_{n*}|.$$

COROLLARY. *If* $\mathbf{A} = (a_{rs})$ *is any complex* $n \times n$ *matrix and* $\rho = \max_{1 \leqslant r,s \leqslant n} |a_{rs}|$, *then*

$$|\det \mathbf{A}| \leqslant \rho^n n^{\frac{1}{2}n}.$$

This inequality is an immediate consequence of Theorem 13.5.3. It will be recalled that it was derived from a different source in § 10.4.2.

The next result we prove is due to Minkowski.

THEOREM 13.5.4. *If* $\mathbf{A}$ *and* $\mathbf{B}$ *are positive definite hermitian matrices of order* n, *then*

$$|\mathbf{A}|^{1/n} + |\mathbf{B}|^{1/n} \leqslant |\mathbf{A}+\mathbf{B}|^{1/n}, \tag{13.5.9}$$

with equality in (13.5.9) *if and only if* $\mathbf{B}$ *is a scalar multiple of* $\mathbf{A}$.

In the proof of this theorem we shall use a special case of Hölder's inequality† which states that if $\alpha_1,\dots, \alpha_n$, $\beta_1,\dots, \beta_n$ are positive numbers, then

$$(\alpha_1 \dots \alpha_n)^{1/n} + (\beta_1 \dots \beta_n)^{1/n} \leqslant \{(\alpha_1+\beta_1) \dots (\alpha_n+\beta_n)\}^{1/n}, \tag{13.5.10}$$

with equality if and only if there exists a number κ such that

$$\beta_i = \kappa\alpha_i \qquad (i = 1,\dots, n). \tag{13.5.11}$$

We know that the positive definite hermitian forms $\bar{\mathbf{x}}^T\mathbf{A}\mathbf{x}$, $\bar{\mathbf{x}}^T\mathbf{B}\mathbf{x}$ can be reduced simultaneously to diagonal form. In fact,

† See Hardy, Littlewood, and Pólya, op. cit. 21–24.

by Exercise 13.4.6 (p. 415), there exists a matrix $\mathbf{P}$ such that $|\det \mathbf{P}| = 1$ and

$$\bar{\mathbf{P}}^T\mathbf{A}\mathbf{P} = \mathbf{dg}(\alpha_1,..., \alpha_n), \qquad \bar{\mathbf{P}}^T\mathbf{B}\mathbf{P} = \mathbf{dg}(\beta_1,..., \beta_n),$$

where all α's and β's are, of course, positive. This implies that

$$\bar{\mathbf{P}}^T(\mathbf{A}+\mathbf{B})\mathbf{P} = \mathbf{dg}(\alpha_1+\beta_1,..., \alpha_n+\beta_n).$$

Now

$$|\mathbf{A}| = \alpha_1 \dots \alpha_n, \quad |\mathbf{B}| = \beta_1 \dots \beta_n, \quad |\mathbf{A}+\mathbf{B}| = (\alpha_1+\beta_1) \dots (\alpha_n+\beta_n),$$

and the inequality (13.5.9) now follows by (13.5.10). Moreover, there is equality in (13.5.9) if and only if (13.5.11) is satisfied, i.e. if and only if $\mathbf{dg}(\beta_1,..., \beta_n) = \kappa\, \mathbf{dg}(\alpha_1,..., \alpha_n)$. This means that $\bar{\mathbf{P}}^T\mathbf{B}\mathbf{P} = \kappa\bar{\mathbf{P}}^T\mathbf{A}\mathbf{P}$, i.e. $\mathbf{B} = \kappa\mathbf{A}$.

Corollary. *If* $\mathbf{A}$ *and* $\mathbf{B}$ *are positive definite hermitian matrices of order greater than* 1, *then*

$$|\mathbf{A}|+|\mathbf{B}| < |\mathbf{A}+\mathbf{B}|.$$

Exercise 13.5.1. Show that the inequality (13.5.9) still holds when the matrices $\mathbf{A}$ and $\mathbf{B}$ are either positive definite or positive semi-definite.

We next deduce an inequality discovered by E. Fischer in 1908.

Theorem 13.5.5. *If* $\mathbf{A} = (a_{rs})$ *is a positive definite hermitian* $n \times n$ *matrix, then, for* $r = 1,..., n-1$,

$$\begin{vmatrix} a_{11} & \cdot & \cdot & \cdot & a_{1r} \\ \cdot & \cdot & \cdot & \cdot & \cdot \\ a_{r1} & \cdot & \cdot & \cdot & a_{rr} \end{vmatrix} \begin{vmatrix} a_{r+1,r+1} & \cdot & \cdot & \cdot & a_{r+1,n} \\ \cdot & \cdot & \cdot & \cdot & \cdot \\ a_{n,r+1} & \cdot & \cdot & \cdot & a_{nn} \end{vmatrix} \geqslant \begin{vmatrix} a_{11} & \cdot & \cdot & \cdot & a_{1n} \\ \cdot & \cdot & \cdot & \cdot & \cdot \\ a_{n1} & \cdot & \cdot & \cdot & a_{nn} \end{vmatrix}. \quad (13.5.12)$$

Denote by $\mathbf{B}$ the matrix obtained from $\mathbf{A}$ when the first r rows and then the first r columns of the latter are multiplied by -1. Then $\mathbf{B}$ is again positive definite by Exercise 13.1.1 (p. 396), and $|\mathbf{B}| = |\mathbf{A}|$. Moreover

$$|\mathbf{A}+\mathbf{B}| = 2^n\Delta,$$

where Δ denotes the value of the left-hand side in (13.5.12). Substituting in (13.5.9), we obtain $|\mathbf{A}| \leqslant \Delta$, which is the required inequality.

Exercise 13.5.2. Deduce the inequality of Theorem 13.5.2 from Theorem 13.5.5.

If $\mathbf{A} = (a_{rs})$, $\mathbf{B} = (b_{rs})$ are two given matrices, we shall denote by $\mathbf{A}\times\mathbf{B}$ the matrix whose (r,s)th element is $a_{rs}b_{rs}$.

THEOREM 13.5.6. *If* $\mathbf{A}$, $\mathbf{B}$ *are positive definite or positive semi-definite hermitian* $n\times n$ *matrices, then the (hermitian) matrix* $\mathbf{A}\times\mathbf{B}$ *is also positive definite or positive semi-definite, and*

$$|\mathbf{A}\times\mathbf{B}| \geqslant b_{11}\ldots b_{nn}|\mathbf{A}|. \tag{13.5.13}$$

The fact that $\mathbf{A}\times\mathbf{B}$ is positive definite or positive semi-definite was noted by Schur, while the inequality (13.5.13) is due to Oppenheim.

We know that there exists a non-singular matrix $\mathbf{P} = (p_{rs})$ such that $\mathbf{B} = \overline{\mathbf{P}}\mathbf{\Lambda}\mathbf{P}^T$, where $\mathbf{\Lambda} = \mathbf{dg}(\beta_1,\ldots,\beta_n)$ and $\beta_1,\ldots,\beta_n$ are non-negative numbers. Writing $\mathbf{B} = (b_{rs})$, $\phi(x_1,\ldots,x_n) = \bar{\mathbf{x}}^T\mathbf{A}\mathbf{x}$, we have

$$b_{rs} = \sum_{k=1}^{n}\beta_k\bar{p}_{rk}p_{sk} \qquad (r,s = 1,\ldots,n),$$

and therefore

$$\bar{\mathbf{x}}^T(\mathbf{A}\times\mathbf{B})\mathbf{x} = \sum_{r,s=1}^{n} a_{rs}b_{rs}\bar{x}_r x_s = \sum_{r,s,k=1}^{n} a_{rs}\beta_k\bar{p}_{rk}p_{sk}\bar{x}_r x_s$$
$$= \sum_{k=1}^{n}\beta_k\phi(p_{1k}x_1,\ldots,p_{nk}x_n).$$

Hence $\bar{\mathbf{x}}^T(\mathbf{A}\times\mathbf{B})\mathbf{x} \geqslant 0$ for all $\mathbf{x}$, and the first part of the theorem is proved. We have shown, in particular, that

$$|\mathbf{A}\times\mathbf{B}| \geqslant 0. \tag{13.5.14}$$

To prove (13.5.13), we write $\mathbf{C} = \mathbf{A}\times\mathbf{B}$ and use induction with respect to n. For $n = 1$ the assertion is true trivially. Assuming that it holds for $n-1$, where $n \geqslant 2$, we have

$$C_{11} \geqslant b_{22}\ldots b_{nn}A_{11}. \tag{13.5.15}$$

Now the matrix $\mathbf{A}'$ (defined on p. 416) is positive semi-definite by Theorem 13.5.1. Hence, by (13.5.14), $|\mathbf{A}'\times\mathbf{B}| \geqslant 0$, i.e.

$$|\mathbf{A}\times\mathbf{B}| - \alpha b_{11}C_{11} \geqslant 0,$$

and so, by (13.5.15),

$$|\mathbf{A}\times\mathbf{B}| \geqslant \alpha b_{11}b_{22}\ldots b_{nn}A_{11} = b_{11}\ldots b_{nn}|\mathbf{A}|.$$

From Theorems 13.5.6 and 13.5.2 we deduce at once a further consequence.

COROLLARY. *If* $\mathbf{A}$ *and* $\mathbf{B}$ *are positive definite or positive semi-definite hermitian matrices, then*

$$|\mathbf{A}\times\mathbf{B}| \geqslant |\mathbf{A}|\,|\mathbf{B}|.$$

EXERCISE 13.5.3. Interpret this corollary for the case $\mathbf{B} = \mathbf{I}$.

EXERCISE 13.5.4. Show that, under the same conditions as in Theorem 13.5.6,

$$|\mathbf{A}\times\mathbf{B}| \geqslant a_{11} \dots a_{nn}|\mathbf{B}|.$$

EXERCISE 13.5.5. Show that if $\mathbf{A}$ and $\mathbf{B}$ are positive definite, then so is $\mathbf{A}\times\mathbf{B}$.

Schur sharpened Theorem 13.5.6 by showing that, under the same conditions,

$$|\mathbf{A}\times\mathbf{B}|+|\mathbf{A}|\,|\mathbf{B}| \geqslant b_{11} \dots b_{nn}|\mathbf{A}|+a_{11} \dots a_{nn}|\mathbf{B}|.$$

In view of Theorem 13.5.2 this inequality clearly implies (13.5.13).

PROBLEMS ON CHAPTER XIII

1. Show that, if $\mathbf{A}$ is a positive definite hermitian matrix, then so is $\mathbf{A}^*$.

2. Let $\mathbf{A}$ and $\mathbf{B}$ be positive definite or positive semi-definite hermitian matrices. Show that $|\mathbf{A}+\mathbf{B}| \geqslant 0$.

3. Determine from first principles the value class of the binary quadratic form $ax^2+2bxy+cy^2$.

4. Let (a_{ij}), (b_{ij}) be two real symmetric $n\times n$ matrices and, for $1 \leqslant r \leqslant n$, write

$$\Delta_r(\lambda,\mu) = \begin{vmatrix} \lambda a_{11}+\mu b_{11} & . & . & . & \lambda a_{1r}+\mu b_{1r} \\ . & . & . & . & . \\ \lambda a_{r1}+\mu b_{r1} & . & . & . & \lambda a_{rr}+\mu b_{rr} \end{vmatrix}.$$

Show that, if $\Delta_r(1,0) > 0$ and $\Delta_r(0,1) > 0$ for all $r = 1,\dots,n$, then $\Delta_r(\lambda,\mu) > 0$ for all $r = 1,\dots,n$ and all $\lambda > 0$, $\mu > 0$.

5. Let $\mathbf{G}$ be the Gram matrix of the $m\times n$ matrix $\mathbf{A}$. Show that $\mathbf{G}$ is a hermitian positive definite or a hermitian positive semi-definite matrix according as $R(\mathbf{A}) = n$ or $R(\mathbf{A}) < n$.

6. Show that a real symmetric matrix $\mathbf{A}$ is positive definite if and only if it can be written in the form $\mathbf{A} = \mathbf{P}^T\mathbf{P}$, where $\mathbf{P}$ is some non-singular matrix.

7. Show that, if the real symmetric matrix $\mathbf{A}$ is either semi-definite or indefinite, then the region specified by the inequality $\mathbf{x}^T\mathbf{A}\mathbf{x} \leqslant 1$ is unbounded.

8. $\mathbf{A}$ and $\mathbf{B}$ are positive definite hermitian matrices. Show that all roots of the equation $|\mathbf{A}-\lambda\mathbf{B}| = 0$ are positive. Show further that all roots are equal to 1 if and only if $\mathbf{A} = \mathbf{B}$.

9. Let $\mathbf{A}$ be a hermitian matrix and write

$$|\lambda\mathbf{I}-\mathbf{A}| = \lambda^n-d_1\lambda^{n-1}+d_2\lambda^{n-2}+\dots+(-1)^n d_n.$$

Show that $\mathbf{A}$ is positive definite if and only if $d_1 > 0,\dots, d_n > 0$.

10. If d is the value of the determinant

$$\begin{vmatrix} \cos\alpha\sin\alpha & \cos\alpha & \sin\alpha & 1 \\ \cos\beta\sin\beta & \cos\beta & \sin\beta & 1 \\ \cos\gamma\sin\gamma & \cos\gamma & \sin\gamma & 1 \\ \cos\delta\sin\delta & \cos\delta & \sin\delta & 1 \end{vmatrix},$$

show, by Hadamard's inequality, that $|d| \leqslant 81/16$.

11. Show that the quadratic form

$$n(x_1^2+...+x_n^2)-(x_1+...+x_n)^2$$

is positive semi-definite, and that it vanishes if and only if $x_1 = ... = x_n$.

12. Determine the range of values of λ for which the quaternary quadratic form

$$\lambda(x^2+y^2+z^2)+2xy-2yz+2zx+t^2$$

is positive definite. Also discuss the case $\lambda = 2$.

13. Determine the rank, signature, and value class of the ternary quadratic form

$$ax^2+by^2+cz^2+ayz+bzx+cxy,$$

where a, b, c are real numbers, not all zero.

14. Determine, for every (real) value of λ, the value class of the ternary quadratic form

$$(5\lambda+3)x^2+(12\lambda+5)y^2+5z^2+(8\lambda+2)yz+(4\lambda+6)zx-2xy.$$

15. Show that Theorem 13.4.5 (p. 412) becomes false if either the condition of reality or of positive definiteness is omitted.

16. Show that Hadamard's inequality (Theorem 13.5.3, p. 418) reduces to an equality if and only if the columns of $\mathbf{A}$ form an orthogonal set of vectors. Also find necessary and sufficient conditions for the equality

$$|\det \mathbf{A}| = |\mathbf{A}_{1*}| \dots |\mathbf{A}_{n*}|$$

to be valid.

17. (i) Show that the inequality given by the corollary to Theorem 13.5.3 (p. 419) is best possible for every value of n.

(ii) Show that this inequality becomes an equality if and only if the columns of $\mathbf{A}$ form an orthogonal set and the elements of $\mathbf{A}$ are all equal, in modulus, to ρ.

18. Let $a_{r1},..., a_{rn}$ $(r = 1,..., m)$ be m given sets of n complex numbers each. By considering the norm of the vector

$$\sum_{r=1}^{m} (a_{r1},..., a_{rn})t_r,$$

where $t_1,..., t_m$ are arbitrary complex numbers, obtain *Gram's inequality*:

$$\begin{vmatrix} \bar{a}_{11}a_{11}+...+\bar{a}_{1n}a_{1n} & . & . & . & \bar{a}_{11}a_{m1}+...+\bar{a}_{1n}a_{mn} \\ . & . & . & . & . \\ \bar{a}_{m1}a_{11}+...+\bar{a}_{mn}a_{1n} & . & . & . & \bar{a}_{m1}a_{m1}+...+\bar{a}_{mn}a_{mn} \end{vmatrix} \geqslant 0.$$

Show also that equality occurs if and only if the vectors $(a_{r1},..., a_{rn})$, $r = 1,..., m$ are linearly dependent.

Prove the same result by considering the Gram matrix of

$$\begin{pmatrix} a_{11} & . & . & . & a_{m1} \\ . & . & . & . & . \\ a_{1n} & . & . & . & a_{mn} \end{pmatrix}.$$

Discuss, in particular, the case $m = 2$ of Gram's inequality.

19. Find a real linear non-singular transformation which reduces one of the two quadratic forms

$$3x^2+5y^2+5z^2+2yz+6zx-2xy, \qquad 5x^2+12y^2+8yz+4zx$$

to $\xi^2+\eta^2+\zeta^2$ and the other to $\lambda\xi^2+\mu\eta^2+\nu\zeta^2$, where λ, μ, ν are suitable constants.

20. Find a real linear non-singular transformation which reduces one of the two quadratic forms

$$6x^2+6y^2+6z^2+8zx+4xy, \qquad 3x^2+3y^2+3z^2-2yz+2zx+2xy$$

to $\xi^2+\eta^2+\zeta^2$ and the other to $\lambda\xi^2+\mu\eta^2+\nu\zeta^2$.

21. Prove Theorem 13.5.2 (p. 417) by writing the positive definite hermitian matrix $\mathbf{A}$ in the form

$$\mathbf{A} = \begin{pmatrix} \mathbf{A}_1 & \mathbf{b} \\ \bar{\mathbf{b}}^T & a_{nn} \end{pmatrix},$$

(where $\mathbf{A}_1$ is a square matrix of order $n-1$) and using the result of Problem I, 16.

22. Show that the quadratic form $Q(x_1,..., x_n)$ is positive definite if and only if it can be written as

$$Q(x_1,..., x_n) = c_1 y_1^2+...+c_n y_n^2,$$

where $c_1,..., c_n$ are positive numbers and, for $1 \leqslant k \leqslant n$, y_k is a linear form in $x_k, x_{k+1},..., x_n$, with 1 as the coefficient of x_k.

If $Q(x_1,..., x_n)$ is expressible in the above form and if

$$Q(x_1,..., x_n) \leqslant x_1^2+...+x_n^2$$

for all values of $x_1,..., x_n$, show that $c_k \leqslant 1$ $(1 \leqslant k \leqslant n)$, and deduce that the determinant of $Q(x_1,..., x_n)$ does not exceed 1.

23. Using Theorem 13.2.1 (p. 398), show that any positive definite hermitian matrix $\mathbf{A} = (a_{rs})$ can be represented in the form $\mathbf{A} = \bar{\mathbf{Q}}^T\mathbf{Q}$, where $\mathbf{Q}$ is triangular. Deduce that $|\mathbf{A}| \leqslant a_{11}a_{22}...a_{nn}$.

24. $\mathbf{A} = (a_{ij})$ and $\mathbf{B} = (b_{ij})$ are symmetric $n \times n$ matrices, and the equation $|\lambda\mathbf{A}-\mathbf{B}| = 0$ has n distinct roots $\lambda_1,..., \lambda_n$. Prove that there exists an $n \times n$ matrix $\mathbf{T} = (t_{ij})$ such that, for all $i, j = 1,..., n$,

$$\lambda_j \sum_{k=1}^{n} a_{ik} t_{kj} = \sum_{k=1}^{n} b_{ik} t_{kj},$$

and such that $\mathbf{T}$ has no column of zeros.

Prove that the matrices $\mathbf{T}^T\mathbf{A}\mathbf{T}$, $\mathbf{T}^T\mathbf{B}\mathbf{T}$ are both diagonal.

25. Let the non-singular quadratic forms $\mathbf{x}^T\mathbf{A}\mathbf{x}$ and $\mathbf{x}^T\mathbf{B}\mathbf{x}$ be simultaneously reducible to diagonal forms $a_1x_1^2+...+a_nx_n^2$ and $b_1x_1^2+...+b_nx_n^2$ respectively. Show that the n roots of the equation $|\mathbf{A}-\lambda\mathbf{B}| = 0$ are $a_1/b_1,..., a_n/b_n$.

Show that the quadratic forms x^2+2xy and $2xy$ are not simultaneously reducible to diagonal form.

26. Let $\mathbf{A}$ be a real positive definite matrix. Show that the region R of (real) n-dimensional space specified by the inequality $\mathbf{x}^T\mathbf{A}\mathbf{x} \leqslant 1$, where $\mathbf{x} = (x_1,..., x_n)^T$, is bounded, i.e. there exists a positive constant K such that, for every point $(x_1,..., x_n) \in R$, we have $|x_1| \leqslant K,..., |x_n| \leqslant K$.

If the volume V of R is defined as the n-fold integral

$$V = \int\limits_{\mathbf{x}^T\mathbf{A}\mathbf{x} \leqslant 1} dx_1...dx_n,$$

show that

$$V = \frac{\pi^{\frac{1}{2}n}}{\Gamma(\frac{1}{2}n+1)} |\mathbf{A}|^{-\frac{1}{2}}.$$

Hence obtain the area enclosed by the ellipse $ax^2+2hxy+by^2 = 1$.

27. Let $\mathbf{\Omega}$ be a real, positive definite matrix, and let a real matrix $\mathbf{A}$ be called $\mathbf{\Omega}$-*orthogonal* if the substitution $\mathbf{x} = \mathbf{Ay}$ leaves the quadratic form $\mathbf{x}^T\mathbf{\Omega x}$ invariant. Show that the most general form of an $\mathbf{\Omega}$-orthogonal matrix is $\mathbf{A} = \mathbf{M}^{-1}\mathbf{PM}$, where $\mathbf{M}$ is a certain fixed matrix and $\mathbf{P}$ an arbitrary orthogonal matrix.

Find the most general $\mathbf{\Omega}$-orthogonal matrix when $\mathbf{\Omega} = \mathbf{dg}(a_1,..., a_n)$ and $a_1 > 0,..., a_n > 0$.

28. Let $\mathbf{\Omega}$-orthogonality be defined as in the preceding question. Show that the characteristic roots of any $\mathbf{\Omega}$-orthogonal matrix are all of unit modulus. Does this statement remain true if the positive definiteness of $\mathbf{\Omega}$ is not insisted on?

29. Show that, if $\mathbf{H}$, $\mathbf{K}$ are positive definite hermitian matrices and $\mathbf{H}^2 = \mathbf{K}^2$, then $\mathbf{H} = \mathbf{K}$.

30. Show that, if $\mathbf{A}$ is a non-singular matrix, then there exist positive definite hermitian matrices $\mathbf{H}_1, \mathbf{H}_2$ and a unitary matrix $\mathbf{U}$ such that

$$\mathbf{A} = \mathbf{UH}_1 = \mathbf{H}_2\mathbf{U}.$$

Verify that this representation of $\mathbf{A}$ is unique. (It is known as the *polar representation* of $\mathbf{A}$.) Show further that $\mathbf{H}_1 = \mathbf{H}_2$ if and only if $\mathbf{A}$ is normal.

31. Show that a matrix $\mathbf{A}$ is similar to a diagonal matrix if and only if there exists a positive definite hermitian matrix $\mathbf{H}$ such that $\mathbf{H}^{-1}\mathbf{AH}$ is normal.

32. $\mathbf{A}$ and $\mathbf{B}$ are positive definite hermitian matrices, and α is a number such that $0 < \alpha < 1$. Show that

$$|\alpha\mathbf{A}+(1-\alpha)\mathbf{B}| \geqslant |\mathbf{A}|^{\alpha}|\mathbf{B}|^{1-\alpha},$$

and that there is equality if and only if $\mathbf{A} = \mathbf{B}$. Extend this result to any number of matrices.

33. Show that, if $\mathbf{A}_1,..., \mathbf{A}_k$ are positive definite hermitian matrices of order n, then $$|\mathbf{A}_1+...+\mathbf{A}_k|^{1/n} \geqslant |\mathbf{A}_1|^{1/n}+...+|\mathbf{A}_k|^{1/n},$$
with equality if and only if $\mathbf{A}_1,..., \mathbf{A}_k$ are all scalar multiples of a single matrix.

34. From the result of Problem VIII, 34 deduce that every positive definite hermitian matrix $\mathbf{H}$ can be expressed in the form $\mathbf{H} = \bar{\mathbf{\Delta}}^T\mathbf{\Delta}$, where $\mathbf{\Delta}$ is triangular.

35. By considering the integral $\int\limits_0^\infty F(x_1 e^{-t}, x_2 e^{-2t},..., x_n e^{-nt})\, dt$ for a suitable quadratic form $F(x_1,..., x_n)$, prove that the determinant $|a_{ij}|_n$ is positive, where $a_{ij} = 1/(i+j)$.

Show further that, if the matrix (a_{rs}) is positive definite, then $(a_{rs}/(r+s))$ is also positive definite.

36. Let $F(x_1,..., x_n) = \sum\limits_{i,j=1}^{n} a_{ij}x_i x_j$ be a quadratic form, and denote by A_r the determinant formed by the first r rows and r columns of the matrix (a_{ij}). Show that, if $A_{n-1} \neq 0$, there exists a transformation $x_i = y_i + t_i y_n$ $(i = 1,..., n)$ such that

$$F(x_1,..., x_n) = \sum_{i,j=1}^{n-1} a_{ij}y_i y_j + \frac{A_n}{A_{n-1}} y_n^2.$$

Deduce that, if $A_r > 0$ $(r = 1,..., n)$, then $F(x_1,..., x_n)$ is positive definite.

37. If $\mathbf{x}^T\mathbf{A}\mathbf{x}$ is a positive definite quadratic form and $\mathbf{x}^T\mathbf{A}\mathbf{x} \leqslant \mathbf{x}^T\mathbf{x}$ for all real vectors $\mathbf{x}$, show that $|\mathbf{A}| \leqslant 1$. Show further that, if $\mathbf{x}^T\mathbf{B}\mathbf{x}$ and $\mathbf{x}^T\mathbf{C}\mathbf{x}$ are positive definite quadratic forms such that $\mathbf{x}^T\mathbf{B}\mathbf{x} \leqslant \mathbf{x}^T\mathbf{C}\mathbf{x}$ for all real $\mathbf{x}$, then $|\mathbf{B}| \leqslant |\mathbf{C}|$.

If $\mathbf{x}^T\mathbf{A}\mathbf{x}$ is a positive definite quadratic form and

$$\mathbf{A} = \begin{pmatrix} \mathbf{A}_1 & \mathbf{V} \\ \mathbf{V}^T & \mathbf{A}_2 \end{pmatrix},$$

find a real matrix $\mathbf{T}$ of determinant 1 such that

$$\mathbf{T}^T\mathbf{A}\mathbf{T} = \begin{pmatrix} \mathbf{A}_1 & \mathbf{O} \\ \mathbf{O} & \mathbf{A}_3 \end{pmatrix},$$

where $\mathbf{A}_3 = \mathbf{A}_2 - \mathbf{V}^T\mathbf{A}_1^{-1}\mathbf{V}$. Deduce Fischer's inequality (Theorem 13.5.5).

38. Let s_k denote the sum of the kth powers of the roots of an algebraic equation of degree n with real coefficients. Show that the n roots are real and distinct if and only if the real symmetric matrix

$$\begin{pmatrix} s_{2n-2} & s_{2n-3} & \cdot & \cdot & \cdot & s_{n-1} \\ s_{2n-3} & s_{2n-4} & \cdot & \cdot & \cdot & s_{n-2} \\ \cdot & \cdot & \cdot & \cdot & \cdot & \cdot \\ s_{n-1} & s_{n-2} & \cdot & \cdot & \cdot & s_0 \end{pmatrix}$$

is positive definite.

39. Let $\mathbf{H}$ be a positive semi-definite matrix of rank r. Show that there exists a (square) matrix of rank r such that $\mathbf{H} = \bar{\mathbf{A}}^T\mathbf{A}$.

40. Let $\mathbf{A}$ and $\mathbf{B}$ be hermitian matrices and suppose that $\mathbf{B}$ is positive definite. If $\alpha_1 \leqslant \dots \leqslant \alpha_n$ are the characteristic roots of $\mathbf{A}$ and $\gamma_1 \leqslant \dots \leqslant \gamma_n$ those of $\mathbf{A}+\mathbf{B}$, show that $\alpha_k < \gamma_k$ $(1 \leqslant k \leqslant n)$.

41. Let $\mathbf{A}$, $\mathbf{B}$ be real symmetric matrices and suppose that $\mathbf{B}$ is non-singular. Show that the quadratic forms $\mathbf{x}^T\mathbf{A}\mathbf{x}$, $\mathbf{x}^T\mathbf{B}\mathbf{x}$ are simultaneously reducible to diagonal forms if and only if, for any k-fold root ω of the equation $|\mathbf{A}-x\mathbf{B}| = 0$, $R(\mathbf{A}-\omega\mathbf{B}) = n-k$.

42. Let $\theta_1,\dots,\theta_n$ be complex numbers whose moduli do not exceed 1. Show, by means of Hadamard's inequality, that

$$|(\theta_1-\theta_2)(\theta_1-\theta_3)\dots(\theta_{n-1}-\theta_n)| \leqslant n^{\frac{1}{2}n}.$$

Show, further, that the sign '$\leqslant$' cannot be replaced by '$<$'.

MISCELLANEOUS PROBLEMS

Unless the contrary is stated, all numbers, vectors, and matrices are assumed to be complex. Vectors are of order n and matrices of type $n \times n$.

1. Show that, for any matrix $\mathbf{A}$ and any vector $\mathbf{x} \neq \mathbf{0}$,

$$|\mathbf{x}|.|\mathbf{Ax}| \geqslant |\bar{\mathbf{x}}^T\mathbf{Ax}|$$

with equality if and only if $\mathbf{x}$ is a characteristic vector of $\mathbf{A}$.

2. Let $\omega_1,\ldots,\omega_n$ be the characteristic roots of the hermitian matrix $\mathbf{H} = (h_{rs})$. Show that there exists a matrix $\mathbf{A}$ whose elements are real and non-negative, whose row-sums and column-sums are all equal to 1, and such that

$$(h_{11},\ldots,h_{nn})^T = \mathbf{A}(\omega_1,\ldots,\omega_n)^T.$$

3. Show that the characteristic polynomial of the matrix

$$\begin{pmatrix} a_1 & 0 & . & . & . & 0 & b_1 \\ 0 & a_2 & . & . & . & 0 & b_2 \\ . & . & . & . & . & . & . \\ 0 & 0 & . & . & . & a_{n-1} & b_{n-1} \\ b_1 & b_2 & . & . & . & b_{n-1} & b_n \end{pmatrix}$$

is equal to

$$\left(x-b_n-\sum_{k=1}^{n-1}\frac{b_k^2}{x-a_k}\right)\prod_{k=1}^{n-1}(x-a_k).$$

4. By finding a suitable matrix of which $f(x) = x^n+a_1x^{n-1}+\ldots+a_n$ is the characteristic polynomial and applying to it Theorem 7.5.4, show that all zeros of $f(x)$ lie in the union of the circles $|z| \leqslant 1$ and

$$|z+a_1| \leqslant \sum_{k=2}^{n}|a_k|.$$

5. Let $\mathfrak{M}$ be a linear manifold and let $\mathfrak{N}$ be a submanifold of $\mathfrak{M}$. For elements $X, Y \in \mathfrak{M}$ we write $X \sim Y$ if and only if $X-Y \in \mathfrak{N}$. Show that $\sim$ is an equivalence relation. Also show that the equivalence classes to which this relation gives rise constitute a linear manifold if multiplication by scalars and addition are suitably defined.

6. Does there exist a non-singular matrix $\mathbf{T}$ such that, for every matrix $\mathbf{A}$, $(\mathbf{T}^{-1}\mathbf{AT})_{11} = 0$?

7. Let $\mathbf{A}$ be any matrix and denote by $\rho_k^2, \alpha_k, \beta_k$ $(1 \leqslant k \leqslant n)$ the characteristic roots of the matrices

$$\bar{\mathbf{A}}^T\mathbf{A}, \quad \tfrac{1}{2}(\mathbf{A}+\bar{\mathbf{A}}^T), \quad \frac{1}{2i}(\mathbf{A}-\bar{\mathbf{A}}^T)$$

respectively. Establish the relation

$$\sum_{k=1}^{n}\rho_k^2 = \sum_{k=1}^{n}(\alpha_k^2+\beta_k^2).$$

8. Let $\mathbf{A}$ be a given matrix and let $\mathbf{D}$ be the matrix of the same order whose (r,s)-th element is equal to $\operatorname{tr}(\mathbf{A}^{r+s-2})$. Show that the characteristic roots of $\mathbf{A}$ are distinct if and only if $\mathbf{D}$ is non-singular.

9. If $\mathbf{H}$ is a positive definite hermitian matrix and $\mathbf{K}$ is a matrix such that $\overline{\mathbf{K}}^T\mathbf{H}+\mathbf{H}\mathbf{K} = \mathbf{O}$, show that all characteristic roots of $\mathbf{K}$ are purely imaginary.

10. Show that, if $\mathbf{x}$ is a characteristic vector of $\mathbf{A}$, then it corresponds to the characteristic root $\bar{\mathbf{x}}^T\mathbf{A}\mathbf{x}/|\mathbf{x}|^2$.

11. For any matrix $\mathbf{A} = (a_{rs})$, we write $\{N(\mathbf{A})\}^2 = \sum_{r,s=1}^{n} |a_{rs}|^2$. If the characteristic roots of $\mathbf{A}$ are denoted by $\omega_1,..., \omega_n$, show that

$$\inf\{N(\mathbf{S}^{-1}\mathbf{A}\mathbf{S})\}^2 = \sum_{k=1}^{n} |\omega_k|^2,$$

where the lower bound is taken with respect to all non-singular matrices $\mathbf{S}$. Show, further, that the lower bound is attained if and only if $\mathbf{A}$ is similar to some diagonal matrix.

12. Let $N(\mathbf{A})$ be defined as in the preceding question. Show that, for any matrices $\mathbf{A}$, $\mathbf{B}$,

$$\sup_{|z|=1} \{N(\mathbf{A}+z\mathbf{B})\}^2 = \{N(\mathbf{A})\}^2+\{N(\mathbf{B})\}^2+2|\mathrm{tr}(\bar{\mathbf{A}}^T\mathbf{B})|.$$

13. Show that a matrix which is both unitary and hermitian positive definite must be equal to the unit matrix.

14. Let $\theta_1,..., \theta_n$ be the roots of the equation

$$x^n+a_1 x^{n-1}+...+a_{n-1} x+a_n = 0$$

and suppose that $|a_1| \leqslant 1,..., |a_n| \leqslant 1$. Use Schur's inequality (10.4.4) to prove the relation

$$\sum_{k=1}^{n} |\theta_k|^2 \leqslant n+2\sum_{k=1}^{n} (k^{-1}+k^{-2}).$$

15. Let $\mathbf{E}$ be an $n \times n$ matrix such that $\mathbf{E}^2 = \mathbf{E}$, and write $R(\mathbf{E}) = r$. Show that $R(\mathbf{I}-\mathbf{E}) = n-r$. Show, further, that if $\mathbf{x}_1,..., \mathbf{x}_r$ are any r linearly independent columns of $\mathbf{E}$ while $\mathbf{y}_{r+1},..., \mathbf{y}_n$ are any $n-r$ linearly independent columns of $\mathbf{I}-\mathbf{E}$, then the matrix $\mathbf{S}$ with $\mathbf{x}_1,..., \mathbf{x}_r, \mathbf{y}_{r+1}..., \mathbf{y}_n$ as columns (in that order) is non-singular and satisfies the relation

$$\mathbf{S}^{-1}\mathbf{E}\mathbf{S} = \begin{pmatrix} \mathbf{I}_r & \mathbf{O} \\ \mathbf{O} & \mathbf{O} \end{pmatrix}.$$

16. Let $\mathbf{A}$ be a given matrix. Denoting by $\sigma_1,..., \sigma_n$ the non-negative square roots of the characteristic roots of $\bar{\mathbf{A}}^T\mathbf{A}$, show that the characteristic roots of the $2n \times 2n$ matrix

$$\begin{pmatrix} \mathbf{O} & \mathbf{A} \\ \bar{\mathbf{A}}^T & \mathbf{O} \end{pmatrix}$$

are $\pm\sigma_1,..., \pm\sigma_n$.

17. If $\mathbf{P}$ is any permutation matrix and $\mathbf{T}$ is the matrix

$$\begin{pmatrix} 1 & 1 & 1 & . & . & . & 1 \\ 1 & -1 & 0 & . & . & . & 0 \\ 1 & 0 & -1 & . & . & . & 0 \\ . & . & . & . & . & . & . \\ 1 & 0 & 0 & . & . & . & -1 \end{pmatrix},$$

show that $\mathbf{TPT}^{-1}$ is a matrix of the form

$$\begin{pmatrix} 1 & \mathbf{0} \\ \mathbf{0} & \mathbf{B} \end{pmatrix},$$

where $\mathbf{B}$ is of type $(n-1)\times(n-1)$.

18. Let $\omega_1,..., \omega_n$ be the characteristic roots of $\mathbf{A}$. Prove that the following statements are equivalent: (i) $\mathbf{A}$ is normal; (ii) the characteristic roots of $\frac{1}{2}(\mathbf{A}+\bar{\mathbf{A}}^T)$ are $\Re\,\omega_k$ $(1 \leqslant k \leqslant n)$; (iii) the characteristic roots of $\bar{\mathbf{A}}^T\mathbf{A}$ are $|\omega_k|^2$ $(1 \leqslant k \leqslant n)$; (iv) $\mathbf{A}$ and $\bar{\mathbf{A}}^T$ are simultaneously similar to triangular matrices.

19. Let $\mathbf{A} = (a_{rs})$ be an $n \times n$ matrix no column of which consists entirely of zeros. Write

$$p_s^2 = \sum_{r=1}^{n} |a_{rs}|^2 \quad (1 \leqslant s \leqslant n),$$

and denote by $\omega_1,..., \omega_n$ the characteristic roots of the matrix (a_{rs}/p_s). Show that

$$\left|\frac{\det \mathbf{A}}{p_1 \dots p_n}\right|^2 \leqslant \left(\frac{|\omega_1|^2+...+|\omega_n|^2}{n}\right)^n$$

and use Theorem 10.4.4 to establish Hadamard's inequality (13.5.8).

20. Let $\mathbf{A}, \mathbf{B}$ be hermitian matrices, and suppose that $\mathbf{B}$ is positive definite. Show that, for any complex number ω,

$$R(\mathbf{A}-\omega\mathbf{B}) = n-k,$$

where k is the multiplicity of ω as a root of the equation $|\mathbf{A}-x\mathbf{B}| = 0$.

21. Let $a_1,..., a_n$ be given numbers, and write

$$\phi_0(x) = 1, \quad \phi_k(x) = (x-a_1)(x-a_2)...(x-a_k) \quad (1 \leqslant k \leqslant n).$$

Show that the characteristic polynomial of the matrix

$$\begin{pmatrix} a_1 & 1 & 0 & . & . & . & 0 & 0 \\ 0 & a_2 & 1 & . & . & . & 0 & 0 \\ . & . & . & . & . & . & . & . \\ 0 & 0 & 0 & . & . & . & a_{n-1} & 1 \\ b_0 & b_1 & b_2 & . & . & . & b_{n-2} & a_n \end{pmatrix}$$

is equal to

$$\phi_n(x) - \sum_{k=0}^{n-2} b_k\, \phi_k(x).$$

22. Let $\mathbf{Z}$ be a matrix with zero trace. Establish the existence of matrices $\mathbf{M}_k$, $\mathbf{N}_k$ and scalars α_k $(1 \leqslant k \leqslant n^2-1)$ such that

$$\mathbf{Z} = \sum_{k=1}^{n^2-1} \alpha_k(\mathbf{M}_k \mathbf{N}_k - \mathbf{N}_k \mathbf{M}_k).$$

23. Let $\mathbf{H}$ be a hermitian matrix with characteristic roots $\omega_1 \geqslant ... \geqslant \omega_n$, and let $\mathbf{x}_1,..., \mathbf{x}_n$ be an orthonormal system of characteristic vectors corresponding to $\omega_1,..., \omega_n$ respectively. Prove that, for $1 \leqslant k \leqslant n$,

$$\omega_k = \sup \bar{\mathbf{x}}^T\mathbf{H}\mathbf{x},$$

where the upper bound is taken with respect to all vectors $\mathbf{x}$ such that

$$|\mathbf{x}| = 1, \quad (\mathbf{x},\mathbf{x}_1) = ... = (\mathbf{x},\mathbf{x}_{k-1}) = 0.$$

24. Suppose that with every matrix $\mathbf{A}$ is associated a number $\phi(\mathbf{A})$ and

suppose, further, that $\phi(\mathbf{AB}) = \phi(\mathbf{BA})$ for any matrices $\mathbf{A}, \mathbf{B}$. Using the identity $\mathbf{E}_{ij}\mathbf{E}_{kl} = \delta_{jk}\mathbf{E}_{il}$ given in Problem III, 37, prove that $\phi(\mathbf{A})$ is a constant multiple of the trace of $\mathbf{A}$.

25. Let $\mathfrak{D}$ denote the set of all $n \times n$ matrices of type $\mathbf{dg}(\pm 1,..., \pm 1)$, where the number of negative signs is even. Show that, for any $n \times n$ matrix $\mathbf{A}$,

$$\sum \det(\mathbf{I}+\mathbf{DA}) = 2^{n-1}(1+\det \mathbf{A}),$$

where the summation extends over all matrices $\mathbf{D}$ in $\mathfrak{D}$.

26. Let $n \geqslant 3$. Let $\mathbf{A}_1,..., \mathbf{A}_n$ be $n \times n$ matrices such that

$$\mathbf{A}_1 \dots \mathbf{A}_n = \mathbf{I}, \quad |\mathbf{A}_1| = \dots = |\mathbf{A}_n| = 1,$$

and $\mathbf{A}_1 - \mathbf{A}_k$ $(2 \leqslant k \leqslant n-1)$ are distinct non-zero scalar matrices. Prove that $\mathbf{A}_1 + (-1)^n \mathbf{A}_n$ is also a scalar matrix.

27. Show that a real matrix can be represented as a product of two real symmetric matrices one of which is positive definite if and only if it is of the form $\mathbf{SDS}^{-1}$, where $\mathbf{S}$ is real and non-singular and $\mathbf{D}$ real and diagonal.

28. Suppose that $\mathbf{A}$ is similar to a diagonal matrix, and let $\mathbf{E}_1,..., \mathbf{E}_k$ be defined as in Problem X, 40. Show that $\mathbf{A}$ is normal if and only if all $\mathbf{E}$'s are hermitian.

29. Show that a singular matrix is a group matrix if and only if 0 is a simple root of its minimum polynomial.

30. Let $(\mathbf{A})_k$ denote the $k \times k$ submatrix in the top left-hand corner of $\mathbf{A}$. Show that, for $1 \leqslant k \leqslant n$ and any $n \times n$ matrix $\mathbf{A}$,

$$|\det(\mathbf{A})_k|^2 \leqslant \det(\bar{\mathbf{A}}^T\mathbf{A})_k.$$

31. Let $\mu(x)$ be the minimum polynomial of $\mathbf{A}$, let $f(x)$ be an arbitrary polynomial, and denote by $d(x)$ the highest common factor of $\mu(x)$ and $f(x)$. Show that $f(\mathbf{A})$ and $d(\mathbf{A})$ have equal rank.

32. Let $\mathbf{\Omega}$ be the $n \times n$ matrix

$$\begin{pmatrix} 0 & 1 & 0 & . & . & . & 0 & 0 \\ 0 & 0 & 1 & . & . & . & 0 & 0 \\ . & . & . & . & . & . & . & . \\ 0 & 0 & 0 & . & . & . & 0 & 1 \\ 1 & 0 & 0 & . & . & . & 0 & 0 \end{pmatrix}$$

Show that $\quad \mathbf{\Omega}^r = \begin{pmatrix} \mathbf{O} & \mathbf{I}_{n-r} \\ \mathbf{I}_r & \mathbf{O} \end{pmatrix} \quad (1 \leqslant r < n), \qquad \mathbf{\Omega}^n = \mathbf{I},$

and deduce that any matrix $\mathbf{A}$ possesses a unique representation of the form

$$\mathbf{A} = \mathbf{D}_0 + \mathbf{D}_1\mathbf{\Omega} + \dots + \mathbf{D}_{n-1}\mathbf{\Omega}^{n-1},$$

where $\mathbf{D}_0,..., \mathbf{D}_{n-1}$ are diagonal matrices. Hence prove that $\mathbf{A}$ commutes with $\mathbf{\Omega}$ if and only if $\mathbf{A}$ is a polynomial (with scalar coefficients) in $\mathbf{\Omega}$.

33. The *range* of the matrix $\mathbf{A}$ is defined as the set of all vectors $\mathbf{Ax}$, where $\mathbf{x} \in \mathfrak{B}_n$; the *null-space* of $\mathbf{A}$ is defined as the set of all vectors $\mathbf{x} \in \mathfrak{B}_n$ such that $\mathbf{Ax} = \mathbf{0}$.

Show that $R(\mathbf{A}) = R(\mathbf{A}^2)$ if and only if $\mathbf{0}$ is the only vector common to the range and the null-space of $\mathbf{A}$.

34. $\mathbf{E}$ is an $n \times n$ matrix of rank $r \geqslant 1$ and $\mathbf{E}^2 = \mathbf{E}$. Show, by means of

the result in Problem 15, that there exist matrices $\mathbf{A}$, $\mathbf{B}$ of type $n \times r$, $r \times n$ respectively such that $\mathbf{E} = \mathbf{AB}$ and $\mathbf{BA} = \mathbf{I}_r$.

35. Suppose that
$$\mathbf{A} = \begin{pmatrix} \mathbf{A}_1 & \mathbf{B} \\ \overline{\mathbf{B}}^T & \mathbf{A}_2 \end{pmatrix}$$
is a hermitian matrix of even order partitioned in such a way that $\mathbf{A}_1$, $\mathbf{A}_2$, $\mathbf{B}$ are square matrices. If $\mathbf{A}$ is positive definite, establish the inequality
$$|\det \mathbf{B}|^2 < \det \mathbf{A}_1 . \det \mathbf{A}_2.$$
What can be asserted when $\mathbf{A}$ is positive semi-definite?

36. Let Q be a real quadratic form and L a real linear form in $x_1,..., x_n$. Show that $Q > 0$ for all real values of $x_1,..., x_n$, other than $x_1 = ... = x_n = 0$, which satisfy $L = 0$ if and only if $Q+tL^2$ is a positive definite form in $x_1,..., x_n$ for all sufficiently large values of t.

37. Let ω be an m-fold characteristic root of the $n \times n$ matrix $\mathbf{A}$, and let μ be its multiplicity as a root of the minimum polynomial of $\mathbf{A}$. Show that
$$R\{(A-\omega I)^\mu\} = n-m.$$

38. Let $1 \leqslant k \leqslant n$ and denote by $\mathfrak{U}$ the subspace of $\mathfrak{B}_n$ spanned by the linearly independent vectors $\mathbf{x}_1,..., \mathbf{x}_k \in \mathfrak{B}_n$. Prove that, for any $\mathbf{x} \in \mathfrak{B}_n$,
$$\inf_{\mathbf{y}} |\mathbf{x}-\mathbf{y}|^2 = \det G(\mathbf{x}_1,..., \mathbf{x}_k, \mathbf{x})/\det G(\mathbf{x}_1,..., \mathbf{x}_k),$$
where the lower bound is taken with respect to all vectors $\mathbf{y} \in \mathfrak{U}$, and $G(\mathbf{z}_1,..., \mathbf{z}_r)$ denotes the $r \times r$ matrix whose (i,j)-th element is $(\mathbf{z}_i, \mathbf{z}_j)$.

39. Let $\mathfrak{S}$ denote the set of all matrices of the form
$$\begin{pmatrix} 0 & \lambda_1 & 0 & . & . & . & 0 \\ 0 & 0 & \lambda_2 & . & . & . & 0 \\ . & . & . & . & . & . & . \\ 0 & 0 & 0 & . & . & . & \lambda_{n-1} \\ \lambda_n & 0 & 0 & . & . & . & 0 \end{pmatrix}.$$
Show that the product of any n matrices in $\mathfrak{S}$ is a diagonal matrix.

40. If r, s are positive integers and $\mathbf{A}$ is a hermitian $r \times r$ matrix, show that there exists an $r \times s$ matrix $\mathbf{X}$ such that $\mathbf{X}\overline{\mathbf{X}}^T = \mathbf{A}$ if and only if $\mathbf{A}$ is positive definite or positive semi-definite and $R(\mathbf{A}) \leqslant s$.

Deduce that, if $1 \leqslant r, s < n$ and $\mathbf{A}$ is an $r \times s$ matrix, then there exists a unitary $n \times n$ matrix of which $\mathbf{A}$ is the top left-hand submatrix if and only if $\mathbf{I}_r - \mathbf{A}\overline{\mathbf{A}}^T$ is positive definite or positive semi-definite and $R(\mathbf{I}_r - \mathbf{A}\overline{\mathbf{A}}^T) \leqslant n-s$.

41. Making use of the result of Problem X, 40, show that two normal (hermitian) matrices $\mathbf{A}$, $\mathbf{B}$ commute if and only if there exist polynomials f, g with complex (real) coefficients and a hermitian matrix $\mathbf{H}$ such that $\mathbf{A} = f(\mathbf{H})$, $\mathbf{B} = g(\mathbf{H})$.

Also prove that any normal matrix can be expressed as a polynomial (with complex coefficients) in a suitable hermitian matrix.

42. Let $\mathbf{A}$, $\mathbf{B}$ be matrices of type $p \times q$, $q \times r$ respectively. Show that there exist matrices $\mathbf{M}$, $\mathbf{N}$, of type $p \times r$, $q \times r$ respectively, such that
$$\mathbf{B} = \overline{\mathbf{A}}^T\mathbf{M}+\mathbf{N}, \quad \mathbf{AN} = \mathbf{O}.$$

43. Let $\mathbf{x}_1,..., \mathbf{x}_n$ be given vectors and denote by $\mathbf{X}$ the matrix whose

(r, s)-th element is $(\mathbf{x}_r, \mathbf{x}_s)$. Show that if any principal minor of $\mathbf{X}$ vanishes, then $\mathbf{X}$ is singular.

44. Let $\mathbf{H}$ be a negative definite or negative semi-definite hermitian matrix. Establish the existence of a matrix $\mathbf{S}$ such that $\mathbf{S}^2 = \mathbf{H}$ and show that, if $\mathbf{S}$ is normal, then it is skew-hermitian.

45. If $\mathbf{A}$ is any matrix, show that there exist non-singular matrices $\mathbf{P}$ and $\mathbf{Q}$ such that $(\mathbf{PA})^2 = \mathbf{PA}$, $(\mathbf{AQ})^2 = \mathbf{AQ}$.

46. Suppose that (i) the characteristic polynomials of $\mathbf{A}$ and $\mathbf{B}$ have no common factor; (ii) there exists a matrix $\mathbf{T}$ such that $\mathbf{AT} = \mathbf{TB}$. Deduce that $\mathbf{T} = \mathbf{O}$.

47. A *circulant matrix* is a matrix of the form

$$\begin{pmatrix} a_1 & a_2 & . & . & . & a_{n-1} & a_n \\ a_n & a_1 & . & . & . & a_{n-2} & a_{n-1} \\ . & . & . & . & . & . & . \\ a_2 & a_3 & . & . & . & a_n & a_1 \end{pmatrix}.$$

Show that a matrix is circulant if and only if it can be expressed as a polynomial in the matrix

$$\begin{pmatrix} 0 & 0 & 0 & . & . & . & 0 & 1 \\ 1 & 0 & 0 & . & . & . & 0 & 0 \\ 0 & 1 & 0 & . & . & . & 0 & 0 \\ . & . & . & . & . & . & . & . \\ 0 & 0 & 0 & . & . & . & 1 & 0 \end{pmatrix}.$$

Deduce that (i) every circulant matrix is normal; (ii) any two circulant matrices commute; (iii) the product of any two circulant matrices is again a circulant matrix.

Also give an alternative proof of (i) by determining a suitable set of characteristic vectors of a circulant matrix.

48. Let $\mathbf{A}$, $\mathbf{B}$ be real matrices. Show that, if there exists a non-singular complex matrix $\mathbf{C}$ such that $\mathbf{AC} = \mathbf{CB}$, then there exists a non-singular real matrix $\mathbf{R}$ such that $\mathbf{AR} = \mathbf{RB}$.

49. Let $F(z) = a_0 + a_1 z + a_2 z^2 + \ldots$, where $a_0 \neq 0$, be a power series with a positive radius of convergence; and write

$$1/F(z) = b_0 + b_1 z + b_2 z^2 + \ldots.$$

Show that

$$b_n = (-1)^n a_0^{-n-1} \begin{vmatrix} a_1 & a_0 & 0 & 0 & . & . & . & 0 \\ a_2 & a_1 & a_0 & 0 & . & . & . & 0 \\ a_3 & a_2 & a_1 & a_0 & . & . & . & 0 \\ . & . & . & . & . & . & . & . \\ a_n & a_{n-1} & a_{n-2} & a_{n-3} & . & . & . & a_1 \end{vmatrix}.$$

50. A square matrix is said to be *doubly-stochastic* if its elements are non-negative and if the sum of the elements in each row and each column is equal to 1. Prove the following results.

(i) The product of two doubly-stochastic matrices is again doubly-stochastic.

(ii) A matrix commutes with every doubly-stochastic matrix if and only if all its diagonal elements are equal and all its non-diagonal elements are equal.

(iii) Every characteristic root ω of a doubly-stochastic matrix satisfies the inequality $|\omega| \leqslant 1$.

(iv) If all characteristic roots of a doubly-stochastic matrix $\mathbf{D}$ lie on the unit circle, then $\mathbf{D}$ is a permutation matrix.

(v) If $\mathbf{D}$ is a doubly-stochastic matrix, then so is $\exp(\mathbf{D}-\mathbf{I})$.

(vi) The numbers $a_1,..., a_n$ are the diagonal elements of some doubly-stochastic matrix if and only if $0 \leqslant a_k \leqslant 1$ $(1 \leqslant k \leqslant n)$ and

$$\sum_{k=1}^{n} a_k - 2 \min_{1 \leqslant i \leqslant n} a_i \leqslant n-2.$$

(vii) If $\mathbf{P}_1,..., \mathbf{P}_k$ are any permutation matrices and $t_1,..., t_k$ any non-negative numbers with sum 1, then

$$t_1 \mathbf{P}_1 + ... + t_k \mathbf{P}_k$$

is a doubly-stochastic matrix. (It is also true that every doubly-stochastic matrix possesses such a representation; but this fact is much harder to prove.)

BIBLIOGRAPHY

ALL books listed below deal wholly or in part with linear algebra. The books of the first series provide a treatment of this subject which is roughly comparable in extent to that offered here. The second list enumerates books which have less affinity with the present treatment but may be usefully consulted on specific points. The books in both series are arranged approximately in order of increasing difficulty.

1. T. L. WADE: *The Algebra of Vectors and Matrices* (Cambridge, Mass.; 1951).
2. W. L. FERRAR: *Algebra—A Textbook of Determinants, Matrices, and Algebraic Forms* (2nd edition; Oxford, 1957).
3. E. T. BROWNE: *Introduction to the Theory of Determinants and Matrices* (Chapel Hill, N.C., 1958).
4. F. E. HOHN: *Elementary Matrix Algebra* (New York, 1958).
5. J. W. ARCHBOLD: *Algebra* (London, 1958).
6. D. C. MURDOCH: *Linear Algebra for Undergraduates* (New York, 1957).
7. G. BIRKHOFF and S. MACLANE: *A Survey of Modern Algebra* (revised edition; New York, 1953).
8. O. SCHREIER and E. SPERNER: *Introduction to Modern Algebra and Matrix Theory* (New York, 1951).
9. H. SCHWERDTFEGER: *Introduction to Linear Algebra and the Theory of Matrices* (Groningen, 1950).
10. S. PERLIS: *Theory of Matrices* (Cambridge, Mass., 1952).
11. R. R. STOLL: *Linear Algebra and Matrix Theory* (New York–Toronto–London, 1952).
12. C. C. MACDUFFEE: *Vectors and Matrices* (Carus Monographs, 1943).
13. R. M. THRALL and L. TORNHEIM: *Vector Spaces and Matrices* (New York, 1957).

---

14. C. V. DURELL and A. ROBSON: *Advanced Algebra*, Volume III (London, 1937).
15. A. C. AITKEN: *Determinants and Matrices* (9th edition; Edinburgh and London, 1956).
16. M. BÔCHER: *Introduction to Higher Algebra* (New York, 1936).
17. E. BODEWIG: *Matrix Calculus* (2nd edition; Amsterdam, 1959).
18. O. PERRON: *Algebra* (2nd edition; Berlin and Leipzig, 1932–3), 2 volumes).
19. H. HASSE: *Höhere Algebra. Band I. Lineare Gleichungen* (3rd edition; Berlin, 1951).
20. P. R. HALMOS: *Finite Dimensional Vector Spaces* (2nd edition; Princeton, 1958).
21. W. L. FERRAR: *Finite Matrices* (Oxford, 1951).
22. R. A. FRAZER, W. J. DUNCAN, and A. R. COLLAR: *Elementary Matrices* (Cambridge, 1938).

23. W. GRÖBNER: *Matrizenrechnung* (München, 1956).
24. R. BELLMAN: *Introduction to Matrix Analysis* (New York–Toronto–London, 1960).
25. A. LICHNEROWICZ, *Algèbre et analyse linéaires* (Paris, 1947).
26. H. W. TURNBULL: *The Theory of Determinants, Matrices, and Invariants* (London and Glasgow, 1929).
27. H. W. TURNBULL and A. C. AITKEN: *An Introduction to the Theory of Canonical Matrices* (London and Glasgow, 1948).
28. N. JACOBSON: *Lectures in Abstract Algebra. Volume II. Linear Algebra* (New York, 1953).
29. F. R. GANTMACHER: *The Theory of Matrices* (translated from the Russian by K. A. Hirsch; New York, 1959), 2 volumes.
30. H. L. HAMBURGER and M. E. GRIMSHAW: *Linear Transformations in n-Dimensional Vector Space* (Cambridge, 1951).
31. C. C. MACDUFFEE: *The Theory of Matrices* (Berlin, 1933).
32. C. C. MACDUFFEE: *An Introduction to Abstract Algebra* (New York, 1940).
33. A. A. ALBERT: *Modern Higher Algebra* (Cambridge, 1938).
34. G. JULIA: *Introduction mathématique aux théories quantiques*, Première partie (2nd edition; Paris, 1949).
35. W. GRAEUB: *Lineare Algebra* (Berlin–Göttingen–Heidelberg, 1958).
36. H. BOERNER: *Darstellung von Gruppen* (Berlin–Göttingen–Heidelberg, 1955).
37. B. L. VAN DER WAERDEN: *Gruppen von linearen Transformationen* (Berlin, 1935).
38. A. SPEISER: *Die Theorie der Gruppen von endlicher Ordnung* (3rd edition; Berlin–Göttingen–Heidelberg, 1937).
39. J. H. M. WEDDERBURN: *Lectures on Matrices* (New York, 1934).
40. F. D. MURNAGHAN: *The Theory of Group Representations* (Baltimore, 1938).
41. J. DIEUDONNÉ: *Sur les groupes classiques* (Paris, 1948).
42. H. WEYL: *The Classical Groups—Their Invariants and Representations* (2nd edition; Princeton, 1946).

INDEX

A CATALOG OF SELECTED
DOVER BOOKS
IN SCIENCE AND MATHEMATICS

NUMERICAL METHODS FOR SCIENTISTS AND ENGINEERS, Richard Hamming. Classic text stresses frequency approach in coverage of algorithms, polynomial approximation, Fourier approximation, exponential approximation, other topics. Revised and enlarged 2nd edition. 721pp. 5⅜ × 8½.
65241-6 Pa. $14.95

THEORETICAL SOLID STATE PHYSICS, Vol. I: Perfect Lattices in Equilibrium; Vol. II: Non-Equilibrium and Disorder, William Jones and Norman H. March. Monumental reference work covers fundamental theory of equilibrium properties of perfect crystalline solids, non-equilibrium properties, defects and disordered systems. Appendices. Problems. Preface. Diagrams. Index. Bibliography. Total of 1,301pp. 5⅜ × 8½. Two volumes. Vol. I 65015-4 Pa. $12.95
Vol. II 65016-2 Pa. $12.95

OPTIMIZATION THEORY WITH APPLICATIONS, Donald A. Pierre. Broad-spectrum approach to important topic. Classical theory of minima and maxima, calculus of variations, simplex technique and linear programming, more. Many problems, examples. 640pp. 5⅜ × 8½. 65205-X Pa. $13.95

THE MODERN THEORY OF SOLIDS, Frederick Seitz. First inexpensive edition of classic work on theory of ionic crystals, free-electron theory of metals and semiconductors, molecular binding, much more. 736pp. 5⅜ × 8½.
65482-6 Pa. $15.95

ESSAYS ON THE THEORY OF NUMBERS, Richard Dedekind. Two classic essays by great German mathematician: on the theory of irrational numbers; and on transfinite numbers and properties of natural numbers. 115pp. 5⅜ × 8½.
21010-3 Pa. $4.95

THE FUNCTIONS OF MATHEMATICAL PHYSICS, Harry Hochstadt. Comprehensive treatment of orthogonal polynomials, hypergeometric functions, Hill's equation, much more. Bibliography. Index. 322pp. 5⅜ × 8½. 65214-9 Pa. $9.95

NUMBER THEORY AND ITS HISTORY, Oystein Ore. Unusually clear, accessible introduction covers counting, properties of numbers, prime numbers, much more. Bibliography. 380pp. 5⅜ × 8½. 65620-9 Pa. $8.95

THE VARIATIONAL PRINCIPLES OF MECHANICS, Cornelius Lanzcos. Graduate level coverage of calculus of variations, equations of motion, relativistic mechanics, more. First inexpensive paperbound edition of classic treatise. Index. Bibliography. 418pp. 5⅜ × 8½. 65067-7 Pa. $10.95

MATHEMATICAL TABLES AND FORMULAS, Robert D. Carmichael and Edwin R. Smith. Logarithms, sines, tangents, trig functions, powers, roots, reciprocals, exponential and hyperbolic functions, formulas and theorems. 269pp. 5⅜ × 8½. 60111-0 Pa. $5.95

THEORETICAL PHYSICS, Georg Joos, with Ira M. Freeman. Classic overview covers essential math, mechanics, electromagnetic theory, thermodynamics, quantum mechanics, nuclear physics, other topics. First paperback edition. xxiii + 885pp. 5⅜ × 8½. 65227-0 Pa. $18.95

ORDINARY DIFFERENTIAL EQUATIONS, Morris Tenenbaum and Harry Pollard. Exhaustive survey of ordinary differential equations for undergraduates in mathematics, engineering, science. Thorough analysis of theorems. Diagrams. Bibliography. Index. 818pp. 5⅜ × 8½. 64940-7 Pa. $15.95

STATISTICAL MECHANICS: Principles and Applications, Terrell L. Hill. Standard text covers fundamentals of statistical mechanics, applications to fluctuation theory, imperfect gases, distribution functions, more. 448pp. 5⅜ × 8½. 65390-0 Pa. $9.95

ORDINARY DIFFERENTIAL EQUATIONS AND STABILITY THEORY: An Introduction, David A. Sánchez. Brief, modern treatment. Linear equation, stability theory for autonomous and nonautonomous systems, etc. 164pp. 5⅜ × 8¼. 63828-6 Pa. $4.95

THIRTY YEARS THAT SHOOK PHYSICS: The Story of Quantum Theory, George Gamow. Lucid, accessible introduction to influential theory of energy and matter. Careful explanations of Dirac's anti-particles, Bohr's model of the atom, much more. 12 plates. Numerous drawings. 240pp. 5⅜ × 8½. 24895-X Pa. $5.95

ORDINARY DIFFERENTIAL EQUATIONS, I.G. Petrovski. Covers basic concepts, some differential equations and such aspects of the general theory as Euler lines, Arzel's theorem, Peano's existence theorem, Osgood's uniqueness theorem, more. 45 figures. Problems. Bibliography. Index. xi + 232pp. 5⅜ × 8½. 64683-1 Pa. $6.95

GREAT EXPERIMENTS IN PHYSICS: Firsthand Accounts from Galileo to Einstein, edited by Morris H. Shamos. 25 crucial discoveries: Newton's laws of motion, Chadwick's study of the neutron, Hertz on electromagnetic waves, more. Original accounts clearly annotated. 370pp. 5⅜ × 8½. 25346-5 Pa. $8.95

INTRODUCTION TO PARTIAL DIFFERENTIAL EQUATIONS WITH APPLICATIONS, E.C. Zachmanoglou and Dale W. Thoe. Essentials of partial differential equations applied to common problems in engineering and the physical sciences. Problems and answers. 416pp. 5⅜ × 8½. 65251-3 Pa. $9.95

BURNHAM'S CELESTIAL HANDBOOK, Robert Burnham, Jr. Thorough guide to the stars beyond our solar system. Exhaustive treatment. Alphabetical by constellation: Andromeda to Cetus in Vol. 1; Chamaeleon to Orion in Vol. 2; and Pavo to Vulpecula in Vol. 3. Hundreds of illustrations. Index in Vol. 3. 2,000pp. 6⅛ × 9¼. 23567-X, 23568-8, 23673-0 Pa., Three-vol. set $38.85

ASYMPTOTIC EXPANSIONS FOR ORDINARY DIFFERENTIAL EQUATIONS, Wolfgang Wasow. Outstanding text covers asymptotic power series, Jordan's canonical form, turning point problems, singular perturbations, much more. Problems. 384pp. 5⅜ × 8½. 65456-7 Pa. $9.95

AMATEUR ASTRONOMER'S HANDBOOK, J.B. Sidgwick. Timeless, comprehensive coverage of telescopes, mirrors, lenses, mountings, telescope drives, micrometers, spectroscopes, more. 189 illustrations. 576pp. 5⅜ × 8¼. 24034-7 Pa. $8.95

ROTARY-WING AERODYNAMICS, W.Z. Stepniewski. Clear, concise text covers aerodynamic phenomena of the rotor and offers guidelines for helicopter performance evaluation. Originally prepared for NASA. 537 figures. 640pp. 6⅛ × 9¼.
64647-5 Pa. $14.95

DIFFERENTIAL GEOMETRY, Heinrich W. Guggenheimer. Local differential geometry as an application of advanced calculus and linear algebra. Curvature, transformation groups, surfaces, more. Exercises. 62 figures. 378pp. 5⅜ × 8½.
63433-7 Pa. $7.95

INTRODUCTION TO SPACE DYNAMICS, William Tyrrell Thomson. Comprehensive, classic introduction to space-flight engineering for advanced undergraduate and graduate students. Includes vector algebra, kinematics, transformation of coordinates. Bibliography. Index. 352pp. 5⅜ × 8½. 65113-4 Pa. $8.00

A SURVEY OF MINIMAL SURFACES, Robert Osserman. Up-to-date, in-depth discussion of the field for advanced students. Corrected and enlarged edition covers new developments. Includes numerous problems. 192pp. 5⅜ × 8½.
64998-9 Pa. $8.95

ANALYTICAL MECHANICS OF GEARS, Earle Buckingham. Indispensable reference for modern gear manufacture covers conjugate gear-tooth action, gear-tooth profiles of various gears, many other topics. 263 figures. 102 tables. 546pp. 5⅜ × 8½. 65712-4 Pa. $11.95

SET THEORY AND LOGIC, Robert R. Stoll. Lucid introduction to unified theory of mathematical concepts. Set theory and logic seen as tools for conceptual understanding of real number system. 496pp. 5⅜ × 8¼. 63829-4 Pa. $8.95

A HISTORY OF MECHANICS, René Dugas. Monumental study of mechanical principles from antiquity to quantum mechanics. Contributions of ancient Greeks, Galileo, Leonardo, Kepler, Lagrange, many others. 671pp. 5⅜ × 8½.
65632-2 Pa. $14.95

FAMOUS PROBLEMS OF GEOMETRY AND HOW TO SOLVE THEM, Benjamin Bold. Squaring the circle, trisecting the angle, duplicating the cube: learn their history, why they are impossible to solve, then solve them yourself. 128pp. 5⅜ × 8½. 24297-8 Pa. $3.95

MECHANICAL VIBRATIONS, J.P. Den Hartog. Classic textbook offers lucid explanations and illustrative models, applying theories of vibrations to a variety of practical industrial engineering problems. Numerous figures. 233 problems, solutions. Appendix. Index. Preface. 436pp. 5⅜ × 8½. 64785-4 Pa. $8.95

CURVATURE AND HOMOLOGY, Samuel I. Goldberg. Thorough treatment of specialized branch of differential geometry. Covers Riemannian manifolds, topology of differentiable manifolds, compact Lie groups, other topics. Exercises. 315pp. 5⅜ × 8½. 64314-X Pa. $6.95

HISTORY OF STRENGTH OF MATERIALS, Stephen P. Timoshenko. Excellent historical survey of the strength of materials with many references to the theories of elasticity and structure. 245 figures. 452pp. 5⅜ × 8½. 61187-6 Pa. $10.95

THE FOUR-COLOR PROBLEM: Assaults and Conquest, Thomas L. Saaty and Paul G. Kainen. Engrossing, comprehensive account of the century-old combinatorial topological problem, its history and solution. Bibliographies. Index. 110 figures. 228pp. 5⅜ × 8½. 65092-8 Pa. $6.00

CATALYSIS IN CHEMISTRY AND ENZYMOLOGY, William P. Jencks. Exceptionally clear coverage of mechanisms for catalysis, forces in aqueous solution, carbonyl- and acyl-group reactions, practical kinetics, more. 864pp. 5⅜ × 8½. 65460-5 Pa. $18.95

PROBABILITY: An Introduction, Samuel Goldberg. Excellent basic text covers set theory, probability theory for finite sample spaces, binomial theorem, much more. 360 problems. Bibliographies. 322pp. 5⅜ × 8½. 65252-1 Pa. $8.95

LIGHTNING, Martin A. Uman. Revised, updated edition of classic work on the physics of lightning. Phenomena, terminology, measurement, photography, spectroscopy, thunder, more. Reviews recent research. Bibliography. Indices. 320pp. 5⅜ × 8¼. 64575-4 Pa. $7.95

PROBABILITY THEORY: A Concise Course, Y.A. Rozanov. Highly readable, self-contained introduction covers combination of events, dependent events, Bernoulli trials, etc. Translation by Richard Silverman. 148pp. 5⅜ × 8¼. 63544-9 Pa. $5.95

THE CEASELESS WIND: An Introduction to the Theory of Atmospheric Motion, John A. Dutton. Acclaimed text integrates disciplines of mathematics and physics for full understanding of dynamics of atmospheric motion. Over 400 problems. Index. 97 illustrations. 640pp. 6 × 9. 65096-0 Pa. $17.95

STATISTICS MANUAL, Edwin L. Crow, et al. Comprehensive, practical collection of classical and modern methods prepared by U.S. Naval Ordnance Test Station. Stress on use. Basics of statistics assumed. 288pp. 5⅜ × 8½. 60599-X Pa. $6.00

WIND WAVES: Their Generation and Propagation on the Ocean Surface, Blair Kinsman. Classic of oceanography offers detailed discussion of stochastic processes and power spectral analysis that revolutionized ocean wave theory. Rigorous, lucid. 676pp. 5⅜ × 8½. 64652-1 Pa. $16.95

STATISTICAL METHOD FROM THE VIEWPOINT OF QUALITY CONTROL, Walter A. Shewhart. Important text explains regulation of variables, uses of statistical control to achieve quality control in industry, agriculture, other areas. 192pp. 5⅜ × 8½. 65232-7 Pa. $6.95

THE INTERPRETATION OF GEOLOGICAL PHASE DIAGRAMS, Ernest G. Ehlers. Clear, concise text emphasizes diagrams of systems under fluid or containing pressure; also coverage of complex binary systems, hydrothermal melting, more. 288pp. 6½ × 9¼. 65389-7 Pa. $10.95

STATISTICAL ADJUSTMENT OF DATA, W. Edwards Deming. Introduction to basic concepts of statistics, curve fitting, least squares solution, conditions without parameter, conditions containing parameters. 26 exercises worked out. 271pp. 5⅜ × 8½. 64685-8 Pa. $7.95

DE RE METALLICA, Georgius Agricola. The famous Hoover translation of greatest treatise on technological chemistry, engineering, geology, mining of early modern times (1556). All 289 original woodcuts. 638pp. 6¾ × 11.
60006-8 Clothbd. $17.95

SOME THEORY OF SAMPLING, William Edwards Deming. Analysis of the problems, theory and design of sampling techniques for social scientists, industrial managers and others who find statistics increasingly important in their work. 61 tables. 90 figures. xvii + 602pp. 5⅜ × 8½. 64684-X Pa. $15.95

THE VARIOUS AND INGENIOUS MACHINES OF AGOSTINO RAMELLI: A Classic Sixteenth-Century Illustrated Treatise on Technology, Agostino Ramelli. One of the most widely known and copied works on machinery in the 16th century. 194 detailed plates of water pumps, grain mills, cranes, more. 608pp. 9 × 12.
25497-6 Clothbd. $34.95

LINEAR PROGRAMMING AND ECONOMIC ANALYSIS, Robert Dorfman, Paul A. Samuelson and Robert M. Solow. First comprehensive treatment of linear programming in standard economic analysis. Game theory, modern welfare economics, Leontief input-output, more. 525pp. 5⅜ × 8½. 65491-5 Pa. $13.95

ELEMENTARY DECISION THEORY, Herman Chernoff and Lincoln E. Moses. Clear introduction to statistics and statistical theory covers data processing, probability and random variables, testing hypotheses, much more. Exercises. 364pp. 5⅜ × 8½. 65218-1 Pa. $8.95

THE COMPLEAT STRATEGYST: Being a Primer on the Theory of Games of Strategy, J.D. Williams. Highly entertaining classic describes, with many illustrated examples, how to select best strategies in conflict situations. Prefaces. Appendices. 268pp. 5⅜ × 8½. 25101-2 Pa. $5.95

MATHEMATICAL METHODS OF OPERATIONS RESEARCH, Thomas L. Saaty. Classic graduate-level text covers historical background, classical methods of forming models, optimization, game theory, probability, queueing theory, much more. Exercises. Bibliography. 448pp. 5⅜ × 8¼. 65703-5 Pa. $12.95

CONSTRUCTIONS AND COMBINATORIAL PROBLEMS IN DESIGN OF EXPERIMENTS, Damaraju Raghavarao. In-depth reference work examines orthogonal Latin squares, incomplete block designs, tactical configuration, partial geometry, much more. Abundant explanations, examples. 416pp. 5⅜ × 8¼.
65685-3 Pa. $10.95

THE ABSOLUTE DIFFERENTIAL CALCULUS (CALCULUS OF TENSORS), Tullio Levi-Civita. Great 20th-century mathematician's classic work on material necessary for mathematical grasp of theory of relativity. 452pp. 5⅜ × 8½.
63401-9 Pa. $9.95

VECTOR AND TENSOR ANALYSIS WITH APPLICATIONS, A.I. Borisenko and I.E. Tarapov. Concise introduction. Worked-out problems, solutions, exercises. 257pp. 5⅜ × 8¼. 63833-2 Pa. $6.95

TENSOR CALCULUS, J.L. Synge and A. Schild. Widely used introductory text covers spaces and tensors, basic operations in Riemannian space, non-Riemannian spaces, etc. 324pp. 5⅜ × 8¼. 63612-7 Pa. $7.00

A CONCISE HISTORY OF MATHEMATICS, Dirk J. Struik. The best brief history of mathematics. Stresses origins and covers every major figure from ancient Near East to 19th century. 41 illustrations. 195pp. 5⅜ × 8½. 60255-9 Pa. $7.95

A SHORT ACCOUNT OF THE HISTORY OF MATHEMATICS, W.W. Rouse Ball. One of clearest, most authoritative surveys from the Egyptians and Phoenicians through 19th-century figures such as Grassman, Galois, Riemann. Fourth edition. 522pp. 5⅜ × 8½. 20630-0 Pa. $9.95

HISTORY OF MATHEMATICS, David E. Smith. Non-technical survey from ancient Greece and Orient to late 19th century; evolution of arithmetic, geometry, trigonometry, calculating devices, algebra, the calculus. 362 illustrations. 1,355pp. 5⅜ × 8½. 20429-4, 20430-8 Pa., Two-vol. set $22.90

THE GEOMETRY OF RENÉ DESCARTES, René Descartes. The great work founded analytical geometry. Original French text, Descartes' own diagrams, together with definitive Smith-Latham translation. 244pp. 5⅜ × 8½. 60068-8 Pa. $6.00

THE ORIGINS OF THE INFINITESIMAL CALCULUS, Margaret E. Baron. Only fully detailed and documented account of crucial discipline: origins; development by Galileo, Kepler, Cavalieri; contributions of Newton, Leibniz, more. 304pp. 5⅜ × 8½. (Available in U.S. and Canada only) 65371-4 Pa. $8.95

THE HISTORY OF THE CALCULUS AND ITS CONCEPTUAL DEVELOPMENT, Carl B. Boyer. Origins in antiquity, medieval contributions, work of Newton, Leibniz, rigorous formulation. Treatment is verbal. 346pp. 5⅜ × 8½. 60509-4 Pa. $6.95

THE THIRTEEN BOOKS OF EUCLID'S ELEMENTS, translated with introduction and commentary by Sir Thomas L. Heath. Definitive edition. Textual and linguistic notes, mathematical analysis. 2500 years of critical commentary. Not abridged. 1,414pp. 5⅜ × 8½. 60088-2, 60089-0, 60090-4 Pa., Three-vol. set $29.85

GAMES AND DECISIONS: Introduction and Critical Survey, R. Duncan Luce and Howard Raiffa. Superb non-technical introduction to game theory, primarily applied to social sciences. Utility theory, zero-sum games, n-person games, decision-making, much more. Bibliography. 509pp. 5⅜ × 8½. 65943-7 Pa. $10.95

THE HISTORICAL ROOTS OF ELEMENTARY MATHEMATICS, Lucas N.H. Bunt, Phillip S. Jones, and Jack D. Bedient. Fundamental underpinnings of modern arithmetic, algebra, geometry and number systems derived from ancient civilizations. 320pp. 5⅜ × 8½. 25563-8 Pa. $7.95

CALCULUS REFRESHER FOR TECHNICAL PEOPLE, A. Albert Klaf. Covers important aspects of integral and differential calculus via 756 questions. 566 problems, most answered. 431pp. 5⅜ × 8½. 20370-0 Pa. $7.95

CATALOG OF DOVER BOOKS

CHALLENGING MATHEMATICAL PROBLEMS WITH ELEMENTARY SOLUTIONS, A.M. Yaglom and I.M. Yaglom. Over 170 challenging problems on probability theory, combinatorial analysis, points and lines, topology, convex polygons, many other topics. Solutions. Total of 445pp. 5⅜ × 8½. Two-vol. set.
Vol. I 65536-9 Pa. $6.95
Vol. II 65537-7 Pa. $6.95

FIFTY CHALLENGING PROBLEMS IN PROBABILITY WITH SOLUTIONS, Frederick Mosteller. Remarkable puzzlers, graded in difficulty, illustrate elementary and advanced aspects of probability. Detailed solutions. 88pp. 5⅜ × 8½.
65355-2 Pa. $3.95

EXPERIMENTS IN TOPOLOGY, Stephen Barr. Classic, lively explanation of one of the byways of mathematics. Klein bottles, Moebius strips, projective planes, map coloring, problem of the Koenigsberg bridges, much more, described with clarity and wit. 43 figures. 210pp. 5⅜ × 8½. 25933-1 Pa. $4.95

RELATIVITY IN ILLUSTRATIONS, Jacob T. Schwartz. Clear non-technical treatment makes relativity more accessible than ever before. Over 60 drawings illustrate concepts more clearly than text alone. Only high school geometry needed. Bibliography. 128pp. 6⅛ × 9¼. 25965-X Pa. $5.95

AN INTRODUCTION TO ORDINARY DIFFERENTIAL EQUATIONS, Earl A. Coddington. A thorough and systematic first course in elementary differential equations for undergraduates in mathematics and science, with many exercises and problems (with answers). Index. 304pp. 5⅜ × 8¼. 65942-9 Pa. $7.95

FOURIER SERIES AND ORTHOGONAL FUNCTIONS, Harry F. Davis. An incisive text combining theory and practical example to introduce Fourier series, orthogonal functions and applications of the Fourier method to boundary-value problems. 570 exercises. Answers and notes. 416pp. 5⅜ × 8½. 65973-9 Pa. $8.95

THE THOERY OF BRANCHING PROCESSES, Theodore E. Harris. First systematic, comprehensive treatment of branching (i.e. multiplicative) processes and their applications. Galton-Watson model, Markov branching processes, electron-photon cascade, many other topics. Rigorous proofs. Bibliography. 240pp. 5⅜ × 8½. 65952-6 Pa. $6.95

AN INTRODUCTION TO ALGEBRAIC STRUCTURES, Joseph Landin. Superb self-contained text covers "abstract algebra": sets and numbers, theory of groups, theory of rings, much more. Numerous well-chosen examples, exercises. 247pp. 5⅜ × 8½. 65940-2 Pa. $6.95

Prices subject to change without notice.

Available at your book dealer or write for free Mathematics and Science Catalog to Dept. GI, Dover Publications, Inc., 31 East 2nd St., Mineola, N.Y. 11501. Dover publishes more than 175 books each year on science, elementary and advanced mathematics, biology, music, art, literary history, social sciences and other areas.